Environmental Science, Engineering and Technology

Monitoring of Water Quality

(Scientific and Applied Aspects)

Environmental Science, Engineering and Technology

Additional books in this series can be found on Nova's website under the Series tab.

Additional e-books in this series can be found on Nova's website under the e-book tab.

Water Resource Planning, Development and Management

Additional books in this series can be found on Nova's website under the Series tab.

Additional e-books in this series can be found on Nova's website under the e-book tab.

ENVIRONMENTAL SCIENCE, ENGINEERING AND TECHNOLOGY

MONITORING OF WATER QUALITY

(SCIENTIFIC AND APPLIED ASPECTS)

ANATOLY MAXIMOVICH NIKANOROV

nova publishers
New York

For permission to use material from this book please contact us:
Telephone 631-231-7269; Fax 631-231-8175
Web Site: http://www.novapublishers.com

Library of Congress Cataloging-in-Publication Data

Monitoring of water quality (scientific and applied aspects) / editor, Anatoly Maximovich Nikanorov.
pages cm.
Includes index.
ISBN: 978-1-62257-684-5 (hardcover)
1. Water quality--Measurement. 2. Freshwater ecology. I. Nikanorov, Anatoly Maximovich.
TD367.M659 2013
577.6--dc23

2012040501

Published by Nova Science Publishers, Inc. † New York

CONTENTS

PREFACE

This monograph is the first attempt in Russian scientific literature to generalize separate and uncoordinated works in accordance with the scientific aspects of surface water quality monitoring. It does not include scientific and methodological projects for ensuring the implementation of the public system for monitoring the composition of these waters. There have been several dozens of these methods that have been recently developed by employees (including the author) of the Institute of Water Chemistry.

The substantive meaning of the term "monitoring," which has remained to the present day, was introduced into the Russian scientific literature by Academician Y.A. Izrael, who made an outstanding contribution to the establishment and development of the system for monitoring environmental changes. He was the head of the Soviet Union Hydrometeorological Service for several decades. In the 1960s and 1970s he laid the foundation for environmental monitoring throughout the former Soviet Union. He developed the scientific rationale for the monitoring of atmospheric air, soil, sea and inland waters, including biomonitoring. The monitoring system utilized by the Russian Federal Service for Hydrometeorology and Environmental Monitoring has been successfully implemented for the past 30 years.

Several studies conducted by the author and his hydrochemical colleagues served as the basis for this book. Most of this work has been published and is listed in the references.

The author thanks his co-authors, colleagues and assistants, who took part in the joint studies and prepared certain sections of this monograph for publishing.

ABSTRACT

This monograph by Associate Member of the Russian Academy of Sciences A.M. Nikanorov evaluates the scientific and applied aspects of inland water monitoring. It describes the different monitoring systems of developed countries (USA, Canada, European states) and the Global Environmental Monitoring System for Water (GEMS/Water). The author examines the structure of the Russian State Surface Water Monitoring System which includes the subsystems for surveillance, operational and special types of monitoring. The majority of the monograph focuses on the results of full-scale environmental modeling and the study of intrawaterbody processes as a basis for environmental regulation. The methodology for complex water quality assessment is analyzed and a new method of complex water contamination assessment is described.

The author proposes several fundamental propositions of mathematical dynamic hydroecology and mathematical methods for assessment of anthropogenic impact on freshwater ecosystems.

The book is intended for scientists and experts in the field of water quality, contamination and environmental protection.; This includes hydrochemists, ecologists and practitioners of network divisions involved in the supervision and monitoring of surface waters and their contamination.

In 2010, this book was awarded the Academic F.P. Savarensky Prize by the Russian Academy of Sciences for distinguished works in the field of inland water research.

INTRODUCTION

Since ancient times, people took a great interest in the composition of natural water. The transition from qualitative to quantitative assessments, however, began after the introduction of physical and chemical methods of analysis, which made it possible to determine the component composition of natural waters. Prior to the middle of the 20th century, analytical control of surface and ground waters was performed by different organizations and authorities for the benefit of these government institutions. There was no unified system for interdepartmental control and exchange of hydrochemical information.

Serious anthropogenic impacts on the environment, including freshwater ecosystems, led to a decrease in the self-regulating capabilities of inland waters, accumulations of pollutants, and changes in the composition of hydrobionts. Progressive environmental pollution was the main reason for the 1972 establishment of the USSR State Committee for Hydrometeorology (Russian Federal Service for Hydrometeorology and Environmental Monitoring since 1992), which was the first government service for environmental monitoring. Under the direction of the Institute of Water Chemistry, the Monitoring Service has been successfully performing its main function—inland surface water quality monitoring.

The following monitoring systems are utilized:

- Monitoring changes of physical, chemical and biological parameters and processes under the influence of natural and anthropogenic factors, which may occur in the natural environment
- Studying the contamination levels of atmospheric air, soil and water bodies
- Assessing the impact of the foregoing processes on the flora and fauna
- Providing concerned organizations and population with current and emergency information related to changes in the natural environment, forecasting environmental conditions and warning of potential hazards

Monitoring the natural environment is an integral part of the quality management system. It also provides feedback for this system. It determines whether existing changes conform to the designated management objectives. In other words, monitoring is designed to assess the effectiveness of environmental measures.

Information provided by the monitoring system and related to insufficiency of certain measures is considered to be a reason for revising control actions.

Until recently, environmental issues focused on the surface water issues and were limited to the construction of sewage treatment facilities for the removal of specific suspended matters and organic substances. These measures were very effective. The reason for the limited monitoring at that time was because the only requirement was that water treatment should meet the established requirements.

The volume of water that has been affected and the complexity of the environmental issues have increased dramatically. There are thousands of pollutants that are emitted into the environment. The number of pollutants has increased annually. The majority of these pollutants is difficult to identify and are highly toxic. No studies have been conducted on the cross-media migration process and the transformation of these compounds in real water bodies. As a result of modern technological limitations, sewage treatment facilities are incapable of removing traces of environmentally hazardous substances.

Due to this situation, it is necessary to further develop water quality monitoring standards and to implement an improved monitoring system. This would include the applicable maximum permissible concentrations (MPCs) with the rate of the ecosystem's response to anthropogenic impact which are environmental MPCs.

The implementation of environmental MPCs in surveillance and special supervision makes it possible to transform the existing environmental pollution monitoring into a process that is guided by scientifically grounded environmental standards.

The development of environmental MPCs, however, is an extremely difficult task. It is first necessary to examine external influences such as sewage discharge, atmospheric precipitation, drainage basin pollutants incoming from drainage basins, etc., on natural water bodies. It is also necessary to study the regional characteristics of water bodies and watercourses to determine the behavior of pollutants in the ecosystem itself (transformation, absorption by biota and dissolved organic substances, cross-media migration, etc.).

It is recommended that environmental standards are differentiated by water body and area with consideration of their individual characteristics. These standards will protect the water body's ecosystem as well as its stability and structural and functional integrity.

Another problem is the difficulty in obtaining the most complete data and information (hydrobiological, chemical, physical, etc.). Unfortunately, in practice it is never possible to get complete and comprehensive data. *Under conditions of modern anthropogenic impact, ecosystems of water bodies change faster than information about these changes can be accumulated.*

Understanding and obtaining information about intra-waterbody processes is crucial for the improvement of the pollution and environmental systems. This monograph describes the research results related to the substantiation of the methodology for studying the migration of pollutants using models of natural water ecosystems (mesocosms). The results have been achieved by the author based on numerous field expeditions to water bodies in southern Russia with minimal human interference on the natural conditions.

Based on the modeling results, the basic fundamental notions indispensable for extending the experimental results to real water bodies were defined. These concepts were similarity of models and criterion of similarity; the relevance of the models to the objectives of modeling; the reproducibility of the results; and the buffer capacity as a calculation basis for maximum permissible environmental load. It is important to emphasize that *the improvement of monitoring is impossible without specially planned experimental studies of particular water bodies used for monitoring.* This condition is technically hard to meet. In our opinion, however, such a transition of monitoring from area principle to selective testing of aquatic ecosystems is absolutely necessary.

We have not identified the only possible approach. It is necessary to search for other ways to identify quantitative criteria for the assessment of aquatic environmental conditions and to propose a general theory of hydroecosystems functioning under conditions of technogenesis.

Many developed European and American countries have national monitoring systems. Each of these systems is unique. These monitoring systems were compared to the Russian State Water Quality Monitoring System (State Service for Monitoring of Environmental Conditions). The first study published in 1994 was the monograph entitled "Systems for Surface Water Quality Monitoring" (by A.M. Nikanorov and V.V. Tsirkunov), which included a comparative analysis of the status of these studies in the USA and Canada. This study drew considerable interest from numerous specialists and experts. Several parts of the book were revised and included in this publication. It was also supplemented with information about monitoring systems employed by EU countries. A separate section of this document describes the structure of the Global Environmental Monitoring System for Water (GEMS/Water), which includes most countries in the world and has been jointly implemented by Russia and other countries.

This system, managed by the World Health Organization (WHO), has several distinctive features compared to national monitoring systems. There are several similarities and differences in the first and second sections. In Russia, in 1989, the Hydrochemical Institute began performing the functions of the national scientific and methodological GEMS center related to the processing and analysis of water quality data obtained through the GEMS/Water observation network.

An important part of the monograph is the section on the theoretical aspects of environmental regulations and mathematical methods for the assessment of the anthropogenic impact on freshwater ecosystems. Many of these methods are isolated from the specific practical problems of environmental regulations. These methods should be developed and extended so that in the future they can be used with the results of the experimental modeling and data that is obtained from natural water bodies. The following key problem was discovered: the identification of key parameters that determine an ecosystem's functional conditions (8 major components). These particular parameters change as they are affected by natural and anthropogenic forces and form the fundamental basis of theoretical calculations within the scientific field of *mathematical dynamic hydroecology*.

While discussing mathematical methods it is important to remember the figure of speech used by Henri Poincaré: "Mathematics has no symbols for confused ideas." Several recent studies have applied full-scale modeling on real water bodies. This has made it possible to clarify their physical and chemical environment and to obtain data that can be used in mathematical calculation to characterize aquatic ecosystems. It is anticipated that additional development of this approach can be used to successfully characterize these ecosystems. It is necessary to validate the transition from qualitative to quantitative methods in order to describe the chemical and biological intrawaterbody processes in the monitoring of surface waters. Based on the reliable experimental data, it is necessary to be cautious in deciding which of the numerous, complex chemical and biological processes can be described using mathematical methods. Obtaining this data is likely to be the most difficult task in the next stage of development in experimental hydrochemistry.

SECTION I. MONITORING AND CLASSIFICATION OF WATER QUALITY IN FOREIGN COUNTRIES

Chapter 1

MONITORING OF INLAND SURFACE WATER QUALITY IN FOREIGN COUNTRIES

Surface water quality monitoring which ensures supervision, assessment and forecasting is an integral part of the system for environmental quality management, its information foundation for selecting objectives, and optimal strategies for control actions.

Under the circumstances of generally observed changes in the natural composition of water when exposed to anthropogenic factors, the scale of environmental issues is growing; the complexity of research and implementation of measures related to elimination (minimization) of negative consequences under the limited resources of society is increasing and the significance of the surface water quality monitoring is rising dramatically. The necessity to improve scientific and organizational methods is becoming the most important task.

This chapter will consider the modern systems and programs for surface water quality monitoring that have been established and applied by several foreign countries with the purpose of assessment and specification of further prospects of development for these systems in Russia.

1.1. SYSTEMS FOR SURFACE WATER QUALITY MONITORING

In the 1970s, developed countries expanded their work in monitoring inland surface water quality, including the design and establishment of new national monitoring networks and the revision of current programs in view of urgent problems. This is related to the activities of the Stockholm Conference on Human Environment (1972), which emphasized the importance of monitoring for the purposes of environmental protection [1].

National systems of water monitoring have a number of similarities determined by the object observed and the problems that are solved. They still differ considerably, however, due to natural and climatic peculiarities; sources of pollution; specific character of their management structures; and the financial, scientific and engineering capabilities of different countries, including the approaches applied to water quality regulations.

Most countries (including the United States, Canada, Japan, Netherlands, Great Britain, etc.) employ national systems of monitoring, within the framework of which major problems related to supervision, assessment and water quality forecasting are solved. They usually consist of several levels including: supervision network, territorial subdivisions and central divisions.

In general, the goals of different national water quality monitoring systems are very similar. They include:

a) The assessment of water quality based on the observance of national standards and regulations and the detection of water bodies that are suffering from the most serious water quality problems;
b) Detection of long-term tendencies (trends) in surface water quality.

The basis of most national monitoring systems is a unified supervision network with stationary posts using manual sampling. Observations conducted by stationary networks are supplemented with occasional regional inspections of water bodies and intensive monitoring of certain areas.

In all countries, the location of stationary observation posts is chosen based on the significance of the water bodies and their level of contamination. The locations of these observation posts are in urbanized areas, near major sources of pollution; on the outlet points of large rivers; the shores of water bodies; and in areas of spawning and wintering of valuable commercial organisms, etc. The number of posts and their density varies by country. For example, the national water quality monitoring network of Great Britain includes 250 stationary posts (one post per 1,000 km^2) [2], the Netherlands—260 (one per 160 km^2) [3], in the USA (Geological Survey)—4,200 posts (one per 2100 km^2) [4], in Japan—over 4,200 posts (one per 84 km^2) [5]. In some countries (particularly in the Netherlands) there are several types of posts, which differ by the number and composition of the measurements as well as the frequency and timing of sampling.

In all countries, programs for surface water quality monitoring utilize a variety of physical, chemical and biological parameters (usually exceeding one hundred). As the number of identified components characterizing the most general requirements for the quality of water (such as dissolved gases, principle ions, chemical oxygen demand (COD), biochemical oxygen demand (BOD), suspended matters, etc.) gradually decreases, it is possible to observe a simultaneous rise in the specific weight of observations of organic and inorganic micropollutants found in the water, bottom sediments and biota. Systems for biomonitoring surface water quality are also being rapidly developed [6, 7].

Water quality monitoring usually includes monthly sampling. This is adequate for determining the long-term trends in the vast majority of parameters. The sampling frequency, however, for different parameters varies by the type of post. As an example, in the Netherlands, major posts (first category) measure 70-100 parameters with a frequency of 6 to 52 times per year. Third category posts only measure primary water quality indices from 6 to 13 times per year every several years.

Several national monitoring systems have more than one supervision network. For example, the national monitoring systems of Finland and Sweden are composed of a number of interrelated subsystems (subprograms). In Sweden, the following subprograms are utilized: monitoring the washout of substances by rivers into adjacent seas; monitoring of water quality and chemical load of large lakes and lakes exposed to liming; soil leaching under the influence of surface water acidulation; monitoring of water quality in the integrated fields of research; state of water bodies on national park territory; and the study of water quality by means of satellites.

Each of the subprograms has its own network of observation posts used for measuring various water quality indices with different frequencies (from 2-4 to 12-24 times per year). Moreover, the quality of discharged industrial and municipal waters is supervised on a regular basis and is supplemented by a number of studies.

Many countries are involved in the intensive development of automated water quality monitoring systems. For example, in the 1980s, Japan implemented automatic control systems on 120 river and lake posts, which functioned, together with control posts on factories, plants and sewage treatment facilities, as part of over ten telemetry systems. In the long-term, control will be ensured by a fully integrated telemetry system for water quality monitoring.

Samples collected by observation networks of national monitoring systems are sent to laboratories for the measurement of parameters required by the program (25-40 on an average). After a variety of data quality tests (results, assessments), they are sent to the Central Data Bank. Special software makes it possible to process and present data in a variety of standard forms (tables, diagrams, charts, maps, etc.). From any location in the country, a user can access current and historical data via a computer terminal.

Integration of data related to the national monitoring of water quality (in the form of reports, reviews and surveys) is carried out by the central division.

Existing national surface water quality monitoring systems have had some drawbacks and controversies due to the complexity of these systems, especially with the use of a single database for water quality management. There have been several positive developments in monitoring pollution sources, including taking into account the characteristics of particular types of production. The integration of pollution data with the results of water quality bioindication and biotesting enables the optimization of monitoring programs and serves as the basis for the next step in environmental monitoring.

There are a number of common drawbacks to these monitoring systems including:

- The most important shortcoming is the contradiction of the goals and objectives of the monitoring programs. Many monitoring networks that are intended for the assessment of long-term trends don't allow the description of short-term changes in water quality. The optimization of a network's structure that is applied to one or more parameters may not suit the other parameters.
- Most programs for national monitoring lack flexibility. This hampers the optimal correlation of the existing standard network with new supervision programs due to changes in goals and conditions.
- The national representative network for water quality assessment is not always suitably chosen because most observation posts are located in the areas with the highest pollution levels.
- Spot sampling is not very representative due to: a) the incorrect placement of background and guaranteed displacement points and b) the lack of regard for the cross heterogeneity of pollutants' concentration patterns.

The optimizing time (terms) for sampling is challenging due to the differences in objectives of the monitoring networks. The detection of trends requires sampling with the same frequency. This is often unacceptable when solving problems related to the detailed study of the periods with the worst water quality problems.

There are a number of problems related to the component and space-time aspects of a network. These problems are primarily caused by the lack of prior special inspections of water bodies during the establishment of national monitoring systems.

The need to address these issues as well as supporting efficient system performance should take place in the project stage with a systematic approach and detailed planning for the acquisition of water quality data and its delivery to the consumer. Special attention should be given to new approaches for determining the optimal correlation between surface water monitoring and the monitoring of stationary and non-stationary pollution sources. This should also include surveillance and the monitoring of physical, chemical and biological parameters.

There is a focus on surface water quality monitoring systems that are utilized in the USA and Canada. Many of these systems are comparable to Russia in their size, diversity of physical and geographical conditions, economic potential and complexity of environmental issues. Detailed analysis of these systems and evolutionary trends are described by A.M. Nikanorov and V.V. Tsirkunov in their monograph titled “Systems for Surface Water Quality Monitoring” [1]. This work examines the fundamental provisions and primary characteristics of these monitoring systems.

1.1.1. Brief Overview of the US System for Surface Water Quality Monitoring

Routine observations of surface water chemical composition in the United States were instituted at the beginning of the 20th century by the US Geological Survey of the United States Department of the Interior (USGS).

With the establishment of the Environmental Protection Agency (EPA) and adoption of the Clean Water Act (1972), the scope of work related to water quality monitoring increased significantly. Several monitoring networks were established, and national programs were implemented, including those which were connected with detection of the major classes of toxic substances (heavy metals, pesticides, etc.) in natural and waste waters.

In the 1980s, the existing national monitoring system underwent a series of revisions based on feedback and recommendations from the National Academy of Sciences, the US General Accounting Office, the Council on Environmental Quality and the Water Quality Act (1987). As a result, a number of programs were developed including: the National Water Information System (USGS), the Environmental Monitoring and Assessment Program (EPA) and others.

Several national monitoring networks are currently being utilized with the United States. Most of the natural water quality monitoring activities is performed by the individual states in cooperation with the USGS. Scientific and methodological management is carried out by the EPA.

The US national system for surface water quality monitoring has a number of levels that are distinguished by their complexity and diversity.

The primary objectives of the surface water quality monitoring program are as follows:

- Description of water resources and detection of water bodies with poor water quality
- Provision of information for the development of high-priority trends, plans and programs related to water quality
- Assessment of pollution prevention measures

The specific purposes of monitoring programs include:

- Checking conformity of water quality with national standards
- Determining the presence or absence of long-term water quality trends
- The establishment of the primary factors having an influence on the water quality
- A comparison of ecosystems before and after the application of water protection measures

The following two main types of water quality monitoring are distinguished [8]:

1) monitoring of pollution sources, and
2) monitoring of natural waters.

The monitoring of pollution sources includes determining the composition of industrial and municipal waste waters discharged into water bodies and the study of natural and waste water mixing zones. It can be carried out by means of "self-monitoring" employed by industrial (municipal) objects discharging waste waters or by sample test inspections for the purpose of validation of "self-monitoring" data, etc.

The monitoring of natural waters comprises all types of monitoring carried out outside the zone of direct influence of sewage discharge (water column, bottom sediments, biota) and may be performed via a network of stationary posts, special surveys or statistical research. Surface water quality monitoring is regarded as the most reliable source of information for carrying out the main functions of the program for water quality management on the territory of each state and the country as a whole.

The main sources of data on natural water quality are the water protection organizations of states, who usually conduct their own monitoring programs. States carry out work mainly at the regional (state's territory) and local (water body or its part) levels under short (season, year) and medium (several years) term programs. They also participate in national programs whenever possible.

An important source of information on the quality of US water resources are the federal departments, with the USGS and the EPA playing a major role.

The USGS is a primary supplier of monitoring information on the quality of water and its functions most closely correspond to those of the Russian Federal Service for Hydrometeorology and Environmental Monitoring.

The USGS is responsible for several national monitoring networks: the Hydrologic Benchmark Network, the National Trends Network and the National Stream Quality Accounting Network.

The first network was established to monitor changes in water quality in small portions of benchmark drainage basins that are free from direct human influence [9]. The second network collects and analyzes weekly liquid precipitation samples for the assessment of the environmental impact of acid precipitation on water bodies [10]. The goals of the third network include: to keep records of the amount and quality of water flowing within and arriving to the territory of the United States; to assess long-term changes in the quality of watercourses; and to prepare the groundwork for further forecasts [11].

The EPA currently has no natural water quality monitoring networks of its own. Therefore, they play a relatively minor role in the acquisition of surveillance data. The EPA has provided scientific and methodological management of the national monitoring system and has concentrated on the following:

1) Development of national policy and new approaches for monitoring based on "the best available scientific knowledge"
2) Management and supervision of monitoring of pollution sources in conformity with the established criteria and water quality standards
3) Development of methods for the processing, interpretation and presentation of information on the quality of aquatic ecosystems, and the maintenance of data banks related to water quality and pollution sources

In addition to the EPA and the USGS, national water quality monitoring is carried out by the National Oceanic and Atmospheric Administration, the Fish and Wildlife Service, the Department of Agriculture and other agencies.

Professional, regional, environmental and public organizations also play a vital role in the acquisition and analysis of data on the quality of water resources at the national and regional levels.

The majority of water quality monitoring work in the United States focuses on projects that study the most urgent problems or are concerned with the most important water bodies. These programs include: the EPA's Dioxin Study [12], the National Pesticides Monitoring Program [13], the Toxic Substances Hydrology Program [14, 15], the National Acid Precipitation Assessment Program [10], the National Biomonitoring Program [16] and others.

Information technology data banks (databases) have become an essential element of the existing monitoring systems. These databases facilitate the consolidation of surface water quality information on a nationwide scale.

Assessment of environmental activities (including surface water quality monitoring) is conducted in the USA with the participation of the federal agencies—the Council on Environmental Quality, the General Accounting Office, etc. [17, 18], and different scientific and professional organizations.

Water quality monitoring involves hundreds of laboratories in four regional and several dozen district USGS divisions, EPA's regional departments, thousands of private sector laboratories, and companies engaged in studying the composition and properties of natural and waste waters, bottom sediments and biota.

The main element of the interaction between monitoring and water resources quality management is *water quality assessment criteria*. The status of the water bodies is evaluated (monitored) and is compared with current laws and regulations. This involves the implementation of specific measures for violations (management). In recent years, principles of determining water quality criteria and their significance have experienced dramatic changes [19] due to the application of risk assessment in aquatic ecosystems.

The United States of America has a *two-tier water quality regulation system*. It includes water quality *criteria* and *standards*. *Water quality criteria* are values of concentrations or verbal descriptions of a component's contents connected with the level of environmental impact. They allow scientifically grounded conclusions to be made. *Standards* represent regulations legal requirements for a specific water body. Standards are primarily based on water quality criteria as a basis for regulation and executive control. These standards are influenced by specific natural conditions (such as the content of natural organic acids) the priority of water bodies and safety considerations.

State water quality standards are based on water quality of water and are strongly influenced by biological characteristics.

The United States has developed guidelines for the assessment of the environmental potential of specific water bodies for the preservation of hydrobionts and has established procedures that modify the national criteria for individual areas. *Biocriteria* have been developed [20] that provide a quantitative representation of water quality standards on the basis of the functional and structural characteristics of aquatic ecosystems [21]. Biocriteria are measures of "biological integrity" which can be used in the assessment of the total environmental impact of numerous pollution sources and other adverse effects [22].

Water quality criteria have been developed by the EPA as national recommendations for those states establishing their own standards. In the cases of risk assessment, the most frequently used parameters are criteria for specific pollutants and waste water toxicity criterion in general. These two types of criteria include the following three components (the first one serves as a *risk assessment index*, the last two are used as an *impact assessment*):

1) Numerical value—determines which concentration of a pollutant (or pollution index, for example, toxicity) is the maximum allowable
2) Duration—the period of time within which the forecasted concentrations in water bodies come to an average in order to make further comparison with the criterion's concentration (this component limits the duration of concentrations exceeding the criterion)
3) Frequency—how often the criterion may be exceeded without having an unacceptable impact on the ecosystem

Threat assessments for particular substances should be performed in accordance with the EPA guidelines [23]. Risks from point sources in the case of the evaluation of waste water toxicity, however, shall be assessed pursuant to the Technical Support Document [24].

National monitoring systems are for solving national and large-scale regional problems and are not intended for implementation at the local level. At the same time, national programs are designed to provide required source information for the monitoring conducted at the local and regional levels.

New national programs (systems) of monitoring have several common functional elements:

- A nationally and regionally unified means of sampling and methods of chemical analysis
- A coordinated list of identified physical, chemical and biological parameters
- Systems of data control and quality assurance

- A uniform system of data (data bank) acquisition, storage and processing
- A consistent system for the consolidation and presentation of information to consumers

New national monitoring programs are characterized by *flexibility and complexity*. *Flexibility* provides an opportunity to rearrange the monitoring system depending on changes in priorities and the retargeting of a program to solve other issues. *Complexity* is expressed in an endeavor to describe as completely as possible the complex nature of water pollution by means of the observation of the aquatic environment, bottom sediments and biota on the basis of measuring certain physical, chemical and biological parameters. Increased attention is paid to the biological parameters which may characterize environmental changes.

The United States emphasizes the development of integrated and coordinated national monitoring systems. These systems should integrate, whenever possible, data from active regional and local programs and additional information on the physical and geographical characteristics of the study area, sources of pollution, objects of land and water use, etc. The coordination of work is carried out at the local, regional and national levels by means of establishing corresponding interdepartmental groups of scientists, managers, representatives of environmental organizations and members of the public.

The main targets of national monitoring are physically (geographically) detached objects—river basins and their drainage areas or individual water bodies. As a rule, however, subnational administrative boundaries are not taken into account.

The characteristic features of national monitoring systems may include a long duration (decades) of supervision at certain intervals. A specific water body usually undergoes relatively short periods of intensive observations alternating with longer periods of less intensive supervision.

Common features of large-scale national monitoring systems include the wide use of statistical methods at the time of their development and the consolidation of data with the purpose of attaining reliable results.

The programs that were analyzed illustrate that national monitoring systems vary based on a number of specific features of the assigned tasks, national peculiarities, physical and financial resources, traditional types of activities, modern scientific concepts, etc.

1.1.2. Brief Overview of the Canadian System for Surface Water Quality Monitoring

Widespread development of water quality monitoring in Canada belongs to the International Hydrology Decade. Established in 1970, Environment Canada's Water Quality Branch ensured the accomplishment of large-scale works on reconnaissance surveys of bodies of water across the whole country. After the execution of monitoring programs by the efforts of the provinces, the branch has focused mainly on national problems.

At the present time, the Water Quality Branch (Water Survey—Environment Canada) is engaged in surface water quality monitoring at the national level. The branch is responsible for designing monitoring networks for the assessment and management of river basins. The National Water Research Institute and the National Hydrology Research Institute render assistance in conducting research works.

The Water Quality Branch is the central division located in Ottawa. Five regional departments (Quebec, Ontario, Yukon and Pacific Region, Atlantic Region and Western and Northern Regions) are subordinate to the branch.

Canada's existing system for water quality monitoring (based on agreements between the federal government and the provinces' administrators) is more rational and flexible than in the United States [1, 25, 26]. Two major levels of the system (federal and provincial) are clearly identified. According to Canadian law, monitoring is an integral part of the water quality management system. Canada has established the same types (forms) of monitoring as the USA but on a smaller scale.

The primary objectives of the system for water quality monitoring include:

a) Assessment of water quality in the country and observance of national standards and regulations in particular
b) Identification of long-term trends in the changes of surface water quality

In accordance with the Canadian Federal Water Policy, the monitoring system shall provide the following:

- Acquisition of information on water quality
- Identification of water quality trends at the national and regional levels
- Determination of whether the stated objectives related to water quality were achieved
- Assessment of control actions aimed at achieving desired water quality
- Establishment of a basis for the revision of requirements regarding waste waters control whenever it may be necessary

Urgent national problems connected with the quality of water require reliable information for monitoring networks. Such problems include the contamination of water bodies used for recreation; the impact of different types of land use on water quality, including water management projects; the effect of long-range transport of pollutants on water quality and the impact of (toxic) pollutants on aquatic ecosystems, just to name a few.

The necessity of solving these foregoing issues focuses the Water Quality Branch's attention toward the challenge of designing perfect monitoring networks and assessing water quality with consideration of diverse consumer requirements.

Monitoring the quality of river basins is done in order to study nature in the observed river systems: its cause and effect relations, the detection of its sources, the consequences of pollution, and for the establishment of existing and arising problems in the sphere of water quality. This type of monitoring is carried out by means of both stationary posts and surveys. Thus, two main kinds of monitoring networks are distinguished: networks of indicator posts and repeated surveys (networks) of river basins.

The network of indicator (stationary) posts is distributed among national territories within river basins and along borders. Samples are taken on a regular basis (quarterly and monthly) for a long period of time. Some of these posts also act as a composite element during the survey of river basins, which enables the evaluation of water quality problems through the use of a more dynamic method.

Canada, as well as the USA, employs a complex, though flexible, *two-tier water quality regulation system*. It includes *guidelines* and *objectives*. Guidelines represent values of concentrations and verbal descriptions of a component's content associated with the degree of environmental impact about which scientifically grounded propositions may be made. Objectives concerning water quality constitute regulations that have legal force and refer to a specific water body or its part. Objectives have water quality guidelines as the basis for regulation and executive control; objectives, however, may differ substantially from guidelines.

This fact may be caused by individual natural conditions of water quality formation (for example, those that determine high concentration of organic acids), priority of water bodies and safety advisable for a specific ecosystem, economic and other factors. Correlations of guidelines and objectives with the Canadian water quality management systems are shown in the Table 1.1.

The development of objectives is necessary in order to solve existing problems as well as in cases of industrial applications for a new production project (object).

Most monitoring networks have controversies (or paradoxes) such as the unreliability of information in cases where excess data has been obtained. The National Benchmark Network for Monitoring Systems was established by the Water Quality Branch to overcome this drawback and for consolidation at the national level. The network is comprised of several independent networks and was designed to provide a constant flow of information on specific national problems related to the quality of water, such as the:

1) Identification of background conditions for all ecological regions of Canada
2) Impact of agricultural pesticides on aquatic ecosystems
3) Influence of pollutants from urban and industrial areas on adjacent aquatic ecosystems
4) Long-range transport of air pollutants (acid rains and toxicants)
5) Eutrophication of aquatic ecosystems

Each problem should be solved independently. This requires an arrangement of several networks, specific for each problem, designed to meet their own specific challenges. Observation posts should be used only for certain purposes, thus preventing acquisition of redundant (superfluous) data. Some details on the organization of the network connected with the second problem are given below in order to illustrate this approach.

Application of pesticides in Canadian agriculture has beneficial effects but is accompanied by a number of deleterious environmental consequences, which is typical for other countries as well. It has been established that less than 1% of pesticides reach target organisms.

Unfortunately, potential risks attributed to the presence of pesticides in surface waters are hard to evaluate, because information concerning location, transformation and migration paths of these compounds is not always properly documented. Basic information on the presence and concentration of this group of compounds in high-risk drainage basins is required to ascertain whether the ecology of the observed water bodies is deteriorating and whether the established standards will be exceeded.

Table 1.1. The role of water quality guidelines and objectives in the Canadian system for water quality management

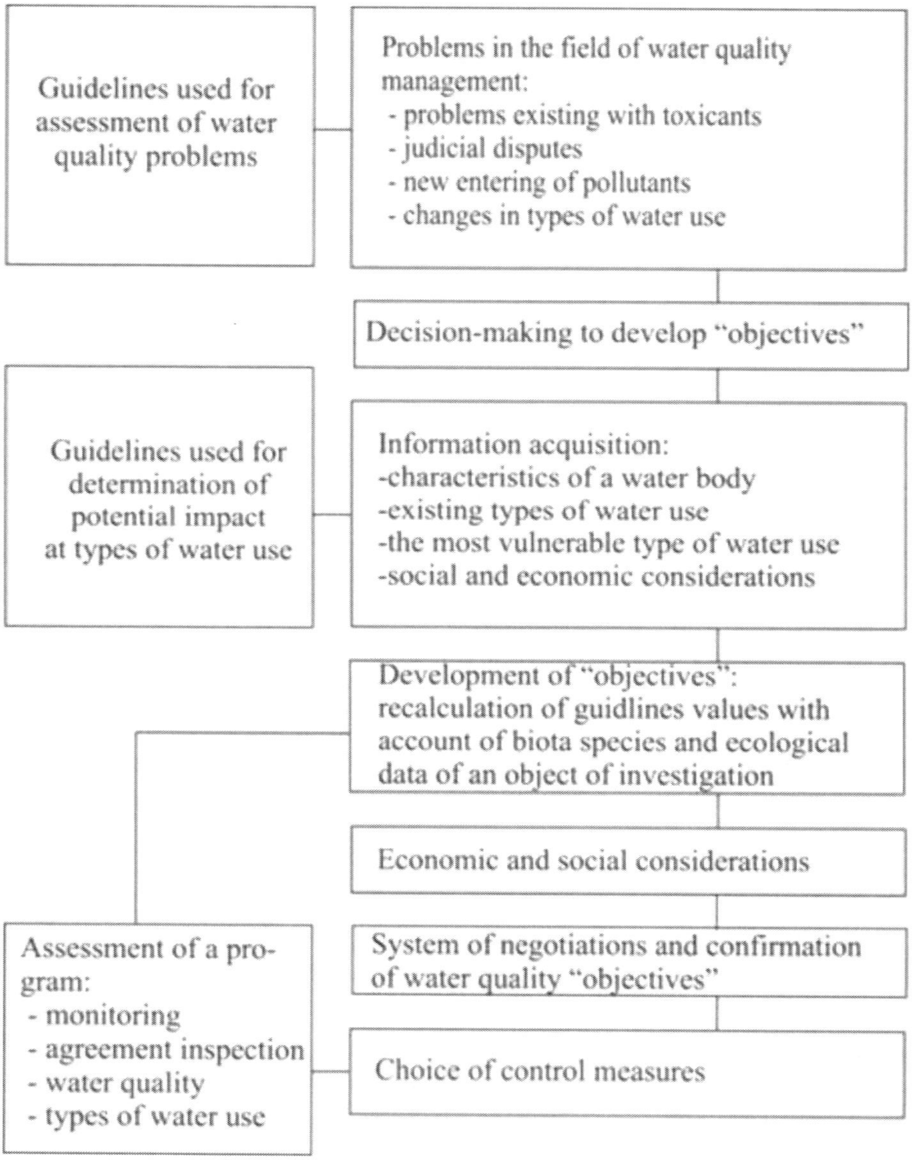

The management objective set for the discussed problem shall "facilitate maintenance and improvement of surface water quality as required by the usage of these water bodies." This objective is converted into the following monitoring purposes: to determine whether detected concentrations of agricultural pesticides contained in water and bottom sediments of bodies of water located in the areas of high agricultural activity are the consequence of applying (using) pesticides and whether such concentrations are harmful for aquatic ecosystems.

The microcomputer geographical information system, SPANS, is used for developing predictive models for each Canadian province. These models are used for determining high-risk areas where the application of pesticides may affect the quality of water. SPANS provides a unique opportunity to unite different types of information, presenting it in a visual form (charts).

Processing of additional diverse information (sketch maps of provinces, charts of water basins, agricultural capability of lands, etc.) with overlay of pesticide application maps is performed in several steps. As a result, the most troubled (tending to be considerably affected) parts of drainage basins are revealed, sampling sites are designated and lists of identified parameters, sampling frequencies and other matters are specified.

This approach, applied to each of the foregoing problems, ensures the interrelation of all monitoring programs.

REFERENCES

[1] Nikanorov A.M., Tsirkunov V.V. Systems for Surface Water Quality Monitoring. Part 1. — St. Petersburg: *Hydrometeoizdat,* 1994. —108 p.

[2] Brown C, Rodda J. C, Williams M.A countrywide study of river water quality for Great Britain // *IAHS Publ.* — 1982. — N 139. — P. 195—205.

[3] Barrows B.H. Analysis and integration of ecological monitoring data // *Proc. Intern. Symp. on the Design of Water Quality Information Systems.* — Colorado Water Research Institute (CWRRI). Information Series. — 1989. — N 61. — P. 276—286.

[4] Peck D.L. Water resources activities of the U. S. Geological Survey // *Water Resources Bulletin.* — 1985. — V. 21, N 6. — P. 901—907.

[5] Fujiwara M. Water quality monitoring In Japan // *Water Quality Bulletin.* — 1985. V. 10, N 3. — P. 10—126.

[6] Cairns J. Evaluating the options for water quality management //*Water Resources Bulletin.* — 1985. — V. 21. — P. 1—6.

[7] Herricks E.E. Aspects of monitoring in river basin management // *Water Science and Technology.* — 1984. — V. 16, N 5—7. — P. 259—274.

[8] Surface Water Monitoring: A Framework for Change. — Wash., D. C: US EPA. Office of Water, Office of Policy, Planning and Evaluation. — 1987. — 41 p.

[9] Hirsch R.M., Alley W.M., Wilder W.G. Concepts for a National Water Quality Assessment Program // *U. S. Geological Survey Circular.* — 1988. — N 1021. — 42 p.

[10] Robertson J. K., Wilson J. W. Design of the National Trends Network for monitoring the chemistry of anthropogenic precipitation. — *U. S. Geological Survey Circular* 964. — 1985. — 46 p.

[11] Steele T.D. A regional analysis of water quality in major streams of the United States // *IAHS Publ.* — 1983. — N 141. — P. 131 —143.

[12] National Water Quality Inventory. 1986. Report to Congress. EPA— 440/ 4—87—008. — Wash. DC: U.S. EPA, Office of Water, 1987. — 141 p.

[13] Gilliom R.J., Alexander R.B., Smith R.A. Pesticides in the Nation- s Rivers, 1975—1980, and implications for future monitoring. — *US Geological Survey Water-Supply.* — Paper 2271. — 1985. — 34 p.

[14] Ragone S.E. Program overview and selected papers from the Toxic Waste Program technical meeting, Tucson, Arizona, 1984. — *U. S. Geological Survey Open-File Report* 86—324. — 1987. — 116 p.

[15] Mallard G.E. et al. U. S. Geological Survey Toxic Substances Hydrology Program — Surface-water contamination: Proceedings of technical meeting. February 2—4, 1987. — Denver, Colorado: *U. S. Geological Survey Open-File Report* 87—764, 1988. — 164 p.

[16] A Review of the USGS National Water Quality Assessment Pilot Prof ram (Water Science and Technology Board, Commission on Geosciences, Environment, and Resources, National Research Council). — Wash., D. C: *National Academy Press,* 1990. — 153 p.

[17] National Water Quality Inventory. 1988 Report to Congress. EPA— 440/4—90—003. — Wash. DC: U. S. EPA, Office of Water, 1990. — 189 p.

[18] Report to the Congress of the United States: Better Monitoring Techniques are Needed to Assess the Quality of Rivers and Streams. CED—81— 30. — V. I, II (April 1981). — N. Y.: U. S. General Accounting Office, 1981. — 211 p.

[19] Quality Criteria for Water 1986. EPA 440/5—86—001. — Wash. D. C: US EPA, Office of Water Regulation and Standards, 1986. — 353 p.

[20] Biological Criteria for the Protection of Aquatic Life: I—III. — Columbus, Ohio.: EPA, Division of Water Monitoring and Assessment. — 1987. — 200 p.

[21] Proceedings of the First National Workshop on Biological Criteria. EPA— 905/9—89/003. — Chicago, Illinois.: US EPA, Region 5. — 1988. — 89 p.

[22] Hughes R.M., Larsen D.P. Ecoregions: an approach to surface, water protection // *J. of Water Pollution Control Federation.* — 1988. — V. 60. — P. 486—493.

[23] Guidelines for Deriving Numerical Water Quality Criteria for the Protection of Aquatic Organisms and Their Uses. — Wash., D.C.: US EPA, 1986. — 26 p.

[24] Technical Support Document for Water Quality Based Toxics Control. — Wash., D. C: US EPA, Office of Water. — 1985. — 86 p.

[25] Dafoe T.J., Watt E.R., Stevens R. Water Quality Branch monitoring activities — a dynamic approach to evolving issues // *Proc. Intern. Symp. on the Design of Water Quality Information Systems, CWRRI. Information Series.* — 1989. — N 61. — P. 47—53.

[26] Canadian Water Quality Guidelines Updates (April 1991). Appendix IX. A protocol for the derivation of water quality guidelines for the protection of aquatic life. — Ottawa, Canada: Environmental Quality Guidelines Division, Water Quality Branch, Environment Canada. — 1991. — 7 p.

Chapter 2

INTERNATIONAL CLASSIFICATIONS AND PROGRAMS FOR SURFACE WATER QUALITY ASSESSMENT

It is commonly known that water quality is evaluated by two primary methods: physicochemical and biological, each having its own merits and demerits.

The first one allows for the so-called point estimate of the quality of water to determine the type of pollution and nature of pollutants along with some side effects of such pollution.

The major advantage of the physicochemical method lies in the fact that analysis results obtained therewith are distinguished by accuracy and data quantity, which makes this method suitable for the evaluation of water quality conformance with the established standards.

The second method enables the assessment of the generalized impact of pollutants on an aquatic ecosystem to determine the degree of its disturbance and to make an integral estimate of the water quality.

National systems for surface water quality monitoring function not only in the USA and Canada, but also in many other countries [1]. The complexity of monitoring programs, increased volumes of data, necessity of its correlation (juxtaposition) to water quality standards/criteria and comparison with data from other bodies of water (or their parts in cases of international border crossings) have triggered the development of general classification schemes allowing these difficulties to be overcome.

2.1. EEC CLASSIFICATION OF SURFACE WATER QUALITY

All of the classification schemes developed by EEC countries are primarily based on the application of physical, chemical and biological methods; they differ considerably, however, by their degree of complexity, perfection, applicability and applied approaches [2].

Meanwhile, most EEC classification schemes have several similarities:

- They include three basic parameters that describe contaminations: dissolved oxygen content, biochemical oxygen demand and ammonia nitrogen content.
- Many biological schemes represent modifications of the Trent index.
- Almost the same color codes are used for mapping water quality.

In spite of the large number of developed and applied EEC classification schemes for water quality assessment, none may be recognized as completely perfect [2].

In *Belgium*, each water sample taken within the national monitoring system is analyzed on approximately 40 parameters of which the most important groups are water oxygen balance and heavy metals [3]. Responsibility for performing this procedure is delegated to the Institute of Hygiene and Epidemiology under the Ministry of Public Health and Family.

According to the Benelux Treaty (economic union of Belgium, the Netherlands and Luxemburg) the quality of surface water in each of these countries is estimated based off of the water oxygen balance (WOB) criterion comprising three key parameters:

- Dissolved oxygen content/saturation
- Biochemical oxygen demand
- Ammonia nitrogen content

Assessment of the WOB in a given point of a body of water is carried out in the following manner. The foregoing parameters are measured in water samples. Then, as provided by the scale (Table 2.1), each parameter is assigned a certain number of points that are thereafter summed up for the overall assessment of the WOB in the sample.

WOB values may vary from 3 to 15. Water quality class at the given point is determined on the basis of the overall index calculated above (Table 2.2).

Color codes provided in the table are used for plotting WOB data on water quality maps. Such a visual method for the consolidation of information is widely applicable by all EEC countries.

Research into the issue of water contamination by heavy metals includes the analysis of samples for the presence of mercury, cadmium, chromium, copper, zinc and lead. Belgium has designed a map labeled with average annual values of heavy metals content, which once exceeded 80% of the mean value, and a map of water contamination with cadmium. Mapping is made according to five classes: < 20%, < 40%, < 60%, < 80% and > 80 % from the national median of all recorded values of cadmium concentrations.

The quality of water is assessed using the hydrobiological method based on the Belgian Biotic Index (BBI) [4, 5], which was used in an attempt to unite comparative advantages of river biotic indices of Woodiwiss [6], Tuffer and Verneaux [7].

This method relies on the analysis of two effects provoked by pollution:

- The decrease in the number of communities
- The decrease of species diversity and extinction of clean water specimens

Assessment of water quality with this index aides in the study of the species composition of macrozoobenthos with consideration for population. The biotic index is calculated by means of standard tables (Table 2.3). The results are classified in accordance with a six grade scale reflecting relations between biotic index values and water quality (Table 2.4). The main advantage of assessment with the hydrobiological method is that it may be based on a single survey, while WOB assessment requires not less than 20 measurements taken within several years.

Table 2.1. WOB assessment table

Points	Dissolved oxygen, saturation %	BOD_5, mg/dm^3	NH_4^+ — N, mg/dm^3
1	91-110	< 3	< 0.4
2	71-90	3.1-6.6	0.5-1.0
	111 - 120		
3	51-70	6.1-9.0	1.1-2.0
	121-130		
4	31-50	3.1-15.0	2.1-5.0
5	< 30 and > 130	> 15	> 5

Table 2.2. Water quality and WOB assessment correlation

WOB points	Quality class	Color code
3-4	1 - very good	Blue
5-7	2 - good	Green
8-10	3 - medium	Yellow
11-13	4 - poor	Orange
14-15	5 - very poor	Red

Table 2.3. Standard table for determining Woodiwiss Biotic Index

Faunistic group	Number of systematic units observed in this faunistic group	Total number of present systematic units				
		0-1	2-5	6-10	11-15	16 and more
ridae (+ Heplageniidae)	Several	—	7	8	9	10
ridae (+ Heptageniidae)	Only one	5	6	7	8	9
Trichoptera in shelters	Several	—	6	7	8	9
	Only one	5	5	6	7	8
Ancylidae or Ephemer optera	More than two	—	5	6	7	8
(excluding Ecdyonuridae)	Two or fewer	3	4	5	6	7
Aphelocheirus or Odo nata or Gammarldae or Mollusca (excluding Sphaeridae)	All of the foregoing systematic units are absent	3	4	5	6	7
Asellus or Hirudinea or Sphaeridae or Hemiptera (excluding Aphelocheirus)	Same	2	3	4	5	—
Tubificidae or Chironoinidae, thummi-plumosus group	»	1	2	3	—	—
Eristalinae (— Syrphi-dac)	»	0	1	1	—	—

Table 2.4. Relations between biotic index values and water quality

Quality class	Biotic index	Color code	Water quality
1	10-9	Blue	Lightly polluted and unpolluted
2	8-7	Green	Slightly polluted
3	6-5	Yellow	Moderately polluted—critical situation
4	4-3	Orange	Heavily polluted
5	2-0	Red	Very heavily polluted
6	0	Black	Impossible to analyze, total absence of bioindicators

In *Denmark,* the Danish National Agency of Environmental Protection has published guidelines for surface water quality planning [8]: for watercourses and lakes (Part I) and coastal waters (Part II).

The planning process includes three stages:

- Determination of water characteristics for certain watercourses depending on the type of use or the type of the object
- Calculation of discharge for industrial and household waste water
- Monitoring and control of water quality in certain watercourses

The guidelines single out groups of water objects (watercourses, lakes or coastal waters) and areas with diverse characteristics.

Quality criteria (biological or chemical) are established for each type of water use determining its quality. These criteria serve as a benchmark for achieving and preserving water quality corresponding to a certain purpose.

The quality of Danish river waters (especially rivers taking water from sewage treatment facilities) is evaluated by utilizing *pollution level by organoleptic, physicochemical and biological parameters* [8]. In this case, general pollution level is determined in accordance with a classification scheme (Table 2.5). Water quality assessed with this method may be represented by color codes.

Table 2.5. Pollution level classification scheme

Pollution level	Saprobity zone	Water quality	Color code
I	Oligosaprobic	Practically unpolluted	Blue
I -II	Same	Same	Green blue
II	B-mesosaprobic	Light pollution	Green
II - III	Same	Same	Yellow green
III	cx-mesosaprobic	Quite high level of pollution	Yellow
IV	Polysaprobic	Severe pollution	Red
	Not classified		Red with shading

In Denmark, there is no generally accepted classification of water quality for lakes yet. Application of elemental classification schemes enables the assessment of water quality of lakes into three classes of quality. In France, systematic monitoring is performed by Basin

Agencies under the Ministry of the Environment. Moreover, the quality of all surface water is examined once every five years. Results thereof are published in the Inventory of Pollution of Surface Waters, Rivers and Channels [9]. The survey (dated 1981) included 1,250 posts on water objects in the country, which were used for taking and analyzing samples with application of the Woodiwiss biological system for stream classification [6]. The primary classification scheme used in *France* for rivers and channels is the Multipurpose Scale [10], which is widely applied by monitoring organizations, including Basin Agencies and the Ministry of the Environment (Table 2.6).

The Multipurpose Scale was established to determine whether the quality of water meets the requirements related to a certain type of use (Table 2.7). This scale may be classified as an interpretive rather than a rating scale. Attempts to describe water quality more accurately have resulted in the development of additional classification schemes (scale II and scale III [10], consisting of six and seven classes respectively, are known as high quality water classification scales). Only the Multipurpose Scale with color coding is used out of all the classification schemes established in France, however, for mapping the quality of surface water.

Quality of lake waters in *Germany* is classified by trophic state and is assessed in accordance with the German standard methods of water analysis by mapping with the use of color coding. The classification scheme is comprised of four classes of lake waters (Table 2.9).

Table 2.6. Multipurpose Scale

Class	Quality criteria for specific types of water use	Quality	Color code
1A	Clean waters suitable for purposes with the highest quality requirements	Excellent	Blue
1B	Waters of slightly worse quality though suitable for any type of use	Good	Green
2	Waters of satisfactory quality conforming to the requirements of irrigation and industrial use, suitable for drinking only after thorough treatment	Satisfactory	Yellow
3	Waters of moderate quality suitable for irrigation, cooling and navigation. Fish may live in such waters, but during certain periods, for example draught or high temperature, they are unlikely to be encountered	Moderate	Orange
Beyond classification	Waters which exceed one or several parameters of the third class. Considered unsuitable for most types of use and may endanger human health and the environment	Bad	Red

Table 2.7. Values of parameters corresponding to quality classes of the Multipurpose Scale

Parameter	Class			
	1A	1B	2	3
Electric conductivity, mmho/cm (at 20°C)	< 400	400-750	750-1,500	1,500-3,000
Temperature, °C	< 20	20-22	22-25	25-30
pH	6.5-8.5	6.5-8.5	6-9	5.5-9.5
Suspended matters, mg/dm^3	< 30	< 30	< 30	30-70
Dissolved oxygen, mg/dm^3	> 7	5-7	3-5	Aerobic environment must be constantly maintained
Dissolved oxygen, saturation %	> 90	70-90	50-70	
BOD_5, mg/dm^3	< 3	3-5	5-10	10-25
Oxidizability, mg/dm^3	< 3	3-5	5-8	—
COD, mg/dm^3 of atomic oxygen	< 20	20-25	25-40	40-80
NH_4^+, mg/dm^3	< 0.1	0.1-0.5	0.5-2	2-8
NO_3^-, mg/dm^3	—	—	< 44	44-100
Total Kjeldahl N	< 1	1-2	2-3	> 3
Fe, mg/dm^3	< 0.5	0.5-1	11.5	—
Parameter	Class			
	1A	1B	2	3
Mn, mg/dm^3	< 0.1	0.1-0.25	—	0.25-0.50
F^-, mg/dm^3	< 0.7	0.71.7	0.7-1.7	> 1.7
Cu, mg/dm^3	< 0.02	0.02-0.05	0.05-1	> 1
Zn, mg/dm^3	< 0.5	0.5-1	1-5	> 5
As, mg/dm^3	< 0.01	< 0.01	0.01-0.05	> 0.05
Cd, mg/dm^3	< 0.001	< 0.001	< 0.001	> 0.001
Cr, mg/dm^3	< 0.05	< 0.05	< 0.05	< 0.05
Cn^-, mg/dm^3	< 0.05	< 0.05	< 0.05	< 0.05
Pb, mg/dm^3	< 0.05	< 0.05	< 0.05	< 0.05
Se, mg/dm^3	< 0.01	< 0.01	< 0.01	> 0.01
Hg, mg/dm^3	< 0.0005	< 0.0005	< 0.0005	> 0.0005
Phenols, mg/dm^3	—	< 0.001	0.001-0.05	0.05-0.5
Synthetic surfactants, mg/dm^3	< 0.2	< 0.2	0.2-0.5	> 0.5
CCE, mg/dm^3	< 0.2	0.2-0.5	0.5-1	> 1
Coliforms per 100 ml	< 50	50-5,000	5000-50,000	—
E. Coli per 100 ml	< 20	20-2,000	2000-20,000	—
Fecal streptococcus per 100 ml	< 20	20-1,000	1000-10,000	—
Difference between biotic and normal indices	1	2-3	4-5	6-7

Notes:

[1] Each parameter characterizes the given quality of water if the most unfavorable changes reach the corresponding lower limit in at least 10% of measurements.

[2] Overall assessment of water quality is made on the basis of the worst quality parameters.

Table 2.8. German surface water quality classification

Class	Organic load	Saprobity zone	Saprobity index	BOD_5, mg/dm^3	NH_4^+ — N, mg/dm^3	Dissolved oxygen, mg/dm^3 (minimal value)
1	None or very low	Oligosaprobic	1.0- < 1.5	1	Traces	> 8
1-2	Low	Transitional from oligosaprobic to β-mesosaprobic	1.5- < 1.8	1-2	< 0.1	> 8
2	Moderate	β- mesosaprobic	1.8 - < 2.3 73 <2.3	2-6	< 0.3	> 6
2-3	Critical	α-β- mesosaprobic boundary	2.3- < 2.7	5-10	< 1.0	> 4
3	High	α- mesosaprobic	2.7- < 3.2	7-13	0.5-1.0	> 2
3-4	Very high	Transitional from β-mesosaprobic to polysaprobic	3.2- < 3.5	10-20	Several units	< 2
4	Excessive	Polysaprobic	3.5- < 4.0	> 15	Same	< 2

Table 2.9. Lake water quality classification scheme

Trophic state	Description of water	Color code
Oligotrophic	Clean, free from biogenic matters, small planktonic products; oxygen saturation more than 70% at the end of stratification period	Dark blue
Mesotrophic	Small biogenic loads, medium planktonic products, transparency more than 2 m; oxygen saturation 30-70% even at great depths at the end of stratification period	Blue
Eutrophic	High content of biogenic matters at the end of stratification period; oxygen saturation 0-30% at great depths, surface is oversaturated with oxygen; in most cases transparency is less than 2 m; large planktonic products	Green
Polyeutrophic	High biogenic load within the whole period; in summer, oxygen is absent at great depths, formation of hydrogen sulfide is observed from time to time; surface is highly oversaturated with oxygen; transparency is very low; intensive phytoplankton products	Yellow

Germany applies such methods for classification of river and lake water quality based on German standard methods of water analysis [11] and provides for the carrying out of biological and environmental research together with the determination of physicochemical parameters. The combination of these different assessment principles allows reliable information to be obtained on the quality of water.

The foregoing system includes seven classes of water quality. Relations between them and the level of contamination of watercourses with organic substances, the saprobity index and corresponding chemical parameters (BOD_5, ammonia nitrogen and dissolved oxygen content) is shown in Table 2.8.

The quality of river water is also classified in accordance with the German standard methods of water analysis through mapping with the use of color coding presented in Table 2.10.

In *Ireland*, information on water quality is mainly based on hydrobiological observations, while the study of physicochemical parameters is quite limited.

The method used for biological assessment of water quality was developed by Toner (Foras Forbartha, National Institute for Physical Planning and Construction Research) on the basis of the Trent Biotic Index [12, 13]. The method stipulates field inspection of microinvertebrate communities inhabiting river and stream bottoms. The method is used for examining two types of pollution influence:

- Decreased microinvertebrate species diversity
- Increased extinction of certain groups typical for clean water fauna as a result of pollution

Biological information is presented in an easy-to-interpret form by means of a five-grade scale (Q-value) of an arbitrary system uniting community composition and water quality (Table 2.11).

In addition, the following intermediate evaluations may be used if necessary: Q 1-2, Q 2-3 and Q 4-5.

Table 2.10. River water quality classification scheme

Water quality class	Contamination level	Color code
1	From zero to light	Dark blue
1-2	Light	Blue
2	Medium	Dark green
2-3	Critical	Light green
3	High	Yellow
3-4	Very high	Orange
4	Excessive	Red

Table 2.11. Water quality indices (five-grade scale)

Quality rating (Q-value)	Water quality
5	Good
4	Satisfactory
3	Suspicious
2	Low
1	Poor

In cases when biological characteristics indicate the presence of toxic pollutants, the corresponding Q-value is added by O indicator (for example, Q 1/O, Q 2/O or Q 3/O). For the purpose of simplification, three main classes of water quality are distinguished: A, B and C. These classes are connected with the five-grade scale and describe contamination levels (Table 2.12).

Table 2.12. Water quality classification (three-grade scale)

Quality rating (Q-value)	Quality class	Contamination level	Color code
Q5; Q4-5; Q4	A	Unpolluted	Blue
Q3-4; Q3; Q2-3	B	Light-medium	Yellow
Q2; Q1-2;Q1	C	Serious	Red

Quality classes are primarily characterized by relative proportions of four groups of water organisms: sensitive, less sensitive, persistent and most persistent species.

Relations shown in Table 2.12 were later updated and presented in the form of a biotic classification of water quality. In Ireland, a classification scheme (Fixed Limits) based on international studies on eutrophication of lakes is used. The studies are coordinated by the Organization for Economic Cooperation and Development (OECD) [14].

Unpolluted lakes are classified as unproductive and productive, the difference between which is determined by the quantity of delivered inorganic biogenic matters:

- The first type (oligotrophic) is fed by small quantities.
- The second type (eutrophic) is well fed.
- An intermediate level of nutrition is specific to mesotrophic lakes.

It is generally known that natural eutrophication is quite a slow process. If the delivery rate of biogenic substances increases due to the discharge of pollutants, then the process of eutrophication quickens considerably and productiveness level sharply rises. The process of eutrophication of such lakes is compared with an unpolluted lake exposed to the same conditions. This comparison enables the evaluation of the degree of pollution present and may be used for classification of lake water quality. The difference was established between oligotrophic and eutrophic conditions by comparing typical, easily observed biological and physicochemical parameters. The OECD has made an attempt to develop quantitative classification acceptable for application within water quality management systems (Table 2.13).

Table 2.13. Classification scheme based on studies by OECD Working Group

Trophic level	P_{tot}, mg/m^3	Chlorophyll "a", mg/m^3		Transparency, m	
		Average	Minimal	Average	Minimal
Ultraoligotrophic	< 4	< 1	< 2.5	> 12	> 6
Oligotrophic	< 10	< 2.5	< 8.0	> 6	> 3
Mesotrophic	10-35	2.5-8.0	8-25	6-3	3-1.5
Eutrophic	35-100	8-25	25-75	3-1.5	1.5-0.7
Hypereutrophic	> 100	25	> 75	< 1.5	< 0.7

At the same time, the foregoing classification scheme includes not only three major trophic levels, but also two additional levels—ultraoligotrophic and hypereutrophic. The classification suggested by the OECD is quite relative because transition from one level to another is gradual [14]. This system, however, is quite useful because it allows the standardization of the approach toward determining trophic state and excess anthropogenic eutrophication caused by contamination of lakes.

In the *Netherlands,* a number of classification schemes are applied, including the one that is used in Belgium: surface water quality assessment by water oxygen balance criterion (WOB) [15].

A separate classification of waters (especially lakes and reservoirs) concerns total phosphorus content due to its limiting role in the processes of eutrophication. MPC of P_{tot} in the Netherlands equals 0.2 mg/dm^3. Allowing for the phenomenon of eutrophication during six months of the warm season, the reference value was set at 0.3 mg/dm^3. The elemental scale for comparing obtained content values with maximum values was chosen as the classification scale.

Thus, the three following gradations were established: lower than standard, of the same order and higher than standard. The classification is based on color coding (Table 2.14). Within the framework of the National Monitoring System, contents of the following six priority metals are controlled: mercury, cadmium, copper, lead, chromium and zinc.

Table 2.14. Relation of P_{tot} content with classification color code

Color code	P_{tot}, mg/dm^3 (from April to September)
Blue	0.10
Green	0.11-0.20
Yellow	0.21-0.30
Orange	0.31-0.74
Red	0.75 and more

Three levels of quality are intended for each metal. In accordance with the established standards [15] the following color codes are used: blue—up to 20% of MPC, yellow—from 20% to 100% of MPC and red—exceeding MPC. A circle divided into six segments representing each metal is used for mapping convenience. This circle is located on the map at the control point and colors of the corresponding sector represent the level of contamination produced by the given metal.

Water quality classification by level of contamination with organic substances is performed in accordance with the following groups: polycyclic aromatic hydrocarbons, organochlorine pesticides, chlorinated hydrocarbons, anionic surfactants and oil products.

The classification principle is the same as for metals. Color coding has two gradations: blue—less than MPC and red—equal to or exceeding MPC. If the parameter was not measured, then the sector is painted yellow.

In *Great Britain* (England, Wales and Northern Ireland), the scheme by the British National Water Council (NWC) is used for the classification of river and channel water quality. This scheme is based on the determination of quality criteria necessary for specific types of water use and consists of four main quality classes (the highest quality class is divided into several subclasses) (Table 2.15).

Table 2.15. Classification of rivers by British NWC

Quality class	Water quality	Quality criteria	Notes	Consumption
1A	Good	Dissolved oxygen more than 80% of saturation $BOD_3 \leq 3$ mg/dm^3 $[NH^+] \leq 0.4$ mg/dm^3 The water must meet the requirements of A2 class (EEC) if it is used for drinking Nontoxic for fish in accordance with EIFAC criteria	Average value of $BOD_3 \leq 1.5$ mg/dm^3 There must be no external features of contamination	Suitable for drinking Commercial fishery for precious species, recreation
1B	Good	Dissolved oxygen more than 60% of saturation $BOD_5 \leq 5$ mg/dm^3 $[NH_4^+] \leq 0.9$ mg/dm^3 The water must meet the requirements of A2 class (EEC) if it is used for drinking	Average $BOD_5 \leq 2$ mg/dm^3 Average value of $[NH_4^+] \leq 0.5$ mg/dm^3 High quality waters which may not be attributed to 1A class due to impact of sewage, low stream slope and eutrophication	Same in whole
		Nontoxic for fish in accordance with EIFAC criteria	1A and 1B form the waters of the first class	
2	Satisfactory	Dissolved oxygen more than 40% of saturation $BOD_5 \leq 9$ mg/dm^3 The water must meet the requirements of A3 class (EEC) if it is used for drinking Nontoxic for fish in accordance with EIFAC criteria	Average value of $BOD_5 \leq 5$ mg/dm^3 Waters having no other physical signs of contamination except for humus coloring and light foaming below floodgates	Suitable for drinking after pretreatment Commercial fishery for small fish
3	Low	Dissolved oxygen more than 20% of saturation Aerobic conditions are unlikely $BOD_5 \leq 17$ mg/dm^3	—	Suitable for recreation
4	Poor	All criteria below the third water quality class	—	

Notes:

[1] Usually classification is based on the values of quality criteria which should be expected in 95% of samples taken. These restrictions are also related to the limits established by the EIFAC (European Inland Fisheries Advisory Commission).

[2] The definition of "nontoxic" for fish is in accordance with EIFAC criteria based on approaches to the classification stated in Water Quality Criteria for Freshwater Fish.

[3] Under extreme weather conditions (flood, drought, freezing) or during growth/decomposition of aquatic vegetation, values of BOD_3, dissolved oxygen or NH^+_4 in waters of 1-3 classes may be exceeded. When this happens, the causes should be ascertained by means of analyses.

[4] In most cases the foregoing chemical classification will be acceptable. The classification, however, is limited by the number of determined chemical parameters, thus chemical compounds not covered by the scheme may decrease the quality of water considerably. In such cases, water class must be lowered on account of actual biota and causes affecting it.

The quality of river waters related to biological parameters is estimated in accordance with the counting system, which was developed for water quality research conducted in 1980 [7]. Sampling posts were located down the stream in order to describe tributaries better than specific sources of pollution. Sample water quality was determined in the following order:

1) All present species communities were described
2) Each of these communities were given points in accordance with Table 2.16
3) Points corresponding to all species communities were summed up to calculate the final grade for the given observation post.

The higher the water quality by biological indices is, the higher the grade.

In *Scotland*, the Development Department together with the Agency for Cleaning Rivers, Clyde, Solway and Tweed, have developed water quality classification schemes. One of these schemes—water quality index (WQI)—includes the following water analyses: dissolved oxygen, BOD, ammonia and ammonia ions, pH, total nitrogen, phosphates, suspended matters, temperature, electrical conductivity and Escherichia Coli. Obtained data is represented as a number characterizing the quality of water, which is calculated on the basis of a method close to the one used in Germany [11, 17]. This approach is not applied by all agencies and is seen as useful for internal operational use.

Due to diverse geographical conditions and types of river use in Scotland, England and Wales, it was considered unacceptable to apply a unified system for chemical classification of river waters. Under these circumstances, a classification scheme (1974) was retained. The scheme is much like the one used in England and Wales, which presents the overall picture of water conditions. The Scottish scheme (system for chemical classification of rivers) is shown in Table 2.17.

Biological classification is based on the system for the distribution of points between biological species used in England and Wales (Table 2.15).

Table 2.16. Distribution of points between biological species

Species	Points
Siplilonuridae Heptageniidae Leptophlebiidae Ephemerellidae Potamanlhidae Ephemerldae Taeniopterygidae Leuctridae Capniidae Periodidae Perlidae Chloroperlidae Aphelocheiridae Phryganeidae Molannidae Beraeidae Odontoceridae Leptoceridae Goeridae Lepidostomatidae Brachycentridae Sericostomatidae	10
Astacidae Leslidae Agriidae Gomphidae Cordulegasteridae Aeshnidae Corduliidae Libellulidae Psychomyiidae Philipolamidae	8
Caenidae Nemouridae Rhyacophilidae Polycentropodidae Limnephilidae	7

Species	Points
Neritidae Viviparidae Ancylidae Hydroptilidae Unionidae Corophiidae Gammaridae Platychemididae Coenagriidae	6
Mesovelidae Hydrometridae Gerridae Nepidae Naucoridae Notonectidae Pleidae Corixidae Haliplidae Hygfobiidae Dytiscidae Gyrinidae Hydrophilidae Clambidae Helodidae Dryopidae Elminihidae Chrysomelidae Curculionidae Hydropsichidae Tipulidae Simuliidae Planariidae Dendrocoelidae	5
Baetidae Sialidae Piscicolidae	4
Valvatidae Hydrobiidae Lymnaeidae Physidae Planorbidae Spliacriidae Glossihoniidae Hiruididae Erpobdellidae Asellidae	3
Chironomidae	2
Oligochaeta	1

Table 2.17. Chemical classification of rivers

Quality class	Water description	Color code
1	Unpolluted rivers and rivers recovered from pollution Sections of rivers: clean water, no significant discharges; clean water, BOD < 3 mg/dm^3>, high oxygen saturation, no significant discharges of toxic and suspended matters affecting the state of river bottoms	Blue
2	High quality rivers Sections of rivers: not included in the first class on account of BOD values, considerably decreased dissolved oxygen content, presence of polluting discharges containing toxic substances; moreover, such substances cannot either affect fish or be removed by means of natural processes	Orange
Quality class	Water description	Color code
3	Low quality rivers Sections of rivers: not included in the fourth class on account of BOD values, content of dissolved oxygen within prolonged periods is not less than 40% of saturation, water contains concentrations of substances that may reach toxic levels	Red
4	Heavily polluted rivers Sections of rivers: water has repulsive appearance or odor, under normal conditions BOD values equal 12 mg/dm^3 or more, poorly saturated with oxygen, water contains concentrations of substances that reach toxic levels from time to time, knowingly uninhabitable for fish	Black

2.2. USGS National Water-Quality Assessment Program

Full-scale implementation of the National Water-Quality Assessment Program (NAWQA) began in 1991; prior to that, it was developed and tested by the USGS.

Primary objectives of NAWQA consist of the following:

- To provide nationwide description of modern water quality for most of the country's water resources
- To determine the presence or absence of long-term water quality trends
- To establish, describe and explain, if possible, main factors affecting the states and tendencies of change in the quality of water

With due account to NRC's recommendations, NAWQA provides for the following:

1. The study of 120 areas around the country forming approximately 80% of total water use in the USA. Studied areas (objects/units under research) include river basins and the main focus is on the quality of surface/ground waters and aquifers with an area of several thousand miles to several tens of thousands square miles. National assessment of water quality includes the study of rivers, reservoirs, estuaries, and, in the first place, lakes, which have a major impact on the quality of river and ground waters.

2. The study of each object of research is carried out by a small group of experts familiar with a certain river basin or aquifer.

3. Objects of research are interconnected in several ways and form a unified national program. The program provides for the use of a previously approved set of approaches and sampling methods, order for carrying out sampling, laboratory analysis and monitoring of performance. Data is acquired and interpreted in accordance with a unified, national set of water quality parameters. Written reports are prepared for each object of research. Observation data is stored in the national database.

4. Much attention is paid to the assessment of changes in water quality; therefore, NAWQA is intended to be practiced for a very long-term. It should evolve with time as a result of changes in scientific knowledge in the fields of hydrology, hydrochemistry and hydrogeology; improvement of measuring methods and emergence of new types of pollutants. The program, however, is focused on measurements repeated over time and thorough documentation of both data acquisition/analysis methods and location/characteristics of sampling posts.

5. For each object of research, works related to quality assessment are carried out on a recurring, but not permanent, basis. Only a certain objects of research are intensively studied in the given period of time. For each object of research, 3-5 year periods of intensive acquisition and analysis of data are alternated with longer periods of less intensive assessment works.

6. The program is concentrated on the study of large-scale or long-term issues of water quality.

7. Measurement of chemical parameters is focused on a number of goal variables. The national list of goal variables consists of a general set of physical measurements and the detection of inorganic and organic compounds included in the sample analysis for all objects

of research. These parameters allow the assessment of water quality on the national level. An additional set of targets is also selected for each object of research.

8. Measurement of biological parameters is used for the following purposes:

- For the detection of occurrence rate and distribution of waters contaminated with feces
- For the detection of occurrence rate and distribution of potentially toxic substances
- For the evaluation of the relationship between physical/chemical characteristics of rivers and functional/structural parameters of biological communities
- For the determination and quantitative description of biological processes affecting physical and chemical parameters of water quality

9. The program provides for wide use of modeling on all stages of water quality assessment, including the planning of the monitoring system, data analysis and evaluation of management decisions.

10. The program employs data acquired on the quality of water in addition to the results of field surveys.

11. The main types of work related to the sampling of river basins include:

a) sampling at stationary posts,
b) synoptic sampling, and
c) the study of selected river sections.

Sampling at stationary posts is done for the:

- Description of temporary changes and recurrence of selected water quality parameters
- Evaluation of washout of substances by river run-offs at the studied areas and in an attempt to establish mass balance for chosen targets between posts
- Determination of long-term water quality trends

Synoptic sampling is mainly used in order to ensure a more detailed (as compared to the capabilities of stationary networks), special resolution while describing certain types of water quality conditions.

A ***synoptic survey*** is generally carried out by measuring unified water quality parameters under specific hydrological conditions at most posts of the researched area within a short period of time.

Study of selected river sections is conducted to solve more specific problems of water quality revealed in the process of research.

The main objective of this research is to provide detailed study of sources for selected river sections. Results of mathematical modeling performed in the process of this research are used for a more complete understanding of the cause-and-effect relations and for the assessment of alternative management strategies. Modeling results may be checked in subsequent research cycles in order to determine whether there are any changes in conditions and to what extent such changes correspond to previous expectations.

Implementation of the National Assessment Program enables researchers to obtain unique information on the quality of water on regional and national levels and the importance of such information can scarcely be overestimated.

1. Information obtained by NAWQA is useful for people involved in developing policies, elaborating rules, setting priorities or managing water resources. Data on water quality conditions and tendencies, including information on its determinant sources, is especially important for agencies engaged in:

- Detecting key substances subject to regulations that require the implementation of studies related to toxicity and impact on humans, and ascertaining treatment possibilities
- Distributing financial resources between competitive types of problems in the sphere of water quality
- Determining the successfulness of solved issues related to water quality
- Designing monitoring programs for different parts of the country (from the perspective of selecting analyzed parameters, places, frequency and terms of sampling)
- Determining the impact of different types of point and nonpoint sources on the quality of water
- Establishing the water quality of aquifers, which require protection
- Assessing management types from the viewpoint of their large-scale impact on the quality of waters of river basins and aquifers

2. Statistical description of water quality conditions and their changes over time is made for most of its components (pesticides, synthetic organic substances, biogenic matters, metals and trace elements). Statistical descriptions include the following information:

- The occurrence rate of water quality parameters, rate of standard water qualities transcendence, concentrations of water quality parameters in key points of rivers and on typical sections of aquifers
- The washout of key substances by river run-offs in selected points of rivers
- The long-term trends of changes in water quality at key points of rivers and on typical sections of aquifers

3. Information on geographical distribution of different pollutants for primary USA river basins and aquifers is used to establish the areas with poor and usually good water quality. This information may be useful for the allocation of water quality management and protection resources. Depending on the spatial distribution of indices, geographical overview includes the following:

- Maps with locations of sampling posts and results of water quality analyses
- Descriptions and maps with marked sections of rivers and zones demarcating differences in average concentration values or in the rate of standard water qualities transcendence
- Descriptions and maps of areas with water quality anomalies spreading over vast territories (tens of river miles or tens of square miles in aquifers)

- Descriptions of water quality changes in ground water depth under different hydrological conditions

4. Acquisition of information related to the main factors affecting the quality of water. This knowledge enables the establishment of correlations between the occurrence rate/concentration of key parameters and the different types of conditions, land use and anthropogenic activities.

Understanding these interrelations provides the basis for prediction of the place and time of probable occurrences of different types of pollution. It is also useful in determining the large-scale impact of alternative water quality management strategies. This information includes:

- Statistical models of expected contamination levels or the probability of exceeding certain levels, which are based on easily calculated variables connected to soil, hydrogeology, land use, population, river run-offs and climate
- Explanation of causes for observed space-time changes in the quality of water
- Information on washout of substances by river run-offs from different point and nonpoint sources
- Description of factors affecting the protection of aquifers from pollution

REFERENCES

[1] Nikanorov A.M., Tsirkunov V.V. Systems for Surface Water Quality Monitoring. Part 1. — *St. Petersburg: Hydrometeoizdat,* 1994. —108 p.

[2] Kimstach V.A. Classification of Surface Water Quality in the European Economic Community countries.

[3] Brown R.M., McLelland N.J., Deininger R.A., Fozer R.Z. A water quality index — do we dare? // *Water and Sewage Works.* — 1970.

[4] A review of water pollution in Ireland // *A report to the Council by An Foras Forbartha. — Irish Water Pollution Advisory Council,* 1983.

[5] De Pauw N., Vanhooren G. Method for biological assessment of watercourses in Belgium // *Hydrobiologia.* — 1983. — V. 100.

[6] Woodwiss F. S. The biological system of stream classification used by the Trent River Board // *Chemistry and Industry.* — 1964. — V. 11.

[7] Tuffer G., Verneaux J. Methode de determination de la qualite biologiques des eaux courantes. Exploitation codifiee des inventaires de la faune du fond. Ministere de l'Agriculture (France) Centre National d'Etudes techniques et de recherches technologiques pour l'agriculture les forets et l'equipment rural CERAFER Section Peche et Pisciculture, 1968.

[8] Guidelines for surface water quality planning. — Danish National Agency of Environmental Protection. — Part I. *Water coastal and lakes.* — 1983. — Report N 1; Part II. *Coastal waters.* — 1983. — Report N 2.

[9] Bureau M., Duport L. Investigation of the results of national inventory of water pollution using equivalence factor analysis method. — French Ministry of Agriculture, 1975.

[10] Duport L., Margat J. Measurement of the quality of waters — methods used to show the quality. — French Ministry of the Environmtnt, 1983. — Report N DPP/ SEB /LD/ SB.

[11] Landwehr J.M. Water quality indices — construction and analysis. — Ph. D. dissertation. — The University of Michigan, 1974.

[12] Sladecek V. System of water quality from the biological point of view // *Arch. Hydrobiol. Beih.* — 1973. — V. 7.

[13] The National Survey of Irish Rivers 1982/83. Biological investigations of quality in selected rivers and streams. — *An Foras Forbartha. National Institute for Physical Planning and Constraction Research.* — 1984.

[14] Eutrophication of waters, monitoring, assessment and control. — *OECD,* 1982.

[15] Tittizer T.G., Miiller D. Comparative study of biological water assessment methods. Practical demonstration of the River Trend and its tributaries // *2-nd Technical Seminar of EC Working Group «Biological monitoring for assessment of surface water quality».* — 1976.

[16] Development of a water quality index. — Scottish Development Department, 1976. — *Report N ARD-3.*

[17] Water Action Programmer for the Netherlands. 1980—1984. — Dritch Ministry for Transport and Public Works, 1980.

SECTION II.
RUSSIAN STATE SURFACE WATER MONITORING SYSTEM

Chapter 3

SURFACE WATER MONITORING IN THE RUSSIAN FEDERATION

In the present time, environmental monitoring shall be understood as a complex system of observations over environmental conditions, assessment and forecast of their changes under the influence of natural and anthropogenic factors [1—5].

3.1. HISTORICAL NOTE

On the territory of the Union of Soviet Socialist Republics (USSR) and the Russian Federation (RF), as a constituent part of the USSR, monitoring of surface waters was carried out from the mid-thirties of the XX century by the Russian Federal Service for Hydrometeorology and Environmental Monitoring (Roshydromet) (formerly named Glavhydrometservice, USSR Hydrometservice, Roskomhydromet, etc.). Development of the service may be divided into several stages. The *first stage* (from approx. 1936) was devoted to observations concerning the chemical composition of natural waters. During the initial period, the observations were irregular with regards to space and time, and during the War (1941-1945) they were reduced on the European part of the country and increased on the Asian part. In the postwar period, the observations were continued on the European part of the country. The most important observations were distinguished from the rest and were used as the basis for the establishment of a surveillance monitoring network. At most posts, hydrochemical observation points matched hydrological points, because observations were organized on the basis of the hydrological network.

In the initial period, the observation program provided for the measurement of physical water properties (odor, taste, color, transparency), its temperature, pH value, solids content, concentration of dissolved gases (oxygen, carbon dioxide, hydrogen sulfide), principle ions (chloride, sulfate, hydrocarbonate, calcium, magnesium, sodium, potassium, ion sum), biogenic matters (ammonium, nitrite and nitrate ions) and organic substances resistant to oxidation (permanganate value). The program was eventually complimented with measurements of total iron, silicon, and phosphates present, while the measurement of organic substances resistant to oxidation by permanganate value was complimented with or replaced by dichromate oxidizability. The following program was named "Standard" for convenience.

In the *second stage* (from approx. 1965), contamination of water bodies drew more attention. First, it became apparent in the observation program that in addition to the standard program, a program named "Pollution" needed to be introduced. This program was intended for the measurement of water pollutants including phenols (volatile), oil products (ether-extractable substances), metal compounds, anionic synthetic surfactants (ASS) and other substances. Later on, the list of substances was extended.

The number of observation posts established in the areas of increased anthropogenic impact was growing.

The beginning of the *third stage* is connected with the USSR Government Order No. 898 "On Enhanced Protection of Nature and Improved Use of Natural Resources," dated December 29, 1972 and the Order (1977) on the establishment of the State Service for Observation and Monitoring of Pollution of the Objects of Nature (SSOMPN), which was further renamed to the State Service for Monitoring of Environmental Conditions (SSMEC). The SSOM was established mainly on the basis of the monitoring network developed by the Federal Service for Hydrometeorology and Environmental Monitoring, including a small number of observation posts pertaining to other ministries and agencies (Ministry of Public Health, Ministry of Land Reclamation and Water Management, Ministry of Agriculture, etc.).

Basic organizational and methodological principles of SSOM's functioning [3] were defined in the third stage of the development of the water quality monitoring system. These principles were further updated by the regulatory documents [6—8] and other governing documents.

In the following years, new principles were introduced and improved, the network of observation posts was rearranged and the monitoring of hydrochemical, hydrobiological and hydrological parameters was integrated. Systems of other ministries and agencies involved in surface water monitoring were developing simultaneously (Ministry of Land Reclamation and Water Management, Ministry of Public Health, Ministry of Fishery and Ministry of Environment). Each of the foregoing ministries had its own monitoring system and carried out its own nature protection policy.

The first attempt to unite the efforts of all the different ministries and agencies was made during the establishment of the SSOM. The attempt, however, turned out to be unsuccessful due to interdepartmental barriers. The scope of activities wasn't clearly distributed between different ministries and agencies. Water users began to control the quality of incoming water for processing needs and discharged into water bodies (measurements were made "in the pipe" and in a water body above and below sewage discharge within a half-kilometer zone in order to control the overall performance of sewage treatment facilities). The Ministry of Land Reclamation and Water Management was in charge of the activities of water users, of keeping the cadaster of waste waters and, if necessary, they were in charge of monitoring certain areas of water bodies. The State Committee on Sanitary and Epidemiological Surveillance was involved in monitoring the quality of water near sources of centralized water supply, drinking water and recreation zones. The Hydrometeorological Service fulfilled the function of monitoring surface water contamination within the framework of the SSOM.

Over the last decades (80-90s of the XX century), the structures and names of ministries and agencies changed and their functional duties related to surface water monitoring and water management were redistributed. These changes and new regulatory documents were aimed at strict coordination of activities in the field of monitoring and at the introduction of monitoring as an integral part of the water management system.

Apart from agencies ensuring the operation of permanent monitoring systems, the *Academy of Science* (USSR/RF) and *institutions of higher education* carried out a lot of work connected to various studies of water bodies. A significant amount of water quality observations was performed by *water users* (in the form of monitoring waste waters and water objects in the areas of sewage discharge).

3.2. Legal and Regulatory Framework of Monitoring

The Constitution of the Russian Federation preserves the right of every citizen to have a safe environment and its duty is to preserve nature/environment and to treat natural resources with care [9]. Nature preservation and environment improvement are rated as governmental and public priorities by the Ecological Doctrine of the Russian Federation [10]. Basic provisions of the Doctrine include the assessment of modern environmental conditions and their influence on the quality of Russian citizens' lives.

A significant legal and regulatory framework has been established in recent years. The most important part of the framework includes the Law on Environmental Protection (2002) [1] and the Water Code of the Russian Federation (1995) [11], both of which determine legal aspects of government policy in the sphere of interaction between society and nature arising from economic and other human activities affecting the natural environment.

Environmental monitoring carried out by governmental and local authorities of the Russian Federation is referred to as *state monitoring* [1, 11]. State monitoring is performed by the *State Service for Monitoring of Environmental Conditions (SSMEC)* under the control of the Russian Federal Service for Hydrometeorology and Environmental Monitoring (Roshydromet) and the Ministry of Natural Resources of the Russian Federation, with the participation of other federal and local executive authorities of the Russian Federation through a system of observation posts [12]. State surface water monitoring is a constituent part of environmental monitoring [11]. The order for the establishment and management of state monitoring, and the distribution of functions and duties are determined by the Order of the Russian Federation Government on Establishment of a Unified State System for Environmental Monitoring [13] and the Statute on State Monitoring of Water Bodies [14]. Other regulatory documents in the field pertaining to the use and protection of federal water resources include References [1, 2, 9—33].

A large base of national (state standards) and departmental (management directives, manuals, guides, etc.) regulatory documents has been developed. These documents determine the order and principles of implementation and the functioning of monitoring systems.

3.3. Structure of Monitoring

The Russian Federation, as well as some other foreign countries (see previous chapter), employs a sophisticated system for environmental monitoring, which includes several levels: administrative, departmental, component, spatial, purpose, etc.

Monitoring involves all *components of the natural environment*: land, subsoil, soil, water, atmospheric air, flora and fauna, including the ozone layer and circumterrestrial space.

The Federal Law of the Russian Federation on Environmental Protection provides for state environmental monitoring, control in the sphere of environmental protection (state, industrial, municipal and public) and scientific research in the field of environmental protection [1].

Monitoring is carried out by different *ministries and agencies* under the rules corresponding to their functional duties and in the order conditioned by their organizational management.

The *objects of monitoring* are natural resources and the sources of impact thereon. *The spatial component of monitoring stipulates observations on local, territorial, regional (basin) and federal levels* [14].

The following types of monitoring are distinguished in accordance with the objectives thereof: *surveillance, operational (under hazardous conditions) and special.*

This complex structure of environmental monitoring is also inherent in its constituent part—the monitoring of water bodies. Moreover, monitoring of water bodies presupposes:

- monitoring of surface water bodies on land and seas, and
- monitoring of underground water bodies.

In its own turn, regulations for monitoring of surface waters on land is applied to water and bottom sediments.

3.4. Grounds of Modern Monitoring Systems

During the last decade of the XX century, environmental monitoring provoked great interest due to the deterioration of environmental conditions caused by both a decrease in the effectiveness of nature protection measures and the increase in the number of industrial/transport accidents and disasters.

The system for surface water monitoring is related to information and measurements and is considered to be a system of regulated observations (with programmed space, time and component resolution). It is used for the assessment and forecast of the quality of water in water bodies.

The process of monitoring represents a sequence of interrelated operations from determining information needs to the use of information products. All these procedures comprise a monitoring cycle (Figure 3.1).

Consecutive operations of the monitoring cycle must be determined and planned with due account of the required information product and the specifications of the previous component in the cycle. Assessment of the obtained information may result in the occurrence of new information needs or in the adjustment of former needs and the initiation of a new sequence of operations for the purpose of improving the monitoring process.

Surface water monitoring is a constituent part of environmental monitoring, monitoring of water bodies and the water management system.

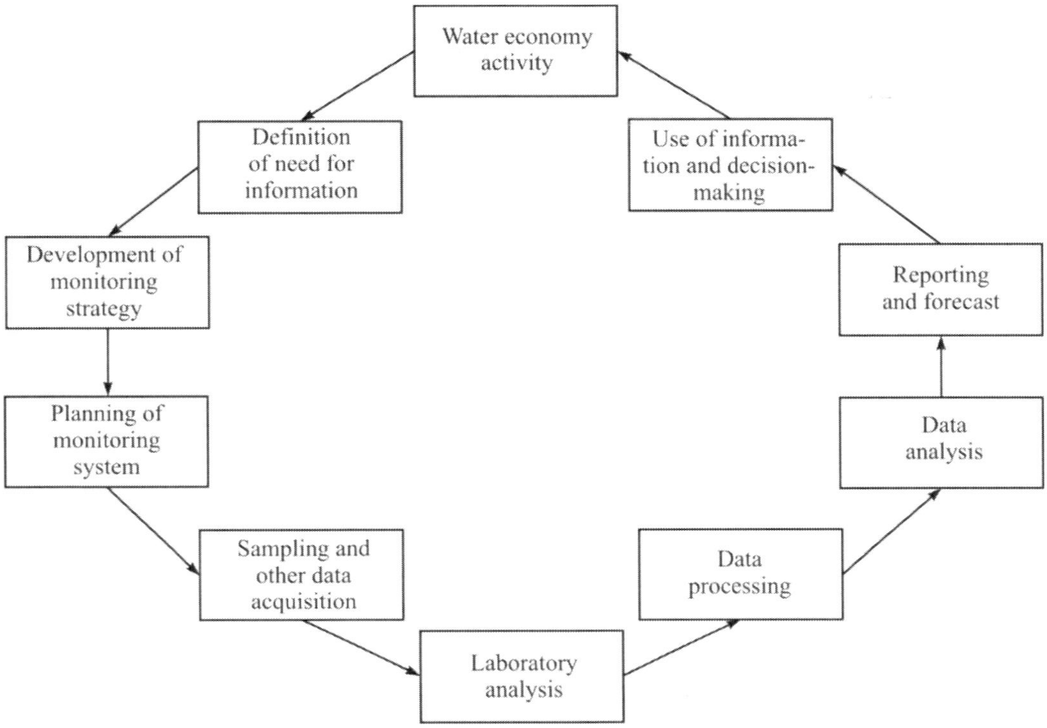

Figure 3.1. Monitoring cycle.

In accordance with the Statute on State Monitoring of Water Bodies, the monitoring is carried out for the following *purposes* [14]:

- Prompt detection and forecast of negative processes affecting the quality of water and conditions of water bodies
- Development and implementation of measures for preventing harmful consequences of such processes
- Assessment of water protection measures
- Information support and control in the sphere of use and protection of water bodies

The strategy for formation of the monitoring system is determined by the information needs of the water management system and the requirements of water users related to the quality of used water. Meanwhile, the integration of different approaches is also necessary; each of these approaches allows for the solution of a limited scope of problems regarding the management of water resources (basin approach), protection of ecosystems (ecosystem approach) or efficient control of pollution (approach based on the interaction between water quality and discharge into water bodies).

The monitoring system must comply with the following *requirements*:

- To ensure the acquisition of complete, reliable and comparable information on the state of a water body and the sources which may affect it

- To ensure the acquisition of information on a time scale allowing for immediate control actions, including cases of hazardous situations related to emergency contamination of water bodies
- To ensure the acquisition of information for the satisfaction of water management needs on local, territorial (basin) and federal levels
- To ensure timely disclosure of information to the consumer
- To be formed under unified methodological and systematic principles

The monitoring systems must *provide*:

- Prompt acquisition and disclosure of information for taking emergency, routine and long-term measures related to the management of water resources, the detection and prevention of environmental ill-being and calamity
- Databases containing information on water resources and ecosystems and their harmonization with international systems and accessibility of such information to a large consumer
- Implementation of unified scientific and technical policy in the field of water monitoring

As it was mentioned above, the following types of monitoring are distinguished as:

- Surveillance
- Operational, including cases of hazardous situations
- Monitoring of anthropogenic impact sources
- Special monitoring

While designing a modern, complex monitoring system, it is economically reasonable to form it on the basis of already operating systems of organized, methodological and systematic approaches. Meanwhile, one of the primary objectives is the establishment of a mechanism for interaction between systems and agencies without disturbing the existing administrative and departmental structures. A system should be aimed at minimizing material and financial costs and the effective use of available information on water bodies and scientific studies.

The system should be open and should provide an unhampered exchange of information with the possibility of applying other departmental systems and objects of monitoring, including other components of the natural environment. Formation and development of a monitoring system should rely upon a proper legal and regulatory framework, which determines the legal status for the system, its constituent parts, its structural elements and for the obtained information. Such a system should further describe the legal and natural responsibilities in the case of concealment, non-disclosure and deliberate misrepresentation of such information. It is also necessary to determine the rights and obligations of those who are involved in the monitoring process and to determine the levels of their responsibilities for implementation, improvement and development of monitoring.

3.5. Principles for Establishing Russian Operating Systems of Monitoring

State environmental monitoring is provided for by the Federal Law of the Russian Federation on Environmental Protection adopted in 2002 [1].

The allocation of functions and duties between ministries and agencies is determined by the Russian Federation Government Order on the Statute on State Monitoring of Water Bodies (1997) [14], in accordance with which state monitoring is carried out by the Ministry of Natural Resources of the Russian Federation (MNR) in cooperation with the Roshydromet (concerning surface water bodies) and other specially authorized public agencies involved in environmental protection, which include:

- The Russian State Fisheries Commission (state monitoring of fauna with regard to the assessment of conditions of water bodies as a habitat for aquatic animals and plants)
- The Ministry of Public Health (social and hygienic monitoring as related to the assessment of the quality of water used for drinking and domestic water supply, including the assessment of water bodies containing natural curative resources and water bodies used for recreation)

Functions of the Russian State Environmental Protection Committee (monitoring of anthropogenic impact on the environment concerning the assessment of anthropogenic impact on water bodies) prescribed by the Statute [14] after its integration with the Russian MNR were placed upon the latter.

Different ministries and agencies employ similar organizational principles of monitoring systems. All systems are based on a multilevel hierarchy. The base is formed of a network of observation posts controlled by committees, centers and departments. Each agency has its own dedicated scientific and research institutions carrying out scientific and methodological management of the system. Organizational structure is shown in the example of the Roshydromet (Figure 3.2).

Work within the systems is carried out under a defined cycle common for all agencies and similar to foreign systems (see Figure 3.1).

Activities of all systems are subordinate to a single objective stated in the Federal Law on Environmental Protection [1]: "State environmental monitoring is carried out in accordance with the federal and local laws of the Russian Federation for the purpose of observing environmental conditions, including environmental conditions of areas with sources of anthropogenic impact and influence of such sources in order to meet the national, legal and private demands in reliable information essential for the prevention and/or reduction of the unfavorable consequences of environmental changes." Each agency ensures the solution of problems connected with its functional duties.

These goals determine the differences in space, time and component parameters of the systems' functioning. Table 3.1 shows these differences and generalities in the example of parameters identified in surface waters of the Lower Don Basin in the Rostov Region. Names of organizations and agencies given below may be obsolete now due to the fact that the data was collected and analyzed in 1997.

Each agency provides its own subordinate production organizations with regulatory documents concerning the management and operation of monitoring systems.

The Roshydromet and State Sanitary and Epidemiological Surveillance (SSEM) have the most developed base of regulatory documents. Since a considerable part of SSEM's information products is supplied by the Roshydromet we shall describe features of the regulatory framework employed by this agency.

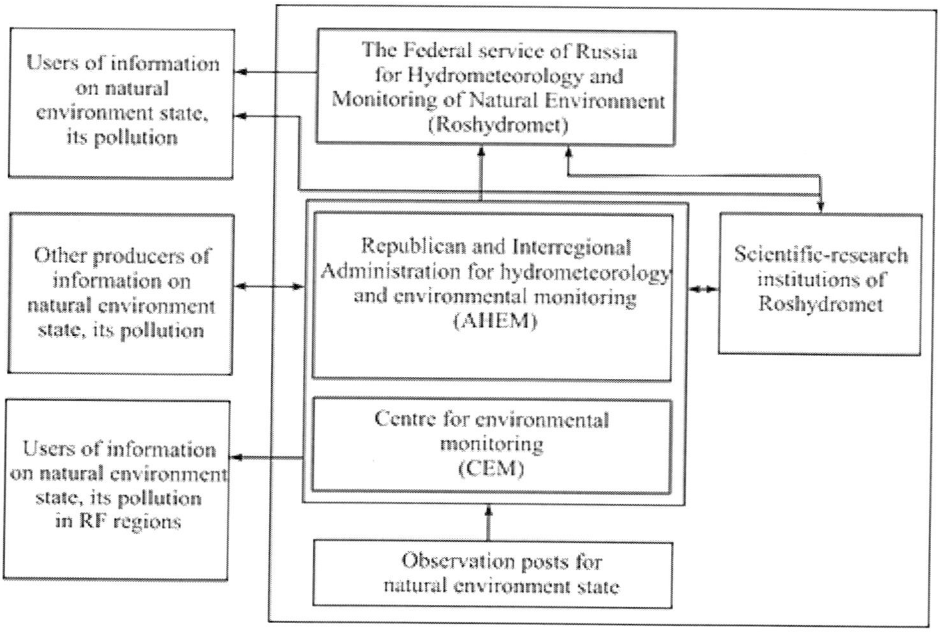

Figure 3.2. Organizational structure of environmental monitoring system employed by Roshydromet.

The major document initially developed by the SSOM was the "Practical Guidelines on the Organization and Implementation of Observations and Control over Surface Water Contamination within the System of the State Service for Observation and Monitoring of Pollution of the Objects of Nature (SSOM)." These guidelines were later revised into the State Standards [7] and the "Practical Guidelines on Organizational Principles of Water Quality Observation and Monitoring System for Water Bodies and Aquifers of Roshydromet's Network within the Framework of the SSOM [35]," and then into Regulatory Document 52.24.309—92 [36].

Table 3.1. Parameters identified in surface waters of the Lower Don Basin on the territory of the Rostov Region (1997)

Parameters	Monitoring and control systems and services				
	NCAHEM	DBWMA	RREPC	SSEPC	AFSRI
HYDROCHEMICAL					
Physical properties					
Temperature	+	+	-	-	+
Chromaticity	+	+	-	-	-
Color	-	-	-	+	-

Parameters	Monitoring and control systems and services				
	NCAHEM	DBWMA	RREPC	SSEPC	AFSRI
Transparency	+	+	-	-	-
Turbidity	-	-	-	+	-
Odor	+	+	-	-	-
Taste	-	-	-	+	-
Floating matters	-	-	-	+	-
pH value	+	+	+	+	
Specific conductance	-	+	-	-	-
Suspended matters	+	+	+	+	-
Dissolved gases					
Oxygen	+	+	-	+	+
Carbon dioxide	+	-	-	+	+
Hydrogen sulfide	+	-	-	+	+
Minerals					
Chlorides	+	+	+	+	+
Sulfates	+	+	+	+	+
Hydrocarbonates	+	+	-	-	+
Carbonates	-	+	-	-	
Calcium ions	+	+	-	+	+
Magnesium ions	+	+	-	+	+
Sodium ions	-	-	-	-	+
Potassium ions	-	-	-	-	+
Total sodium and potassium ions	+	-	-	+	+
Hardness	+	+	-	+	-
Alkalinity	-	+	-	+	+
Acidity	-	+	-	-	
Total ions	+	-	-	-	+
Solids content	-	+	+	+	-
Overall indices of organic substances					
COD	+	+	-	+	+
Permanganate value	-	-	-	+	+
BOD_5	+	+	+	-	+
BOD_{total}	-	-	-	+	-
Biogenic matters					
Nitrogen ammonia	+	+	+	+	
Nitrogen nitrite	+			+	+
Nitrogen nitrate	+	+	+	+	+
Phosphates	+	+	-	+	+
Total iron	+	+	-	-	+
Silicon	+	-	-	-	+
Total phosphorus	+	-	-	-	-
Polyphosphates	-	-	-	+	-
Orthophosphates	-	-	-	+	
Widespread organic pollutants					
Phenol	+	+	-	+	+
Oil products	+	+	+	+	+
Resins and asphaltenes	+	-	-	-	-
ASS	+	+	+	+	+
Pesticides	+	+	-	+	+
Metal compounds					

Table 3.1. (Continued)

Parameters	Monitoring and control systems and services				
	NCAHEM	DBWMA	RREPC	SSEPC	AFSRI
Copper	+	+	+	+	+
Zinc	+	+	+	+	+
Iron	-	-	-	+	+
Chromium	+	+	+	+	
Manganese	+	+	-	+	+
Mercury	+	-	-	-	+
Aluminum	+	+	+	+	+
Nickel	+	+	-	+	+
Cobalt	+	-	-	+	+
Tin	+	-	-	-	+
Titanium	+	-	-	-	+
Bismuth	+	-	-	-	-
Molybdenum	+	-	-	+	+
Vanadium	+	-	-	-	+
Lead	+	+	+	+	+
Silver	+	-	-	-	+
Barium	-	-	-		-
Cadmium	-	+	+	+	
Arsenic	-	+	-	+	+
Tungsten	-	-	-	-	+
Other pollutants					
Sulfides	-	+	+	+	-
Fluorine	-	-	+	+	-
Polyacrylamide	-	-	-	+	-
Fats	-	-	-		-
Residual chlorine	-	-	-	+	-
HYDROLOGICAL	+	-	-	-	+
HYDROBIOLOGICAL	+	+	-	-	+
MICROBIOLOGICAL	-	-	-	+	+
TOXICOLOGICAL (biotesting)	-	+	+	-	-
ICHTHYOLOGICAL	-	-	-	-	+
PHYSIOBIOCHEMICAL	-	-	-	-	+
ASTACOLOGICAL	-	-	-	-	+

Notes:

NCAHEM—North Caucasus Administration of Hydrometeorology and Environmental Monitoring (Roshydromet), DBWMA—Don Basin Water Management Association (Russian Ministry of Natural Resources), RREPC—Rostov Regional Environment Protection Committee (Russian State Environment Committee), SSEPC—State Sanitary and Epidemiological Supervision Center of the Rostov Region (Russian Ministry of Public Health), AFSRI—Azov Fisheries Scientific and Research Institute (Fisheries Department, Russian Ministry of Agriculture and Food Products)

Among the regulatory documents of other agencies, we should note the Sanitary Rules and Regulation on Environmental Protection of Surface Waters [37], List of Fishery Standards [38] and Sanitary and Hygienic Standards [39].

Observations of surface waters carried out by Roshydromet are divided into several types depending on their purposes: *surveillance, operational, background* and *special*.

The purpose of background observations is to study slow changes in the surface waters of the country. The obtained data represents a background for surveillance observations determining seasonal, annual and interannual changes. Local changes, as a rule, provoked by discharges of anthropogenic pollutants, occur against the background of surveillance changes. The purpose of special observations is to determine the content of works even though they are individual for each subgroup of these observations.

General information on the quality of surface waters is currently obtained within the framework of surveillance observations. Background observations are carried out practically under the program of surveillance observations, thus this type may be characterized as a special block on the basis of surveillance observations. In general, such organization of work is acceptable, though it results in a certain redundancy of these type of observations. Operational observations are poorly developed. Today, they are also based on surveillance observations with additional complex works carried out in case of emergency situations. This type of monitoring requires further development.

3.6. Practical Implementation of Monitoring in the Russian Federation

Different systems and services for observation and monitoring of water quality are functioning on the territory of the Russian Federation. As previously mentioned, state systems and services include the Roshydromet, the Ministry of Public Health, the MNR and the State Fisheries Commission. Production control is provided by water users. Observations are also conducted by organizations of other agencies and municipal institutions.

In general, a great number of observations over the quality of surface waters are carried out in the country. The systems of observation, however, are not united by a common objective of integration. Their departmental disconnection and poor coordination on the interdepartmental level lead to the following drawbacks:

- Fragmented systems for monitoring conditions and use of water resources lead to their poor commitment to management problems
- Absence of methodological and metrological software and hardware connections of operating services
- Wasteful use of funds due to partial overlap of observations
- Incomparability of data obtained by different agencies due to uncoordinated procedures for quality assurance and other reasons
- Absence of integrated systems for collection, analysis, assessment and forecast of water resources as a main link of management
- Problems with access to data obtained by different departmental systems

Monitoring systems and services observe physical, chemical, biological, microbiological and toxicological parameters of water with identification of hydrological characteristics less frequently than they observe bottom sediments and biota. Most observations is concerned with physical and chemical parameters with a minor emphasis on the toxicological.

Frequency of observations varies in each system and service, with monthly sampling prevailing.

The Roshydromet conducts departmental policy in the sphere of environmental monitoring within its boundaries and on the national level, provides acquisition of information on environmental conditions and pollution on the territory of the whole country. It also discloses this information to different consumers.

Among the existing systems of monitoring, extensive experience was accumulated by Roshydromet's system for surface water monitoring. Its advantages include its systematic character and the sufficient duration of surveillance observations, which allows them to carry out long-term monitoring and to detect trends in the level of water pollution. The system enables the identification of priority water bodies, their sections and the tracking of tendencies in the quality of water. An important advantage of the system is the integration of observations and, primarily, the provision of most observation posts with quantitative characteristics of water bodies (water flow rate, level of water and current velocity). It is the only system which continuously provides consumers and the Administration of Hydrometeorology Service with generalized and systematic information related to basins of major rivers and the country as a whole. Technically, however, the system is not intended to solve issues of regional monitoring because it fulfills functions of national surface water monitoring and partially those functions of the State Service for Monitoring of Environmental Conditions. The system has some shortcomings: the list of identified parameters is not optimized; its effectiveness and flexibility are not high enough; and the lack of integration of observations in the system of water—bottom sediments and of the environment as a whole (ground waters, sea waters, soils and atmospheric air), etc.

The next chapters deal with different types of monitoring—surveillance, operational (remote sensing methods and computerized systems) and special.

REFERENCES

[1] Federal Law of the Russian Federation No. 7-FZ on Environmental Protection adopted on January 10, 2002.

[2] Federal Law of the Russian Federation No. 113-FZ on Hydrometeorological Service adopted on July 19, 1998.

[3] Izrael Y.A., Gasilina N.K., Rovinsky F.Y., Filippova L.M. Monitoring of Environmental Pollution in the USSR. — Leningrad: Hydrometeoizdat, 1978. —111 p.

[4] Izrael Y.A. Environment and Control of Environmental Conditions. — Moscow: *Hydrometeoizdat,* 1984. —560 p.

[5] Kimstach V.A., Friedman M.D., Dmitriev E.C., Yazvin L.S., Neumann E.Y. Concept of Russian Environmental Monitoring System // *Meteorology and Hydrology.* 1992. — No.10. — P. 5-8.

[6] State Standard 17.2.3.01—77. Nature Preservation. Atmosphere. Rules for Controlling Air Quality of Inhabited Localities. — Moscow: *Publishers of Standards,* 1978. —4 p.

[7] State Standard 17.1.3.07—82. Nature Preservation. Hydrosphere. Rules for Controlling Water Quality of Water Bodies and Aquifers. — Moscow: *Publishers of Standards,* 1986. —12p.

[8] State Standard 17.1.3.080—82. Nature Preservation. Hydrosphere. Rules for Controlling Quality of Sea Waters. — Moscow: *Publishers of Standards,* 1982. —8 p.

[9] Constitution of the Russian Federation (Fundamental Law). — Moscow, 1993.

[10] Order of the Russian Federation Government No. 1225-R on Ecological Doctrine of the Russian Federation dated August 31, 2002. — *Collection of Government Orders,* 2002.

[11] Federal Law of the Russian Federation No. 167-FZ on the Water Code of the Russian Federation adopted on November 16, 1995.

[12] Order of the Russian Federation Government No. 662 on the Statute of State Service for Monitoring of Environmental Conditions dated August 23, 2000.

[13] Order of the Council of Minister of the Russian Federation Government on Establishment of Unified State System for Environmental Monitoring dated November 24, 1993.

[14] Order of the Russian Federation Government on the Statute on State Monitoring of Water Bodies dated March 14, 1997.

[15] Federal Law of the Russian Federation No. 187-FZ on Continental Shelf of the Russian Federation adopted on November 30, 1995.

[16] Federal Law of the Russian Federation No. 109-FZ on Safe Handling of Pesticides Agricultural Chemicals adopted on July 19, 1997.

[17] Federal Law of the Russian Federation No. 117-FZ on Safety of Hydrotechnical Structures adopted on July 21, 1997.

[18] Federal Law of the Russian Federation No. 71-FZ on Payment for Use of Water Bodies adopted on May 6, 1998.

[19] Federal Law of the Russian Federation No. 155-FZ on Inland Sea Waters, Territorial Sea and Adjacent Zone of the Russian Federation adopted on July 31, 1998.

[20] Federal Law of the Russian Federation No. 191-FZ on Exclusive Economic Zone the Russian Federation adopted in 1998.

[21] Federal Law of the Russian Federation No. 94-FZ on Protection of Baikal Lake adopted on May 1, 1999.

[22] Order of the Russian Federation Government No. 625 on the Statute of State Sanitary and Epidemiological Service dated June 5, 1999.

[23] Order of the Russian Federation Government No. 1097 on the Order for Development, Adoption and Implementation of Integrated Use and Protection Schemes for Water Resources dated September 13, 1996.

[24] Order of the Russian Federation Government No. 1403 on Introduction of State Water Cadaster dated November 23, 1996.

[25] Order of the Russian Federation Government No. 1404 on Adoption of the Statute on Water Protection Zones for Water Bodies and their Coastal Protection Belts dated November 23, 1996.

[26] Order of the Russian Federation Government No. 1504 on the Order for Development and Adoption of Indices of Maximum Permissible Harmful Impact on Water Bodies dated December 12, 1996.

[27] Order of the Russian Federation Government No. 383 on the Rules for Allowing Water Bodies Owned by the Government, for Establishment and Revision of Water Use Limits, for Issuing Water Use and Distribution Licenses dated April 3, 1997.
[28] Order of the Russian Federation Government No. 716 on the Statute on Implementation of State Control over Use and Protection of Water Bodies dated June 16, 1997.
[29] Order of the Russian Federation Government No. 1320 on Organization of State Surveillance over Safety of Hydrotechnical Structures dated 1997.
[30] Order of the Russian Federation Government No. 818 on Adoption of Minimum and Maximum Payment Rates for Use of Water Bodies of Certain Basins of Rivers, Seas and Economic Regions dated July 22, 1998.
[31] Order of the Russian Federation Government No. 555 on the Statute on Russian Federal Service for Hydrometeorology and Environmental Monitoring dated May 20, 1999.
[32] Order of the Russian Federation Government No. 128 on the Statute on Provision of Information Related to Environmental Conditions, Pollution and Emergency Situations of Anthropogenic Nature, which Have Made, are Making or Likely to Make Negative Impact on the Environment dated February 14, 1999.
[33] Order of the Russian Federation Government No. 726 on the Statute of Ministry of Natural Resources of the Russian Federation dated September 25, 2000.
[34] Nikanorov A.M., Tsirkunov V.V. Systems for Surface Water Quality Monitoring. Part 1. // Surface Water Quality Monitoring Systems of the USA and Canada. Their Evolutionary Trends. — St. Petersburg: *Hydrometeoizdat,* 1994. —107 p.
[35] Practical Guidelines on Organizational Principles of Water Quality Observation and Monitoring System for Water Bodies and Aquifers of Roshydromet's Network within the Framework of the SSOM — Leningrad.: *Hydrometeoizdat,* 1984. —39 p.
[36] Regulatory Document 52.24.309—92. Practical Guidelines. Protection of Nature. Hydrosphere. Organization and Carrying Out of Surveillance Observations over Contamination of Surface Water of Roshydromet's/Roskomhydromet's Network. — St. Petersburg: *Hydrometeoizdat,* 1992. — 64p.
[37] Sanitary Standards and Regulations No. 4630—88. Sanitary Standards and Regulations for Protection of Surface Waters from Contamination/USSR Ministry of Public Health. — Moscow, 1988. —69 p.
[38] List of Fishery Standards: maximum allowable concentrations (MPC) and safe reference level of impact (SRLI) of hazardous substances for water bodies of commercial fishing importance. — Moscow: Publishers of Russian Scientific and Research Institute for Fishery and Oceanography, 1999. —304 p.
[39] Sanitary Standards and Regulations 42-123-5319—91. Sanitary and epidemiological standards on maximum allowable concentrations (MPC) and safe reference level of impact (SRLI) of pesticides in the water of water bodies used for utility, drinking, cultural and household purposes, including methods of their measurement (list)/Ministry of Public Health. — Moscow, 1991. —92 p.

Chapter 4

SURVEILLANCE MONITORING OF SURFACE WATERS

The unfavorable environmental situation chiefly conditioned by pollution of the biosphere is a characteristic of most Russian regions and in some of their parts, changes in environmental conditions assumed a steady nature. Certain areas experienced profound irreversible environmental changes followed by the destruction of ecosystems and the deterioration of population health. Such territories (areas and sections) are rated as zones of environmental emergency in accordance with scientifically grounded criteria for assessment of environmental situation. The maldistribution of water resources on the territory of the country, overregulation of major rivers' run-off and increase of its irretrievable removal, high levels of pollution under conditions of insufficient effective control and imperfection of current standards along with several other causes, have led to deep, destructive damage of aquatic ecosystems and aggravation of the problem of providing the population with clean water.

Industry, agriculture, transport, power engineering and other spheres within the economy are frequently characterized by an extremely low environmental potential of production, which fails to provide a sufficient level of treatment for all types of emissions and waste disposal. The aquatic environment is the most exposed to anthropogenic impact. Modern conditions of water resources show that the problem of their rational use and protection hasn't been solved yet.

Of all the types of consequences of anthropogenic factors, contamination of water bodies is the most dangerous and is extremely hard to eliminate. The major part of pollutants arrives with waste waters from different sources of production. At the same time, water bodies are exposed to other types of stress coming both from concentrated (local) and from dispersed sources of pollution, urbanized territories, etc. Under these conditions, the essential significance of surface water monitoring for the general assessment of water quality and establishment of trends and changes thereof is unquestionable. At the present time, acquisition of general information on the conditions of water bodies and their pollution on the territory of the Russian Federation is provided by the system for surface water surveillance monitoring within the framework of the State Service for Monitoring of Environmental Conditions (SSMEC).

The objectives of surveillance monitoring as a part of surface water monitoring include:

- Acquisition of information related to conditions of surface waters and their pollution
- Provision of public authorities and organizations concerned with systematic information and forecasts regarding contamination of water bodies and aquifers, including emergency information on dramatic changes in the pollution of water.

The procedure of *surveillance observations* consists of the following: carrying out observations under established programs, collection and accumulation of data related to conditions of water bodies and aquifers, forming and replenishment of a data bank, provision of information to consumers in accordance with the established order and the development and preparation of forecasts concerning the conditions of water bodies.

The network of surveillance monitoring is based on approximately 1,750 posts located on 1,150 water bodies. The density of the network of observation posts is not high and is equal to 1 post per 9.8 thousand km^2. Approximately 2% of the posts are used for both surveillance and special observations.

Twenty-three thousand water samples and about 100 samples of bottom sediments are taken within the framework of surveillance observations. These samples are analyzed by 94 laboratories. In general, about 110-120 physical and chemical water parameters and 7-9 sediment parameters (organochloride pesticides, oil products, resins and asphaltenes) are measured around the country.

The organization and carrying out of surveillance observations rely on the following basic principles: the integration and systematic character of observations, the coordination of their duration with specific hydrological situations and the identification of composition and property of water by a common or accurate method.

The systematic character of observations is conditional based upon observations of physical, chemical, hydrological, biological and toxicological (biotest) indices in water and of a certain number of pollutants in bottom sediments.

The system of surveillance observations is governed by regulatory documents [1—17], among which the most important is Regulatory Document 52.24.309—92. This document has established the requirements for the organization and carrying out of surveillance monitoring.

Observation program. Samples are taken on observation posts, which are installed with due account of current use of a water body or aquifer and prospective plans for the development of areas exposed to anthropogenic impact. Observation posts are also built in unpolluted sections of water bodies for background observations. Posts may be set up in the areas of: city and settlement locations, discharge of waste waters by detached major enterprises, spawning and wintering of the most valuable species of commercial organisms, near dam sections of rivers, national borders, outlet points of major and medium rivers and mouths of contaminated tributaries.

With consideration of the hydrometeorological and morphometric peculiarities of water bodies or aquifers; location of pollution sources; amount, composition and properties of waste waters; and concerns of water users, an observation post may be equipped with one or several points containing one or several gage points depending on the mixing conditions of natural waters with waste or tributary waters. Depending on the depth of a water body, samples may be taken from near-surface level or from several levels.

The category of a post determines the component and temporal parameters of observations (Tables 4.1 and 4.2).

Measurements of physical, chemical (except for pesticides) and hydrological parameters of water are obligatory for all posts. Pesticides in water and bottom sediments are identified by some posts selected with due account of the peculiarities of pesticide delivery, migration and transformation in the aqueous medium. Measurements of biological and toxicological (biotest) parameters are currently of advisory nature, because possibilities are limited.

Table 4.1. Observed hydrochemical parameters

Index	Program			
	obligatory	reduced 1	reduced 2	reduced 3
Water flow rate at watercourses	+	+	+	+
Water current velocity (on watercourse)[1]	+	-	-	+
Water level (in water bodies)	+	+	+	+
Visual observations	+	+		+
Temperature	+	+	+	+
Coloration	+	-	-	-
Transparency	+	-	-	-
Oxygen	+	+	+	+
Carbon dioxide	+	-	-	-
Suspended matters	+	-	+	
pH value	+	-	+	+
Ox-redox potential	+			
Specific conductance	-	+	+	-
Chloride ions[2,4]	+	-	-	-
Sulfate ions[2,4]	+	-	-	-
Hydrocarbonate ions[2,4]	+	-	-	-
Calcium ions[2,4]	+	-	-	-
Magnesium ions[2,4]	+	-	-	-
Sodium ions[2,4]	+	-	-	-
Potassium ions[2,4]	+	-	-	-
Total ions[2,4]	+	-	-	-
Ammonium ions[4]	+	-	-	-
Nitrite ions[4]	+	-	-	-
Nitrate ions[4]	+	-	-	-
Phosphate ions[4]	+	-	-	-
Total iron[4]	+	-	-	-
Silicon[4]	+	-	-	-
BOD_5[5]	+	-	+	+
COD	+	-	+	+
Oil products[4]	+	-	-	-
Phenol (volatile)[4]	+	-	-	-
Heavy metals[3,4]	+	-	-	-
Pollutants[3,4,6]	-	-	+	+

[1] Information on water current velocity is provided if sampling is combined with supporting measurements of water flow rate.

[2] In cases of the presence of pollution source in several points, concentrations of principle ions and total ions are measured only in the first point located after the waste waters discharge point.

[3] List of heavy metals and pollutants characteristic of a certain post, which are to be determined by reduced programs 2 and 3, is established on the basis of observations over waste waters and preliminary studies of the water body.

[4] Typical pollutants may include principle ions, biogenic matters and widely spread pollutants. In this case, these parameters are measured by reduced programs 2 and/or 3.

[5] The first point located after waste waters discharge point is used for observations under reduced program 1 and sampling of water volumes of not less than 5 liters stored for the period of 5 days in case of a necessity to make analyses in an emergency (mortality, killing of fish and other aquatic organisms, emergency discharges of pollutants).

[6] Reduced program 3 is used for detection of all pollutants typical to a specific post; reduced programs 2 and 3 are used for detection of the most typical pollutants.

Table 4.2. Frequency of observations and types of hydrochemical observation programs

Frequency of observations	Category of observation post			
	I	II	III	IV
Daily	Reduced program 1	Visual observations	-	-
Decennially	Reduced program 2	Reduced program 1	-	-
Monthly	Reduced program 3			-
During main phases of water conditions	Obligatory program			

The list of identified chemical parameters of water is determined with consideration of the intended use of a water body or aquifer, the composition of discharged waste waters and the requirements of information users. The list of identified pesticides is established with due account of the list of priority pesticides recommended for detection in water, volumes and scales of the use of pesticides and primary agricultural crops on the territory of the observed area, information on companies producing these pesticides and the level of pollution within the observed water body. Observations are carried out under obligatory and reduced programs (see Table 4.2), which include the identification of obligatory parameters and substances that contaminate water in the given observation post.

Organochlorine pesticides are detected in bottom sediments. Identification of oil hydrocarbons, resin components and metals is also provided for.

The list of measured biological and toxicological parameters is determined with due consideration of regional peculiarities and level of water body pollution. In this respect, various observations may be carried out for phytoplankton, zooplankton, zoobenthos, periphyton, macrophyte, microbiological indices, phytoplankton photosynthetic and organic substance destruction rates, organic substances consumption rate, alkaline phosphatase activity, Daphnia magna biotesting of acute and chronic toxicity and algal biotesting.

As to the results of long-term observations of moderately and partially contaminated water bodies, it is acceptable to reduce the number of controlled parameters to one or two indices that change notably under anthropogenic stress.

Temporary parameters are characterized by a periodic nature and terms of observations.

Frequency of observations is determined in accordance with the category of a post:

- Hydrochemical indices (except for pesticides)—in accordance with Table 4.2
- Content of pesticides in water—3-7 times per year with consideration of their persistence and use
- Biological and toxicological parameters—4-6 times per year (annual frequency) or with an interval of 2-3 years (interannual frequency)
- Content of pollutants in bottom sediments—2-3 times per year

The terms of observations of hydrochemical parameters of water should be coordinated with the hydrological conditions of water bodies and observations of pesticides—with hydrological conditions and terms of reclamation, observations of biological and toxicological parameters—with hydrological conditions.

Roshydromet employs the following frequency and terms of sampling: 4-8 times per year under the obligatory program for all posts, and for posts of 1-3 categories, the decennially, monthly and daily sampling for reduced analysis is added.

Sampling terms may be combined due to the requirement of integration of observations for all types of parameters.

Background observations are carried out primarily on surveillance posts and under the surveillance program. Observations are also done on a limited number of stations for complex background monitoring located in biosphere reserves [18]. These stations are used for measuring pH value, sulfate, chloride, calcium, magnesium, potassium, sodium, ammonium, nitrate, phosphorus (total), nitrogen (total), lead, mercury, cadmium, arsenic, DDT and its metabolites, HCH isomers and 3,4-benzpyrene.

Requirements for observations are determined by the requirements for separate phases of work and are governed by the corresponding documents:

- Collection of representative samples and pollution control during sampling [3]
- Measurements (measurement methods [4—8]; measurement tools [13, 14]; reference standards [15]; teaching of a specific measurement method to personnel, techniques of laboratory work and analytical analysis)
- Constant interlaboratory and occasional outside control of information quality (includes control of all work phases) [16, 17]

The requirements for information are determined by the requirements of all previous work phases and by the stage of acquisition, processing and disclosure of hydrochemical information, control of which is regulated by the Regulatory Document [16].

The generalization of acquired hydrochemical data is carried out with the application of individual and complex assessments. Individual assessments compare measured parameters with fishery, sanitary and hygienic water quality standards. Complex assessments consider formalized (rated) parameters: water pollution indices and specific water pollution combinatorial indices. Values of the foregoing indices allow the classification of water quality. Both methods of complex assessment are based on water quality standards as an assessment criterion.

For certain classes of water quality, biological and toxicological conditions are assessed by comparing them with regulatory values.

Contamination of bottom sediments is assessed by comparing concentrations of pollutants in samples of bottom sediments taken from observation and background points.

References

[1] Regulatory Document 52.24.309—92. Practical Guidelines. Protection of Nature. Hydrosphere. Organization and Carrying Out of Surveillance Observations over Contamination of Surface Waters of Roshydromet's Network. — St. Petersburg: *Hydrometeoizdat,* 1992. — 64 p.

[2] Regulatory Document 52.24.609—99. Practical Guidelines. Organization and Carrying Out of Observations over Concentration of Pollutants in Bottom Sediments. — St. Petersburg: *Hydrometeoizdat,* 2001. — 23 p.

[3] Regulatory Document 52.24.353—94. Sampling of Surface and Waste Waters. — St. Petersburg: *Hydrometeoizdat,* 1995. — 28 p.

[4] Regulatory Document 52.18.595—96. Federal List of Measuring Methods Authorized for Use in the Field of Monitoring of Environmental Pollution. — St. Petersburg: *Hydrometeoizdat*, 2001. — 97 p.

[5] State Register of Methods for Quantitative Chemical Analysis and Assessment of Environmental Conditions Acceptable for State Environmental Control and Monitoring Carried Out by the Russian Ministry of Natural Resources. — Moscow, 1995.

[6] Guide on Chemical Analysis of Surface Waters / Edited by A.D. Semenov — Leningrad: *Hydrometeoizdat,* 1977. —541 p.

[7] Guide on Hydrobiological Monitoring of Freshwater Ecosystems / Edited by V.A. Abakumov — St. Petersburg: *Hydrometeoizdat,* 1992. — 318 p.

[8] Methods for Bioindication and Biotesting of Natural Waters. Issue No. 2 / Edited by V.A. Bryzgalo, T.A. Khoruzhey. — Leningrad: *Hydrometeoizdat,* 1989. — 276 p.

[9] Instructions for Hydrometeorological Stations and Posts. Hydrological Observations and Works on Large and Medium Rivers. Issue No. 6. Part I. — Leningrad: *Hydrometeoizdat,* 1978. — 384 p.

[10] Instructions for Hydrometeorological Stations and Posts. Hydrological Observations and Works on Small Rivers. Issue No. 6. Part II. — Leningrad: *Hydrometeoizdat*, 1972. — 256 p.

[11] Instructions for Hydrometeorological Stations and Posts. Hydrological Observations and Works on Lakes and Reservoirs. Issue No. 7. Part I. — Leningrad: *Hydrometeoizdat,* 1973. — 476 p.

[12] Regulatory Document 52.10.324—92. Regulatory Document. Practical Guidelines. Hydrological Observations and Works within Hydrometeorological Network in Mouth Areas of Rivers. — Moscow: *Hydrometeoizdat,* 1993. — 187 p.

[13] Rules 50.2.006—94. State System for Ensuring Uniform Measurement. Order for Inspection of Measurement Tools. — Moscow: *Publishers of Standards,* 1994. — 50 p.

[14] Rules 50.2.009—94. Order for Testing and Approval of Types of Measurement Tools. — Moscow: *Publishers of Standards,* 1994. 51. State Standard 8.315—91.

[15] Reference Standards. Main Provisions. Order for Development, Attestation, Approval, Registration and Application. — Moscow: *Publishers of Standards,* 1988. —9 p.

[16] 52.24.268—88. Practical Guidelines. System of Accuracy for Observation of Pollution Parameters in Controlled Environment. — Leningrad: *Hydrometeoizdat,* 1986. — 30 p.
[17] 53. Regulatory Document 52.24.509—96. Practical Guidelines. Order for Controlling of the Quality of Hydrochemical Information. — St. Petersburg: *Hydrometeoizdat,* 1999. — 55 p.
[18] Izrael Y.A., Rovinsky F.Y., Cherkhanov Y.P. Study of Background Environmental Pollution in the USSR: Functional Structure of Background Monitoring / Monitoring of Background Pollution of the Natural Environment. Issue No. 2. — Leningrad: *Hydrometeoizdat,* 1984. — P. 3 — 10.

Chapter 5

OPERATIONAL MONITORING OF SURFACE WATERS

Great damage to environmental conditions of water bodies and the economy of water users (water consumers) is caused by emergency and volley (including illegal) discharges of contaminated waste waters. Accidents occurring in the course of production, transportation and storage of different chemical compounds (including toxic compounds) are accompanied by dangerous consequences. On oil refineries all over the world, about 1,500 different accidents happen every year. Risk of accidents increases when diverse enterprises are located in one place due to overlapping accident risks generated by different productions.

The list of hazardous cargoes transported by water includes over 3,000 items, 300 of which are transported in bulk. The most frequent disasters in the world are spills of oil and oil products caused by accidents on tankers, pipelines and drilling platforms. The most noticeable sign of accidental pollution is the mass mortality of fish and bottom organisms. Ecosystems of water bodies have been ruined and damages may reach tens of thousands or millions of roubles.

The untimely detection of extreme situations occurring on water bodies due to different types of accidents, inefficient inspection and forecast of pollution zone's movements hampers adoption of urgent water protection measures and control actions, which results in significant economic and environmental losses (costs).

All changes in surface waters induced by natural and anthropogenic factors are divided into three classes of space-time variability of characteristics (processes), in accordance with which three types of monitoring are implemented. These three types of monitoring include their own methodology of observations, methods of analysis and characteristics of research programs, and are referred to as *background, surveillance* and *operational* monitoring.

Due to different frequencies of processes studied by these types of monitoring, certain requirements are imposed on system parameters (frequency of monitoring, number of monitored components, area embraced by the system, distance between sampling posts, etc.).

In the first place, the state system for monitoring of surface waters (and other environments) employed by the Russian Federation meets the goals and objectives of surveillance monitoring described in the previous chapter.

Established terms of sampling, expressly fixed observation posts (points), common methods and practical guidelines focus on the measurement of a comparatively large number of water parameters and components by permanent laboratories—all of these provide general assessment of surface water pollution on the territory of the whole country. The system, however, is limited by coverage even within separate industrially developed regions, which does not allow surveillance monitoring to solve the problems of *operational* identification/detection of hazardous situations provoked by accidental pollution on a national scale. Information on the level and nature of water pollution obtained by the network of surface water surveillance monitoring does not fully reflect the real conditions of water bodies, because it covers only a small part of the pollutants delivered by various flows from concentrated and dispersed sources. In particular, less than ten out of many dozens (or even hundreds) of agricultural pesticides are monitored, and certain classes of chemical compounds remain completely outside the scope of monitoring. On the other hand, monitoring programs generally include analysis of substances dissolved in water and sometimes provide for observation of other elements of ecosystems (suspended matters and bottom sediments), in spite of the fact that a great number of widely spread and hazardous toxic substances migrate primarily in sorbed state and are accumulated in bottom sediments and hydrobionts. Due to the foregoing reason, analysis of such incomplete information may give a distorted idea of the real level of water pollution and may poorly correlate with data on conditions of biological resources (decrease of fish capacity, etc.).

Furthermore, it should be noted that the time required to take a sample, its transportation to and analysis by a permanent laboratory under conditions of the current system for surveillance monitoring, as a rule, noticeably exceeds (even by several times) the development time of a hazardous phenomenon provoked by a discharge of contaminated waste waters. Investigations of numerous cases of hazardous environmental situations prove this proposition. If a hazardous phenomenon has been detected, experts of territorial network divisions and scientific research institutes belonging to different departments (State Committee for Nature Protection, State Committee for Hydrometeorology, Ministry of Fisheries, etc.) would have been invited to investigate the causes of the phenomenon, the extent of damage and to establish the persons responsible for the pollution. Expeditionary units would be sent to accident areas and a great amount of work related to sampling and analysis of water, bottom sediments and biological material would be performed.

The field work conducted by the foregoing authorities, however, usually yields poor results, because they are performed under conditions of deficiency of time and source information (related to conditions of the water body before the accident, potential pollution sources, their nature, waste treatment methods, etc.), limited financial and technical support, a poorly coordinated program of mutual actions and an irregular exchange of observational data. Another substantial cause is the *delay factor* of investigation commencement (from several to ten or more days). In cases where the primary or only sign of a hazardous water phenomenon was dead fish, the delay period could be even longer because, for example, dead sturgeons rise to the surface after 7-10 days. In this case, the identification of a pollution source is extremely difficult and would only be possible if emergency discharge continued for a long period of time.

5.1. Purposes, Objectives and Structure of Operational Monitoring, Methods and Means of Support

Intricate problems in the prompt detection of emergency (including illegal) discharges; the investigation of hazardous water situations induced by them or other causes; the effective control and forecast, including establishment of persons responsible for pollution and in the provision of recommendations related to urgent water protection measures demanded that the SSMEC develop the *system (subsystem) for operational monitoring* of surface waters with its own parameters, methods, methodologies and hardware appropriate for the subject of research [1-2]. In the 1970-80s, by the order of the State Committee for Science and Technology, research institutions and public authorities (USSR Academy of Science, State Committee for Hydrometeorology, Ministry of Land Reclamation and Water Management, Ministry of Higher Education, Ministry of Electrical and Tool Engineering, etc.) launched a project to design and establish an operational monitoring system and its constituent components on the basis of a systemic approach to the problem that used the whole scope of methods for effective control, unified methodology of observations and technical and automatic means of support.

Occurring simultaneously as the new developments, on the basis of Roshydromet's territorial departments, certain elements of operational monitoring were gradually implemented in major industrial centers. Posts of the first category began to be used for daily monitoring of water conditions under reduced programs, and operational control groups were organized for carrying out urgent works in emergency areas. A little later, automated systems for monitoring and control of major water bodies began to be developed and express methods of analytical and biological control were introduced.

The problem of establishing a system for operational control servicing all water bodies under conditions of vast national territories and diverse potential pollution sources is extremely difficult and practically impossible to solve.

The complexity of operational monitoring of water bodies is aggravated by the occurrence of hazardous situations differing by location and time, diverse causes and compositions of pollutants and the short time of impact to which the aquatic environment is either poorly adapted or not adapted at all. At the same time, exposure and ecological adaptation rates become unsynchronized. This type of impact is highly dangerous and is accompanied by major damages.

Highly concentrated, short-lived pollutants of surface waters, entailing major and profound changes in conditions of ecosystems or their separate parts, are usually localized in space and characteristic of areas with violent anthropogenic stress. A high probability of hazardous water situations determines the appropriateness of immediate operational monitoring of such areas.

The effective control of water pollution provoked by emergency discharges should be carried out on a real time scale. Timely recovery measures in their own turn facilitate reduction of environmental and economic damages inflicted by the accident (Figure 5.1).

In order to reduce damages and to choose the most effective recovery method, it is necessary to use comparative analysis of information based on conditions of a water body in the period during the accident and in the period before the accident occurred (background). Such information is usually collected by network surveillance monitoring.

During the last decade, in accordance with regulatory and technical documentation governing the system for environmental monitoring, criterial terms such as “disastrous phenomenon (DP)” and “extremely high pollution (EHP)” were widely used in order to reflect the essence of controlled phenomenon. *EHP actually denotes concentrations of pollutants, the short-term impact of which causes or is likely to cause damage to the national economy, population health and environmental conditions.*

Hazardous water phenomena falling under these categories emerge as a consequence of volley emergency and illegal discharge of large amounts of waste waters, material wastes and industrial wastes containing polluting and toxic substances.

Consequences of these phenomena are manifested in the deterioration of aquatic ecosystems and water quality, the increase of water purification and treatment costs, the decrease of the recreational value of a water body and the loss of biological resources, including fish.

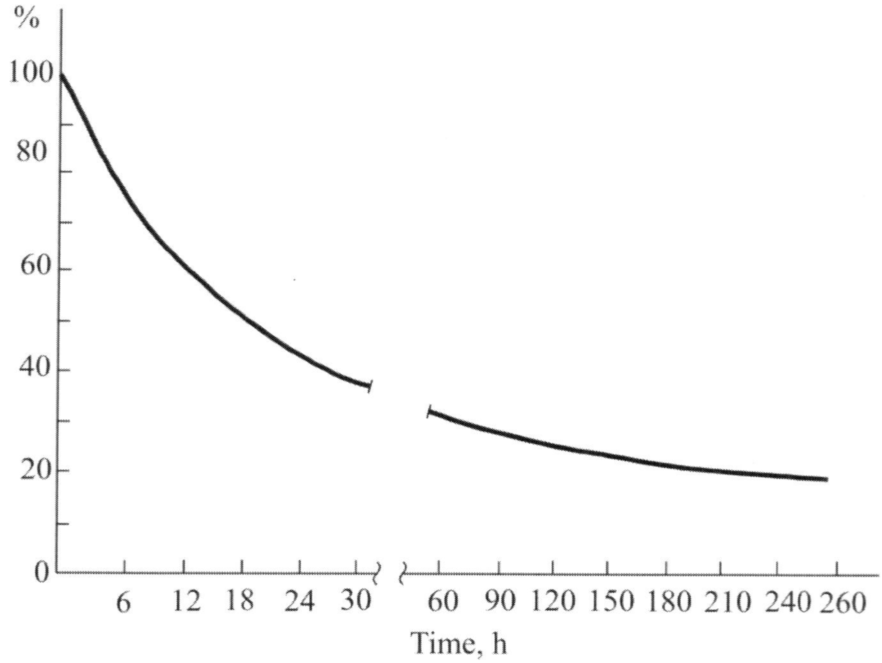

Figure 5.1. Relation between effectiveness of control actions (%) and delay of hydrochemical information on detection of high and extremely high levels of water pollution.

Pollution of water bodies may be determined by both regular (systematic, chronic) and volley discharges of pollutants. Regular discharges of waste waters resulting in recurrent damages are the subject of surveillance monitoring and are compensated for by regular (planned) water protection measures.

Volley discharge of waste waters is an emergency discharge of pollutants largely exceeding target discharge standards (by 100 times or more). This type of discharge is of a spontaneous nature as related to both time and location; it can occur on the territory of an enterprise or near it. Industrial wastes, finished and semi-finished products, reagents, raw materials, etc., may serve as pollutants.

Maintaining constant stress on a water body's regular discharges constitutes a background on which consequences of an accident evolve and on which it is necessary to control the impact of volley discharges, e.g. regular discharges create noise that disguises the signals of volley discharges. Systematic pollution by itself leads to unfavorable environmental conditions of a water body, e.g. to such level of background when even a slight additional volley discharge may lead to the profound imbalance of an ecosystem. The greater the regular discharge at a constant rate of volley, the smaller the relation of signal to noise determining the reliable control of volley discharges.

Summarizing the above, we can draw the conclusion that the *subject of surveillance monitoring* is not the level of water pollution, because it is the subject of observations carried out within surveillance monitoring and does not reflect parameters of short-term anthropogenic impact, *but is instead the sudden changes in conditions of a water body as a result of such impact.*

The main characteristic features of emergency volley discharges that determine the requirements for operational monitoring include the following:

- Difficulty in forecasting the nature of arising emergency situations
- Diversity in their origins
- High variability of pollutants' composition
- Short time of impact

The major objectives of operation monitoring are:

- Timely detection of sudden changes in conditions of water bodies or their sections, which may entail substantial economic and environmental damages
- Preliminary assessment of the extent of changes, their causes and possible consequences
- Provision of the latest information on hazardous phenomenon and recommendations on effective measures aimed at the protection of an aquatic ecosystem

The goals aimed at achieving the major objectives are:

- Detection of places and spatial parameters of zones with dramatic changes in conditions of aquatic ecosystem, and the assessment of extent of the damage
- Operational assessment of the impact on an ecosystem (degree of damage)
- Preliminary establishment of the causes and sources (polluter) of a hazardous water phenomenon
- Preliminary identification and assessment of the water pollution level affecting the ecosystem
- Preliminary forecast for possible expansion of extremely polluted zones
- Signaling/warning of EHP (taking into account preliminary forecasts of expansion)
- Update of the nature of water pollution and its sources (originators of EHP)
- Update of the forecast for expansion of polluted zones and its consequences
- Elaboration and provision of recommendations for effective measures for the protection of ecosystems of the monitored water body and reduction of negative consequences

- Development of suggestions for the organization with control over the consequences of accidental pollution of a water body
- Preparation of recommendations on the improvement of the system for operational monitoring, its constituent components, methods of control, etc.

Thus, *operational monitoring of surface waters represents information, measurement* and *preventive systems* [1]. Its functional scheme (Figure 5.2) is based on the optimal principles of known information systems, the perfection of which is expressed to the highest possible degree. Such systems include, for example, human and animal systems of sensor information.

System for operational monitoring of anthropogenic changes in natural environments (including water) should include two subsystems—*remote sensing (noncontact)* and *contact observations*. The first one (analogous to peripheral vision) is intended for the detection of a hazardous water phenomenon resulting from an accident and is implemented promptly through a complex set of methods and means but does not specifically respond to *changes in the image of a water body*.

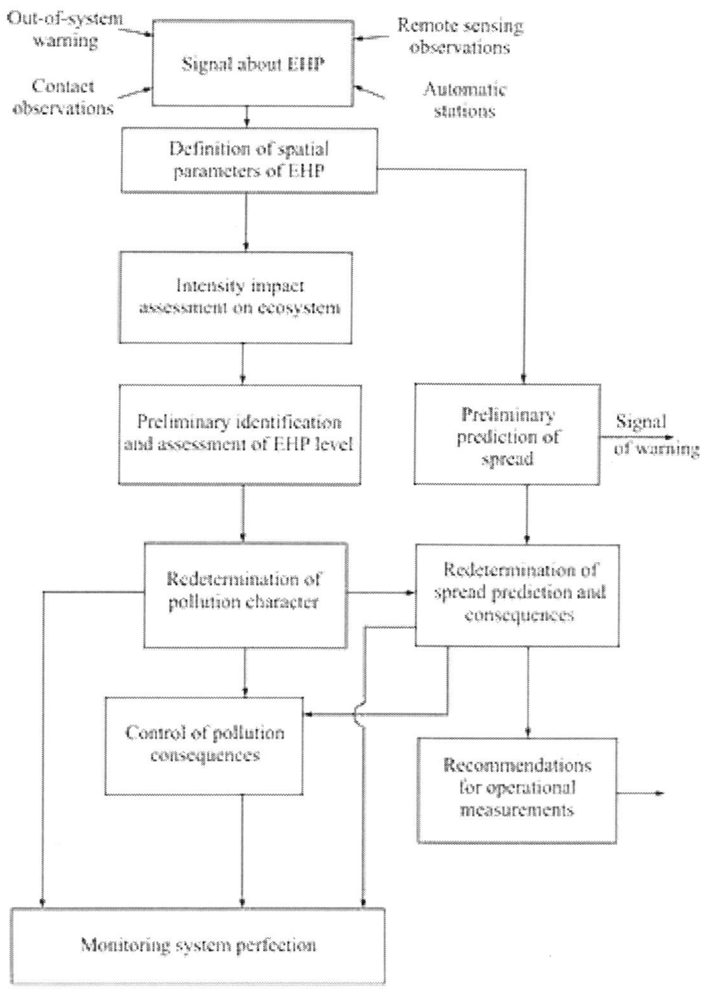

Figure 5.2. Functional scheme of operational monitoring.

From this viewpoint, the best means are those that allow not only for the establishment of the facts concerning dramatic changes of environmental conditions, but that also allow for the assessment of their spatial parameters. The *subsystem of noncontact observations* includes a remote sensing component (Figure 5.3) with its own methods and technical means of control presented by multi-zone video information arriving from a space station or an aircraft (airplane or helicopter) and from an observation post located on a vessel.

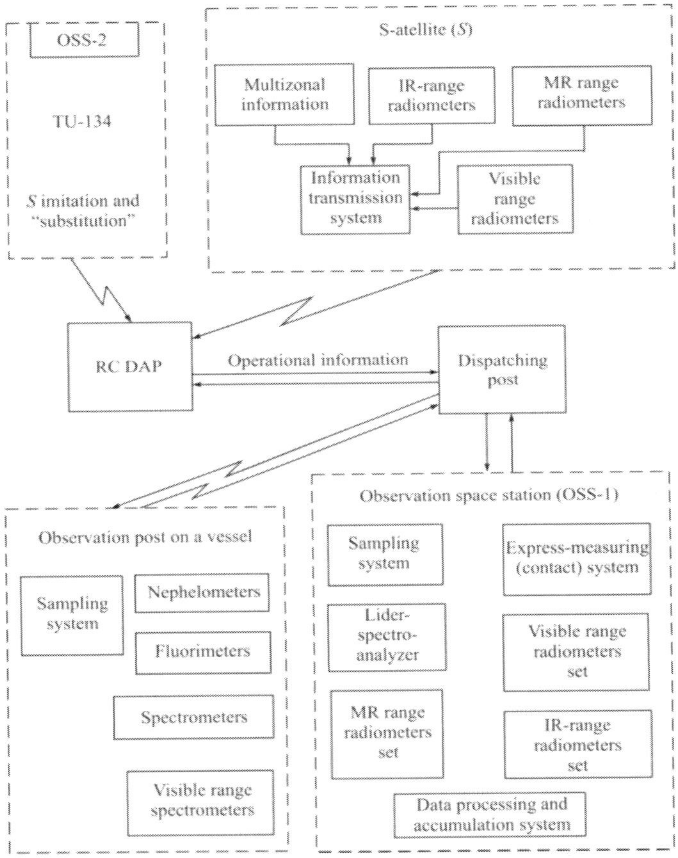

Figure 5.3. Remote sensing component of operational monitoring.

Observation stations located on a helicopter or a vessel, as compared with air and space stations, are a more informative instrument from the viewpoint of the component resolution of a changed water body image due to the possibility of carrying out additional contact observations; they lose to the latter, however, with regard to space-time resolution.

An aerial observation station equipped with a two-channel spectrometer (parallel visible range photometer) and installed on a multipurpose helicopter makes it possible to solve operational tasks related to the inspection of a water body, establish water area zones that optically differ by water quality indices and to take samples of water for direct contact, onboard measurements [3].

Developed by the Hydrochemical Institute, methods for the indication of spatial heterogeneity of distribution fields of organic and mineral suspended matters by brightness of radiation upwelling from water are used for the remote measurement of relative and absolute concentrations of chlorophyll "a" and suspended matters under conditions of considerable variations regarding their concentrations on the background of a relatively low variability of concentrations of other optically active components.

The functional capabilities of remote sensing controllers, which enable the determination of changes of image (space-time parameters), are currently limited in their detection of such changes. In other words, not every change in water conditions provoked by an emergency situation may be deciphered by means of the remote sensing component. Due to this fact, EHP shall be determined not only by the remote sensing component, but also by means of other components comprising the second subsystem of operational monitoring, a *subsystem of contact observations*.

Stationary surveillance observation posts serviced by SSMEC's staff observers were used during the first stages of development of the operational monitoring.

The urgency of obtaining information on EHP increased considerably while using other means of contact operational observations. These means include automatic control stations (multi-parameter analyzers) installed on water bodies at the most crucial points—places of probable occurrence of accident consequences. The stations carry out constant automatic control of physical and chemical water parameters (20 or more) on a periodic, predetermined basis [2]. Collection of samples and their express analysis (up to 25-30 parameters) directly near the water body is performed by mobile hydrochemical laboratories sent to the areas with EHP.

Automatic control stations as well as mobile hydrochemical (hydrobiological) laboratories (mounted on automobiles, helicopters and vessels) function within another important component of operational monitoring—the automated subsystem of observations and control over environmental pollution of surface waters, also known as ANKOS-VG (Figure 5.4). These systems provide prompt detection of volley and emergency discharges, timely measurement, processing and assessment of water pollution, operational short-term forecasting of pollution level and transfer of information to nature protection, controlling and other authorities. Effectiveness of operational monitoring may be achieved by simultaneous use of noncontact (remote sensing) and contact systems of observations (Figure 5.5).

Application of mathematical methods establishing connections between separate water quality parameters allows for the increased informativeness of measurement data by amplifying it with calculation data. Time variability of the foregoing connections determines the necessity of using adaptive models occasionally adjusted on the basis of analytical measurements of all parameters controlled on the given body of water.

The least operational type of receiving signals for hazardous water situations are the so-called non-system warnings from industrial enterprises, sanitary/fishery inspections and individual persons.

Contact methods and means as compared with remote sensing have considerably smaller capabilities for the effective establishment of spatial parameters for water pollution zones. In this case, it is possible to use calculation methods to forecast the spread of hazardous emergency pollution zones, including different schemes for advisory forecasts with due account of the completeness of initial information on the discharges of waste waters and distribution diagrams of their emissions into a water body.

The foregoing schemes of predictive calculations provide a number of important operational characteristics, including the time period for when the front, central and tail parts of a polluted zone will reach a specified river point and the maximum concentration of major pollutants at the specified point when the zone comes through it [4].

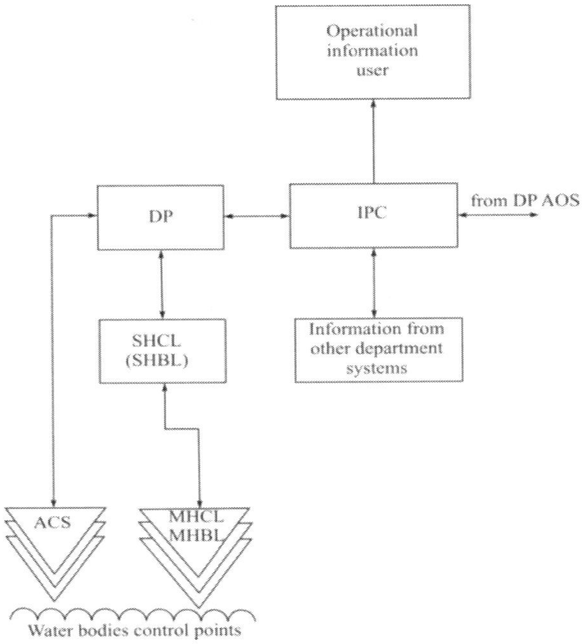

ACS—automatic control station, MHCL (MHBL)—mobile hydrochemical (hydrobiological) laboratory, SHCL (SHBL)—stationary hydrochemical (hydrobiological) laboratory, DP—dispatching post, AOS—automatic operation system, IPC—information processing center.

Figure 5.4. Structural scheme of the automated system of observations and control over ANKOS-VG type of surface waters.

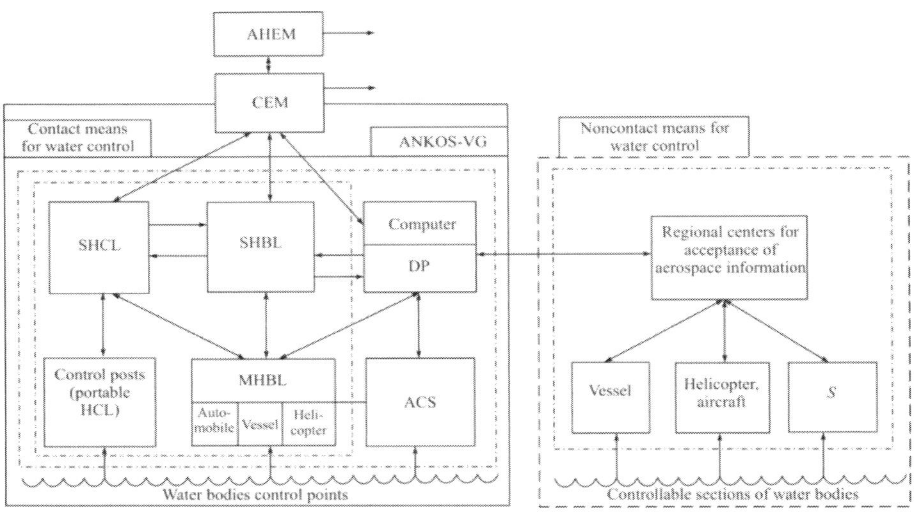

ACS—automatic control station, MHCL (MHBL)—mobile hydrochemical (hydrobiological) laboratory, SHCL (SHBL)—stationary hydrochemical (hydrobiological) laboratory, DP—dispatching post, AOS—automatic operation system, IPC—information processing center.

Figure 5.5. General structure of the operational monitoring of surface waters based on remote sensing (noncontact) and contact observations.

Among the methods and means of operational control, indirect methods deserve particular consideration. For example, developed by the Hydrochemical Institute, *trace methods* give a practical solution to problems related to the timely detection (establishment) of the fact that an accident has occurred, identification of the pollution source, forecasting of the evolution of a water pollution zone, travel speed of the plume, travel time of the front edge, etc. These methods are based on an original approach connected with a special way of marking pollution masses coming into a water body with tracer markers [5].

Simplicity, efficiency, high sensitivity (tracer may be detected even if water is diluted by 10^{13} times) and quickness are the main advantages of the trace method.

Studies of different types of tracer marker detection in surface water indicate that the most promising type is laser probing by observation stations installed on an airplane or vessel.

In addition, effective use of tracers is possible under the control of discharges generally made by major concentrated pollution sources and is almost inapplicable for controlling dispersed sources.

In the process of developing observation methods and selecting the most informative (and fastest) methods for operational monitoring, particular attention was paid to *biological (hydrobiological) methods* for controlling hazardous environmental situations.

For assessment of conditions and damages of hydrobiocenosis resulting from extreme emergency impact on a water body, it is reasonable to include the following three activities based on application of classical hydrobiological methods of bioindication and biotesting [6]:

- Assessment of biocenosis structure on the basis of the analysis of the number and species composition within the aquatic community
- Assessment of functional composition of aquatic organisms on the basis of physiobiochemical parameters—primary products of phytoplankton, concentration of chlorophyll “a” in phytoplankton and activity of extracellular enzymes (alkaline phosphatase, esterase)
- Assessment of water toxicity by Daphnia and algal biotest (fluorescent parameters)

As opposed to bioindication, i.e. traditional hydrobiological methods biotesting doesn’t examine conditions of biota, but rather water properties—its toxicity dependent on accidental or chronic pollution, with application of laboratory biotests.

The foregoing methods are notable for simplicity, quickness, opportunity for use in the field, comparability of analysis time and time scale of an emergency and they are also notable for good reproducibility.

An important outcome of these works is represented by the analysis and comparison of toxicological and hydrobiological data with the results of hydrochemical works.

The specific character of the objectives of operational monitoring has established the necessity to develop its own practical guidelines and methodological support intended for use under conditions of both individual and joint operation of its components. The development of these fundamentals includes the elaboration and testing of analytical methods corresponding to the purposes and objectives of operational monitoring, coordination and approval of regulatory and technical documentation, selection of methods for a unified set and complex assessments according to criteria pertaining to their effectiveness, accuracy, economy, etc.

Practical guidelines and methodological support for the purposes of operational monitoring are developed within a tight schedule and with phased selection and improvement of methods, taking into account the priority of their selection determined by the specific character of particular objectives of this type of monitoring, the most limiting stages of the whole process of operational monitoring, accumulated experience in the investigation of real emergency situations and material and technical security, etc. Since the problems of preliminary identification and assessment of pollution by remote sensing methods are only partially solved, complete fulfillment of this function shall be ensured by contact methods implemented by other elements of the operational monitoring system (subsystem of contact observations) in the field on the basis of elemental technical means and in permanent laboratories equipped with modern measurement tools.

In this relation, *test methods* attract great attention for determining priority pollutants with high concentrations (> 10 MPCs) on the basis of available means and methods, for example, reactive test papers, visual colorimetry and ionometry. These principles are also used by portable multi-parameter analyzers widely applied in both the field and laboratory work (for example, Aqua Test, ACV 106). The impossibility of conducting a detailed field study of chemical composition has determined the importance of preliminary preparation of water samples for their further transportation to a permanent laboratory for complete analysis. Furthermore, the necessity of the preliminary preparation of samples for analysis of certain groups of pollutants demanded the design of auxiliary laboratory equipment (for water filtration, concentration, extraction, UV-irradiation, etc.) to suite the requirements of the methods for analysis of different natural objects (water, bottom sediments, hydrobiont) while they were in permanent laboratories.

Effective, highly informative, multi-element (multicomponent) methods are employed for detailed identification of individual pollutants and the calculation of their concentrations in permanent laboratories. For metals, it may be chemical spectral, X-ray fluorescence or emissive with inductively coupled plasma and neutron activation types of atomic absorption. The most effective method for the identification and quantitative determination of organic pollutants is chromatographic mass spectrometry; for measurement of specific polycyclic aromatic hydrocarbons (PAH), high-efficiency liquid chromatography with high resolution capacity allows the measurement of PAH in the elements of aquatic ecosystems. From the viewpoint of economy, the use of expensive devices is rational only if they are fully loaded. This fact should be taken into account while deploying permanent laboratories in the regions, thus fulfilling objectives of both surveillance and operational monitoring. Performance of permanent laboratories engaged in analysis of a considerable number of water samples taken under programs of operational and surveillance monitoring may be increased by the introduction of automated flow lines uniting measuring analyzers with computer equipment for data processing.

Briefly described issues and objectives of operational monitoring, including methods and means of their solution within contact (ANKOS-VG, trace, hydrobiological methods, etc.) and noncontact (remote sensing) observations, are discussed in detail in Sections 5.2 and 5.3 of this chapter and in other chapters as well.

5.2. Trace Markers in Operational Monitoring

The multiplicity and diversity of potential pollution sources (point and dispersed), and the unpredictability of place, time and scale of accidents causes difficulties in solving major problems of operational monitoring—prompt detection of emergency discharges, identification of polluters, operational observations and forecasting of the evolution of a polluted zone in a water body.

The complexity of these objectives is aggravated by the fact that it is unknown which substances, composition and media may be discharged by a potential polluter, which methods (express methods) should be prepared and used under scarce time conditions for hydrochemical and hydrobiological control of a hazardous situation and movement of a polluted zone. In addition, it should be noted that it is quite difficult to organize operational observations and carry out analytical works directly near a water body.

Foreign and domestic experiences in the investigation of such situations shows that even sufficiently evident emergency discharges of waste waters, illegal discharges in particular, are registered by monitoring agencies with a delay, often significant. As it was previously demonstrated, timely detection of an emergency discharge reduces costs of recovery by ten times. Under conditions of unavailability of prior information on possible accidental pollution and the complexity of discharged waters' composition, however, fulfillment of the foregoing objective is only possible upon the application of highly informative, analytical methods, which by themselves are rather expensive and sophisticated.

The aforementioned causes have motivated the Hydrochemical Institute to execute works related to the establishment of a system of rather simple methods (measures) for marking pollution masses with inexpensive, affordable, environmentally safe and stable trace markers, which may be detected even if dilution exceeds 10^{10}-10^{13} times [7—13].

Detailed study of this issue revealed that the major part of known tracer substances cannot be unfailingly detected in contaminated waters if the emergency current is diluted by a water body. In particular, maximum possible dilution of fluorochromes (rhodamine and others) used in hydrology equals 10^6 times.

Tracer indicators based on short-lived radioisotopes (iodine-131, bromine-82, aurum-196) ensure quantitative registration if the dilution is low. The high cost of specimens and special conditions required for tests, however, hampers the wide use of isotopes for marking emergency discharges in water bodies.

Use of bacterial viruses (bacteriophages) as indicators, which may be quantitatively detected in dilutions of up to 10^{14} times, has also turned out to be unacceptable for operational monitoring due to the high complexity level of the method and its limited range of application.

Moreover, all of the former indicators do not allow for the simultaneous marking of a large number of potential pollution sources.

In order to solve different practical and theoretical issues of the monitoring, including operational monitoring, the Hydrochemical Institute has developed an essentially new type of tracer (marker) for studying movements of fluid phase, dissolved and suspended matters in a water body. To be effective markers, it is suggested that a trace substance be used in the form of solid, spherical microscopic particles made of brightly fluorescent, inert (polymeric) materials with specific densities. Particles of such tracers should meet the following initial requirements:

- Hold a spherical form
- Come in microscopic sizes that are considerably smaller than the sizes of the least turbulent masses in water bodies
- Density of particle material should equal water density
- Easily detectable, identifiable and registrable particles in the presence of interfering foreign particles

The most suitable for these purposes were daytime fluorescent pigments in the form of superfine powders of transparent to visible light polymeric or polycondensation resins colored with luminescent dye.

The initial basis of these pigments is a melamine, formaldehyde, toluenesulfonamide polymer with added luminophores. At the present time, 20 pigments covering the fluorescence spectrum from blue to red and characterized by high fluorescence quantum efficiency are produced or prepared for production.

The new tracers do not dissolve in the marked medium, are not toxic and are resistant to physical, chemical and biological factors. The size of a tracer particle may vary from 0.5 to 2.5 microns; the size of the most frequently used particles spans from 0.5 to 1.0 microns. The regular spherical form and size of trace particles ensure sedimentological and flocculation resistance in a dispersed system.

While designing the markers for operational investigation of emergency pollution discharges, there was a necessity to ensure long-term aggregative and sedimentological stability of highly concentrated disperse systems and to maintain their primary properties—intensity and color of fluorescence. This problem was solved by enclosing concentrated disperse tracer material in a solid phase or matrix and further chopping it into superfine powder, which can be stored in a pressure vessel for an unlimited period of time.

The proposed method for registration of a tracer marker is based on the filtration of a definite volume of water through a membrane filter with a pore size of 0.45 microns and further examination (counting) of particles on the filter surface under the microscope with a mercury-vapor lamp as a light source exciting fluorescence. If foreign luminescent particles are present in the sample, they should be discriminated by intensity and color of fluorescence, form and size.

Quantitative microspectrofluorimetry with luminescent microscopic adapter FMEL-1A is applied if tracer markers with similar fluorescent characteristics are used at the same time. In this case, photomultiplier FEU-79, with a set of interferential color filters, is used as a spectrum analyzer.

This method is distinguished by high selectivity and detectability even when dilution exceeds 10^{13} times.

The use of fluorescent marks injected directly into a pollution zone or contaminated media during an accident simplifies prompt detection of accidental water pollution to a great extent. At the same time, detection criterion remains the same regardless of composition, properties and other characteristics of incoming contamination. The high reliability of detection is provided by the fact that under normal (non-emergency) conditions, such markers are absent in water bodies.

The most effective method is to inject the tracer marker in advance, for example, in cases of oil transportation and oil productions by vessel, or directly into the marked medium during emergency discharge. In this case, all reservoirs and tanks used by industrial, agricultural and transport companies should be equipped with autonomous devices for emergency injection of markers into the front edge of the contaminated current (for example, siphons going off when waste water flow rate exceeds a specified value).

Water may be marked with a combination of luminophores of different colors, if it is necessary to detect an individual polluter of a major river basin that services a number of industrial enterprises. Having a comparatively small number of base luminophores of different luminescent colors, it is possible to obtain the required number of individual markers.

Depending on the implemented scheme of operational investigation in an emergency situation resulting from an emergency discharge, markers in a water body may be registered different ways: by means of frequent water sampling at SSMEC's posts, by mobile laboratories and in the long-term with use of simple automatic monitors installed at the places of potentially high accidental risks. The latter method of registration is the most promising and requires urgent implementation.

Apart from the foregoing techniques, another important method was developed and patented, that is, the quantitative detection of new water tracer markers by a laser lidar located onboard an airplane or vessel.

The most effective use of tracers is achieved while controlling discharges of major concentrated pollution sources, though they are practically inappropriate for monitoring dispersed sources.

Let us consider operational observations and forecasting of the evolution of an accidental pollution zone on a watercourse with the application of fluorescent tracers on the example of a simulation experiment conducted on the Seversky Donets River from the mouth of the Krasnaya River to the point of water intake at Krasitel, Production Department (near Rubezhnoe) (Figure 5.6.)

Volley discharge of pollutants (tracer) was modeled in the Krasnaya River across the entire section of the river before its mouth and was observed at three points located downstream of the Seversky Donets River. Results of the hydrometrical works are provided in Table 5.1.

Estimated and actual travel time of the front edge of the pollution wave and the peak of pollutants concentration are shown in Table 5.2.

Water samples were taken simultaneously across the entire area of passage at two downstream points with due account of the approximate change of travel time based on average flow rate and distance between the point of tracer injection, imitating an emergency discharge of waste waters to a control point.

Obtained results (Figure 5.7) allow us to visualize the evolution of the pollution wave despite the imperfection of simulation models made through the comparison of estimated time both for the arrival of the front edge and for the maximum concentrations of pollutants with corresponding values for the tracer wave. This test determines the degree of waste water dilution at certain points, places of maximum concentrations and appropriateness of the sampling post location.

In the examined case, the evolution of the contamination plume, conditioned mainly by two factors (advection and longitudinal turbulent dispersion), is presented by parameters that are quickly changing from point to point. In actual practice under the influence of a large number of factors, the evolution of a zone may be even more complex, especially with regard to low-flow and still water bodies.

In this case, two main forecasts of evolution are possible. The first one is based on direct observations over the zone marked with a tracer, and requires the sampling and analysis of large amounts of water (at all applicable points and gage points), which allows for the mapping of the space-time evolution of the zone with dilution indices of initial concentrations.

The second one provides for the use of simulation models parameterized on the basis of special, full-scale trace experiment modeling of emergency discharges. In this example, travel time and the factor of pollutants dilution are the main parameters of pollution prevalence in a water body that are determined.

Studies conducted on a number of watercourses, water bodies and ground waters in real cases of accidental water pollution with the application of new tracers (superfine suspensions of fluorescent microspheres) have proved their simplicity, quickness and effectiveness in terms of operational monitoring.

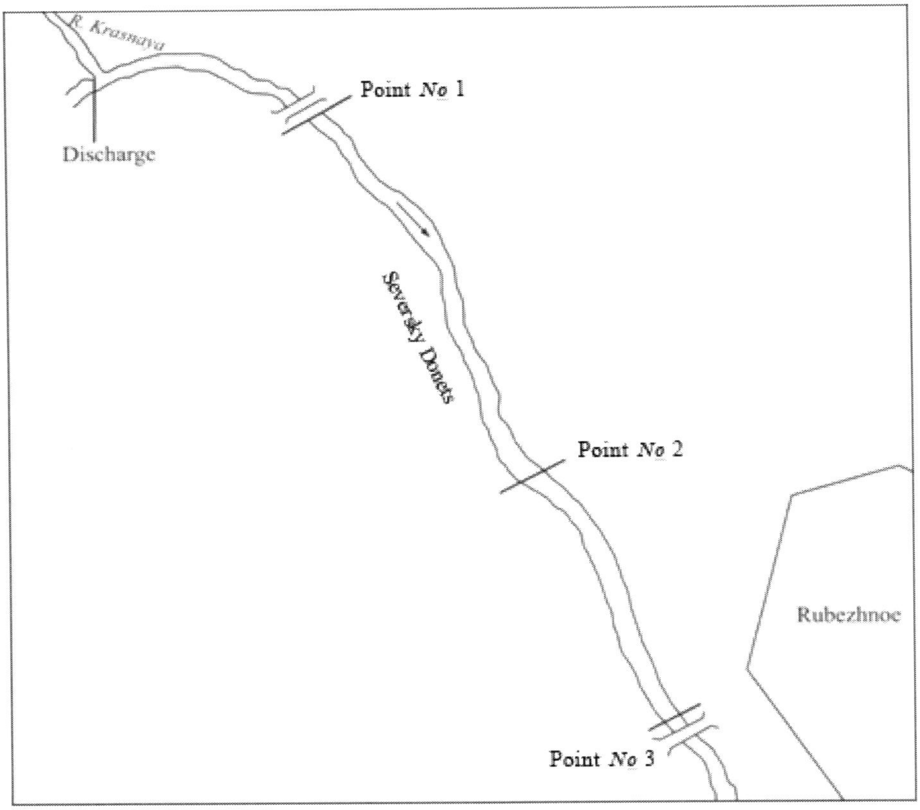

Figure 5.6. Section of the Seversky Donets River used for modeling an emergency discharge of pollutants.

Table 5.1. Primary hydrochemical characteristics of the Seversky Donets River at the studied sections

Parameter	Section between points		
	0-1	1-2	2-3
Flow rate, m^3/sec	68	68	67
Area of passage, m^2	190	130	194
Width, m	53	68	78
Depth, m			
maximum	4.56	3.08	4.48
average	3.6	2.3	2.48
Velocity, m/sec			
maximum	0.57	0.66	0.57
average	0.36	0.52	0.34
Hydraulic slope	0.00035	0.00042	0.00036

Table 5.2. Travel time of the front edge of the pollution wave (numerator) and concentration peak (denominator) at the Seversky Donets River in October 1988

Point No.	Distance between points, km	Travel time, min	
		estimated	actual
1	2	58/93	35/55
2	4.7	106/135	80/120
3	4.2	137/230	85/145

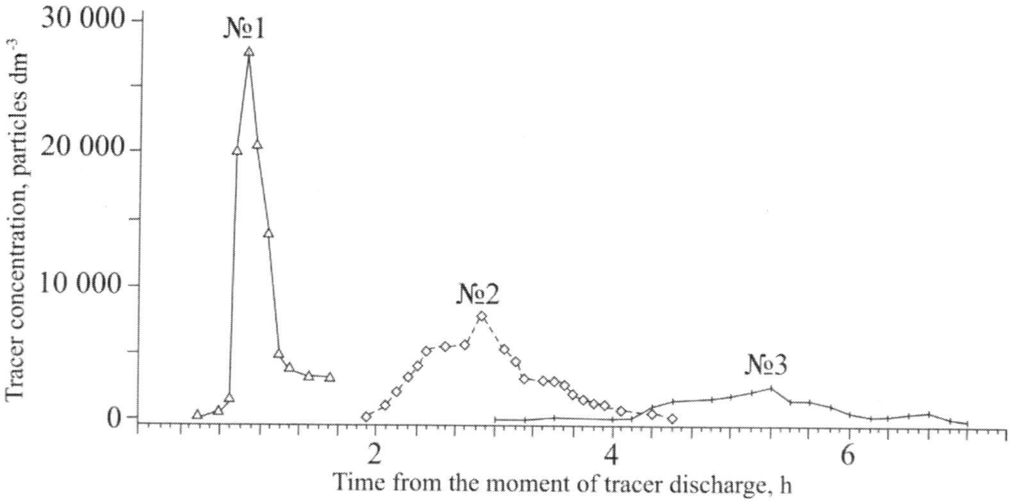

Figure 5.7. Evolution of pollution wave between points No. 1, 2 and 3 at the Seversky Donets River in October 1988.

5.3. Biological Methods in Operational Monitoring

A complex set of objectives for operational monitoring related to the assessment of emergency anthropogenic impact on ecosystems may be successfully fulfilled with hydrobiological methods. The operational component of hydrobiological monitoring based on modern surface water surveillance monitoring [14] is provided mainly through the application of methods that reflect functions of an ecosystem rather than its structure [6, 15].

In this respect, the most promising methods are those that are based on the study of physiological and biochemical parameters. This group includes the following methods of bioindication: strength of seston and benthos enzymes, consumption and mineralization rates of metabolites and the correlation of productional and destructional processes.

An important place in the system for operational biomonitoring is occupied by biotesting methods, which allow for the determination of the toxicity of an aquatic environment. Biotesting denotes assessment of water quality according to the response of aquatic organisms being the test-objects (algae, crustacean, fish, etc.) Express biotests are also useful for further orientation of programs for analytical control (chemical analysis) towards specific classes of pollutants.

Express biotests carried out within the observation system provide information on the presence of a biologically hazardous agent in the aquatic environment and enable the approximate evaluation of the adverse effects of this agent on hydrobionts [14].

Biotesting is intended to establish acute toxic action and to detect unfavorable environmental situations of varying extents [6, 16—18].

Acute toxic action (acute toxicity) is an impact causing the rapid response of a test-object. Acute toxic action is often measured by a test-reaction named "survival rate" for a relatively short period of time. This group of methods includes assessment of water toxicity according to motion activity of benthic invertebrates (leeches, mollusks), algal fluorescence, survival rate and physiobiochemical parameters of Daphnia. These methods are focused on the detection of acute toxic action dependent on a high level of pollution and the presence of hazardous, toxic chemical compounds [6, 17].

Indicators of operational detection of extremely high pollution in water bodies and watercourses consist of the following parameters of aquatic communities: recurrence of community death and injured species, abundance of α-saprobic species in plankton communities and abundance of Oligochaeta in macrozoobenthos communities.

If acute toxic action of water is detected, biotesting is used in order to identify the dilution factor at which acute toxic action does not occur. Quantitative approaches to biotesting cannot be applied yet due to a number of reasons—the most important of which include the following [17]:

- Biotesting methods are not quantitative and don't have numerical expressions for results
- Current frequency of observations over toxicological parameters falls behind the frequency of chemical control by public authorities

It is acceptable to use the assessment of toxicity according to the maximum possible number of analyzed samples provided that, if toxicity is detected in even one sample, then conditions of a water body or its section should be classified as unfavorable.

The use of Table 5.3. [18] is recommended for the assessment of toxicological conditions of a water body or its section from the viewpoint of an aquatic ecosystem's well-being. All classes of the scale are relative, though at the current stage, it can be temporally used for both surveillance and operational observations.

The fourth and fifth positions characterizing critical and disastrous conditions of an aquatic ecosystem (environmental emergency and disaster) don't require systematic observations. These positions may be evaluated after the results of biotesting of one water sample are completed, if they indicate acute toxic action of water detected in even one dilution of water by 50 times and more.

Information on acute and chronic toxic action of natural water may be obtained by surveillance observations and in the course of field work. All five positions provided by Table 5.3 may be used within the system for surveillance monitoring.

In the field work, toxicological conditions are assessed at the specific moment in time. In this case, whether water has a toxic action and whether this action is chronic or acute is established. Then, if the latter is detected, the dilution factor at which acute toxic action does not occur will be identified.

If acute toxic action is detected in water, then the water body or its section is declared environmentally unfavorable. An aquatic ecosystem is assessed by its dilution factor in accordance with the last two positions of Table 5.3.

Table 5.3. Assessment of toxicological conditions of a water body or its section

Toxicological conditions of an aquatic ecosystem	Assessment of toxicological parameters according to biotest data
Favorable	Chronic toxic action is detected in neither survey of surveillance monitoring[1]
Unfavorable	Chronic toxic action is detected in one survey of surveillance monitoring[1]
Extremely unfavorable	Chronic toxic action is detected in two or more surveys of surveillance monitoring[1] or acute toxic action is detected in one survey if such action is eliminated by double dilution
Environmental emergency[2]	Acute toxic action of water is detected if such action is eliminated by a dilution factor from 50 to 100 (per one survey)
Environmental disaster	Acute toxic action of water is detected if such action is eliminated by a dilution factor of 100 and more (per one survey)

[1] The number of surveys (samples taken from the same point) is determined by the program for toxicological parameters established for each department of SSMEC's network [14].

[2] The presence of an emergency environmental situation and environmental disaster is identified by a complex set of criteria [18].

By the year 1991, biotesting had become an obligatory element of surface water monitoring that is provided for by the Rules for Protection of Surface Waters [19]. Biotest parameters of natural waters were included in the list of indices used for the detection of environmental emergency zones and environmental disaster zones [18].

Biotesting of natural waters represents an assessment of toxicity in an aquatic environment of unknown composition and in this relation has several peculiarities. Even if water pollution sources are known and it is possible to suppose which chemical compounds may be present in a sample of natural water, nevertheless, the results of biotesting cannot be predicted accurately. These results depend on the following factors [16]:

- Combined effects of chemical compounds contained in water
- Water temperature

- Transformation and metabolism velocity of chemical compounds in specific aquatic ecosystems
- Hydrochemical conditions, etc.

Thus, only by biotesting water samples is it possible to ascertain the actual toxicity of a natural water sample for hydrobionts and to assess the hazard level of toxic pollution for an aquatic ecosystem. As an example, we shall provide results of studies conducted in 1988-1989 on the Seversky Donets River, which is one of the most contaminated rivers according to hydrochemical parameters [15]. The river is rated as a water body with a dangerous level of pollution, experiencing frequent events of EHP. In the year 1988, toxicity assessment was carried out through a complex expedition organized by the Hydrochemical Institute for the purpose of operational monitoring by helicopter survey [20]. The river was studied along a considerable length, including areas of industrial influence of the following towns and cities: Rubezhnoe, Lisichansk, Kamensk-Shakhtinsky, Belaya Kalitva, Ust'-Donetsky (Figure 5.8 and 5.9). The survey took three full days (from 16 to 18 of May). Biotests were performed right after the helicopter had delivered the samples. The acute toxicity of water was evaluated by the Daphnia biotest. Observations were carried out for 24-48 hours of exposure in the studied water. The immobilization (loss of translational movement) and survival rate of Daphnias were registered. The level of acute toxicity was evaluated in accordance with the five-point scale developed by the Hydrochemical Institute (Table 5.4).

Table 5.4. Scale for assessment of acute toxicity in natural water by 24-hour Daphnia test

Daphnia survival rate, %	Level of water toxicity	Points
0–65	Very high	5
66–80	High	4
81–90	Moderate	3
91–95	Low	2
> 95	Nontoxic[1]	1

[1] Chronic tests may detect toxicity.

Spatial dynamics of water toxicity in the Seversky Donets River were generally characterized by a high toxic level (see Figure 5.8); samples 19 and 29 displayed toxic effect on Daphnias. In most cases, toxicity increased in the river sections located after cities and at the inflow of tributaries.

On May 31, 1989 Lissoda Factory made an emergency discharge of waste waters in the Belenkaya River (tributary of the Seversky Donets) and, consequently, the concentration of ammonia nitrogen exceeded MPC by 30 times. The toxicity and bioindication of enzyme activity were quickly assessed. A survey was carried out within three days in the area of Lisichansk (see Figure 5.9).

It was established that Daphnias were affected by acute toxicity of water from the Belenkaya River (see Figure 5.9, point 4). Toxic action was not detected in the samples taken from other sections of the river. According to the research, the toxicity of even highly contaminated water may be detected by the Daphnia test within first 24 hours of exposure and in certain cases, within 3-4 hours. This time frame is in regard only to acute toxicity, because longer exposure (over 7 days) may result in the occurrence of chronic effects.

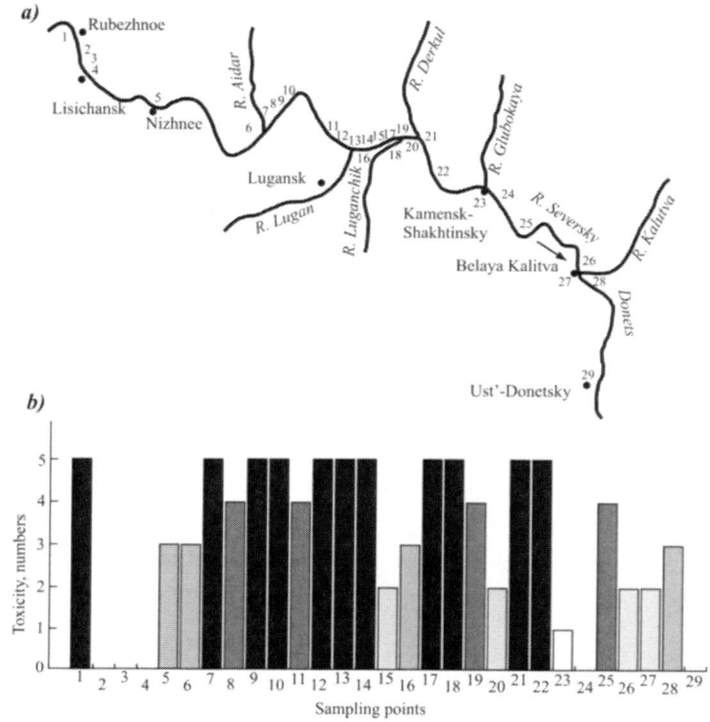

a) Sketch map of the studied section: 1—29 — sample points (13, 17, 20, 23, 27—mouths of tributaries); b) Level of water toxicity at sample points: 1—nontoxic, 2—low toxicity, 3—moderate toxicity, 4—high toxicity and 5—very high toxicity.

Figure 5.8. Spatial distribution of toxicity level in the Seversky Donets River from Rubezhnoe to Ust'-Donetsky.

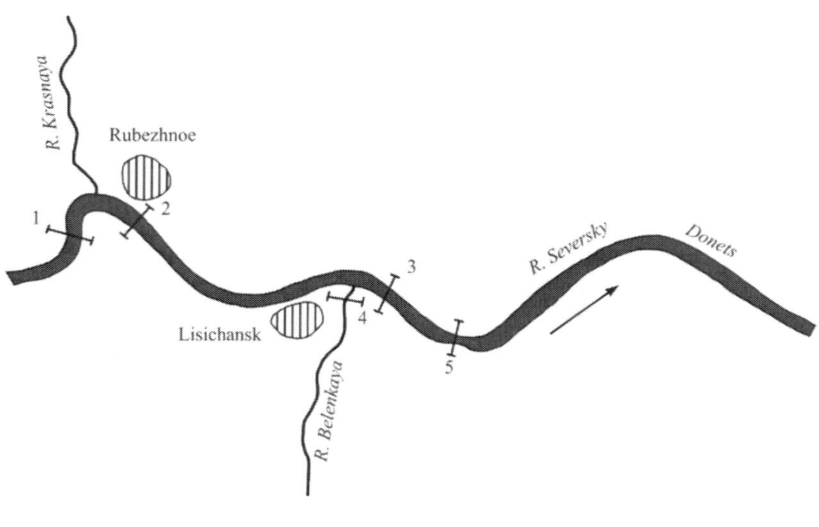

Figure 5.9. Sketch map of the studied section.

REFERENCES

[1] Kimstach V.A., Nikanorov A.M. Methodological Principles for Development of Operational Monitoring of Water Body Conditions / The Institute of Water Chemistry, Volume 100. — Leningrad: *Hydrometeoizdat,* 1991. — P. 3—14.

[2] Belousov A.P., Nikanorov A.M., Zavesa M.P., Gekov V.F. Present State and Prospects of Ground Operational Automated Control of Surface Water Pollution / Contemporary Issues of Regional and Applied Hydrochemistry. — Leningrad: *Hydrometeoizdat,* 1987. — P. 206—217.

[3] Practical Guidelines on Remote Sensing Methods for Monitoring of Surface Water Quality. Issue 1. Preparation and Carrying Out of Express Hydrochemical Survey. — Leningrad: *Hydrometeoizdat,* 1987. —32 p.

[4] Klimenko O.A., Bystrov A.V., Gekov V.F. Operational Forecasting of Expansion of Hazardous Accidental Pollution Zones in Watercourses / The Hydrochemical Materials, Volume 100. — Leningrad: *Hydrometeoizdat,* 1991. — P. 93—113.

[5] Trunov N.M., Nikanorov A.M., Askalepov V.N., Trofimchuk M.M. User of Trace Methods in Operational Monitoring of Accidental Pollution of Water Bodies / Hydrochemical Materials, Volume 100. — Leningrad: *Hydrometeoizdat,* 1991. — P. 85 —92.

[6] Methods of Bioindication and Biotesting of Natural Waters. / Edited by Bryzgalo V.A., Khoruzhey T.A.. Issue 2. — St. Petersburg: *Hydrometeoizdat,* 1989. —252 p.

[7] Trunov N.M., Kubranov V.L., Lastenko I.P., Khopryachkova O.V. Microscopefluorimeter for Counting Microorganisms // *Biological Resources of Water Bodies under Conditions of Anthropogenic Impact.* — Kiev, 1985. — P. 131—132.

[8] Nikanorov A.M., Trunov N.M. Using a New-type of Fluorescent Tracers to Study the Dinaraic Processes in Surface and Ground Water // *Tracers in Hydrology. Proc. of the Yokohama Symposium,* July 1993. — IANS Publication N 219, 1993. —P. 249—251.

[9] Nikanorov A.M., Trunov N.M., Bystrov A.V., Askalepov V.N. et al. A Fluorescent Tracer for Hydrodynamic Process Studies// *Hydrological, Chemical and Biological Process of Transformation and Transport of Contaminants in Aquatic Environments. Proc. of the Rostov-on-Don Symposium,* May 1993. — IANS Publication N 219, 1994. —P. 387—397.

[10] Trunov N.M. Practical Guidelines. Methods of Biotesting for Contaminated Water Bodies // *Regulatory Document 52.44.2—94. Regulatory Document. Practical Guidelines. Protection of Nature. Complex Examination of Contaminated Industrial Areas with Violent Anthropogenic Stress.* — Moscow, 1996. —P. 60—63.

[11] Trunov N.M. Practical Guidelines. Methods of Examination of Water Pollution Level According to State of Phytoplankton Community (Bioindication) // *Regulatory Document 52.44.2—94. Regulatory Document. Practical Guidelines. Protection of Nature. Complex Examination of Contaminated Industrial Areas with Violent Anthropogenic Stress.* — Moscow, 1996. —P. 64—66.

[12] Trunov N.M., Dyagtyarev A.F., Teplyakov Y.V. New Type of Tracer for Study of Dynamic Processes in Natural Environments // *Materials of the 28th USSR Hydrochemical Council,* May 1984 — Leningrad, 1987. —168 p.

[13] Nikanorov A.M., Trunov N.M. Intrawaterbody Processes and Water Quality Control. — St. Petersburg, 1999. —155 p.

[14] Regulatory Document 52.24.309—92. Practical Guidelines. Protection of Nature. Hydrosphere. Organization and Carrying Out of Surveillance Observations over Contamination of Surface Waters of Roshydromet's Network. — St. Petersburg: *Hydrometeoizdat,* 1992. —67 p.

[15] Khoruzhaya T.A., Bryzgalo V.A., Predeina L.M., Korotkova E.I. Methods of Bioindication and Biotesting in Studies of Freshwater Ecosystems under Accidental and Extremely High Pollution // Hydrochemical Materials. T.S. — St. Petersburg: *Hydrometeoizdat,* 1991. — P. 47—68.

[16] Nikanorov A.M., Khoruzhaya T.A., Brazhnikova L.V., Zhulidov A.V. Water Quality Monitoring: Toxicity Assessment. — St. Petersburg: *Hydrometeoizdat,* 2000. —160 p.

[17] Regulatory Document 52.24—94. Methods for Toxicological Assessment of Pollution of Freshwater Ecosystems. — Moscow: *Hydrometeoizdat,* 1994. —130 p.

[18] Criteria for Assessment of Environmental Conditions for Detection of Environmental Emergency Zones and Environmental Disaster Zones // *Green World. Russian Environmental Newspaper,* 1994, No. 12. — P. 8—11.

[19] Rules for Surface Water Protection (Standard Provisions). Introduced on 01.03.91. — Moscow: *Publishers of USSR State Committee for Natural Protection,* 1991. —38 p.

[20] Recommended Practice on Remote Sensing Methods for Surface Water Quality Control. Issue 3. — Leningrad: *Hydrometeoizdat,* 1989. —47 p.

Chapter 6

REMOTE SENSING METHODS OF OPERATIONAL MONITORING

Monitoring of inland water reservoirs is a constituent part of the national system for surface water monitoring. Under conditions of violent anthropogenic stress on the natural environment, society demands a timely (operational) supply of qualitative information on environmental conditions and, consequently, the establishment of a subsystem of operational monitoring. It is obvious that such subsystems could not be designed without development of methods and technical means of remote sensing monitoring. Thus, remote sensing monitoring is an essential constituent part of the subsystem of operational monitoring of the natural environment in general and of inland water bodies in particular.

The system of remote sensing monitoring is a complex, hierarchical structure. Generally, it has two or three levels. The top level is based on satellite information (or information from a high-flying airplane). The purpose of this level is to warn the lower system levels of any sudden changes in environmental conditions and to form the signal of place and time for the detailed examination of a water body. The middle level of observations relies upon information obtained from spectrometers with different spectral ranges and fulfills functions of express, though rather rough, control of environmental conditions. The bottom level determines environmental parameters requiring high component resolution.

Monitoring of inland water bodies is quite specific due to the peculiarities and diversity of objects involved in the first place and, secondly, due to extreme space-time variability conditioned by hydrological instability of these objects, qualitative and quantitative heterogeneity of pollutants and the presence of a wide variety of external pollution sources. Moreover, secondary contamination may be caused by the escape of pollutants from bottom sediments on account of the relative shallowness of the foregoing objects, certain weather conditions and water roughness.

The subsystem of remote sensing monitoring for inland water bodies should be developed with due account of these peculiarities. The first peculiarity connected with typical scales of inland water bodies (except for seas and major lakes) leads to the point that the use of information sent from space is ineffective for monitoring water bodies. For top level information, it is more reasonable to use surveys made by a high-flying airplane equipped with the same devices as a spaceship. In this case, hardware resolution may prove to be sufficient due to lower flight altitude; besides, the lock-in range at this altitude allows an aircraft to fly in the direction of watercourse movement without repeating all details of its bed. If an airplane is used, the influence of weather conditions would not be so substantial, because a survey could be conducted even under certain types of cloudy conditions. The second peculiarity connected with the qualitative diversity of pollutants and hydrological instability of water bodies seriously constrains studies related to monitoring of inland water bodies.

Remote sensing information is based on the intensity of upwelling from the water brightness of radiation in different ranges of the electromagnetic spectrum or brightness ratios (BR) indifferent to illumination. In order to decipher this information and convert it to concentrations of particular components of natural waters, it is necessary to use reference information, which in its own turn requires the introduction of a base component—the sampling of water and, in certain cases, of bottom sediments. This stage of monitoring should be performed by a helicopter or an amphibious aircraft in order to supply the latest information.

These two principles serve as the basis of remote sensing methods developed by the Hydrochemical Institute [1—21]. In particular, the methods were used for designing a subsystem of operational monitoring of inland water bodies and inland surface waters (ISW). The following items were developed:

- Methods for the detection of radioactive components*
- Equipment for the middle level of the subsystem
- Methods of operation for observation space stations (OSS) belonging to the middle level of the subsystem
- Principles for interaction of operational monitoring divisions with traditional analytical services engaged in hydrochemistry and hydrobiology

Works related to the implementation of complex methods for active and passive detection of pollutants are being carried out. Work was commenced on the classification of surface water conditions based on analysis of only remote sensing information, though it was accompanied by the acquisition of detailed reference information in the system's development stage.

6.1. Use of Aerospace Means of Observations in the Operational Survey of Aquatic Ecosystems. Statement of Objectives

The primary purpose for the use of aerospace means of observations in the operational survey of aquatic ecosystems is to *collect information on the water quality parameters* of the observed ecosystem. At that task, the composition and volume of obtained measurement information must provide an opportunity to accomplish the following goals:

* Here the term "radioactive" is equivalent to the term "radiation" with a wavelength range from centimeter (SHF and microwave) to nanometer (visible and infrared).

1) To obtain reference information for the analysis of aquatic ecosystems
2) To assess the current state of the entire aquatic ecosystem
3) To forecast changes of this aquatic ecosystem and to evaluate its possible oscillations

The foregoing objectives of observations may be fulfilled only with the use of certain mathematical models of a water body. Thus, information obtained by observation is used to solve the following issues:

- Development of new and update of the existing mathematical models of aquatic ecosystems
- Assignment of initial and boundary data and data used in simulation
- Checking of simulation data

While solving these problems it is necessary to obtain a great amount of real time measurement information on water quality parameters (or, for the first stage, concentrations of optically active water components).

Special attention should be paid to the *problem of determining sudden changes in the conditions of a water body.*

Such a situation may occur, for example, in the case of emergency volley discharges of pollutants. Here, the accuracy of registration for brightness of upwelling radiation is not of primary significance, because the most important role is occupied by higher requirements of operational observations of a water body and by the detection of water areas with considerably different optical properties. Areas with sudden changes of water conditions must undergo a complete spectrometric survey (with a high registration resolution for the brightness spectrum) and, if applicable, water samples must be taken for further hydrochemical and hydrobiological analyses.

Joint use of remote sensing and contact methods is most effective in the following cases where:

- Remote sensing parameter of water quality expressly describes the spatial structure of water pollution
- Remote sensing parameter of water quality correlates with the pollution parameter's spatial structure, which requires assessments

In particular, remotely sensing concentrations of optically active components (OAC) in water (suspended mineral matters, chlorophyll "a" in phytoplankton, dissolved organic substances) may be used as parameters describing the spatial structure of a polluted area. In this case by using only remote sensing measurements, it is possible to detect water areas with sudden changes in pollution level. Consequently, data from remote sensing of OAC aides in the optimization of contact sampling by coordinating locations of sampling stations with spatial structures of polluted areas.

6.1.1. Use of Space Information on the Visible Range for Monitoring Aquatic Ecosystems

The main types of output information obtained from space means of observation in probing the visible spectrum range include the following:

- Schemes of spatial zoning of water bodies related to the optical properties of water becoming apparent on aerospace images
- Maps of OAC concentration patters in natural waters (suspended mineral matters, chlorophyll "a" in phytoplankton, dissolved organic substances)
- Maps of the spatial distribution of environmental water parameters [7, 22, 23]

Different clusterization algorithms may be used in solving the first problem; meanwhile, no additional information is required except for the identified aerospace images. As a result of clusterization, the obtained schemes related to optical parameters for the spatial zoning of a water area enable researchers:

- To follow changes in water masses and to determine appropriateness of more detailed observations in the area (through a range of consecutive photographs of the same area of a water body)
- To choose the flight route for an OSS (in cases of receiving a photograph of a water body)

The measurement of OAC concentration in natural waters based on deciphering a multi-zone (not less than three representative spectral channels) scanner photograph of a water body was developed in order to map concentration patterns.

As opposed to the way of obtaining information to solve the first issue, in this case, it is necessary to simultaneously collect reference information of a limited volume while the scanning survey of a water body is occurring in order to ensure the proper adjustment of the deciphering algorithm. Expenses on the collection of reference information are repaid by the development of new output products—maps of OAC concentrations. At this stage, it is possible:

- To observe the process of eutrophication in water bodies (and to determine water areas with high levels of anthropogenic impact resulting in eutrophication)
- To control the erosion of shorefaces and distribution of sediments
- To establish the spatial structure of concentration patterns for biogenic elements

Methods for the assessment of environmental conditions of water bodies are currently at the stage of scientific research (see Section 6.5). It should be noted that the appropriateness of this research is unquestionable, because the establishment of quantitative relations between optically active and passive components is a promising way to increase the effectiveness of remote sensing means.

The experience of the Hydrochemical Institute in this sphere has shown that it is impossible to use aerospace means for the development of the foregoing output products without a special database. This database is primarily intended to store and use the following information:

- Aerospace images of water bodies
- Schemes of spatial zoning of water bodies based on clusterization of their aerospace images
- OAC concentrations in test sections (reference information)
- Brightness spectrums and BR in test sections
- Measurements of additional parameters used for the establishment of relations between optically active and inactive components

6.2. Equipment Used

The design and technical parameters of equipment installed on board an OSS are discussed in documents [12—15, 18, 24]. A flow chart of the automated onboard data measurement system (ODMS) is shown on Figure 6.1.

All equipment installed onboard the OSS may be conventionally divided into four blocks:

- Passive remote sensing
- Active remote sensing
- Contact express measurement of water quality parameters
- Flight and navigation equipment including an onboard computer and data transceiver

6.2.1. Equipment for Passive Remote Sensing

Visible range spectrometers are intended for the registration of upwelling from water (B↑) and incident (B↓) radiation as well as irradiance (E). The range of the wavelength should be between 400-750 nm, the spectral resolution should not be less than 3 nm and the spectrum record time should not be more than 10-15 sec. The main element of the spectrometer is the monochromator representing a prism or diffracting screen. The viewing angle of the spectrometer usually varies from 3° to 10°.

There are spectral devices with fixed receiving channels, and interference light filters act as the main dispersing element [18, 24]. A two-channel photometer of visible range is designed for the continuous recording of the B↑/B↓ ratio. Spectral separation of the signal is performed through interference light filters that have a bandwidth of 10-15 nm. Silicon photodiodes serve as detectors. There is a viewing angle of 6-10°. The SHF radiometric complex consists of three radiometers: B2, B18 and B30 for wavelengths of 2.25, 18.0 and 30.0 cm, respectively.

The B2 radiometer is based on the modulation of a superheterodyne receiver equipped with a remote ferrite polarizer. B18 and B30 radiometers are based on the modulation of a tuned-radio-frequency receiver.

The range of measured temperatures for B2 and B18 radiometers equals 10-300 K and 30-1000 K for a B30 radiometer.

The range of an antenna directivity diagram of a B2 radiometer equals 10-15° and 30-40° for a B18 and B30 radiometer. The fluctuating sensitivity of the radiometers is not less than 0.3 K with a time constant of $\tau = 1$ c [14]. A television system is used for convenient review of examined objects and in modern equipment, it is integrated in visible range spectrometers.

6.2.2. Equipment for Active Remote Sensing

The lidar spectrum analyzer, LSB-8, is designed for airborne registration of pollutants in visible and near IR spectral range. For the excitation of fluorescence, ruby and neodymium lasers are used with frequency doubling. The set laser mode is "impulse." Nine channels with replaceable interference input filters are used for registration of radiation.

A registration block records amplitudes of short impulse signals and constant components at the photomultiplier's input. This allows the measurement of spectra that are scattered and reflected solar radiation upwelling from water within intervals between laser impulses.

The lasers are controlled and spectra are registered by a minicomputer.

The RLSBO side-looking radar has not been used up to this point in this ODMS while monitoring surface waters.

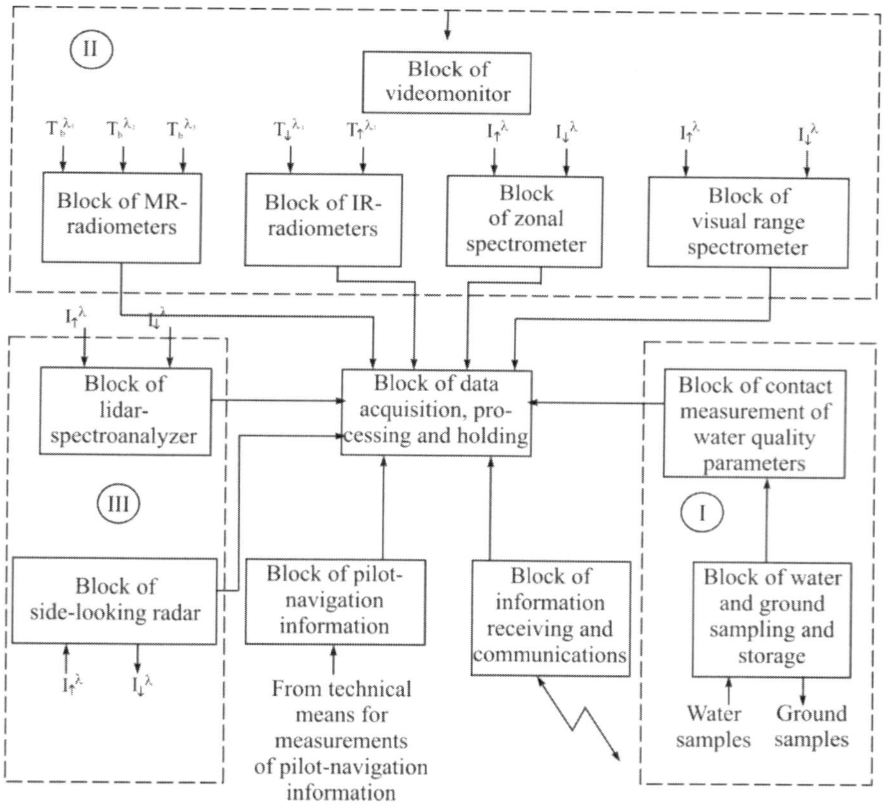

I—Group of blocks for contact measurement of water quality parameters, II—Group of blocks for passive remote sensing of water quality parameters, III—Group of blocks for active remote sensing of water quality parameters.

Figure 6.1. Flow chart of an automated onboard data measurement system (ODMS).

6.2.3. Equipment for Contact Measurement of Water Quality Parameters

The Quantum-5 fluorimeter has a spectral range of fluorescence excitation from 260 to 700 nm and can register fluorescence from 280 to 750 nm. Necessary spectral sections are marked by absorption and interference light filters supplied along with the device.

The PCFD-Eureka portable photocolorimeter and densitometer is used for measuring metal concentrations (Cu, Pb, Zn, etc.) and turbidity (at the level of MPC percentage).

Anion-210 is a portable device for measuring hydrochemical water parameters: pH, Eh, pX, t°C, æ.

The IK-Eureka is an instrument for measuring concentrations of oil products at the level of 0.8 of MPC.

6.2.4. Equipment for Airborne Sampling

The design of a water sampler is described in Figure 6.2. A hovering helicopter drops the sampler on a wire line with an LPG-4 winch equipped with special cable shears. Sampling methods used in hovering mode are described in the "Recommended Practice" document [25].

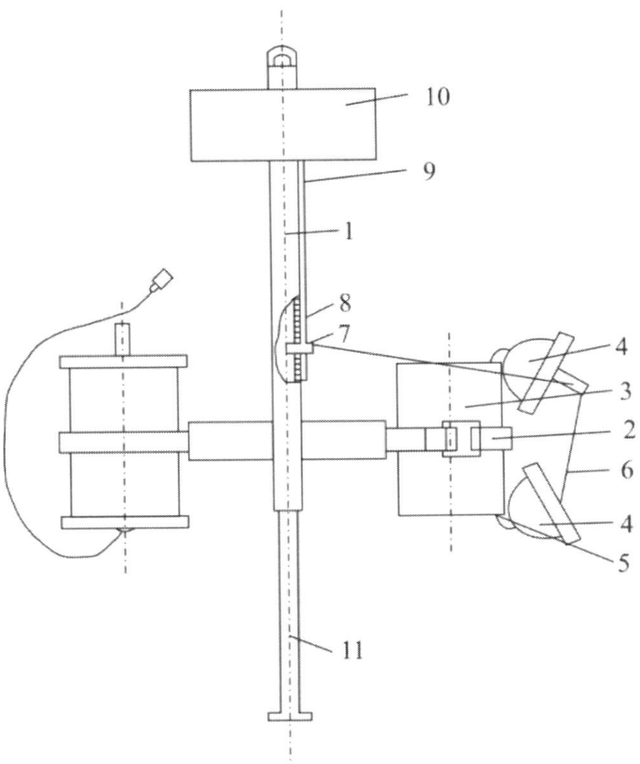

1—supporting bar, 2—clamp, 3—cylinder, 4—cover, 5—elastic tourniquet, 6—capron thread, 7—pin, 8—hole, 9—control rod, 10—float, 11—pusher

Figure 6.2. Scheme of bathometer for taking samples on board the helicopter.

6.2.5. OSS Flight and Navigation Equipment

Modern aircrafts (AC) used for environmental monitoring must be equipped with navigation equipment allowing them to conduct flights under adverse weather conditions at any time of the day or night, to communicate with base airports and to land on austere and unprepared airfields. For these purposes, the aircrafts should be equipped with special devices, such as a multipurpose integrated complex of onboard equipment (MICBO-4).

Primary functions of the complex include:

- Supporting flights in domestic and foreign airspace under any weather conditions, at any time of the day or night and in accordance with the rules of instrument and visual flights based at any airdromes, landing fields and water areas
- Ensuring safe, comfortable and economic flight by providing communication with ground services
- Providing instrument landing approach under meteominimum of the first ICAO category
- Ensuring accurate piloting and navigation

The latter is carried out by the system of satellite communication; in this respect, location accuracy equals 20-30 m, which is of vital importance in the monitoring of major water bodies.

The complex also decreases consumption of electricity from the aircraft's power supply system and optimizes consumption of fuel during the flight.

It is also possible to directly find ultra-short band stations.

In case of an emergency, an alarm signal is sent to the Cospas-Sarsat satellite-based rescue system and another emergency signal is emitted to guide rescue AC.

The complex is able to function under a wide range of operating pressures and temperatures; it is reliable and has a relatively light weight.

6.3. Methods of Interpreting Aerospace Information

6.3.1. Structure of Surface Water BR Spectra

Typical BR spectra obtained in the course of field works (rivers: Don, Seversky Donets, Kuban; reservoirs: Tsimlyanskoe, Rybinskoe; Azov Sea; lakes: Baykal, Balaton) are shown in Figures 6.3-6.5.

The maximum wavelength from 540 to 580 nm is primarily conditioned by optical dispersion on suspended mineral particles. Concentrations of chlorophyll "a" in phytoplankton peculiar to surface waters were at a clear minimum of $\rho(\lambda)$ at a registered wavelength of 670 nm. Their position of maximum change ranged from 665 nm at low values of C_{chl} to over 700 nm at $C_{chl} > 300$ mg/m^3.

On the high end, phytoplankton concentrations at the minimum $\rho(\lambda)$ registered in the range of 630-635 nm and had a maximum range of 650-655 nm. All local extremums of $\rho(\lambda)$ relation were mainly connected with the spectral movement of light absorbance by phytoplankton pigments, $æ_{php}(\lambda)$: $\rho(\lambda)$ minimums at 630 and 670 nm correspond to $æ_{php}(\lambda)$ maximums, while the maximum at 650 nm corresponded to the $æ_{php}(\lambda)$ minimum.

The maximum $\rho(\lambda)$ relation at a wavelength of 685 nm may have had a number of causes—fluorescence of phytoplankton pigments [26] and changes in correlation between absorbance of water, $æ_{в}(\lambda)$, and phytoplankton pigments, $æ_{php}(\lambda)$, at places with high concentrations of phytoplankton [27, 28]. Without ruling out the possibility of the fluorescence signal's contribution to water radiation at λ= 680 nm [29], we should note that the shift of maximum's position to a long-wave range with an increase of C_{php} confirms the appropriateness of the model suggested by document [27]. At C_{chl} = 1 mg/m^3, the contribution of $æ_{php}(\lambda)$ into the total absorbance at λ=670 nm equals less than 40% and is almost unnoticeable in the spectrum.

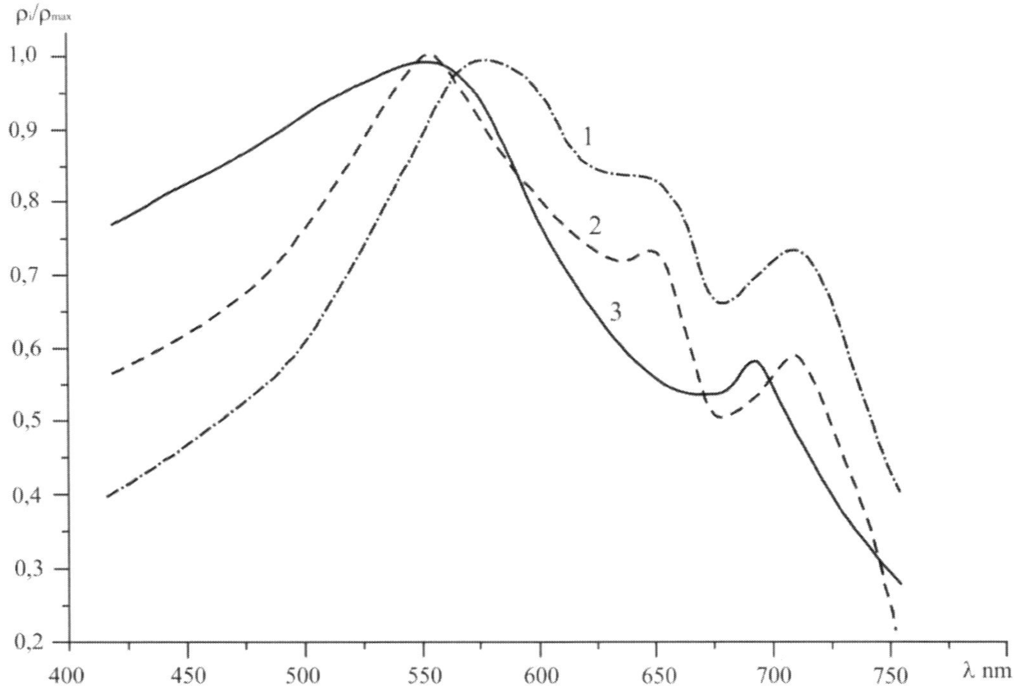

1—Don, 2—Seversky Donets, 3—Gulf of Taganrog, Azov Sea.

Figure 6.3. Typical spectra of brightness ratio in the visible range of wavelengths for different water bodies.

Meanwhile, for the $\rho(\lambda)$ relation, we register only a "shoulder" at 685-690 nm. With an increased $æ_{php}(\lambda)$, the maximum at λ = 670 nm becomes more apparent and the minimum appears on $æ(\lambda)$ at 690 nm. At C_{chl} = 10 mg/m^3 the contribution of $æ_{php}(\lambda)$ into $æ_{\Sigma}(\lambda)$ at λ=670 nm equals 80%, while the absorption minimum shifts from 690 nm into a long-wave range. The maximum wavelength at which $\rho(\lambda)$ maximum was experimentally observed, equaled 725 nm (at C_{chl} = 100 mg/m^3, C_{sus} = 35 mg/l).

Factor analysis of spectrums allows us to understand the conditionality of $\rho(\lambda)$ peculiarities and to find deciphering signs of different ingredients in spectrometric information [30, 31].

These signs should represent a certain function from SBR at different waves, like color index [9], where one of the $\rho(\lambda_i)$ values is maximally changeable at variations of concentrations of the sought ingredient C_x, and the other $(\rho(\lambda_i))$ is minimally dependent on C_x. $(\rho(\lambda_i))$ as well as $\rho(\lambda_i)$, however, should maximally reflect the impact of other optically active components, conditions of the survey and the spread function of the registering device.

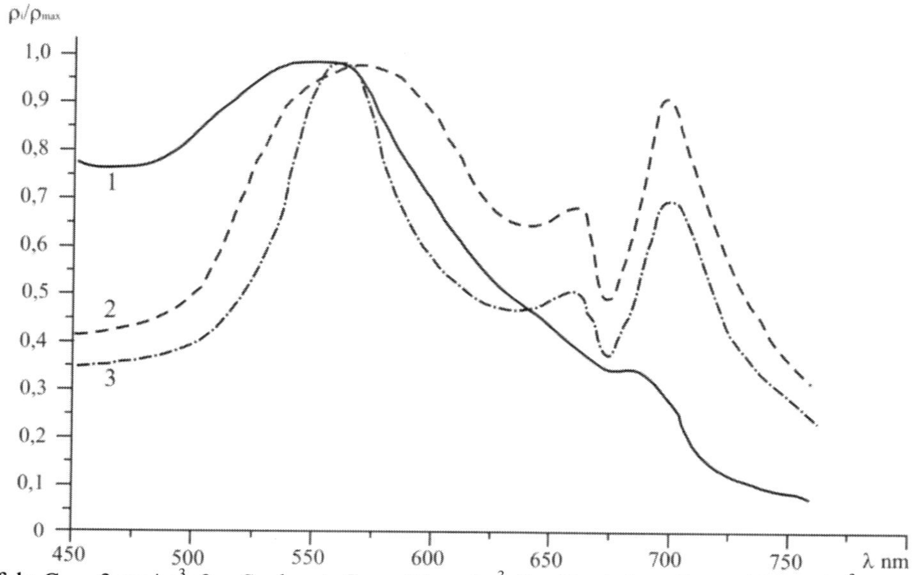

1—Siofok: C_{chl} =3 mg/m^3; 2—Segleget: C_{chl} = 50 mg/m^3; 3—Kestheley: C_{chl} = 100 mg/m^3

Figure 6.4. Typical spectra of brightness ratio in the visible range of wavelengths for Balaton Lake.

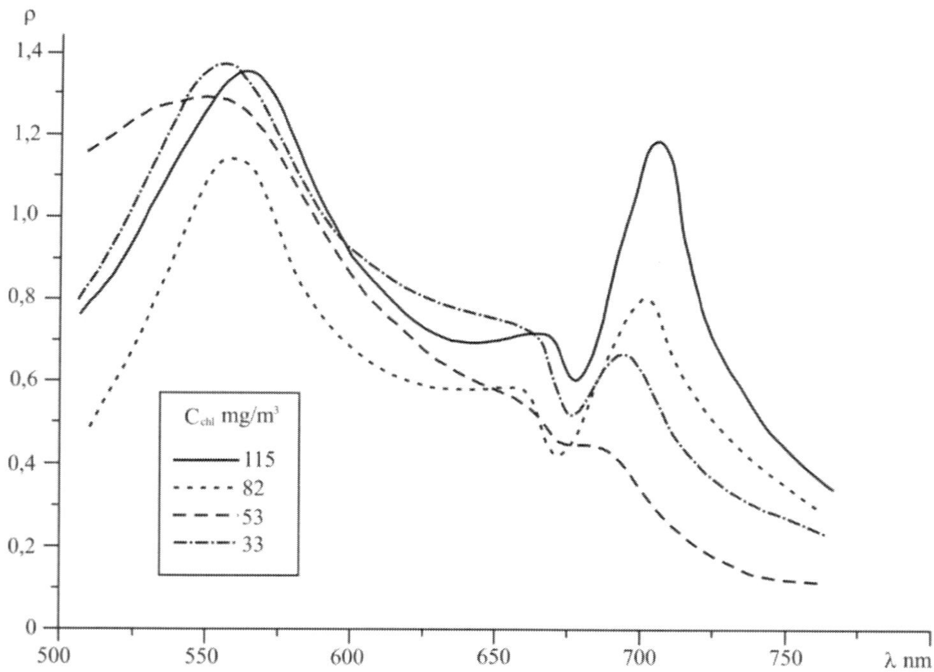

Figure 6.5. Typical spectra of brightness ratio in the visible range of wavelengths for Mueggelsee Lake.

6.3.2. Measurement of Chlorophyll "A" Concentrations in Phytoplankton

A great number of works are devoted to the remote sensing of phytoplankton concentrations (particularly of chlorophyll "a") in sea, ocean and surface waters. We shall mention only some of them—see, for example, documents [6, 7, 10, 21, 24, 27, 29, 33].

Let us discuss the deciphering signs (in some works, this term is an equivalent of "radiation parameters") on spectrometric data in order to determine concentrations of chlorophyll "a". In this case, the spectral movement of initial hydrooptical characteristics is a determinative element. $\rho(\lambda)$ relation changes when C_{chl} is adjusted primarily on account of changes in $æ_{php}(\lambda)$. The contribution of selective scattering of phytoplankton $\sigma_{php}(180°)$ in $\sigma(180°)$ at C_{sus} = 10 mg/l is rather small and equals less than 10% even at C_{chl} = 100 mg/m^3 [27]. Therefore, the ratio of the scattering coefficient $\sigma\lambda_i(180^o)/\ \sigma\lambda_i(180°)$ at variations of C_{chl} remains practically invariable.

In the shortwave range of spectrum, $æ_{\sum}(\lambda)$ depends on $æ_{php}(\lambda)$, $æ_{POB}(\lambda)$ and $æ_{в}(\lambda)$, and in the long-wave (absorbance of dissolved organic substance $æ_{DOM} \approx 0$), it depends only on $æ_{php}(\lambda)$ and $æ_{в}(\lambda)$. If it is increased, $C_{chl}æ_{\sum}(670)$ increases and $æ_{\sum}(700)$ decreases. Thus, a constituent part of deciphering signs may be $\rho(700)$ and $\rho(670)$. The absorption of light by phytoplankton pigments is minimal within the range of 560-580 nm; the relation rate of $æ(C_{chl})$ is also minimal in this range. Therefore, the value of ρ in the range of 560-580 nm may be normalizing in the deciphering sign. We should also take into account that if C_{chl} is increased, the value of SBR decreases along the whole visible spectral range, so that if C_{chl} increases, the value of $\int_{\lambda_{min}}^{\lambda_{max}} \rho(\lambda)d\lambda$ decreases, where λ_{max} and λ_{min} stand for the red and blue boundaries of wavelengths of the visible spectral range.

The presence of the foregoing spectral peculiarities of $\rho(\lambda)$, $æ(\lambda)$ and $\sigma\lambda(180°)$ provides an opportunity to propose and test several deciphering signs:

$$\rho(700)\rho(560)\sim æ(560)/æ(700) \quad (1)$$

$$\rho(560)\rho(670)\sim æ(670)/æ(560) \quad (2)$$

$$\rho(700)\rho(670)\sim æ(670)/æ(700) \quad (3)$$

$$[\rho(700)-\rho(675)]/[\rho(700)+\rho(675)] \quad (4)$$

$$\rho(700)/\int_{\lambda_{min}}^{\lambda_{max}} \rho(\lambda)d\lambda \quad (5)$$

$$[\rho(700)-\rho(675)]/\int_{\lambda_{min}}^{\lambda_{max}} \rho(\lambda)d\lambda \quad (6)$$

$$[\rho(670)/\int_{\lambda_{min}}^{\lambda_{max}} \rho(\lambda)d\lambda]^{-1} \quad (7)$$

The analysis of experimental data showed that there is quite a high correlation of C_{chl} with all radiation parameters (RP) of types (1)-(7). The regression equation showing the relation of C_{chl} to each of the RP has the following form:

$$C = A \cdot RP^B \quad (8)$$

where A and B are parameters of equation.

Table 6.1 shows the numerical evaluation of the parametrization effectiveness of type (8) for the Don River, where $F = r^2(n - 1)/(1 - r^2)$.

$$\delta C_{chl} = [\sum_{i=1}^{n} (C_i - \bar{C})^2 (1 - r)/(n - 2)]^{\frac{1}{2}}$$

n—volume of samples (number of stations), $\bar{C}$—average concentration, δC—relative accuracy of the concentration measurement, r[C, RP]—correlation coefficient.

For the Don River, the most effective radiation parameter is RP = p(700)/p(560). If C_{chl} is changed from 1 to 40 mg/m^3 and C_{sus} from 5 to 40 mg/l, the parameters of model (8) are as follows: A = 67.67 mg/m^3, B = 2.75.

Figure 6.6 demonstrates the correlation between concentrations of C_{chl} (measured in accordance with luminescence of water samples) and the p(700)/p(560) ratio.

Table 6.1. Statistical characteristics of the correlation between concentrations of chlorophyll "a" and radiation parameters type (1)-(7)

Statistical parameter	Radiation parameter						
	(1)	(2)	(3)	(4)	(5)	(6)	(7)
r[C_{chl}, RP]	0.96	0.93	0.94	0.88	0.86	0.90	0.84
F	826	425	491	225	190	285	157
δC_{chl}, mg/m^3	2.4	3.4	3.1	4.1	4.2	3.6	4.6

Deciphering signs were tested on Balaton Lake with a wide range of concentrations: C_{chl} ranged from 4.5 to 100 mg/m^3 and C_{sus} ranged from 5 to 40 mg/l. Radiation parameters (1)-(3) turned out to be the most effective for this ecosystem. The minimum standard evaluation error was achieved in relation to ρ(700)/ρ(560) at n = 103; $r^2 = 0.94$; F > 2000; $\delta C_{chl} = 2.8$ mg/m^3. The parameters for model (8) were as follows: A = 122.85 mg/m^3 and B = 2.3 (Figure 6.7).

For hypereutrophic water bodies with a C_{chl} of up to 400 mg/m^3 at a radiation parameter of ρ(700)/ρ(560) in equation (8), A = 270 mg/m^3 and B =1.2 at $r^2 > 0.94$ and $\delta C_{chl} = 23$ mg/m^3. At a parameter of ρ(700)/ρ(675), $r^2 > 0.98$ and $\delta C_{chl} = 13$ mg/m^3 [3, 10] (Figure 6.8).

Thus, the concentration of chlorophyll "a" is functionally connected with radiation parameters in respect to aquatic ecosystems of different trophicity. Parameters A and B in this equation depend on granulometric composition and the type of prevalent suspended mineral matters, plankton species composition in each group, concentration of dissolved organic substances (humic and fulvic acids), etc.

Therefore, parameters A and B in model type (8) are constant for ecosystems with definite relations between concentrations of optically active substances; they represent a system of the same type in accordance with radiation properties.

Differences in radiation characteristics are determined by the diversity of environmental conditions, which are dependent on a large number of factors. In this respect, it is currently unlikely that researchers will be able to measure the contribution of each factor to parameters A and B of the model and to introduce a unified parametrization for each type of ecosystem.

The extrapolation radius of the designed radiation models is quite large; it includes basins of large rivers, including coastal zones of seas into which they flow, reservoirs built on these rivers, etc. Therefore, the radiation model established for the lower reaches of the Don River allows remote sensing of C_{chl} on relatively large areas (from river head to mouth, including the Gulf of Taganrog and the Tsimlyansk Reservoir) within one hydrological season and with an error of less than 3 mg/m^3.

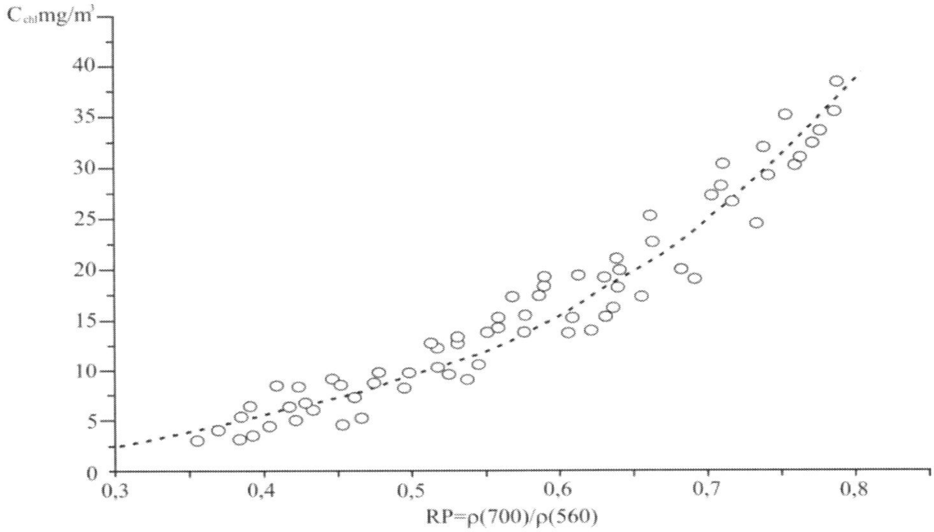

Figure 6.6. The correlation between concentrations of chlorophyll "a" in phytoplankton and RP with a ratio = ρ(700)/ρ(560) for the Don and Seversky Donets Rivers based on spectrometer data with high wavelength resolution.

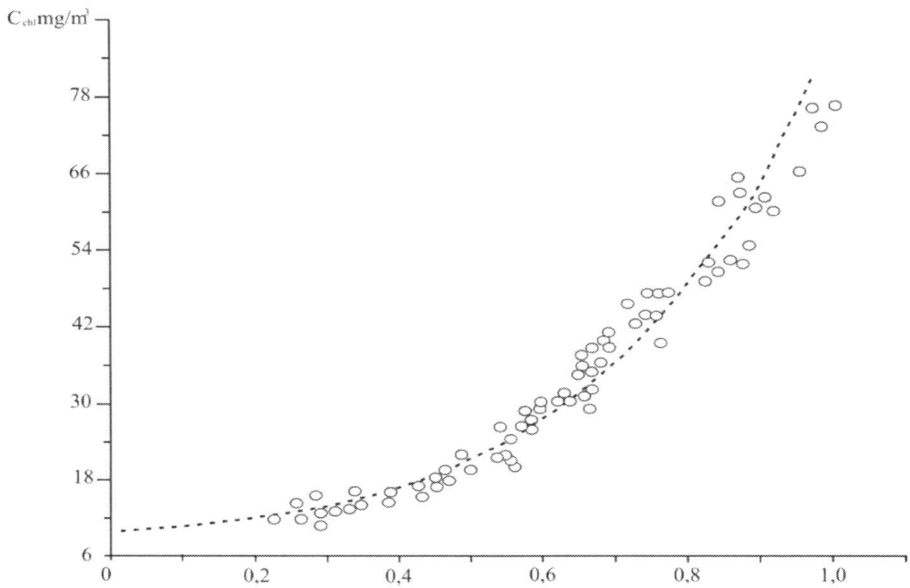

Figure 6.7. The correlation between concentrations of chlorophyll "a" in phytoplankton and RP with a ratio = ρ(700)/ρ(560) for Balaton Lake.

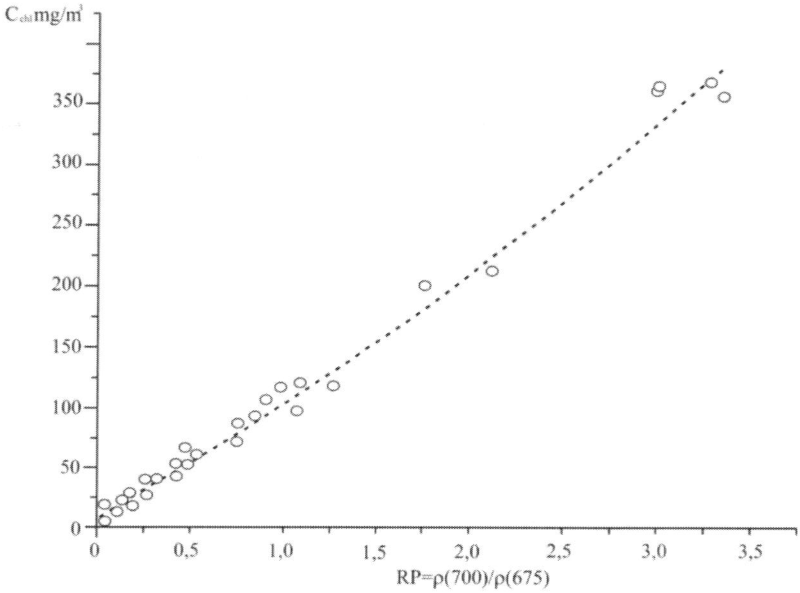

Figure 6.8. The relation between concentrations of chlorophyll "a" in phytoplankton and RP with a ratio = ρ(700)/ρ(675) for Mueggelsee Lake.

6.3.3. The Measurement of Concentrations of Suspended Minerals and Dissolved Organic Matter

Optical characteristics of suspended mineral matters are mainly determined by their scattering coefficient. The highest rate of change for $\rho(\lambda)$ in the variation of $\sigma\lambda(180°)_{sus}$ is observed in the wavelength range from 540 to 580 nm [28, 31]. In this range, absorbance of phytoplankton pigments is minimal. Thus, the selection of one BR comprising a deciphering sign is obvious: $\lambda \cong \rho(560)$. The maximum dependence of $\rho(C_{sus})$ is observed on this wave or near it.

The situation is more complicated when determining the wavelength at which BR is minimally dependent on C_{sus}. Factor analyses of spectrometric data [22, 30] show that in the range of wavelengths from 580 to 650 nm and from 500 to 520 nm, spectral variable makes an insignificant contribution to the dispersion of all three factors, which allows researchers to talk about the minor influence of spectral brightness ratios in these wavelength ranges on the total factor solution. In this way, values of SBR contribution at wavelengths of 500-520 nm in a dispersion of factors equals, for factors with a maximum at 700 and 560 nm, less than 0.5 %, and for factors with a maximum at 430 nm, less than 5%. This fact enables the usage of BR in the foregoing ranges of wavelength for normalization in deciphering signs of C_{sus} on spectrometric information. In accordance with the above, the following radiation parameters (RP) were tested for model (8):

$$\rho(560)/\rho(520) \quad (9)$$

$$\rho(560)/\rho(620) \quad (10)$$

$$[\rho(560) - \rho(520)]/[\rho(560) + \rho(520)] \quad (11)$$

$$[\rho(560) - \rho(620)]/[\rho(560) + \rho(620)] \quad (12)$$

Table 6.2 provides the numerical evaluation of the effectiveness of these parameters for the Don River, Tsymlyansk Reservoir and Azov Sea (n = 63).

Parameter (11),[ρ(560) - ρ(520)]/[ρ(560) + ρ(520)], turned out to be the most effective. If C_{sus} changes from 5 to 40 mg/l parameters, model (8) is equal to: A = 60.68 mg/l and B = 0.49. Figure 6.9 shows the correlation of this parameter to C_{sus} in accordance with survey data for Balaton Lake, dated 1985.

Table 6.2. Statistical characteristics of correlations between concentrations of suspended mineral matter and radiation parameters of type (9)-(12)

Statistical parameter	Radiation parameter			
	$\frac{\rho(560)}{\rho(520)}$	$\frac{\rho(560)}{\rho(620)}$	$\frac{\rho(560)-\rho(520)}{\rho(560)+\rho(520)}$	$\frac{\rho(560)-\rho(620)}{\rho(560)+\rho(620)}$
$r[C_{sus}, RP]$	0.87	0.83	0.93	0.91
F	192	135	405	297
δC_{sus}, mg/l	4.3	4.9	3.2	3.6

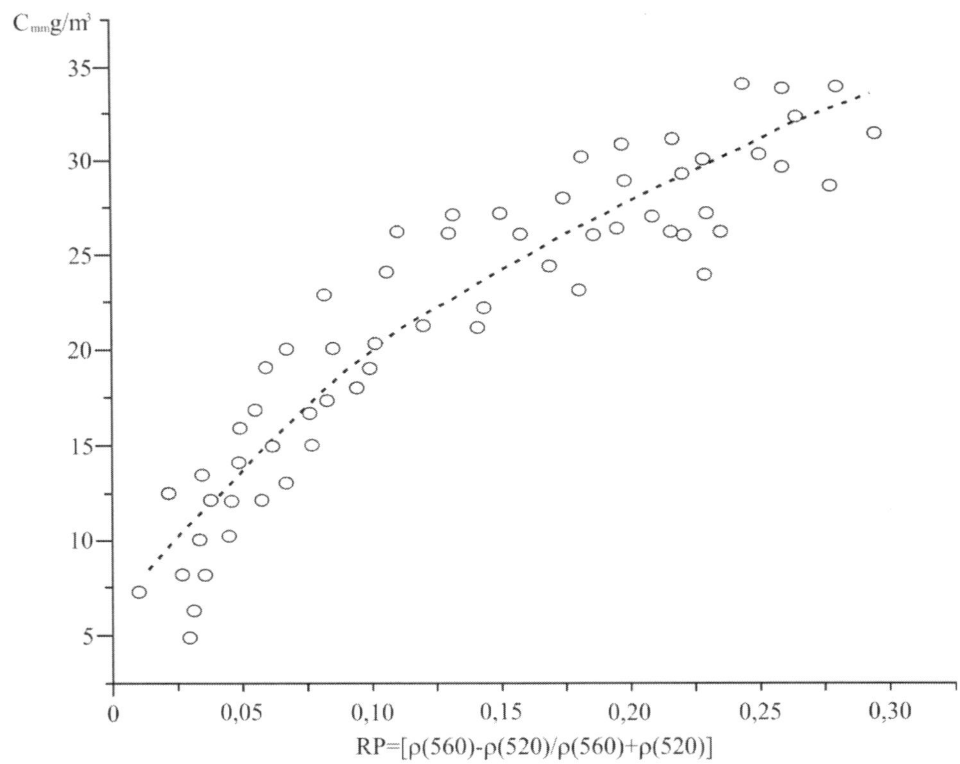

Figure 6.9. The correlation between concentrations of mineral suspended matter and RP with a ratio = [ρ(560) - ρ(520)]/[ρ(560) + ρ(520)] for the Don and Seversky Donets Rivers.

Quite high values of standard errors in the evaluation of C_{sus} in accordance with remote sensing characteristics may result for two reasons—the variation of wavelengths, at which the scattering coefficient, BR, σλ(180°) and ρ(λ) change, and the variation of the granulometric and mineralogical composition of suspended matters, which mostly affect the BR value. In the case of remote sensing assessment of waters, it appears that the most appropriate approach is one that determines not C_{sus} but the parameter that is functionally connected with a concentration of mineral suspension, the average radius of particles and the law of their distribution by size. This integral parameter determining the turbidity of a medium is closely related to the abovementioned factors having maximum loads at wavelength from 540 to 660 nm [22, 30].

We should note that radiation parameters of type (9)-(12) are effective in determining C_{sus} when concentrations of C_{chl} and C_{sus} are independent. When the concentrations of these ingredients are strongly correlated (for example, at Balaton Lake $r[C_{sus}, C_{chl}] > 0.88$), sufficiently accurate evaluations of C_{sus} may be obtained by using radiation parameters (1)-(7).

The selection of a deciphering sign for the work is hampered by the fact that for the most part, the studied objects are strongly connected with C_{chl}. Therefore, for Balaton Lake, the Tsymlyansk Reservoir and the Azov Sea, C_{DOM} was determined by one of deciphering signs developed for C_{chl}. This means that for Balaton Lake, equation type (5) with a correlation coefficient of $r > 0.94$ connects C_{DOM} with RP = $\rho(700)/\rho(560)$ and provides an opportunity to assess C_{DOM} with an error less than 0.5 mgC/l. For the Seversky Donets River, we have managed to check the deciphering signs of C_{DOM} based on the normalization of BR on waves where $æ_{DOM}$ is quite high (in the range of 430-450 nm) and on BR that range from 580 to 670 nm, where $æ_{DOM} = 0$. In these cases, RP = $\rho(620)/\rho(430)$ and RP = $[\rho(430) - \rho(620)]/[\rho(430) + \rho(620)]$. These models allowed us to evaluate C_{DOM} with an error of less than 0.65 mgC/l.

Distribution maps of suspended mineral substances (Figure 6.18) vividly demonstrate the variability of these characteristics, wherefore their remote sensing assessment proves to be quite effective.

In work [34], the concentration of DOM is measured by another method. ΔD is used as a deciphering sign for meso- and eutrophic water bodies:

$$\Delta D_{400\text{-}550} = \lg(1/\rho_{400}) - \lg(1/\rho_{550}) = \lg(\rho_{550}/\rho_{400}), \qquad (13)$$

where ρ_{400} and ρ_{550} are BR on wavelengths of 400 and 550 nm, respectively. The concentration of yellow substance, C_{ys} (in oceanology, the term "yellow substance" is similar to DOM in surface waters), which in most cases comprises the prevalent part of DOM, is determined as follows:

$$C_{YS} = C_{DOM}^{REL} \approx K_{YS}(\lambda) = K_{YS}(\lambda_0)e^{-\mu(\lambda-\lambda_0)},$$
$$K_{YS400} = 0.0955\rho_{550}^2\Delta D_{400-550}. \qquad (14)$$

Here, μ is a coefficient varying across different water bodies from 0.012 to 0.019 nm^{-1} (on average, $\mu = 0.015$ nm^{-1}). In this method, the average relative assessment error for K_{YS400} does not exceed 25%.

6.3.4. Remote Sensing of Concentrations of Dissolved Organic Substances (Active Probing Method)

The spectral-luminescent properties of humic and fulvic acids (HA/FA) were studied in order to ascertain requirements for the systems for the induction (excitation) of a fluorescence signal within dissolved organic substances along with its registration and the determination of deciphering signs of its ingredients in accordance with the obtained information.

Observing the wave, λ_1 = 470 nm maximum on the spectrum where FA fluorescence excitation is detected on waves ranging from 340-350 nm, while observing the wave of 550 nm maximum, FA excitation is detected on waves of 460-470 nm. In the range of wavelengths from 400 to 450 nm, there is a plateau on the excitation spectrum in which excitation intensity equals 0.7-0.8 from the maximum (see insertion on Figure 6.10).

Spectral peculiarities of fluorescence excitation at waves of 470 and 550 nm allow us to carry out remote sensing assessments of FA and HA in solutions. In the excitation of spectrum by radiation with a wavelength, λ_B = 350 nm, FA fluorescence is at maximum observed at a wavelength of λ_H = 470 nm, and HA fluorescence is at λ_H = 550 nm, though the intensity of HA fluorescence is considerably lower than FA. In the excitation of λ_B = 470 nm, the intensity of HA fluorescence is 4-5 times higher than FA (Figure 6.10). Thus, in the excitation of fluorescence by waves of 350 and 450 nm and in the observation of waves of 470 and 560 nm, it becomes possible to measure concentrations of dissolved organic substances. It is reasonable to evaluate concentrations in mixed solutions with due account of the HA fluorescence contribution at λ_{H_1}= 470 nm and for FA, at 560 nm.

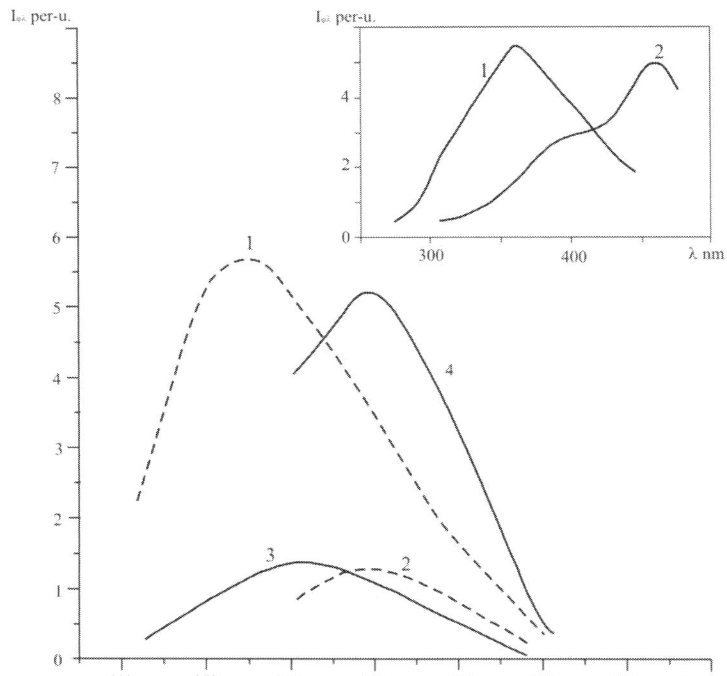

Figure 6.10. Fluorescence spectra of FA (1, 2) and HA (3, 4) in excitation by waves λ = 350 nm (1, 3) and λ = 470 nm (2, 4). Insertion: Spectra of fluorescence excitation for FA is at λ=470 nm (1) and for HA, it is at λ = 560 nm (2).

Laboratory research was carried out with probing distances from 2 to 5 meters. The maximums of FA and HA fluorescence were registered by lidar in the wavelength ranges of 447 ± 10 and 540 ± 10 nm, respectively (Figure 6.11). The detected spectral peculiarities of fluorescence under laser excitation allowed the measurement of concentrations of dissolved organic substances by excitation of one wave at 347 nm. Intensities of re-radiation at λ_{H_1} = 447 nm and λ_{H_2} = 540 nm, respectively, are measures of HA and FA concentrations in a two-component solution. The dependence of the fluorescent parameter, φ_0, (ratio of the fluorescence signal at the foregoing wavelengths (I_φ) to the signal of Raman scattering for water (I_{Rs})—see the insertion on Figure 6.11) on HA and FA concentrations indicates that this parameter may serve as an indicator of a concentration of dissolved organic substances in a two-component solution [12, 13].

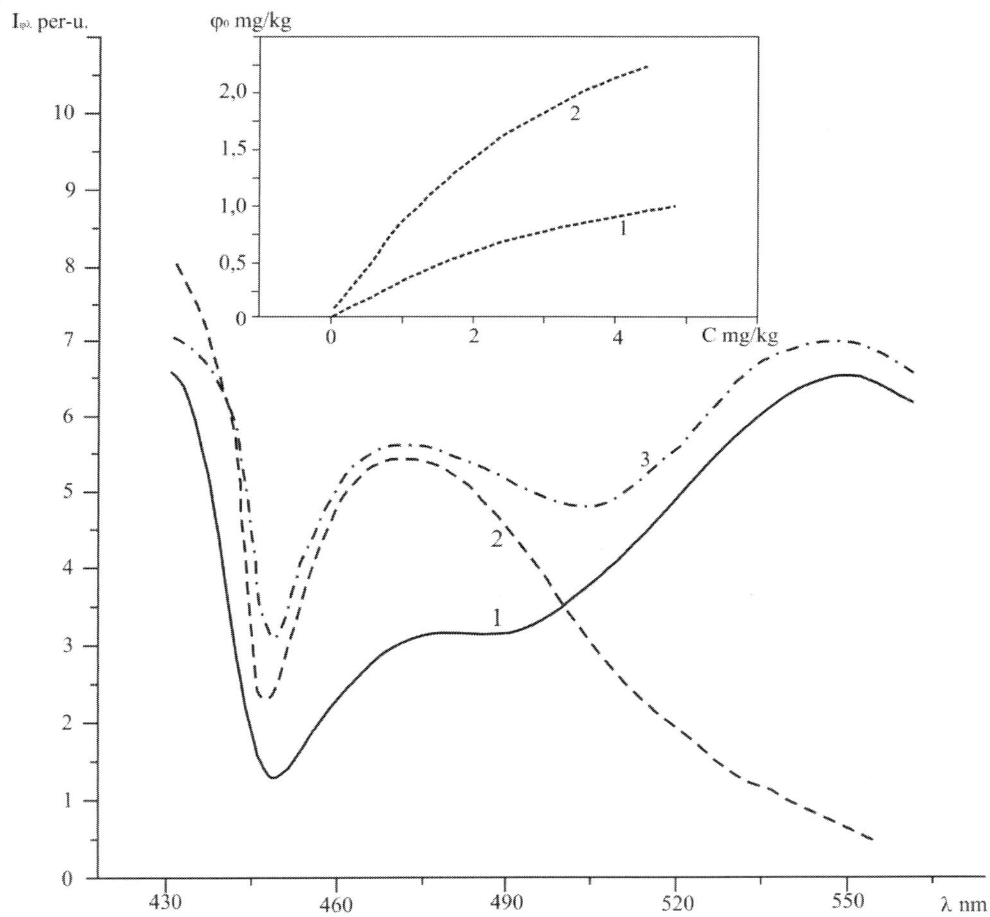

Figure 6.11. Spectra of fluorescence obtained by a lidar spectrum analyzer for HA(1), FA(2) and a two-component solution of HA and FA. Insertion: Dependence of $\varphi_0 = I_\varphi/I_{Rs}$ on the concentration of C for FA(**1**) and HA(2).

The assessment of the fluorescence section of both organic components attributed to an average molecule is also quite interesting. The fluorescence section, σ_φ, is calculated in accordance with the following formula:

$$\sigma_\phi = \varphi_0 \frac{\sigma_{Rs} n_{H_2O}}{n_\varphi}, \tag{15}$$

where σ_{Rs} is the Raman scattering section (m^2/strad), n_{H_2O} is the concentration of water molecules, n_φ is the concentration of FA and HA molecules, $\varphi_0 = I_\varphi/I_{Rs}$ is the fluorescence parameter and I_φ and I_{Rs} represent the intensity of the fluorescence signal and Raman scattering, respectively.

The fluorescence parameter is measured at a low (0.1-0.5 mg/l) concentration of organic substances. At the same time, the relative contribution of fluorescence in the signal of Raman scattering is rather low.

For HA, an average molecular mass would be 35,000 u. $\sigma_{ha} = (1.2 \pm 0.5) \cdot 10^{-23} m^2$/strad. For FA, an average molecular mass would be 1,000 u. $\sigma_{fa} = (1.5 \pm 0.5) \cdot 10^{-25}$ м^2/strad. For such complex multicomponent substances as HA and FA, the fluorescence section may vary significantly depending on the ratio of molecules in the solution to different masses. The effectiveness of FA and HA fluorescence is largely determined by this factor.

The sensitivity threshold of this method, as well as its error, is determined by conditions of measurement, in particular, by probing depth, hydrooptical parameters of a medium, etc. In the experiments described above, the sensitivity threshold did not exceed 5 μg/l at a probing depth of 0.25 m. It is obvious that at greater probing depths, the sensitivity threshold will be lower.

Application of the fluorescence method for surface waters is often hampered by rather high values of the water attenuation coefficient and the complexity of fluorescence contribution of different organic substances into the signal at a Raman scattering frequency. In this respect, surface water bodies require a fundamentally new method for measuring the hydrooptical parameters of a medium and for the normalization of a fluorescence signal during remote laser sensing. It may be based on measurements by a lidar spectrum analyzer in the intervals between impulses of a laser with a brightness ratio equal to the ratio of upwelling and incident radiation brightness. BR conveys information on the absorption and scattering properties of a water environment. By using the obtained deciphering signs of optically active ingredients on spectrometric data and by normalizing the fluorescence signal registered by a lidar for functional BR at different wavelengths, we have managed to consider the influence of absorption and scattering properties on the intensity of the florescence signal registered by a lidar. This fact makes it possible to broaden the application of the laser spectroscopy method and to use it for remote sensing of DOM in surface waters.

6.3.5. Remote Sensing of Mineralization and Water Temperature

The SHF radiometric method for the measurement of mineralization and water temperature is based on the dependence of the intrinsic SHF radiation intensity of water bodies on their physicochemical parameters (PCP). The brightness temperature is a measure of intrinsic radiation [14, 15, 35]. Laboratories measuring the SHF radiometric complex included radiometers with wavelengths of 2.25, 3.0 and 7.77; they were between 18 and 30 cm and fluctuation sensitivity was more than 0.3 K. All measurements on the field were taken by an An-2 aircraft. The aerial systems of this onboard complex provided the following relation between flight altitude (H) and ground resolution (D): D = 0.7H. Some of the results of the conducted experimental studies and calculations are shown on Figures 6.12 and 6.13. Full-scale experiments covered the measurement of the PCP of water bodies with mineralization from 0 to 400 g/kg.

The studies proved the presence of a stable functional connection between the intrinsic radiation field, water temperature and mineralization, which provides an opportunity to measure them by remote sensing. Essential spectral peculiarities in the dependence of radiation on mineralization and temperature were detected. These peculiarities serve as a basis for the solution of a theoretically incorrect, inverse problem, namely, the measurement of the PCP of water bodies by their intrinsic SHF radiation and development of methods for remote sensing of PCP.

In accordance with the model representations and experimental data, radiation properties of water bodies in the SHF range are intricately interrelated with water mineralization, temperature and the state of the water surface. The degree of influence of certain factors on radiation characteristics, however, is significantly different on the spectrum. This characteristic feature determines capabilities of the remote indication of hydrophysical parameters, the assessment of their nature and their intensity in accordance with SHF radiation power (brightness temperature T_b) on certain sections of a spectrum by solving a system of hydrophysical radiation equations [14]:

$$F_{\lambda_i}(Q_{\lambda_1},\ldots, Q_{j_1},\ldots, Q_n) = T_{b_i},$$
$$i = 1, 2, \ldots, m,\ m \geq n. \qquad (16)$$

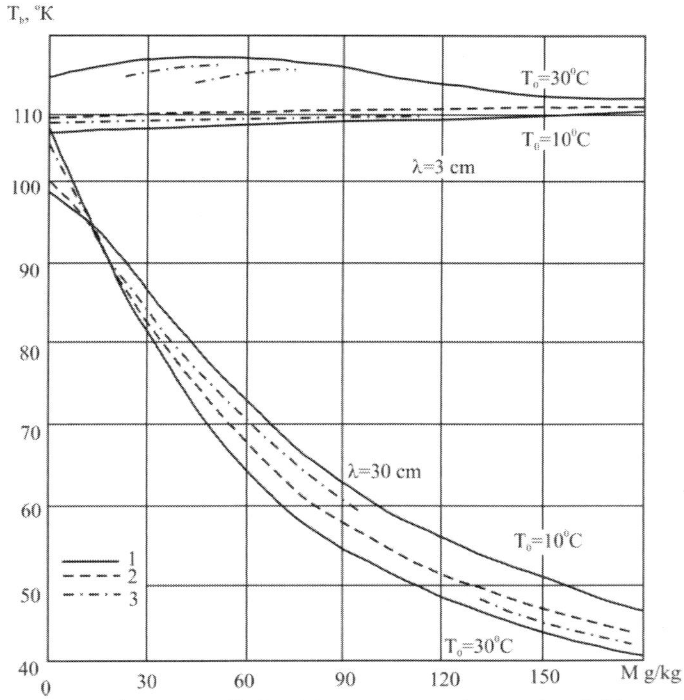

1—estimate (at T_0 = 30°C and T_0 = 10°C), 2—laboratory experiment, 3—field experiment

Figure 6.12. The dependence of water brightness temperature T_b on Kelvins in the SHF range and on mineralization (M) at different values of thermodynamic temperature (T_0).

Due to the fact that the dependence of T_b on water temperature and mineralization is not linear (see Figures 6.12 and 6.13), the system of equations (16) is also not linear in most cases. If target characteristics are not absolute values of temperature, mineralization or other parameters of a_j, but are instead variations of ΔQ_j in a water body, then system (16) may be represented in a linear form:

$$\sum_{i=1}^{n} a_{ij}\ \Delta Qi = \Delta\ T_{b_i}, i = 1, 2, 3, \dots, m \tag{17}$$

where $\Delta\ T_{b_i}$ represents variations of brightness temperature on spectrum sections caused by variations of the PCP of a water body; $a_{ij} = \Delta T_{bi}/\Delta Q_i$ are coefficients defining the sensitivity of the radiation field at a wavelength of λ_i to variations of the Q_i parameter (within the limits of linear dependence $\Delta T_{bi}(\Delta Q_i)$). In centimeter and decimeter ranges of wavelength, the slope of dT_b/dT_0 largely depends on the temperature of the water surface. An exception includes only wavelengths from 5 to 8 cm, where the radiation coefficient is weakly dependent on the PCP of water, while dT_b/dT_0 reaches 0.5 K/°C. The influence of salinity (mineralization) of the water bodies is most vividly expressed in the decimeter spectrum range (see Figure 6.12) where the value of radiation-mineralization dependence slope varies from 0.5 to 1 K.

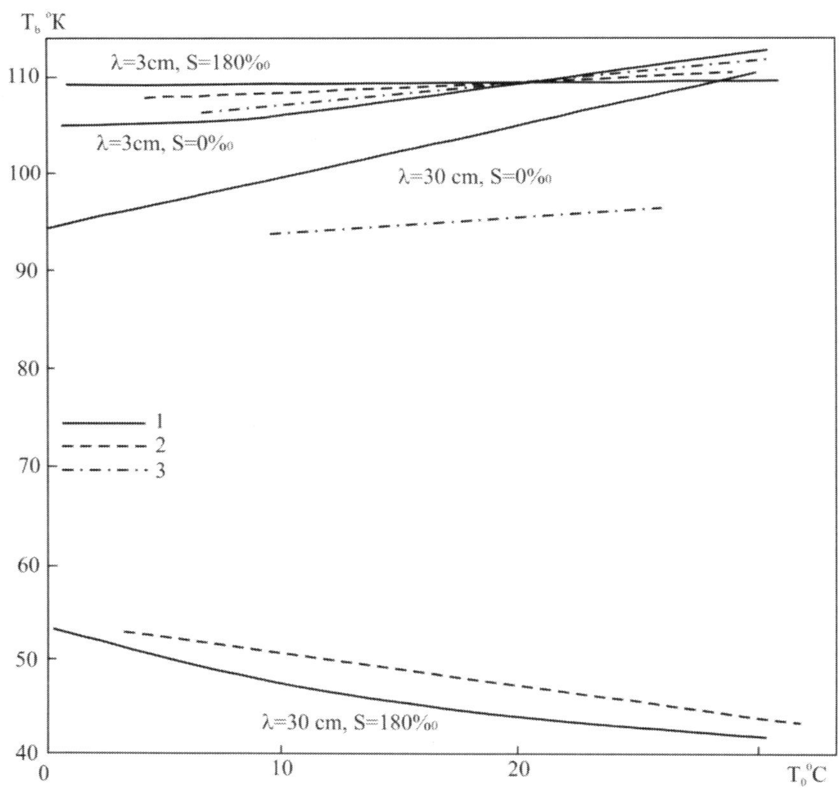

1—estimate (at S = 0‰ and S = 180‰), 2 and 3—laboratory experiment (2) and field experiment (3)

Figure 6.13. Spectral dependence of SHF radiation intensity on variations of thermodynamic temperatures of the water surface.

The capabilities of remotely sensing water's physicochemical parameters through their intrinsic SHF radiation were tested in laboratory and field conditions in a wide range of water mineralization (from 0 to 400 g/kg) and temperature (from 5 to 27°C) variations [12, 15].

It is quite a challenge to determine the accuracy of the method in natural conditions. Due to quite high spatial and time variability of the physicochemical parameters of water, and test fields in particular, joint experiments on synchronous contact and noncontact measurements of PCP were conducted.

Joint synchronous measurements of geophysical parameters by express contact and remote sensing methods allowed the determination of the accuracy of the method in a wide range of PCP measurements.

As a result of tests, experiments and metrological certification of the method, the following metrological, technical and economic indices were established:

- Discrimination threshold for mineralization from 0.2 g/kg for HC1 to 2 g/kg for NaCl
- Mineralization (salinity) detection range from the value of the discrimination threshold to 400 g/kg
- Resolution of mineralization at a radiometer time constant of $\tau = 1$ sec equals 1 g/kg
- Error of determination of area-average mineralization (salinity) at radiometer time constant $\tau = 1$ sec with a confidence probability of 0.95 equals 2 g/kg
- Resolution of water surface temperature at a radiometer time constant of $\tau = 1$ sec equals 0.6 °C
- Error of determination of area-average water surface temperature does not exceed 2 °C with a confidence probability of 0.95 within the range of temperatures from 0 to 35 °C

Spatial resolution equals \ 0.7 from flight altitude. The performance of a hydrochemical survey for a field test at a distance of 1 km between lanes equals 162 km^2/h. The time needed for preparing results of the hydrochemical survey in the form of water mineralization and temperature variation maps equals one hour [36].

Figures 6.14 and 6.15 display the correlation of data on water mineralization and thermodynamic temperature obtained by contact and remote sensing methods with the use of the SHF radiometric complex.

6.4. Use of Remote Sensing Data for Controlling Water Quality Parameters

This section provides examples of applications of remote sensing measurements in studying water bodies that vary in properties and sizes. Surveys were conducted for water bodies exposed to violent anthropogenic stress and areas of environmental emergency.

Information was obtained from different levels of observations: from boards of artificial Earth satellite (AES) and the observation space station (OSS). Due to absence of domestic system for natural environmental space monitoring, the pictures presented only demonstrate an opportunity for the effective use of space information.

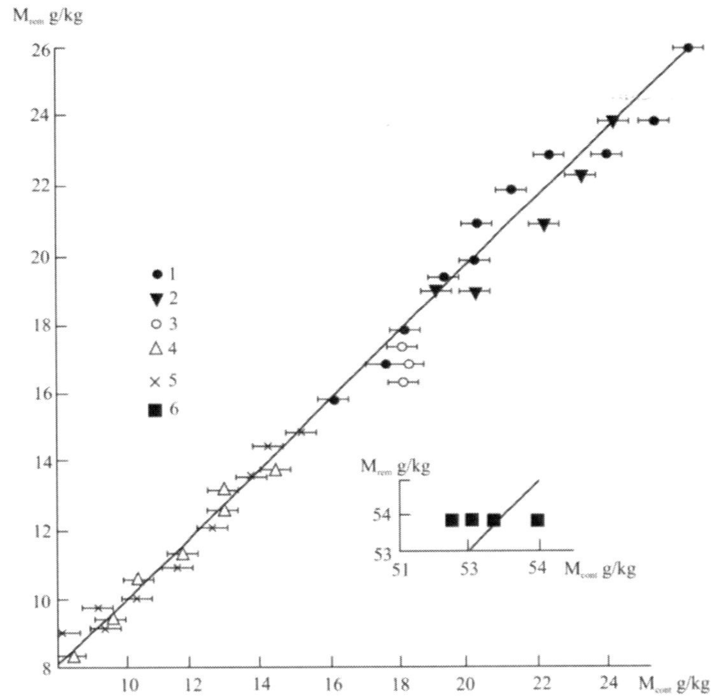

1, 2—Manych-Gudilo Lake, 3—Black Sea, 4—Sea of Okhotsk, 5—Dnepro-Bugsky Estuary, 6—Sivash Gulf

Figure 6.14. Comparison of mineralization measurements made by contact (M_{cont}) and SHF radiometric (M_{rem}) methods.

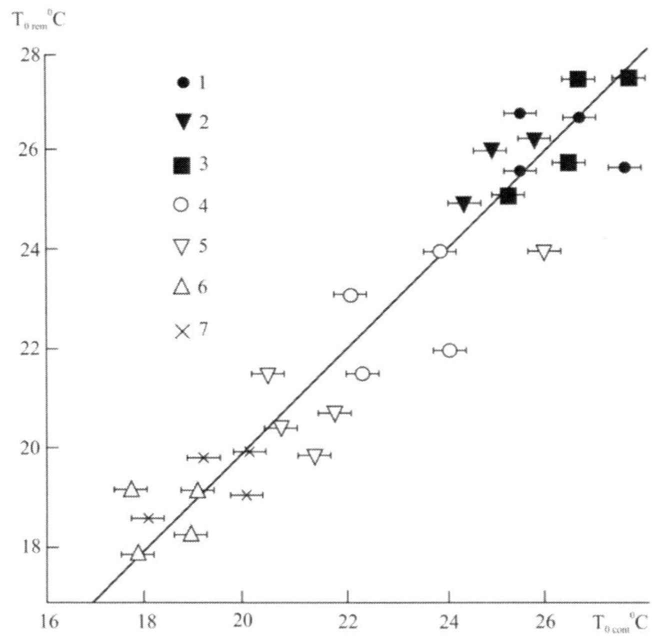

1, 2—Manych-Gudilo Lake, 3—Sivash Gulf, 4-7—Caspian Sea

Figure 6.15. Comparison of thermodynamic temperature measurements made by contact ($T_{0\ cont}$) and SHF radiometric ($T_{0\ rem}$) methods.

6.4.1. Satellite Information

Figure 6.16 shows schemes for the distribution of suspended mineral matters and chlorophyll "a" in phytoplankton for Balaton Lake in accordance with data obtained by a scanner installed on the American satellite, Landsat. The information was processed under the method developed by the Hydrochemical Institute [3, 4]. We should note the mismatch of the spatial structure of suspended mineral matter and phytoplankton distribution fields, which is only possible to obtain by multi-spectral equipment. Figures 6.17 and 6.18 provide schematic maps obtained by an analog of space equipment (Mattra scanner) installed onboard the high-flying aircraft (at the altitude of about 10 km), TU-134 SH. The distribution of optically active components (OAC) in the tail water of the Tsymlyansk Reservoir was scanned twice within a two week interval. It is obvious that within this period of time, the general structure of OAC fields remained almost the same, while areas of intensive phytoplankton bloom increased and shifted, which is explained by a rapid bloom of diatoms at the end of April. Evidently, it is almost impossible to capture this bloom peak with conventional means of surveying (by vessel) or even to make a sufficiently detailed mapping of bloom zones.

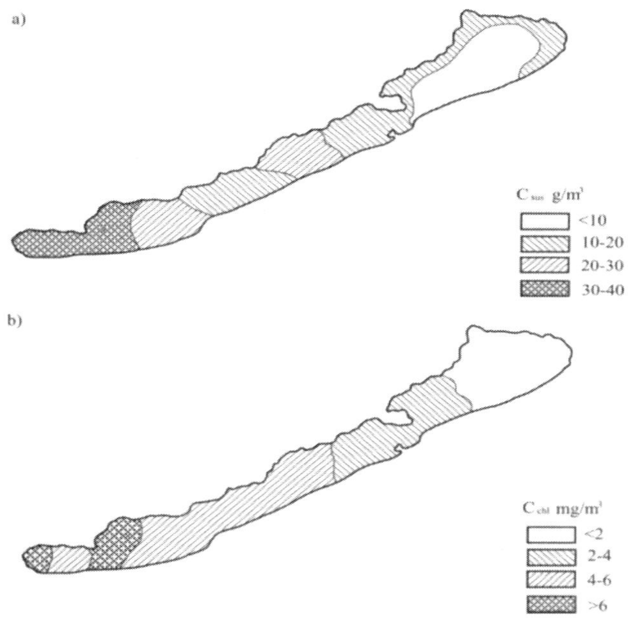

In accordance with a survey conducted by MSS Landsat on August 20, 1983.

Figure 6.16. Distribution map of suspended mineral matters (a) and chlorophyll "a" in phytoplankton (b) for the western part of Balaton Lake.

Figure 6.19 contains a turbidity distribution map for the Azov Sea prepared by AES NOAA (equipment AVHRR) after receiving a signal of environmental emergency in the Temryuk Gulf area. In the center, with a gradation of 20-30 g/m^3, the equipment detected a spot of clean water with turbidity of 5-10 g/m^3. On Figure 6.20, this area is zoomed in. The foregoing images clearly display the contamination plumes (but not the expected areas of pollution prevalence). After obtaining this information, an OSS equipped with spectrometric equipment and a sampler was sent to this area.

6.4.2. Information Obtained by an OSS

An observation space station may be installed onboard any aircraft—whether airplane or helicopter. It is most reasonable, however, to install it onboard a carrier that allows it not only to perform remote sensing of radiation parameters of water quality (which is indispensable for mapping polluted water areas), but also to make samples of water for detailed chemical and biological analysis. It is therefore preferable to install an OSS onboard a helicopter or amphibian aircraft.

Figure 6.21*a* shows the flight course of a helicopter equipped with an OSS through an emergency situation area detected by satellite information. Figure 6.21*b* provides a graph of a two-channel photometer continuously operated along the flight route of an OSS. If the photometer's readings suddenly changed, spectrums of upwelling radiation were recorded and samples were taken for chemical and biological analysis. As a result of the causes of pollution, its composition and sources were established.

a—Tsymlyansk Reservoir, April 28, 1986; b —in accordance with data obtained by the Mattra scanner from an altitude of 10 km, May 12, 1986

Figure 6.17. Distribution map of chlorophyll "a" in phytoplankton.

a—Tsymlyansk Reservoir, April 28, 1986; b —in accordance with data obtained by a Mattra scanner from an altitude of 10 km, May 12, 1986.

Figure 6.18. Distribution map of suspended mineral matters.

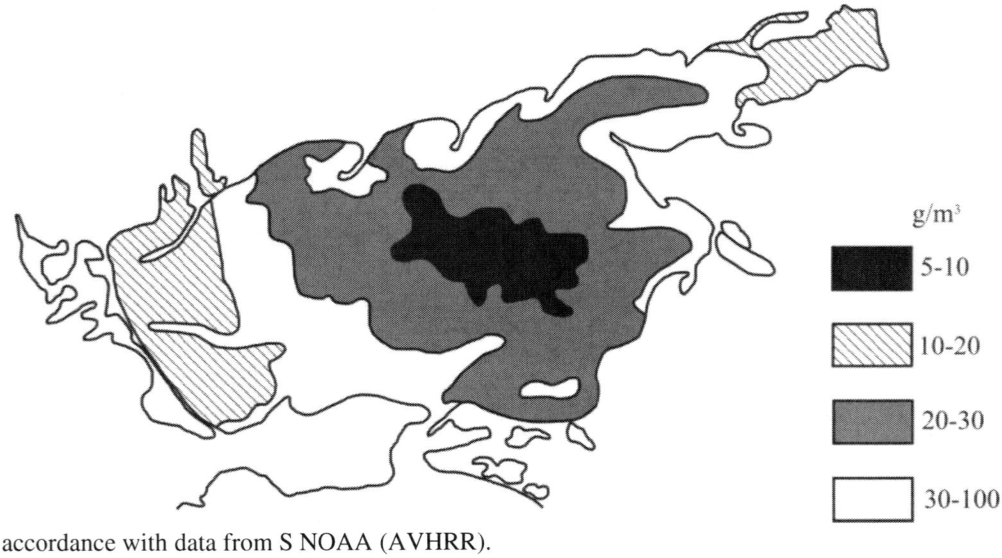

In accordance with data from S NOAA (AVHRR).

Figure 6.19. Turbidity distribution map of the Azov Sea. Overall radiogram.

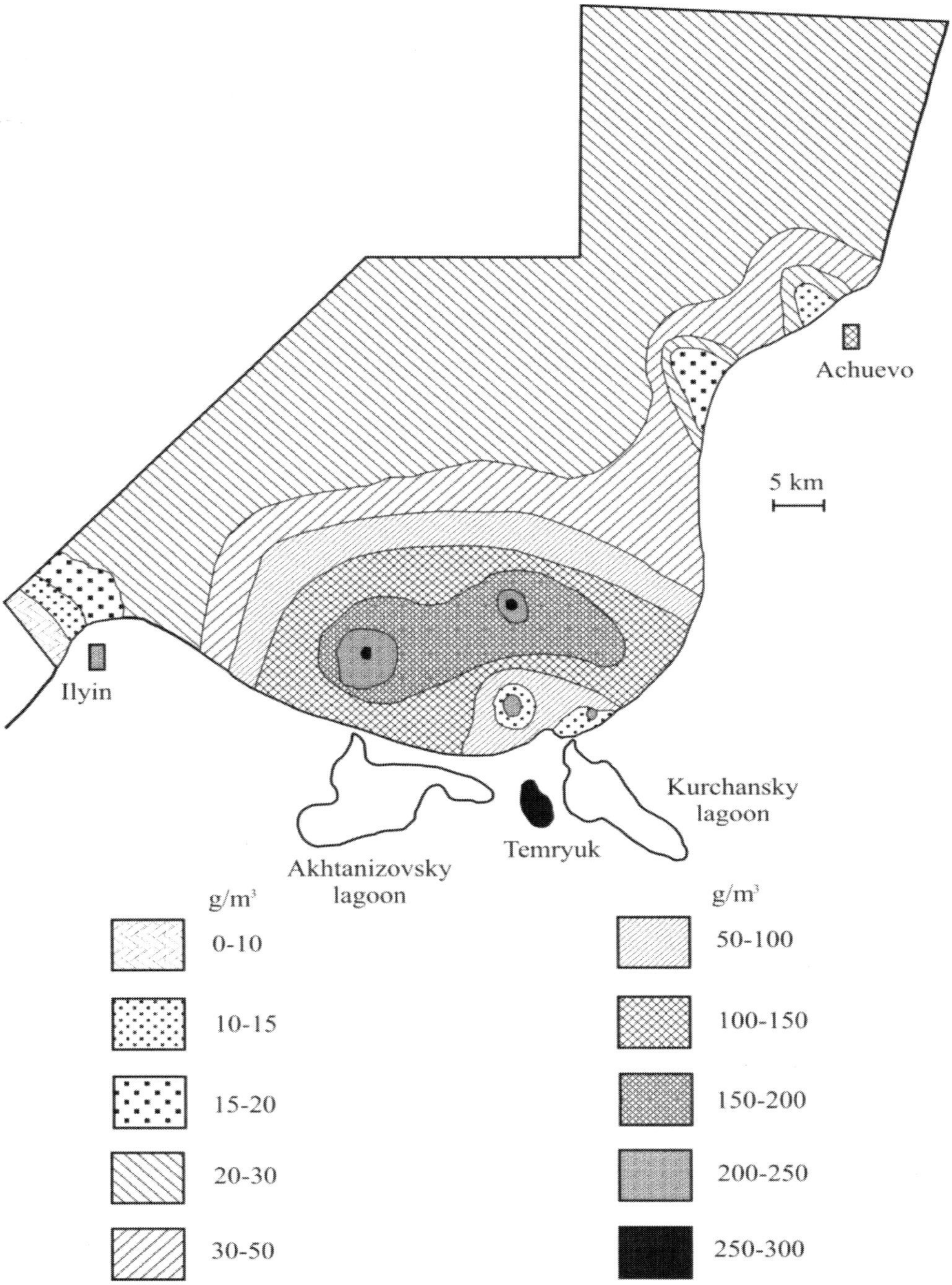

Fragment with increased numbers of gradations organized by optically active component in accordance with data from S NOAA (AVHRR).

Figure 6.20. Turbidity distribution map of the Temryuk Gulf in the Azov Sea.

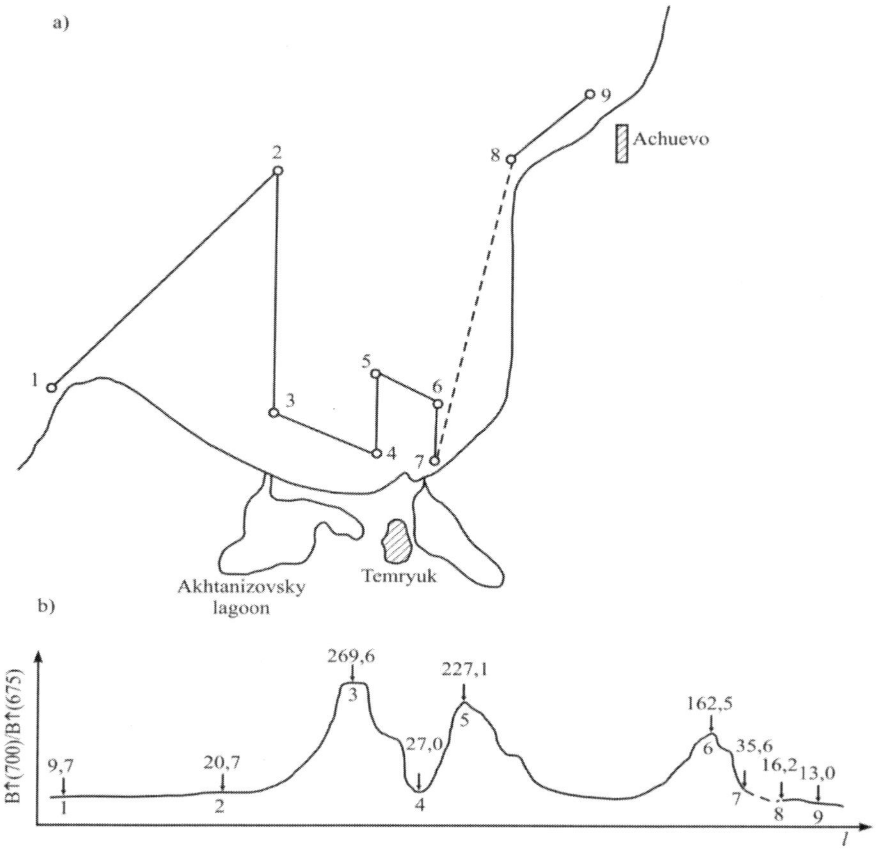

Figures above the curve represent concentrations of chlorophyll "a" in **phytoplankton** (mg/m^3) measured in water samples.

Figure 6.21. The flight course of an OSS and scheme of the emergency situation area (*a*); the graph of the two-channel photometer along the flight route of the OSS (*b*).

Figure 6.22 displays the distribution of chlorophyll "a" in phytoplankton at Lake Ladoga in accordance with spectrometric and fluoremetric measurements taken during emergency situations in the area of St. Petersburg water intake. Consequently, certain measures were recommended for improvement of the environmental situation.

Figure 6.23 presents data on the distribution of suspended mineral matters, chlorophyll "a" in phytoplankton and fulvic/humic acid in the Rybinsk Reservoir. This information was obtained in May 1987 in the course of a complex expedition aimed at studying environmental conditions of this unique water body.

A curious comparison of control and remote sensing measurements are mention in monograph [37] (Figure 6.24). On Figure 6.24, it is clear that the structure of the distribution fields of chlorophyll "a" in phytoplankton of the Mingechaur Reservoir differs considerably between measurements taken from an aircraft and a vessel. The authors [37] rightly note the substantial mismatch of absolute values of concentration, which is not a surprise due to the fact that remote sensing data was recovered by algorithms intended for sea waters. The viewpoint of monitoring a detailed structure of phytoplankton (Figure 6.24*b, c*), however, is the most important, especially in the head and tail waters of the reservoir, though such detailed structure may be obtained only with the use of remote sensing measurements.

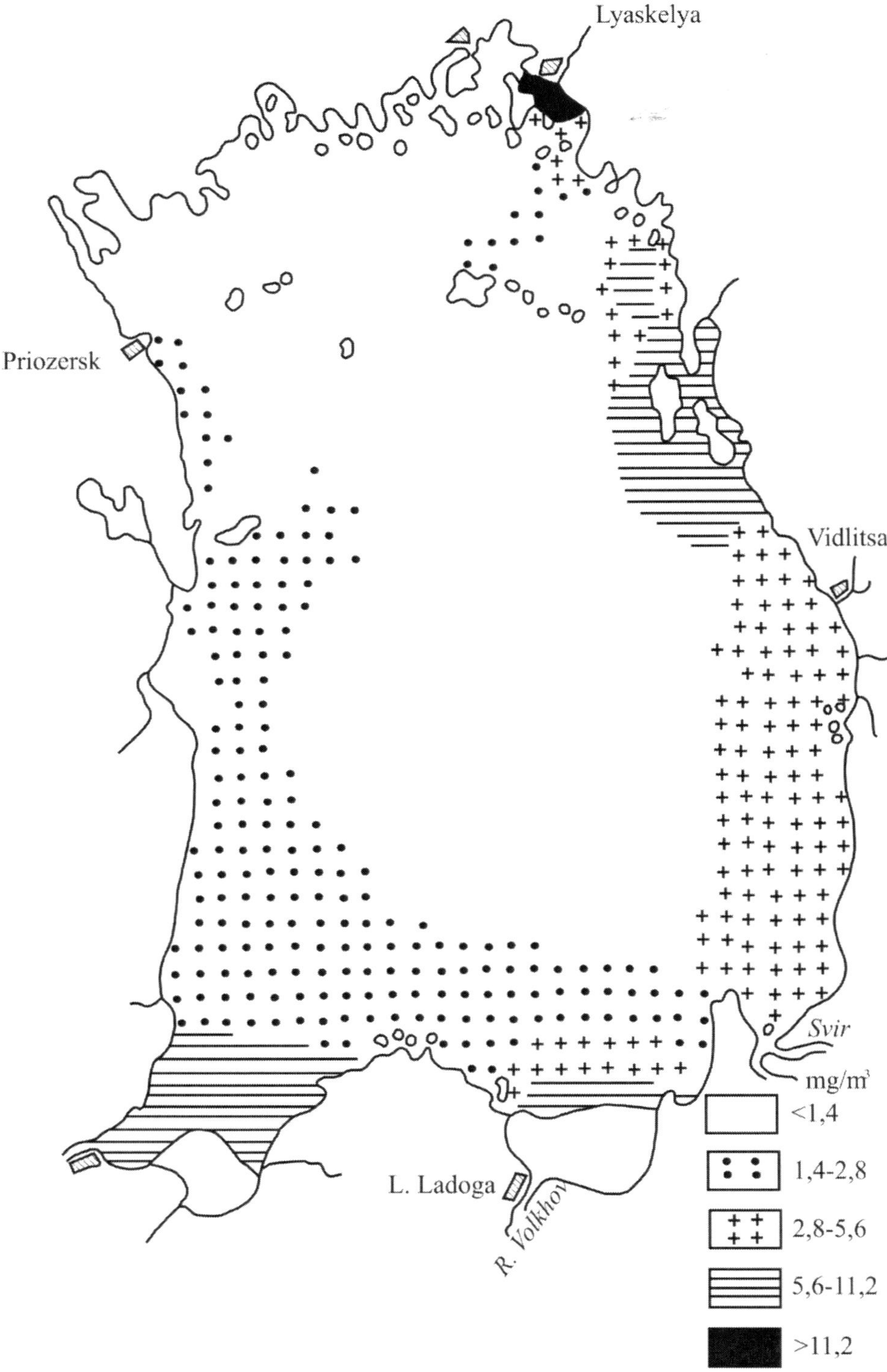

Figure 6.22. Distribution map of chlorophyll "a" in phytoplankton (mg/m^3) in Lake Ladoga on November 8, 1986.

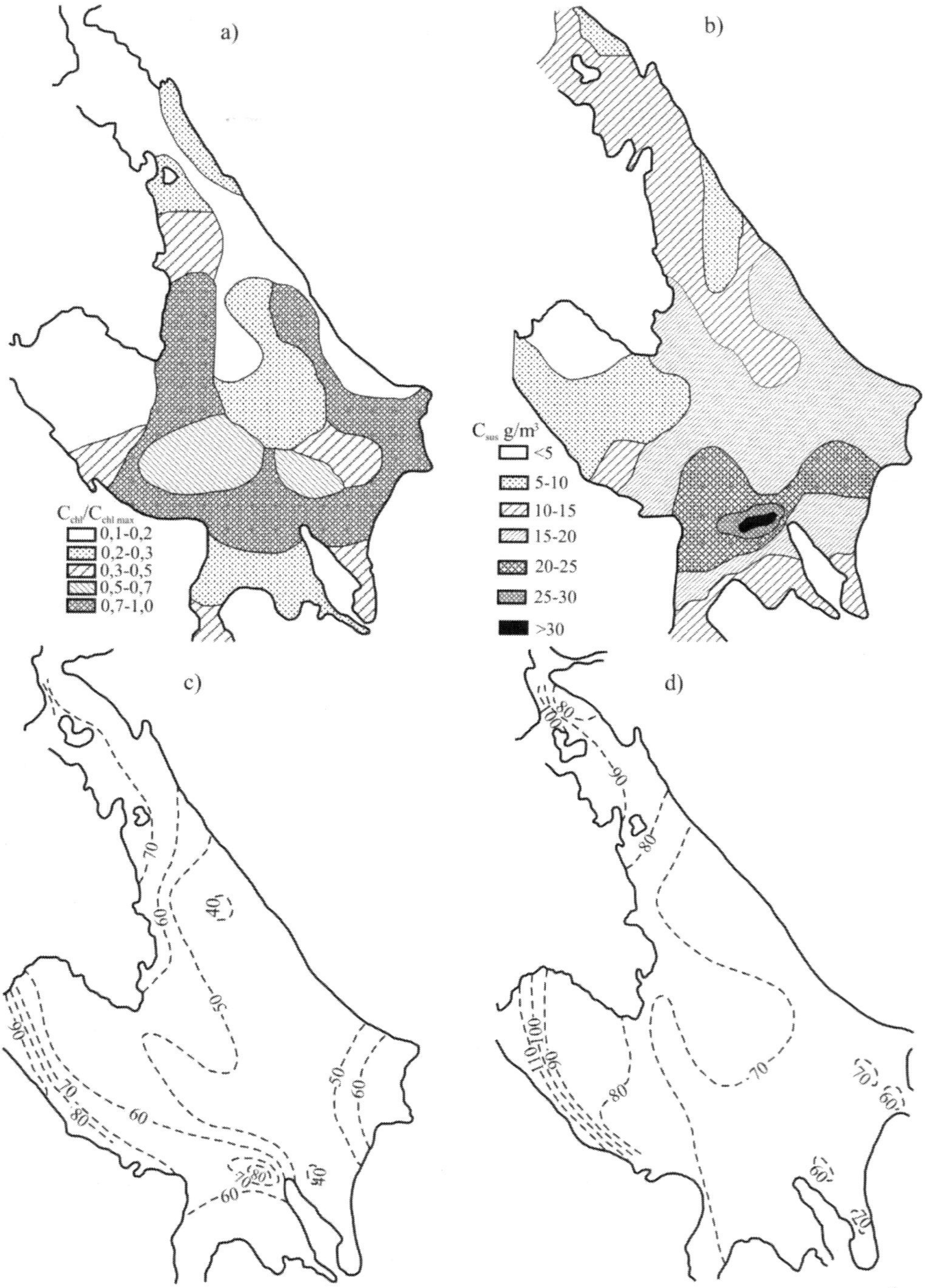

a—chlorophyll "a" in **phytoplankton** (in relative units), b—suspended mineral matters (g/m^3), c—humic acids ((g/m^3 of carbon) · 10^{-2}), d—fulvic acids (g/m^3 of carbon)

Figure 6.23. Distribution of optically active components in the Rybinsk Reservoir, May 28, 1987.

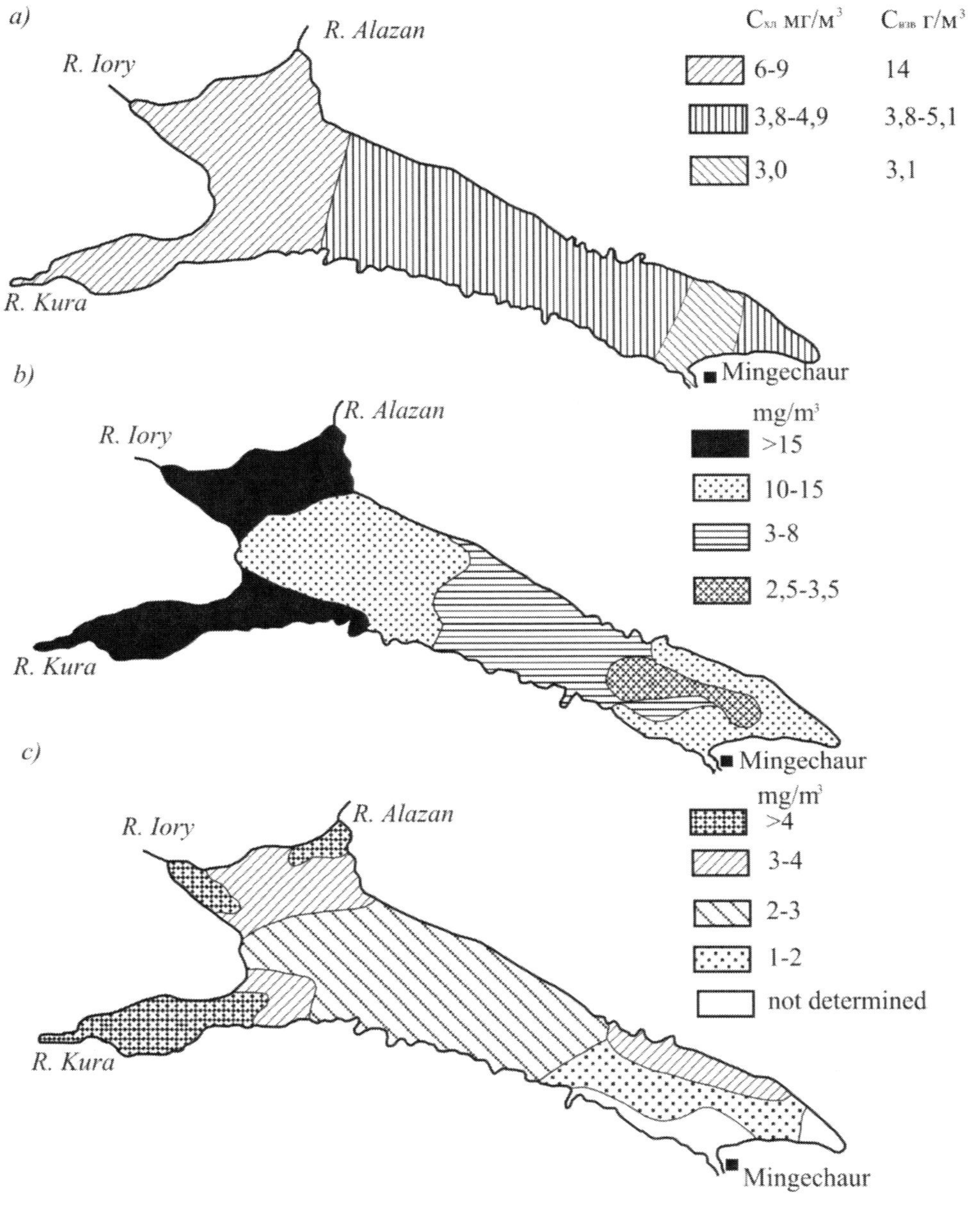

Figure 6.24. Distribution of chlorophyll "a" concentrations in phytoplankton in the Mingechaur Reservoir, dated August 28, 1984 in accordance with contact measurements (*a*) and remote sensing with application of Stepin (*b*) and Morel (*c*) algorithms.

Figure 6.25 describes the distribution of phytoplankton in Lake Sevan obtained by a lidar [37]. It is probably the first map with such spatial resolution. The data provided corresponds to long-term observations of phytoplankton distribution in the surface layer of the lake. Strange as it may seem, Big Sevan Lake is more exposed to eutrophication despite its deep waters. Areas exposed to higher anthropogenic stress are clearly apparent.

We should mention that the OSS survey is carried out in real time and, thus, it may be used for making managerial decisions within 10-40 hours from the moment of survey commencement.

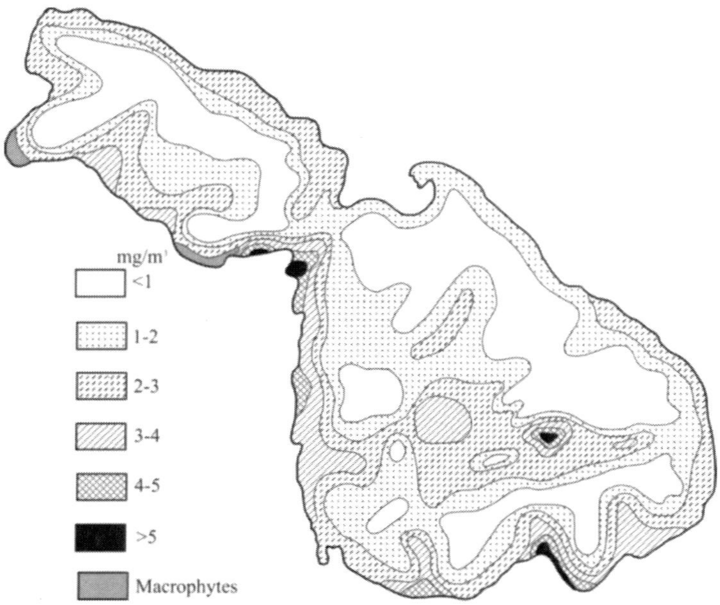

Figure 6.25. The distribution of chlorophyll "a" in phytoplankton of Lake Sevan in accordance with data obtained by lidar probing on September 1–4, 1986.

6.5. Prospects for Further Development

Prospects for the development of operational monitoring of surface waters in relation to the use of remote sensing data consist of improvement of both the equipment (with special attention to active remote sensing) and methods for processing remote sensing data.

The most methodologically productive developments include the elaboration of a general concept for the assessment of environmental conditions of water bodies by remote sensing parameters [38—44]. In order to illustrate the potential capabilities of such an approach, we should examine modern methods for interpreting optical spectrometric information.

Visible range spectrometric data obtained by passive remote sensing of water bodies is traditionally used for determining concentrations of optically active components (OAC). Statistical modeling is usually employed for this purpose through the use of the following regressive dependence:

$$C_{rem} = f(B(\lambda_1)/B(\lambda_2)), \quad (18)$$

where C_{rem} is the assessment of OAC concentrations and $B(\lambda_1)$ and $B(\lambda_2)$ are values of brightness and brightness ratios of upwelling radiation at maximally changeable and stable waves every time the OAC changes. Section 6.3 [6, 18, 20, 32] provides a great number of such dependencies for measuring the concentration of chlorophyll "a" in phytoplankton of power type (8), where $RP = B(\lambda_1)/B(\lambda_2)$. Analogous relations (models) are also used for the measurement of suspended mineral matters [6, 32, 45].

Despite the known advantages of regressive models of type (18) (simplicity, clarity, high efficiency of OAC concentrations assessment achieved in several cases) attempts to use these models in practice have led to certain difficulties connected with the instability of model parameters from one experimental sample to another and from one water body to another—regional dependencies.

In order to illustrate this point, Figure 6.26 contains plane [RP, C_{chl}] (where C_{chl} is an analytically determined concentration of chlorophyll "a" in phytoplankton and RP = B(700 nm)/B(675 nm)) with experimental points corresponding to two surveys of the Taganrog Gulf (Azov Sea) in substantially different phytophases—in April and August of the year 1990. Each remote sensing survey was carried out from the board of a helicopter within several hours [25]. Afterwards, the survey regressions of type (8) were completed with a correlation coefficient of 0.86 (April) and 0.6 (August). Parameters of the corresponding models A and B differ considerably: A = 3.16 and B = 11.16 in April, A = 13.07 and B = 5.17 in August.

A power regression built in accordance with combined samples from Figure 6.26 has the following parameters, A = 5.21 and B = 10.85 at a correlation coefficient of 0.75. It is equally inaccurate for both the April and August (blooming) phases of phytoplankton development. Most experimental points assume the character of hardly interpreted discharges.

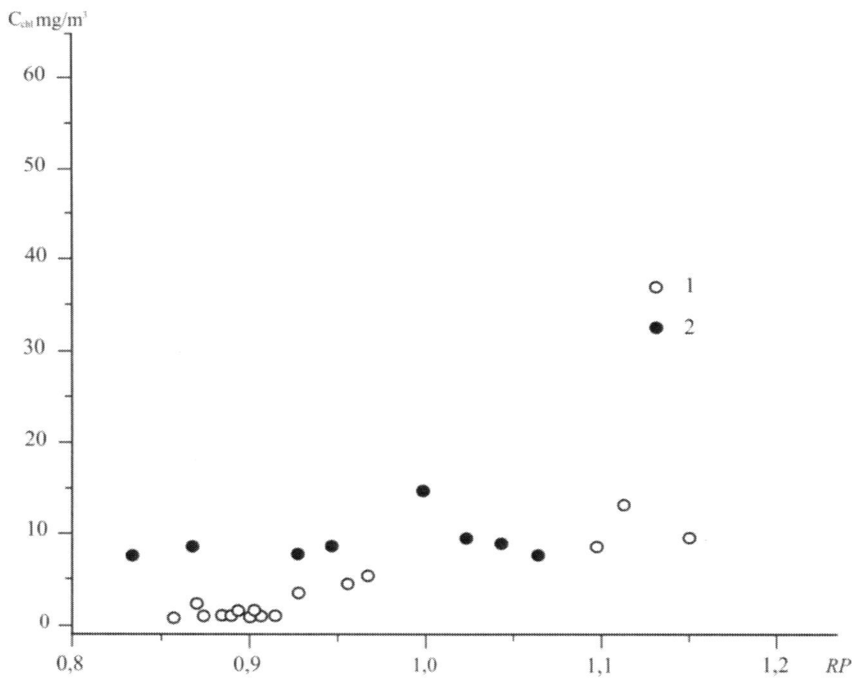

1 — April 1990, 2 — August 1990

Figure 6.26. Relation of C_{chl} to radiation parameter, where RP = B(700 nm)/B(675 nm) for the Taganrog Gulf in the Azov Sea.

The instability of the parameters of the regression models of type (18) fuel the search for other means of practical use of remote sensing data along with plotting relation of radiation to concentration. The above example proves not only the instability of the parameters of the regression model, but also an opportunity to differentiate between seasonal conditions of an aquatic ecosystem on the basis of remote sensing measurements. In this sense, we can pose a more general question than the type associated with (18), a question about the correlation of optical state and environmental conditions within an aquatic ecosystem.

An alternative method of analysis for remote sensing data consists in plotting optical state in relation to environmental conditions. An elemental scale of environmental conditions is found in the following pair: standard-pathology*.

The metric of space within the optical state is determined through the representation method of spectrometric information. The space within an optical state may be built as one with a coordinate axis in which the coordinates are generalized and determined by spectrums of brightness ratios of upwelling from water radiation in the optical range of wavelengths.

The notion of spectrum form is used as a basis, which requires the following preliminary processing of reference spectrometric information.

1) The form of each spectrum is approximated in optical ranges of wavelengths by a number of discrete indications. A sampling increment is determined by a minimal width of spectral peculiarities encountered in experimental samples. The evidence suggests that a sampling increment of 10 nm is enough.
2) For proper comparison of different spectral forms, they are normalized in accordance with the following rule:

$$B^*_{i,j} = \frac{B(i,j)}{B(1,j)+\ldots+B(n,j)}, \tag{19}$$

where i is the number of indication (i =1,...,n, n is the number of spectral indications) and j represents the number of spectrums.

Consecutive transformation (19) designates distribution of points in an n-dimensional attribute space corresponding to the considered experimental samples. Each point of this space corresponds to an optical state of a water body under the field of vision of a remote sensing device. At the same time, it is known that a complete set of descriptive signs may often be substituted by a more concise range of type-generating signs for the purpose of classification [46]. Mathematical procedures for factor analysis are used in order to determine the informativeness of each n sign of the spectrum form and to reduce spatial dimension [47].

Such an approach to the interpretation of remote sensing data requires a special training stage, i.e. simulation experiments with preset and adjustable physicochemical and biological indices. On this stage, experimental works should be carried out on a water body's environmental conditions *a priori*. The methodology of this experiment were implemented by us and our employees on artificial aquatic ecosystems—mesocosms. Below we provide an example of such an experiment and its results, which clearly indicate capabilities of remote sensing that have not been used to their full extent until now.

6.5.1. The Relationship between the Optical State and Environmental Conditions of Water Bodies

In July and August of 1991, the Mertvy Donets River was used for ecological modeling within the course of which ecosystems of mesocosms were distracted by salts of heavy metals (copper and mercury) [39, 40]. The following four mesocosms were examined during the

* "Standard" and "pathology" here refer to aquatic ecosystems unimpaired and impaired by pollutants, respectively.

experiment: 1st and 2nd by adding $CuSO_4$ (through a variety of means), 3rd by adding $Hg(NO_3)_2$ and the 4th one was for reference. In all mesocosms and nearby river sections (hereinafter referred to as, "the background"), the following parameters were measured at approximately equal intervals (2-3 days): contents of Cu and Hg, species composition, number and biomass of phytoplankton, concentration of chlorophyll "a" in phytoplankton, etc. The spectra of brightness ratios of upwelling optical radiation in the range of wavelengths from 490 to 720 nm were registered simultaneously. The equipment for and methods of spectrometric measurements are described in document [24]. The experiment continued for 26 days; the complete set of parameters was measured within ten series of measurements and one series of spectrometric measurements.

First of all, the experimental data was used to check the type (18) regression model. The effectiveness of this model is shown on Figure 6.27 and 6.28. The coordinate plane [RP, C_{chl}], where $RP = B(700\ \text{nm})/B(675\ \text{nm})$, includes points corresponding to each mesocosm and background for every day of the experiment (see Figure 6.27) and particularly for the 1st and 3rd mesocosms (see Figure 6.28). It is evident (Figure 6.27) that building a regression of any type through a number of experimental points is technically possible, even though it is rather unreasonable, because the correlation coefficient between RP and C_{chl} equals less than 0.3. Figure 6.28 clearly demonstrates the imbalance of the regression model; the poorly correlated set of points in Figure 6.27 is formed by a range of subsets, each of which is related to a definite mesocosm and is described by a regressive dependence with unique parameters. The inappropriateness of the regression dependence (18) in the interpretation of remote sensing data obtained from the experiment with mesocosms highlights the limitations of spectrometric information.

A factor analysis of the correlation matrix comprised of experimental correlation coefficients between 24 type (19) signs (with a sampling increment of 10 nm) distinguished three general factors that were subsequently used as generalized coordinates, which altogether explain 93% of the observed total dispersion of these signs; the first factor was 46%, the second was 33% and the third was 14%. The calculations are based on a varimax factor rotation algorithm.

This means that variability in the form of brightness ratio spectra in the experiment with mesocosms may be described with 93% accuracy through only three generalized coordinates. The spectral distribution of the generalized coordinates' factor loadings is described in Figure 6.29.

Computational procedures for factor analysis allow the calculation of values for generalized coordinates of each spectrum from the original samples. The processing of the spectrum array leads to the distribution of points in a three-dimensional space of generalized coordinates, which reproduces a distribution of a set of experimental points in a 24-dimensional space of type (19) signs with an accuracy of 93%. Analysis of the projection of this distribution on planes 1 and 2, 1 and 3, and 2 and 3 factors or, put another way, of the generalized coordinates (GC), shows that the most isolated groups of points are generated by a projection of the 1st and 3rd generalized coordinates on the plane. Figure 6.30 displays the distribution of points in the plane of the 1st and 3rd generalized coordinates with points indexed in accordance with the experimental day number and mesocosm (data on several mesocosms is not available for certain days due to technical problems).

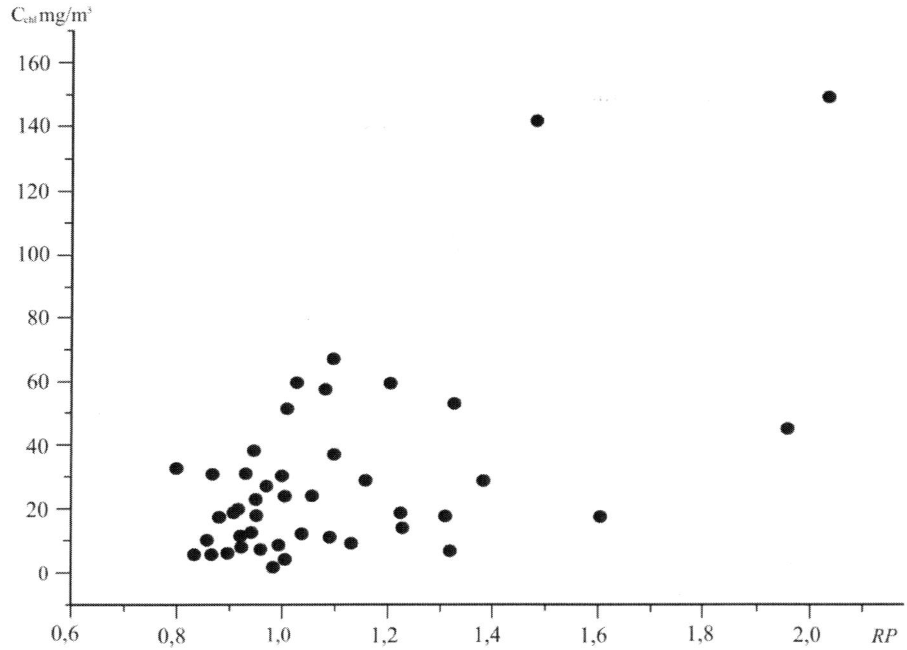

Figure 6.27. The dependence on radiation parameter, *RP* = B(700 nm)/B(675 nm), for the entire set of measurements in the experiment with model ecosystems.

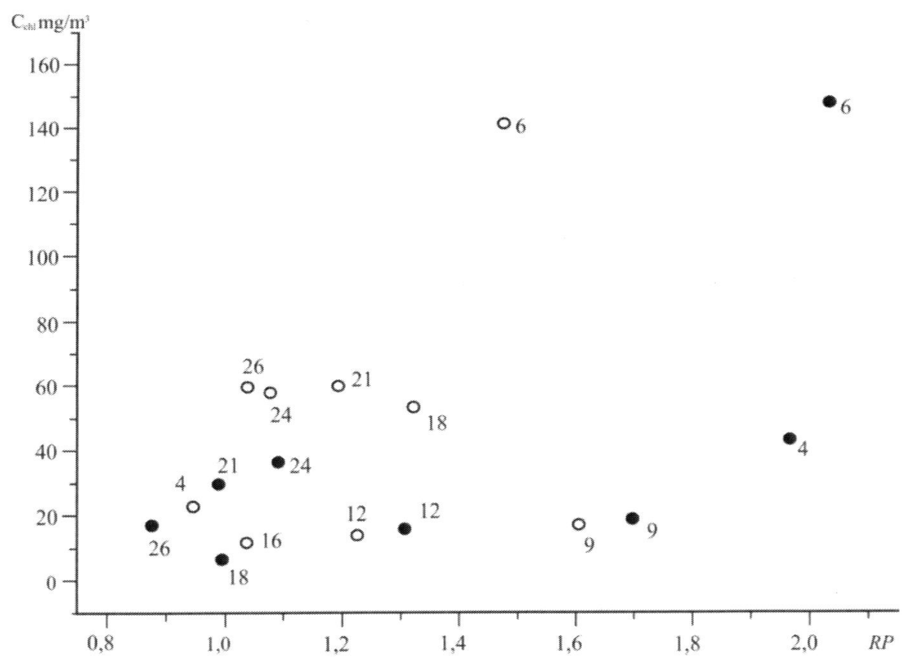

Numbers denote the corresponding day of the experiment.

Figure 6.28. The dependence of C_{chl} on radiation parameter, RP = B(700 nm)/B(675 nm), for the 1st (filled circles) and 3rd (unshaded circles) mesocosms.

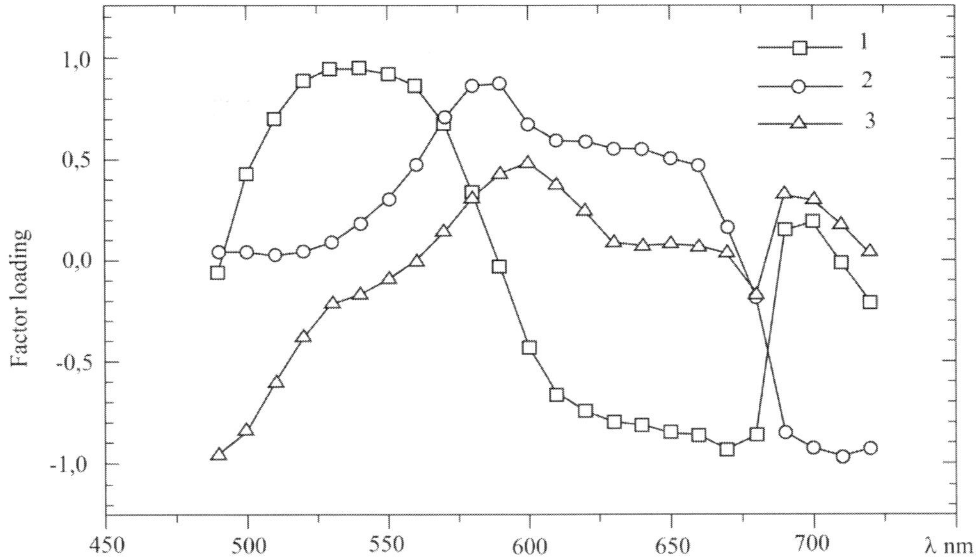

1, 2 and 3 are 1, 2 and 3 factors, respectively.

Figure 6.29. Spectral distribution of factor loadings *R*.

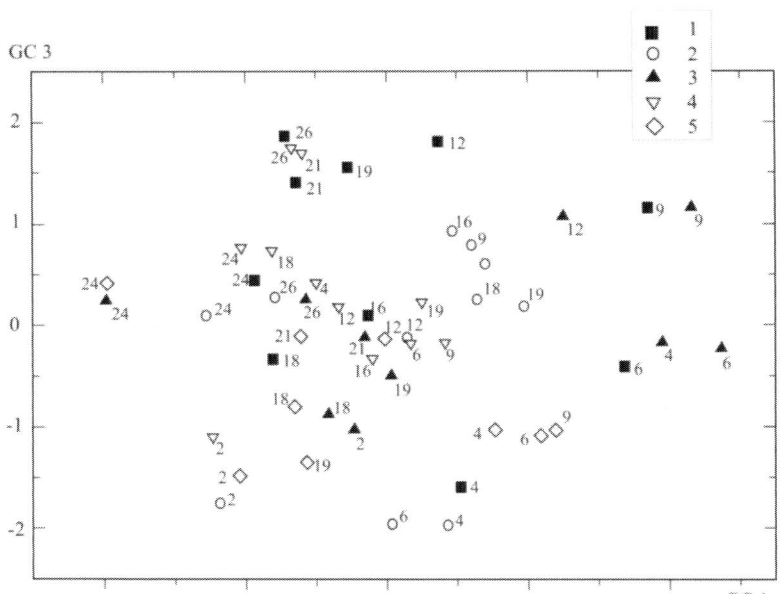

1—1st mesocosm, 2—2nd mesocosm, 3—3rd mesocosm, 4—river (background), 5—reference mesocosm. The numbers denote the corresponding day of the experiment.

Figure 6.30. Projection of a set of experimental points on planes 1 and 3.

Figure 6.30 shows the motion paths of the optical state of each mesocosm in the course of the experiment and areas of factor space associated with pollution. For the sake of simplicity, we shall unite the groups of isolated points on Figure 6.30 into conventional zones *A, B, C, D, E, F, G* and *H* and examine the transition scheme of the mesocosm optical state between the zones on Figure 6.31.

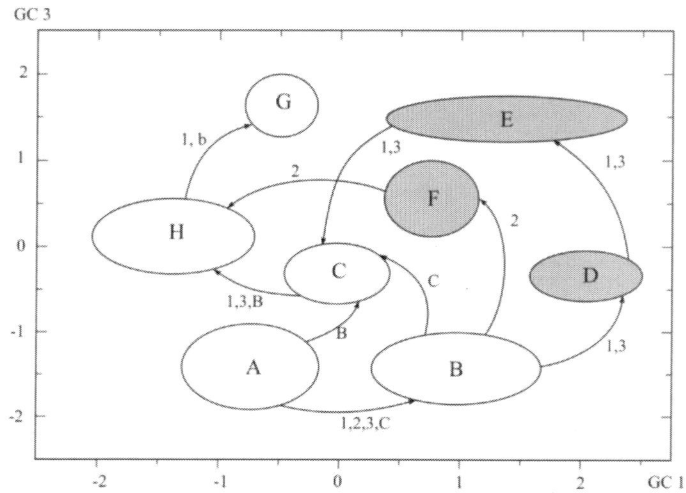

Figure 6.31. Transition scheme of the mesocosms' optical state.

The reference optical state of all mesocosms and the background corresponds to zone *A* in the beginning of the experiment. Organization (partition) of mesocosms leads to a transition of optical state into zone *B*. This transition is accompanied by the increase of GC 1 values, while the GC 3 values remain almost unchanged and correspond to minimal values in the sample. In accordance with the reference data, the transition from zone *A* to zone *B* is characterized by the purification of water; in both zones, the concentration of chlorophyll "a" in phytoplankton is very low.

At the very beginning of the experiment, the river's background optical state moves from zone *A* to zone *C* (with average values of GC 1 and GC 3), where it stays until there is a sudden change of weather conditions on days 16-18 of the experiment (temperature jump, precipitation, downward winds). From zone *B*, the optical states of different mesocosms move in considerably different directions. Optical states of the 1st and 3rd mesocosms (contaminated with the single addition of metals) do not stay for long in zone *B*, but move first to zone *D* and then into zone *E*. In the examined space of GC, these zones are characterized by maximum values of GC 1 and average (zone *D*) and large (zone *E*) values of GC 3.

In accordance with this, reference data zones *D* and *E* correspond to a combination of high concentrations of chlorophyll "a" in phytoplankton (over 100 μg/l) and a rather high transparency of aquatic environment. Such peculiar blooms in the 1st and 3rd mesocosms continued for 12-16 days, whereupon the optical state of these mesocosms was located in the background zone *C*; accordingly, the concentration of chlorophyll "a" in phytoplankton of these mesocosms shrunk to average values (10-20 μg/l). For the same period of time, the concentration of metals in the water of the 1st and 3rd mesocosms reduced from reference values of 145 MPC for copper and 83 MPC for mercury to a few MPC. The period of time within which the optical state of the 1st and 3rd mesocosms moves from zone *A* to zone *C* corresponds to the adaptation period of these ecosystems to a disturbing influence. The specific motion paths of the optical states of the 1st and 3rd mesocosms moving from zone *A* to zone *C*, as well as the velocity of this movement, serve as an intrinsic characteristic of the adaptation processes. In this respect, zones *D* and *E* may be interpreted as specific optical states inherent in ecosystems severely contaminated with heavy metals.

Zone *F* describes the optical state of the 2nd mesocosm being contaminated during the experiment by relatively small (20 MPC), though constant (5 MPC), additions of copper. Having left zone *B* too late as compared with the optical states of the 1st and 3rd mesocosms, the optical state of the 2nd mesocosm stayed in zone *F* almost continually for more than ⅔ of the experiment time and then moved to zone *H* after new weather conditions considerably different from initial ones had become stable.

Relatively, the compact zone *F* is not far from the background zone *C* and is characterized by higher values of GC 1 and 3, which are not as extreme as those in zones *D* and *E*. In accordance with the reference information, this zone does not have any evident peculiarities.

The optical state of the reference mesocosm from zone *B*, in which it stayed for quite a long period, made a direct transition to zone *C*. The reference mesocosm's optical state movement proves the eccentricity of optical states in zones *D*, *E* and *F* and their connection with contamination.

Due to sudden changes of weather conditions in the second part of the experiment, the further movement patterns of optical states was somewhat "noisy," which hampered the reliable interpretation of the mesocosms' optical state transitions in this period (transition from background zone *C* to zones *A*, *G* and *H*). We can only note the different reactions of optical states of mesocosms to intensive weather influence.

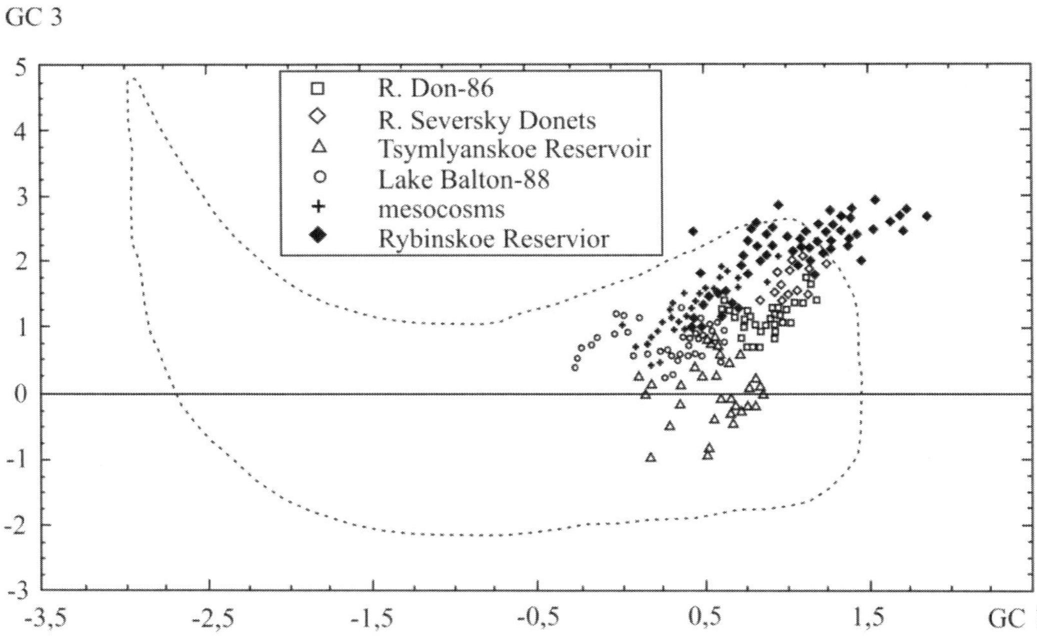

The marks show optical states of real aquatic ecosystems. The dotted line represents a region of space filled with model spectrums.

Figure 6.32. Projection of optical states space on planes GC 1-GC 3.

The experiment with artificial ecosystems showed that simulated influences disturbing the ecosystem are definitely reflected in spectrometric information, which in its own turn distinguishes and interprets spaces of optical states connected with standard and pathology conditions. The limited volume of experimental spectrometric information enabled us to build a limited space of optical states.

A complete space of optical states may be built through the use of spectrums obtained from all water bodies of the Earth during different hydrological seasons. This problem may only be solved theoretically but not in practice. Model spectra of brightness ratios with maximum variation of model parameters are used for building a complete space of optical states in practice. These parameters include, first of all, concentrations of optically active components—mineral suspended matters and dissolved organic substances along with the species composition of phytoplankton with due account of changes in their pigment composition. Such an approach was used in works [46, 49], where the space of the optical states was based on over 15 thousand models and about 500 experimental spectra obtained from different water bodies. The most informative projection of this three-dimensional space (GC 1-GC 3) is provided on Figure 6.32 with a dotted line representing the region of space with generalized coordinates implemented only by model spectrums.

It is obvious that almost all experimental spectrums are found in the region of space of optical states established by theoretical model spectra. Some spectra (optical states) beyond the foregoing region, however, were also observed. This fact proves that the employed model either does not consider all physical processes taking place in an aquatic ecosystem in the scattering of sunlight or doesn't cover all of the parameters (initial hydrooptical characteristics) that are encountered in different aquatic ecosystems. Spectra from the Kuybyshev Reservoir and the Seversky Donets River may serve as examples. These water bodies differ from other surface waters through their increased concentration of dissolved organic substances and high total water contamination.

Normalized spectra of BR in the space of generalized coordinates are optical states of aquatic ecosystems. In this connection, it would be interesting to compare spectrums of different water bodies. Mid-spectra of several water bodies are presented on Figure 6.33. Vertical segments designate ranges of spectrum intensity in the experiments.

Figure 6.34 is similar to Figure 6.32 but apart from mid spectra of different water bodies, it includes the optical states registered on Lake Baikal (marked with numbers). In work [42], it is noted that this array of spectra may be considered unique in relation to diversity of optical and environmental properties in different parts of this water body. A spectrum of very clean water, comparable by its optical properties with spectra of seas and oceans, is marked by the number *19*. Spectrum of a eutrophic section (Lake Arangatuy) is marked by the number *43* and is characteristic of water bodies of southern Russia.

Spectra *41* and *42* are unique due to the fact that turbid waters of this section almost never contain any phytoplankton.

In addition, Figure 6.34 includes areas definitely interpreted by their environmental properties as I—areas of theoretically clean water for real waters, II—areas corresponding to water contaminated with mineral suspended matters without phytoplankton or III—eutrophic water and hypereutrophic water bodies.

The obtained results with respect to spectrometric information are evidence of a single mechanism of formation of optical spectrums in different experimental samples and of a universal approach to the analysis of such information described in works [38—44].

Unfortunately, the multistage formation mechanism of these and other optical states, including those that are connected with a preset, disturbing influence remains unclear. Further development of remote sensing data requires training experiments with model ecosystems that have a more detailed program of reference measurements.

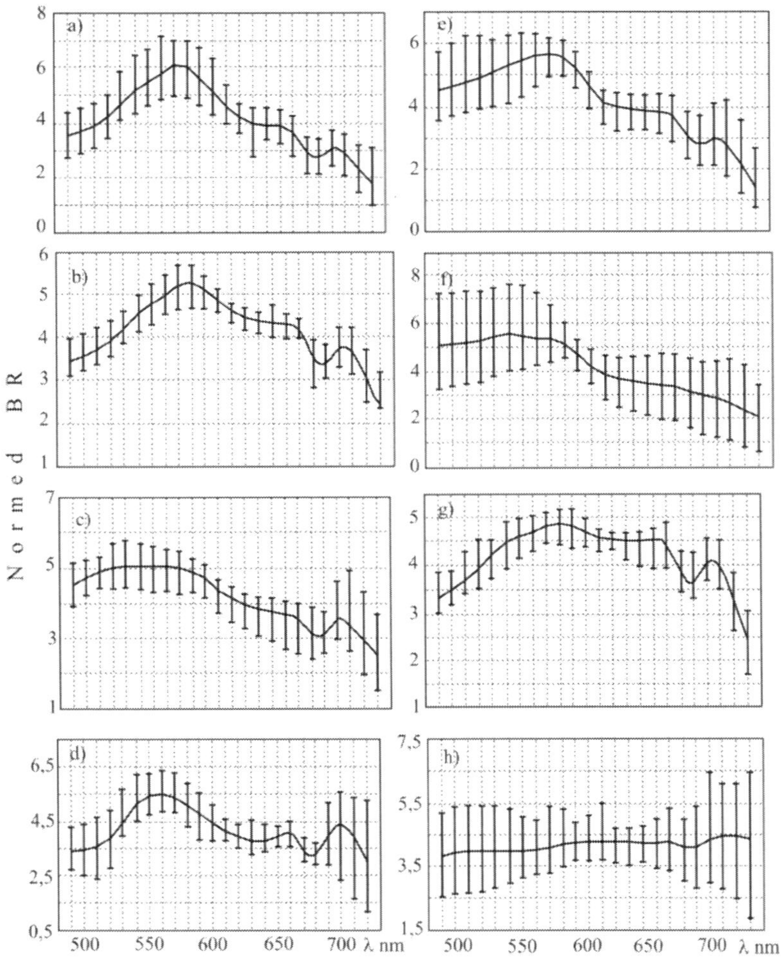

Vertical segments are ranges of spectrum intensity in the experiments.

a—Tsymlyansk Reservoir, b—Don River (1986), c—Mertvy Donets (1991, mesocosms), d—Lake Balaton (1986), e—Lake Balaton (1988), f—Lake Baikal (1989), g—Seversky Donets River (1986), h—Rybinsk Reservoir (1988)

Figure 6.33. Mid spectra of different water bodies.

6.5.2. The Use of Road Structures (Bridges) in Remote Sensing Operational Monitoring of River Ecosystems

Spectrometric information may be successfully interpreted in terms of environmental monitoring [48]. Aspects of the accurate collection of such information from a vessel or helicopter are well known. In this work, we are discussing the collection of information from bridges located mainly in large cities and along the whole territory where a city is contiguous to a river. The height of bridges over navigable rivers is sufficient for the passing of large river-sea vessels and is comparable to the flight altitude of a helicopter during spectrometric survey. Meanwhile, the cost of work is ten times less as compared with the use of a helicopter or vessel.

In this paper, the objective within the framework of the foregoing problem is formulated as a check of the limits of an *efficient tracking system for environmental conditions of a large river running within the bounds of a large (lengthy) city*.

In order to solve this issue, the author together with G.P. Garbuzov and B.L. Sukhorukov [49] have made measurements on the Don River from three car bridges, the location of which is shown on Figure 6.35.

The first bridge (marked as *A*) is located in the place where the river crosses the city line. In this section, the river is not exposed to any loads associated with waste waters discharged in the city. The second bridge (*V*) is located in the middle of the city. Anthropogenic stress before bridge *V* rises but stays minimal. The third bridge (*Z*) is located downstream from all discharge sewers of the city. Consequently, anthropogenic stress on the water body is maximal here. The average velocity of the Don River between bridges *A* and *Z* equals 0.1 m/sec. The time of water travel from the upper to the lower bridge totals 50 hours. Therefore, taking into account that one of the main consequences of water pollution is eutrophication of water bodies (watercourses) and other changes in water conditions, we may expect that the influence of anthropogenic stress at a section of the water body will be apparent both from one survey to another and from one point to another, and that such stress may be detected by remote sensing methods.

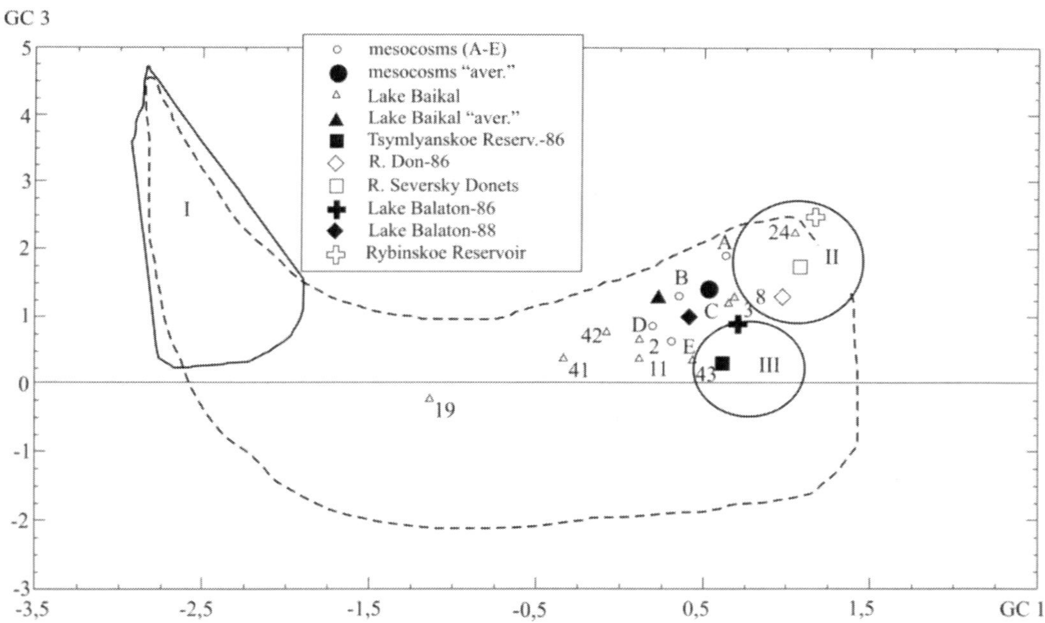

The marks show the average optical states of real aquatic ecosystems. The dotted line represents a region of space filled with model spectrums. The numbers denote the optical states of different areas of Baikal Lake, mentioned herein. Roman figures correspond to regions of space definitively interpreted with due account to the reference information.

Figure 6.34. The projection of optical states on plane GC 1-GC 3.

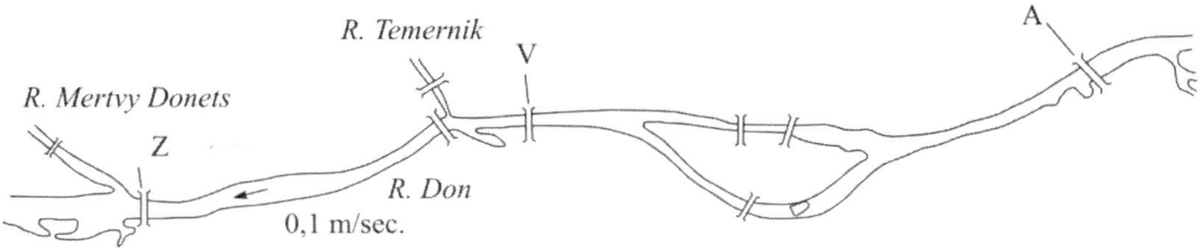

Figure 6.35. Scheme of the Don River running through Rostov-on-Don.

An upgraded spectrometric complex [40] was used in the surveys. The survey time of one instant spectrum is about 3 seconds. At a bridge height of about 30 m and a lens angle of 10°, the probing surface of the river equaled about 20 m^2 and the probing depth was about 1 m. The complex was upgraded in order to have an appropriate size and autonomy. The continuous operation period with light-weight batteries equaled 1.5 hours. The mass of the whole complex with batteries and a monitor used for controlling the registered spectra was about 14 kg. Spectra were recorded on magnetic media.

During the period from summer to fall, we have managed to take six series of measurements of brightness ratio spectra on the 1 and 22 of July*, 21 of August*, 15 and 22 of September, 24 of October*, 1997. Four series of spectrum measurements (marked with an asterisk) were accompanied by the synchronous sampling of water for analysis of species composition. Water samples were taken from the surface by an enameled container and the volume of samples totaled 2 l. The sequence of measurements within one series (one day) was as follows: *Z—A—V*. Spectrums corresponding to each bridge (point) are based on four or five consecutive instant spectrums of the same area near the river's channel.

The spectra were smoothed with a moving average filter to eliminate short-period noise primarily caused by patches on the water surface . Smoothed and normalized [40] spectra for each series of measurements were used for the classification of optical states [48]. The following indexing is used hereinafter; points (bridges) correspond to the letters *A, V, Z* and series of measurements correspond to indices from 1 to 6. Samples were analyzed by the Azov Fisheries Scientific and Research Institute. Reference data is presented in Table 6.3.

Hierarchy classification (agglomerative hierarchy) [52] was applied to both the obtained reference and spectrometric data. The distance between classes S_l and S_n was measured by the principle of average connection, i.e. the arithmetical mean of all possible pair distances between class representatives was calculated:

$$\rho_m(S_l, S_m) = \frac{1}{n_i n_m} \sum_{X_i \in S_l} \sum_{X_j \in S_m} d_E(X_i, X_j) \tag{20}$$

where n_i is the number of objects (vectors) forming a class and d_E is the Euclidean distance between vectors X_i and X_j of observation X. The maximum distance between classes corresponds to 1 or 100%:

$$\rho_{max}(S_l, S_m) = 1, \ \rho_{min}(S_l, S_m) = \rho_{min}(S_l, S_l) = 0 \tag{21}$$

Table 6.3. Number (10^5 m^{-3}) and biomass (mg/m^3) of phytoplankton

Point and series	Number					Biomass				
	Diatomaceous	Green	Blue-green	Euglenales	Pyrrhophyta	Diatomaceous	Green	Blue-green	Euglenales	Pyrrhophyta
A1	1,049	78	91	—	—	24.6	0.9	3.7	—	—
V1	298	—	18	—	—	5Д	—	0.3	—	—
Z1	301	—	—	—	—	6.42	—		—	—
A2	11,375	860	135	10	—	277.0	13.7	17.5	1.5	—
V2	11,991	420	190	5	5	292.8	11.1	9.7	1.2	2.4
Z2	7,580	225	545	—	15	228.7	6.8	27.7	—	4.5
A3	15,860	585	1,200	—	15	288.6	52.7	1,047.1	—	6.3
V3	18,950	240	755	—	10	385.4	11.8	1,052.7	—	4.2
Z3	15,200	720	245	5	60	236.1	18.9	455.6	0.9	10.3
A6	216	584	872	24	—	7.7	8.6	43.3	9.6	—
V6	527	694	1,794	6.7	—	53.9	11.3	54.2	1.5	—
Z6	128	64	512	—	—	12.4	11.5	13.6	—	—

Consequently, the distance between classes describes the proximity of biological conditions of an ecosystem either by number (Figure 6.36) or by biomass (Figure 6.37) of phytoplankton. The space of ***X*** dimension in the classification of reference data corresponds to the number of phytoplankton types determined during the initial hydrobiological analysis, i.e. it equals five.

Figure 6.38 provides the classification of normalized brightness ratio spectra. This classification is based on 115 spectral features evenly distributed between wavelengths from 492 to 705 nm. The use of normalization leads to the classification of spectrum forms, therefore, each class on the dendrogram (Figure 6.38) represents a special class of brightness ratio spectra.

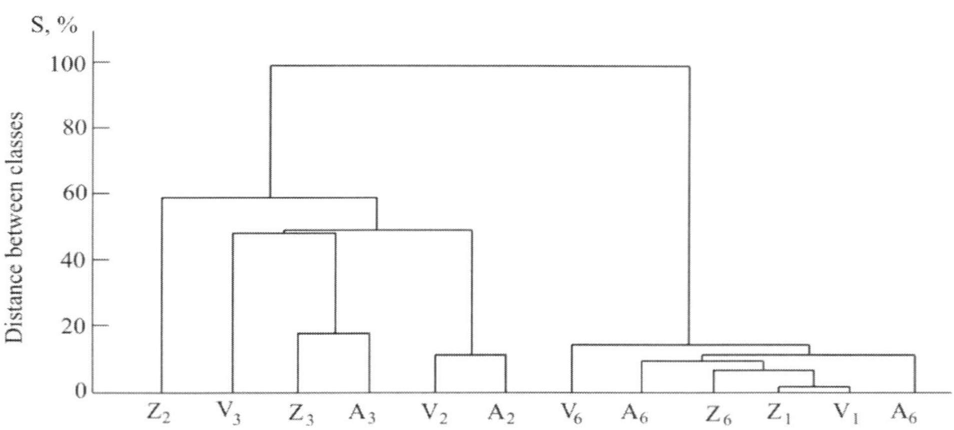

A, V and Z correspond to points shown on Figure 6.35; the digital index corresponds to the survey number.

Figure 6.36. Classification of a river ecosystem by the number of phytoplankton cells identified in water samples.

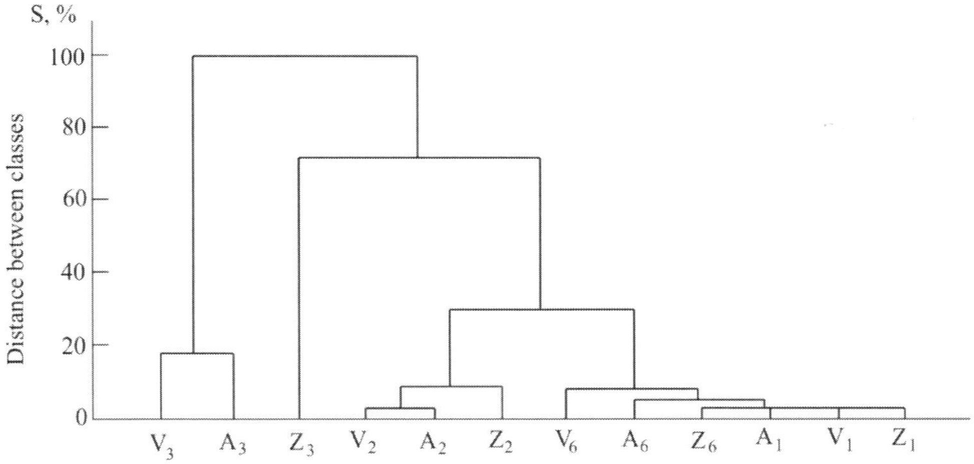

A, V and Z correspond to points shown on Figure 6.35; the digital index corresponds to the survey number.

Figure 6.37. Classification of a water ecosystem by phytoplankton biomass measured in water samples.

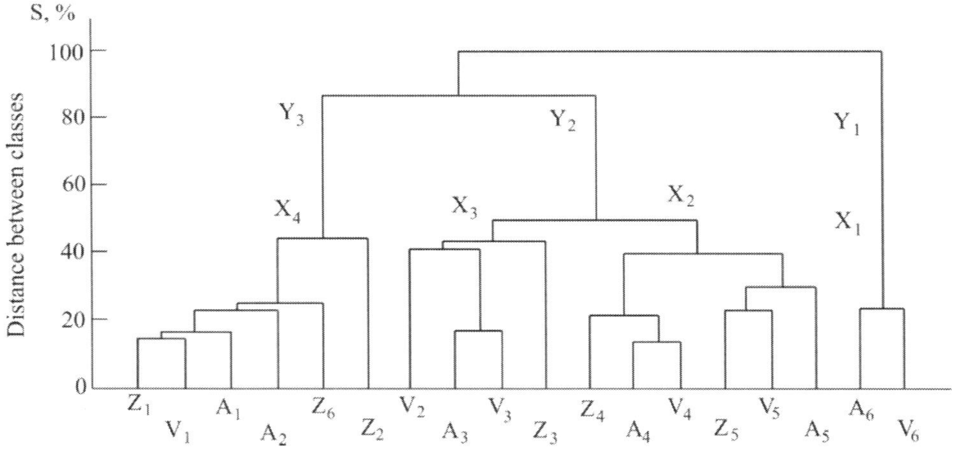

A, V and Z correspond to points shown on Figure 6.35; the digital index corresponds to the survey number.

Figure 6.38. The classification of optical states of the Don River by brightness ratio spectrums: X_1-X_4 and Y_1-Y_4—classes of optical states are formed at different stages of consolidation.

The analysis of the classification shows a distinct influence of the biosystem's seasonal changes on the river's optical state. The dendrogram (Figure 6.38) clearly distinguishes three separate classes of spectra, Y_1, Y_2 and Y_3, which correlate with three seasonal phytophases: before summer blossom of blue-green algae (class Y_1), blossoming (Y_2) and the late fall decline in phytoactivity of plankton (Y_3). Seasonal changes in the conditions of phytoplankton ecosystem are evident in the classification by number; samples from the blossom phytophase are located in one sufficiently diverse class while other samples are located in the phases of pre- and post- blossom. A "finer" structure is seen on the dendrogram along with the distinct seasonal classes. A large class of blossoming biosystem, Y_2, is very heterogeneous, and when the threshold distance between the classes drops below a certain level, the class may be divided into subclasses—X_2 and X_3, which is reflected by the different phases of biosystem blossom: initial and final, knowingly differentiated by their relative concentration of various phytopigments.

It is significant that the transition from quiet development of phytoplankton to blossom occurs gradually from the first to the third survey (interval between surveys equals 50 days). The second intermediate survey, in accordance with the classification of spectrum forms, shows preblossom conditions for bridges *Z* and *A*, and expresses blossom for bridge *V*. Such intermediate and indefinite ecosystem conditions during the second survey (between two phytophases) are confirmed by the classification of the reference information. In the classification by number, the ecosystem is located in the blossom class during the second survey, while in classification by biomass, it is located in the preblossom class with the formation of an expressed subclass.

By comparing structures of the classification (dendrogram) with forms of spectra, and on the basis of the reference data, we may draw a conclusion that remote sensing methods are significantly more informative than analytical measurements (at least those chosen by us). This is proved by a large number of expressed subclasses on the dendrogram formed at a high level of threshold distance between the classes. A very simple structure for the classification of the reference information will most likely confirm the high approximates and averages of our reference data, which allowed us to detect only the most general processes in the ecosystem.

Compared with traditional biological measurements, another resolution of remote sensing is well demonstrated by the last and sixth survey where an aquatic ecosystem in point *Z* is classified by its optical properties as being similar to summer preblossom, a rather active condition of the phytophase, while the other two points pertain to a special type of spectrum form specific to late fall. The difference between spectrum forms considerably exceeds the possible measurement errors. Through a priori information, this observation is found to be quite accurate since point *Z* is located downstream from the inflow location of the extremely contaminated River Temernik, which is rich with biogenic matters; for some time, the river ecosystem at this point could perfectly continue such an existence in different pseudosummer conditions. The results of analytical measurements do not show such noticeable anomalies in the sixth survey, which is rather homogeneous and is related to an amorphous class with low numbers and biomass.

Apart from the marked seasonal changes in the optical states of the river's ecosystem, the conducted surveys detected quite a significant variation of optical states within one survey while moving downstream from one bridge to another. Such variation is especially evident during the transition from one phytophase to another: in the beginning and at the end of blossom (second and sixth survey). Even during relatively stable periods of blossom (surveys 3-5), within one survey, we can observe variations of spectrum forms in separate subranges. The intersurvey variation of spectrum forms becomes apparent in the significant distance between the forms of certain spectrums on the dendrogram (see Figure 6.38). Half of surveys display a certain opposition of optical conditions at point *Z* to optical conditions of the river at other points, which may be caused by the influence of the Temernik River, as was mentioned before.

The conducted research demonstrated real capabilities of an efficiently organized remote sensing system based on the use of road structures. Technically, real time tracking may be carried out through automatic two-channel photometers equipped with recording blocks developed by the Hydrochemical Institute [24]. Each observation point requires two photometers adjusted so that they can measure brightness ratio coefficients in different ranges of wavelengths.

References

[1] Belousov A.P., Gittelson A.A., Kimstach V.A., Nikanorov A.M. Methods of Surface Waters Operational Monitoring // USSR Hydrochemical Council / Water Quality and Scientific Basis of Protection Section. — Leningrad: *Hydrometeoizdat,* 1986. — P. 64—65.

[2] Berestovsky I.F., Gittelson A.A., Sabo D. Experience of Subsatellite Experiments on Water Bodies of USSR and PRH // Study of the Earth from Space. — 1984. — No. 5. — P. 107—111.

[3] Gittelson A.A., Keidan G.P., Shishkin V.N. Surface Water Quality Assessment by Polyzonal Aerospace Information // *Study of the Earth from Space.* — 1965. — No. 6. — P. 26 — 36.

[4] Gittelson A.A., Shishkin V.N. Methods of Operational Aerospace Monitoring of Surface Waters // Noncontact Methods and Means for Measurement of Oceanographic Parameters. — Moscow: *Hydrometeoizdat,* 1986. — P. 23—27.

[5] Gittelson A.A., Zavolokin Y.V., Sukhorukov B.L. Methodology of Reference Information Collection on Grounds of Surface Waters for Interpretation of Remote Sensing Data // Complex Remote Sensing Monitoring of Lakes. — Leningrad.: *Science,* 1987. — P. 239—245.

[6] Gittelson A.A., Siladi F. Radiation Models of Mesotrophic and Eutrophic Water Bodies // *Study of the Earth from Space.* — 1988. — No. 6. — P. 72—82.

[7] Gittelson A.A., Nikanorov A.M. Remote Sensing Monitoring of Environmental Conditions of Aquatic Ecosystems // *Impact of Pollutants on Aquatic Organisms and Ecological Systems.* — Kharkov: Russian Water Protection Research Institute, 1988. — P. 105—122.

[8] Gittelson A.A., Dubovitsky G.A., Keidan G.P. et al. Remote Sensing of Hydrochemical and Hydrobiological Parameters of Aquatic Ecosystems by Radiation in Visible Spectrum. — Rostov-on-Don: Hydrochemical Institute, 1986. — p. 51 — USSR Institute of Scientific and Technical Information, 25.11.86, No. 8010—8016.

[9] Neuymin G.G., Solovyev M.V., Martynov O.V. Some Results of Measuring Water Colour Index of Different Areas of the World Ocean // Optical Methods for Studying Oceans and Inland Water Reservoirs. — Novosibirsk: *Science,* 1979. — P. 27—33.

[10] Mittenzwei K.-H., Gittelson A.A., Lopatchenko L.L., Sukhorukov B.L. Radiation Properties of Eutrophic and Hypertrophic Water Bodies // *Materials of German-Soviet Experiment on Water Bodies of GDR.* — Moscow, 1987. — p. 43 — USSR Institute of Scientific and Technical Information, 11.08.87, No. 58—29—B87.

[11] Kondratyev K.Y., Gittelson A.A., Dubovitsky G.A. Remote Sensing Method for Determining Concentrations of Dissolved Organic Substances in Aquatic Ecosystems // *Reports of USSR Academy of Sciences.* — 1987. — V. 295, No. 3. — P. 569—271.

[12] Gittelson A.A., Dubovitsky G.A. Peculiarities and Capabilities of Laser Spectroscopy of Surface Waters // *Environmental Regulation and Simulation of Anthropogenic Impact on Aquatic Ecosystems.* Issue. 1. — Leningrad.: *Hydrometeoizdat,* 1988. — P. 135—146.

[13] Gittelson A.A., Dubovitsky G.A., Lapin I.A. Determining Concentrations of Humic and Fulvic Acids by Laser Spectroscopy // *Environmental Regulation and Simulation of Anthropogenic Impact on Aquatic Ecosystems.* Issue. 1. — Leningrad.: *Hydrometeoizdat,* 1988. — P. 147—154.

[14] Gittelson A.A., Grankov A.G., Lieberman B.M., Lopatchenko L.L., Shutko A.M. Basic Principles and Results of Using SHF Radiometric Method for Remote Sensing of Physicochemical Parameters of Surface Waters // *Environmental Regulation and Simulation of Anthropogenic Impact on Aquatic Ecosystems*. Issue. 1. — Leningrad.: *Hydrometeoizdat*, 1988. — P. 155—172.

[15] Gittelson A.A., Lopatchenko L.L., Lieberman B.M., et al. Remote Sensing of Physicochemical Parameters by their Study in SHF Range of Wavelengths // *Complex Global Biospehere Monitoring.* — V. 1. — Leningrad: *Hydrometeoizdat,* 1986. — P. 326—535.

[16] Gittelson A.A., Grankov A.G., Lieberman B.M. et al. Metrological Characteristics and Results of Using SHF Radiometric Method for Determining Mineralization and Temperature of Water Bodies // *Meteorology and Hydrology.* — 1987. — No. 5. — P. 106—109.

[17] Kondratyev K.Y., Filatov N.N., Zaycev L.V. Assessment of Water Exchange and Pollution Zones of Water Bodies by Space Survey // *Reports of USSR Academy of Sciences*, 1989. — V. 304, No. 4. — P. 829—832.

[18] Gittelson A.A., Dubovitsky G.A., Keydan G.P. et al. Assessment of Quality of Aquatic Ecosystems by their Radiation in Visible Spectrum // *Environmental Regulation and Simulation of Anthropogenic Impact on Aquatic Ecosystems.* Issue. 1. — Leningrad.: *Hydrometeoizdat,* 1988. — P. 101—135.

[19] Gittelson A., Nikanorov A. Remote monitoring of ecological condition of aquatic ecosystem / Fate and effects of pollutant on aquatic organism and ecosystem athens // Georgia. 1988, EPA /600/9-88/001. — P. 166—182.

[20] Gittelson A.A., Nikanorov A.M., Szabo G., Sziladyi F. Etude de la qualite des eaux de surface par teledetection / Monitoring to detect changes in water quality series. Proc. of the Budapest Symp., July, 1986 // *IAHS Publ.* — 1986. — No 157. — P. 111 — 121.

[21] Mittenzwey K.-H, Gittelson A.A., Lopatchenko L.L., Sukhorukov B.L., Voigt T. Insitu monitoring of water quality on the basis of spectral reflectance // *Int. Revue des Quality Hydrobiol.* — 1988. — V. 73, N 1. — P. 61—72.

[22] Kondratyev K.Y., Garbuzov G.P., Gittelson A.A. New Approach to the Issue of Monitoring of Aquatic Ecosystems by Spectrums of Dispersed Radiation // *Reports of USSR Academy of Sciences.* — 1987. – V. 295, No. 4. — P. 825—827.

[23] Bukata R.P., Jerome J.H., Kondratyev K.Ja., Pozdnyakov D.V. Satellite monitoring of optical-active components of inland waters: an essential input to regional climate change impact studies // *J. Great Lakes Res.* —1991. — V. 17, N 4. — P. 470—478.

[24] Gittelson A.A., Garbuzov G.P., Dubovitsky G.A., Lopatchenko L.L., Makhotenko A.N., Sukhorukov B.L. Methods and Equipment for Remote Sensing Monitoring of

Surface Waters Conditions // *Hydrochemical Materials*. — 1991. — V. 100. — P. 15—46.

[25] Recommended Practice for Remote Sensing Methods of Surface Water Quality Control. Preparation and Carrying out of of Express Hydrochemical Survey. Issue 1. — Leningrad: *Hydrometeoizdat,* 1987. — p. 32.

[26] Gordon H.R. Diffusive reflectance of the ocean: The theory of its augmentation by chlorophyll «a» fluorescence on 685 nm // *Appl. Opt.* — 1979.— V. 18, №8.—P. 1161—1166.

[27] Vasilkov A.P., Kopelevich O.V. About Causes of Maximum Occurrence Near 700 nm in Spectrum of Radiation Upwelling from Sea Water // Oceanology. — 1982. — V. 22, No. 6. — P. 945—950.

[28] Sukhorukov B.L. Model Calculation of Brightness Ratios of Water Bodies. Part 1: Initial Hydrooptical Characteristics of Polydisperse Mineral and Organic Suspension; Part II: Spectrums of Suspension. Dispersion Relations / Rostov-on-Don, Hydrochemical Institute, 1991. — p. 121 — Manuscript at USSR Institute of Scientific and Technical Information 22.08.1991. No. 3537—591.

[29] Sidko F.Y., Aponasenko A.D., Sidko A.F., Lopatin V.N. Assessment of Intensity of Phytoplankton Chlorophyll Fluorescence Excited by Solar Radiation // *Study of the Earth from Space.* — 1990. — No. 2. — P. 11—17.

[30] Garbuzov G.N., Gittelson A.A., Siladi F. Determining Stable Spectral Peculiarities of Dispersed Radiation as Indicators of Conditions of Aquatic Ecosystems. — Rostov-on-Don: Hydrochemical Institute, 1986. — p. 69 — USSR Institute of Scientific and Technical Information, 1.12.86, No. 8661—B867.

[31] Burenkov V.I., Gurevich I.Y., Kopelevich O.V., Shifrin K.S. Spectra of Upwelling Radiation Brightness and their Changes with Observation Altitude // Optical Methods for Studying Oceans and Inland Water Reservoirs. — Novosibirsk: *Science,* 1979. — P. 41—58.

[32] Radiation Mode and Optical Properties of Lakes / Adamenko V.N., Kondratyev K.Y., Pozdnyakov D.V. — Leningrad: *Hydrometeoizdat,* 1991. — p. 300.

[33] Kondratyev K.Ja., Pozdnyakov D.V. Passive and active optical remote sensing of the inland water phytoplankton // *J. of Photogrammetry and Remote Sens.* — 1990. — P. 257—294.

[34] Sidko A.F., Aponasenko A.D., Sidko F.Y. et al. Assessment of Spectral Indices of Dissolved Organic Yellow Substance Absorption in Meso- and Eutrophic Water Bodies by Optical Remote Sensing Methods // *Reports of Academy of Sciences.* — 1990.— V. 315, No. 4. — P. 847—849.

[35] Kondratyev K.Y., Melentyev V.V., Nazarkin V.A. Space Remote Sensing Indication of Water Areas and Water Bodies. — St.Petersburg: *Hydrometeoizdat,* 1992. — p. 248.

[36] Recommended Practice for Remote Sensing Methods of Surface Waters Control. Determining Water Mineralization by SHF Radiometric Complex Installed on an Airplane. Issue 2. — Leningrad: *Hydrometeoizdat,* 1987. — p. 48.

[37] Kondratyev K.Y., Pozdnyakov D.V. Optical Properties of Natural Waters and Remote Sensing of Phytoplankton. — Leningrad: *Science,* 1988. — p. 181,

[38] Garbuzov G.P., Sukhorukov B.L. Investigation of ecological state of water bodies according to the remotely sensing optical spectra // *Proc. 1st Int. airborne remote sensing conf. and exhib.* — Strasbourg, France. — 1994. _ v. 3. — P. 37—45 (GS-94).

[39] Sukhorukov B.L., Garbuzov G.P., Nikanorov A.M. Optical Properties of Contaminated Aquatic Ecosystems // *Meteorology and Hydrology.* — 1997. — No. 11. — P. 5—19 (SGN-97).

[40] Sukhorukov B.L., Garbuzov G.P., Nikanorov A.M. Assessment of Water Bodies by Spectrums of Brightness Ratio // *Water Resources.* — 2000. — No. 5. — P. 579—588 (SGN-00).

[41] Sukhorukov B.L., Garbuzov G.P., Akawiec A.A. Model calculations of the brightness coefficients spectra for the interpretation of the spectrometric data on fresh water quality // Seventh International symposium on atmospheric and ocean optics / G. Matvienko, M. Panchenko, Eds., *Proc. of SPIE* 2000, V. 4341. — P. 344—351.

[42] Sukhorukov B.L., Novikov I.V. The information content of spectra of the brightness coefficient for water ecosystems // Eighth International symposium «Atmospheric and ocean optics. Atmospheric physics» / G. Matvienko, V. Banach, Eds., *Proc. of SPIE* 2001, V. 4678. — P. 503—510.

[43] Sukhorukov B.L., Novikov I.V. Information Content of Brightness Ratio Spectra from Aquatic Ecosystems // *Optics of Atmosphere and Ocean.* — 2001. — No. 10. — P. 944—949.

[44] Sukhorukov B.L., Garbuzov G.P., Akawiec A.A., Nikanorov A.M. Integral Values of Environmental Conditions of Water Bodies in Interpretation of Remote Sensing Spectrometric Data // *Reports of Academy of Sciences.* — 1998. — V. 363, No. 2. — P. 278—281.

[45] Smith R. C, Baker K. S. Optical classification of natural waters. — *Limnol. Oceanogr.* — 1979. — V. 23. — P. 260—267.

[46] Applied Statistics: Classification and Reduction of Dimensions / Edited by S.A. Aivazyan. — Moscow: *Finances and Statistics,* 1987.

[47] Iberla K. Factor Analysis. — Moscow: *Statistics,* 1980.

[48] Aivazyan S.A., Buchstaber V.M., Enyukov I.S, Meshalkin L.D. Applied Statistics. Classification and Reduction of Dimensions. — Moscow: *Finances and Statistics,* 1989. — p. 670.

[49] Garbuzov G.P., Sukhorukov B.L., Nikanorov A.M. Use of Road Structures in the System of Remote Sensing Monitoring of River Ecosystems // *Reports of Academy of Sciences.* — 2000. — V. 373, No. 5. — P. 688—691.

Chapter 7

AUTOMATED SUBSYSTEMS FOR OBSERVATION AND CONTROL OF SURFACE WATER POLLUTION

Different problems of operational water monitoring, including the timely identification (detection) of short-term, extreme anthropogenic impacts (emergency and illegal volley discharges of waste waters, accidents during storage and transportation of hazardous substances, etc.); the effective control and forecast of changes in chemical composition and the contamination of waters may only be solved with the application of modern automated equipment.

The most effective use of automatic/automated means of water analysis and computers is achieved by uniting them in a system with a unified coordinated program and common software for obtaining, processing and outputting information regarding a water body under control for the purpose of taking urgent water protection measures.

Such a system is an automated system (subsystem) for the observation and control of surface water pollution (a variety of a more general automated system for the observation and control of environmental conditions) that fulfills measurement, information and preventive functions.

Through the structure and interaction of its main internal components (automatic control stations, mobile and permanent laboratories, information processing centers, dispatching posts), and through its applied methods and equipment, an automated system (subsystem) represents a new, more complex and higher level of development of the national surface waters monitoring system.

In the beginning of the 1970s, about 250 automatic stations were operated all over the world, 200 of which were located in the USA. The first automated system for control of surface waters contamination was designed in the United States, mounted on the Ohio River and named "Orsanko." It initially consisted of 12, and later on of 20, autonomous water control stations produced by Beckman and included one central station used as a control center. Now, the USA possess over a dozen of these systems, including 8-15 different automatic stations controlling more than ten physical and chemical water parameters. The primary objectives of these systems are to carry out effective control of the impact exerted by waste waters of major pollution sources on a water body, to calculate the amount of fines for violation of regulations (related to the volume and composition of discharge waters) and to develop standards for the discharge of waste waters along with the requirements for their treatment (with due account of the environmental conditions of the water body and its economic capabilities).

Highly industrialized countries like Japan, Sweden, England, Germany, Poland, the Czech Republic, Hungary, etc., employ automated water contamination control systems designed in accordance with the principles used in the USA and intended for similar purposes.

The structure of the foregoing automated systems may be based on two or three levels. The first level includes automatic control stations and mobile hydrochemical laboratories. The second level is comprised of permanent hydrochemical laboratories. The third level represents a post equipped for the collection and primary/secondary processing of information in order to make operational assessments and short-term forecasts.

The facilities for automated systems of these countries are mainly comprised of standard laboratory and portable autoanalyzers, laboratory devices and appliances. Some devices and equipment for dispatching posts were specially designed. Germany, the Czech Republic and Russia (USSR) have a partially established automated system for monitoring major rivers of the country composed of several dozens of domestically produced automatic stations. A unified automated water control system consisting of 50 Nayada automatic stations is being established in the Czech Republic on the basis of existing monitoring systems.

In Russia, the work in the field of automatic water contamination control commenced at the end of the 1960s and in the beginning of the 70s of the previous century.

7.1. Purposes and Objectives of Automated Systems. Principles of Design and Structures of Systems

About twenty research and design institutions and organizations of different ministries and agencies were involved in the work devoted to the establishment of a scientific and methodological basis for automated water control and to the development and production of equipment for automatic and computerized measurements. The Roshydromet played a leading role in this work. Functions of the main institute on this problem were vested on the Hydrochemical Institute (A.M. Nikanorov, V.B. Stradomsky, A.P. Belousov, A.A. Nazarova, M.P. Zavesa, L.I. Minina, V.F. Gekov, A.G. Muzykov et al. [1—20]).

The automation of control over the chemical composition of contaminated surface waters is a difficult scientific and technical issue that includes the following interrelated components:

- Development of design principles for automated control systems
- Establishment of instrumental methods and corresponding equipment for automatic/computerized control
- Preparation of system software

The design and operation of analyzers showed that the metrological characteristics and functional reliability of standard devices intended for other purposes failed to meet the requirements for automated control of such a complex object as natural waters, while some other instruments were not produced at all.

Foreign experience related hereto was studied and taken into account in a domestic, special base complex of technical means for automated systems of observations and control over water pollution.

Fundamental principles of the conducted works included the constant improvement of methodological and engineering solutions used in the base models of equipment for the purpose of enhancing the quality of their technical, metrological and operating performance. Works related to equipment design were preceded by the preparation of a coordinated priority list of controlled water composition and property parameters, which included 130 items from the general list of substances sorted by MPC in accordance with the recommended chemical and analytical support (Table 7.1). A similar American list included 125 items.

Table 7.1. Priority list of pollutants and parameters; recommended methods for analysis of surface waters

No.	Parameter	Recommended methods of analysis		
		Express	Laboratory	Automatic (for stations)
1	Ammonia nitrogen	5	1, 5	5
2	Nitrate nitrogen	5	1, 5, 13	5
3	Nitrate nitrogen	1	1, 13	1
4	Total nitrogen	15	1	0
5	Acrylonitrile	0	11	0
6	Aluminum	0	1, 8, 9, 10	0
7	Aniline	0	11, 12, 14	0
8	Asphaltenes	0	4, 14	0
9	Acetic aldehyde	0	11	0
10	Benzpyrene	0	4, 12, 14	0
11	Benzol	0	1, 11	0
12	Beryllium	0	9	0
13	Borates	0	1, 13	0
14	BOD	0	16	0
15	Vanadium	0	9, 10	0
16	Suspended matters	1, 16	16	1
17	Vinyl chloride	0	11	0
18	Bismuth	0	1, 9, 10	0
19	pH	5	5	5
20	Herbicides	0	11, 12, 14	0
21	Hydroquinone	0	12	0
22	Dimethylamine	0	11	0
23	2, 4-dinitrotoluene	0	11, 12, 14	0
24	1, 2-dichloroethylene	0	11	0
25	1, 2-dichloroethane	0	11	0
26	Diethylamine	0	11	0
27	Dibutyl phthalate	0	11, 12	0
28	Dioctyl phthalate	0	11, 12	0
29	Dithiophosphate	0	12, 13	0
30	Dichlorobenzenes	0	11	0
31	Mercury diethyl	0	15	0
32	Iron (3^+)	1, 7, 16	1, 8, 9, 10	1, 7[1]

Table 7.1. (Continued)

No.	Parameter	Recommended methods of analysis		
		Express	Laboratory	Automatic (for stations)
33	Total iron	1, 7, 16	1, 8, 9, 10	1, 7[1]
34	Potassium ions	5	5, 8	5
35	Cadmium	7, 16	6, 7, 9, 10	7[1]
36	Calcium ions	5	5, 8	5
37	Caprolactam	0	0, 12, 14	0
38	Carbonyl compounds	0	1, 11	0
39	Dissolved oxygen	15	15	15
40	Cobalt	0	9, 10	0
41	M- and p-cresols	0	11, 12	0
42	Silicon	1	1, 3	1
43	Xanthogenate	0	1,12	0
44	Xylols	0	11	0
45	Magnesium ions	5	5, 8	0
46	Manganese	0	1, 7, 9, 10	0
47	Copper	1, 5, 7, 10, 16	7, 8	1, 7
48	Mercaptan	0	11	0
49	Methanol	0	1, 11	0
50	Methyl mercaptan	0	11	0
51	Methyl methacrylate	0	11	0
52	Methylstyrene	0	11	0
53	Turbidity	1	15	1, 15
54	Arsenic	6, 7	6, 7, 9	7[2]
55	Sodium ions	5	5, 8	5
56	Naphthalene	0	11	0
57	Oil products	15, 16	3	0
58	Nickel	0	1, 9,1 0	0
59	Nitrobenzene	0	11, 12	0
60	Nitrosamines	0	11, 12	0
61	Nitrophenol	0	12, 14	0
62	Oxidizability	15	15	0
63	Organic acids		11, 12, 13, 14	0
64	Pesticides	0	11, 12, 14	0
65	Pyridine	0	1, 11	0
66	Polyacrylamide	0	1	0
67	Polyphosphate	0	1	0
68	Polychlorbiphenyl	0	11, 12	0
69	Transparency	15	15	0
70	Rhodanate	0	1	0
71	Mercury	15, 16	15	7[1]
72	Lead	5, 7	6, 7, 9, 10	5[1], 7[1]
73	Selenium	0	9	0
74	Total sulfur	1, 16	1	0
75	Nonsulfate sulfur	1, 16	1	0
76	Carbon disulfide	0	1, 11	0
77	Silver	0	7, 9, 10	0
78	Hydrogen sulfide and sulfides	5	1	0

No.	Parameter	Recommended methods of analysis		
		Express	Laboratory	Automatic (for stations)
79	Turpentine	0	1	0
80	Resins	0	4, 14	0
81	Total heavy metals	1, 16	1	0
82	Synthetic surfactants	1	1, 13	0
83	Alcohols	0	11	0
84	Strontium	0	8	0
85	Sulfates	5	1	5
86	Sulfites	0	1	0
87	Antimony	0	1, 7, 9	0
88	Tellurium	0	9	0
89	Temperature	16	16	16
90	Terpene	0	11	0
91	Tetrachloroethylene	0	11	0
92	Tetrachloroethane	0	11	0
93	Tetraethyltin	0	9	0
94	Tetraethyl lead	0	9	0
95	Titanium	0	9	0
96	Toluene	0	11	0
97	Threebutylmethacrylate lead	0	9	0
98	Trichlorethylene	0	11	0
99	Triethanolamine	0	11	0
100	Carbohydrates	0	1	0
101	Organic carbon	15	15	15
102	Carbon tetrachloride	0	11	0
103	Phenol	1, 16	1, 11	
104	Formaldehyde	0	1, 11	0
105	Mineral phosphorus	16	1	0
106	Total phosphorus	16	1	0
107	Elemental phosphorus	0	1	0
108	Fluorine	5	1, 5	5
109	Furfural	0	11, 12	0
110	Chlorine	0	16	0
111	Chlorbenzene	0	11	0
112	Chlorides	1 ,5	1, 5	5
113	Methylene chloride	0	11	0
114	Chloroprene	0	11	0
115	Chloroform	0	11	0
116	Chlorophenol	0	11, 12	0
117	Chromium (Cr^{6+})	1	7	1, 7[1]
118	Chromium (Cr^{8+})	0	1, 6, 7	0
119	Chromaticity	16	1	0
120	Cyanide	5	1, 5, 13	0
121	Hexahydrophenol	0	11	0
122	Ketohexamethylene	0	11	0
123	Ketohexamethyleneoxine	0	12	0
124	Zinc	7, 16	1, 6, 7, 8	7[2]
125	Alkalinity	16	5	0
126	Electrical conductivity	15	15	15

Table 7.1. (Continued)

No.	Parameter	Recommended methods of analysis		
		Express	Laboratory	Automatic (for stations)
127	Ethylbenzene	0	11	0
128	Ethylene glycol	0	11	0
129	Ethers	0	11, 12	0
130	Molybdenum	0	7, 9	0

Notes:

Dash (—)—no data available, 0—express and automatic methods are not applicable at the current stage, [1]—after establishment of automatic blocks based on inversive voltamperometry for AMA-203.

List of methods mentioned in the table:

1. Visible range photometry
2. UV spectrometry
3. IR spectrophotometry
4. Fluorescence (luminescence)
5. Ionometry
6. Polarography
7. Inversive voltamperometry
8. Flame atomic absorption (with flame photometry)
9. Flameless atomic absorption
10. Emission spectrometry
11. Gas chromatography
12. Liquid chromatography
13. Ion chromatography
14. Thin-layer chromatography
15. Special analyzers
16. Auxiliaries for sample preparation and analysis

Anthropogenic impact on surface waters is mainly caused by twenty major industries and affects approximately 60 parameters of water composition and properties, 20 of which are the most frequently controlled (temperature, dissolved oxygen, pH, electrical conductivity, suspended matters, phenols, synthetic surfactants, heavy metals, nitrogen compounds, etc.). Therefore, the works related to automated control were primarily concentrated on the foregoing parameters.

The most important aspects of the strategy for the development of automated water control included the 1) development of design and operation principles for automated systems; 2) improvement of structure and interrelation of components; 3) optimization of the number of automatic control stations and places of their installation, including control posts serviced by mobile means; 4) selection of the most reasonable frequency for measuring physical, chemical and biological water parameters and 5) the optimization of their information, software and methodological performance.

The first domestic realization of the developed scientific and methodological principles of automated control was the establishment of an experimental, local computerized system for surface waters control of the ASCSW type on the Moscow River (from the Mozhaysk Reservoir to the Oka River in 1976).

Experience from the design and operation of the first automated systems, large amount of scientific and engineering work, and problems solved by developers allowed the formulation of a number of primary provisions (requirements), which had to be taken into account. The new system of the ANKOS-VG type (automated environmental observations and control—water, main) should have:

- Been established on the basis of existing network of Roshydromet's divisions (environmental pollution control centers)
- Organizationally and methodically united all water control divisions within its zone, including observation posts and points and permanent hydrochemical and hydrobiological laboratories located on the territory covered by the system within 180-250 from its center
- Been built in accordance with the principle of hierarchy, reasonable distribution of objectives between its main components and organization of interactions between them providing minimal time and labor costs for all procedures, from sampling to the provision of information

The system of the ANKOS-VG type was intended for solving the following problems:

- Effective measurement, collection and processing of data; assessment of water pollution and the registration, storage and provision of information on conditions of the water body (area, post and point) under control
- Timely detection of substantial volley and emergency waste water discharges
- Operational short-term forecast of pollution level (from several hours to five days)
- Prompt notification to water users about observed and forecasted high and extremely high levels of water pollution
- Provision of information to supervising and managing authorities for taking measurements aimed at the prevention or minimization of negative consequences entailed by an emergency or unauthorized discharge of waste waters

The automated system of the ANKOS-VG type is economically justified when it serves as the source of initial operational information for the general real time water quality management system. The use of such a system for solving problems of surveillance monitoring within the framework of the State Service for Monitoring Environmental Conditions is unreasonable, however, because these problems may be solved by cheaper, conventional means.

Systems of the ANKOS-VG type are to be established for controlling areas of economically important water bodies that are exposed to occasional, violent anthropogenic stress. The establishment of such systems on water bodies with a continually high level of pollution is ill-advised. The economic feasibility of such systems was also proven for the operational monitoring of territories (with areas of 20,000-25,000 m^2) with scarce water balance, which are water bodies that are occasionally exposed to violent anthropogenic stress. The number of water areas controlled by ANKOS-VG systems may vary from 35 to 70 posts (points) located at a distance of not more than 120-130 km from the control center.

Observance of the foregoing system parameters ensures timely detection of the dangerous pollution of a water body, prompt investigation of its causes and proper notification to water users thereof.

In cases with a considerable increase in the area and distance of a control post (point), the advantages of automated systems decreases dramatically. The presentation of hydrochemical information with a delay of over four hours prolongs the decision-making process by a water protection authority by up to 6-8 hours. The ANKOS-VG system is based on a hierarchy principle and includes three levels that are grounded by technical reasons and economic factors. Only this type of system structure may provide settlement of major issues of operational monitoring and organize a rational interaction of its components, including the distribution of work with the use of new and conventional methods along with technical means (Figure 7.1).

The first level of the system is presented by automatic water control stations and mobile hydrochemical (hydrobiological) laboratories.

The main objective of this level is the 24-hour, automatic and computerized collection, processing and transmission of data related to the physical properties and chemical composition of water. Water parameters are measured with a preset frequency (from 1 to 6 hours) and the initial data is processed by an automatic station under the control of a computer. At the appointed time, the computer turns on station pumps and hydraulics for taking samples of water and feeds it to sensing devices. Measurement data is transmitted to the corresponding converter blocks and then it is processed by the computer in accordance with predetermined programs and saved in the form convenient for further transmission to a data collection and processing center. Mobile hydrochemical (hydrobiological) laboratories serve the purpose of 1) taking water samples in the areas of a water body, which are not equipped with automatic control stations; 2) on-site, express analysis of samples and 3) delivering samples to permanent hydrochemical (hydrobiological) laboratories for detailed analysis.

The second level of the system includes permanent hydrobiological laboratories. The major task of this level is to analyze the collected samples in order to obtain more reliable hydrochemical and hydrobiological information on the conditions of the water body under control. This (or the third) level comprises laboratories for the adjustment, repair and metrological support of the system.

The third level of the system is presented as the processing center for data received from automatic stations and mobile and permanent laboratories. This center is comprised of a computer system and dispatching post with a communications center. The major objective of the complex includes 1) providing communication to all components of the system and consumers of information; 2) maintenance of the whole complex of equipment and 3) the collection, processing, storage and provision of information on current and forecasted conditions.

Components of ANKOS-VG systems are equipped with modern technical means, including automatic and computerized laboratories, portable measuring devices, auxiliary facilities and equipment. Description of all components of the system, their purposes, functions, equipment, etc., are provided below. The system is managed by special software (a package of application programs) designed to transmit commands from a dispatching post to automatic control stations and mobile and permanent laboratories.

The level of system's mathematical support and software is determined by the consumer's requirements concerning the information.

Mathematical support and software are intended to solve the following problems:

- Collection, validation, initial processing, registration and storage of information on the pollution of a water body under control; the information is obtained by all components of the system
- Detection of increased MPC (or sharp deviations) of observed parameters and warning thereof
- Short-term forecasts of pollution for a controlled water body
- Preparation and provision in a tabular form of information related to the current and forecasted level of pollution
- Control of technical means

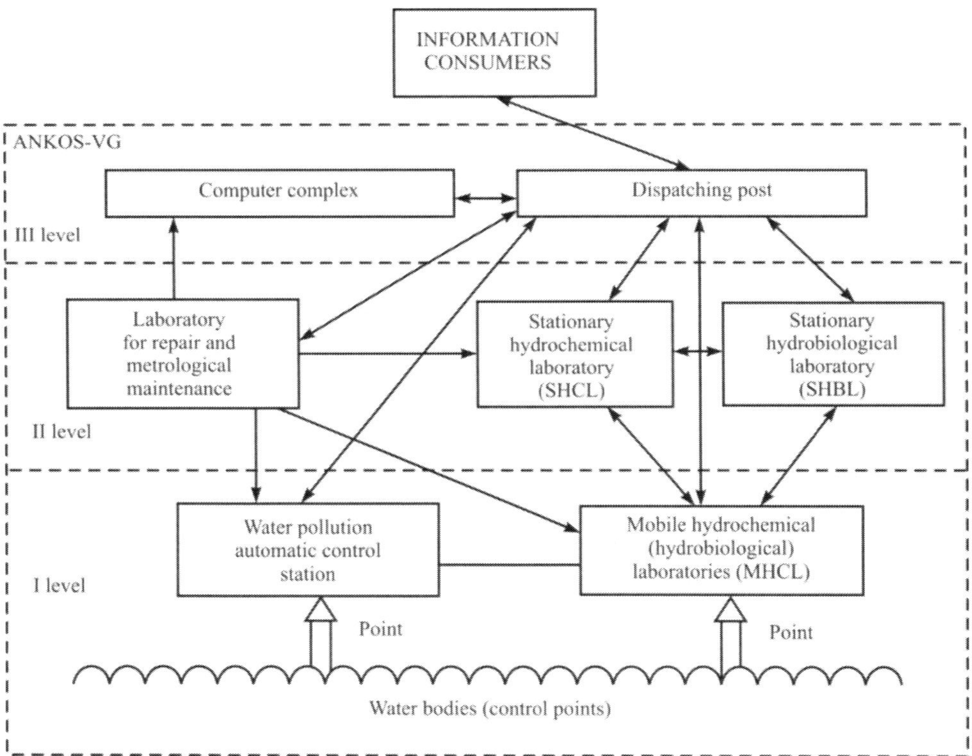

Figure 7.1. A block diagram of the ANKOS-VG automated system.

The ANKOS-VG system with all its components is able to control 50 parameters of water properties and composition. Meanwhile, up to 20 parameters (and in the future 25-30) are measured by automatic stations, up to 30 parameters (and in the future up to 40) by mobile laboratories with quite a high level of automation (up to 60%) and up to 50 parameters by permanent laboratories (Table 7.2).

The monitoring frequency of different system components may vary widely from once a month to 12 times or more per day in accordance with the adopted program. The content of information necessary to solve issues of operational monitoring, assessment and short-term forecast may reach 150-300 thousand measurements spread across 30-60 control posts (points) within the system.

The operational short-term forecast (from several hours to five days) provided by ANKOS-VG has had a success rate of at least 85%. Forecasting is based on the conditional expectation method and its values are calculated by models with a multiplicatively additive structure that describes nonstandard, random processes in a water body. The least squares method is used for occasional adjustment of model parameters. The detection probability of volley discharges lasting at least one hour by posts (points) with automatic stations equals o 85-90%.

According to research from Russian experts, the following contributes to the decrease in damages caused by EHP of a water body: timely detection of emergency volley discharge of waste waters into a water body; prompt survey thereof; adoption of urgent water protection measures, including those that are related to polluters and proper notification of major water consumers (water users).

The economic effect from decreased damages caused by EHP to a water body controlled by ANKOS-VG is 50-60% higher as compared with the effect achieved by conventional methods of observations implemented by the network of surveillance monitoring.

Under conditions of severe anthropogenic impact on the natural environment, there is an objective necessity to develop integrated automated control systems covering surface waters as a natural object and waste waters as a factor of anthropogenic impact on this object. Only in this case will there be an opportunity to detect the following in proper time:

a. Beginning of volley discharges under breach of maintenance procedures or in case of an emergency
b. Particular polluters, which have provoked EHP

This may prevent or reduce damages caused by a volley discharge on account of a shortened response time to the information obtained and an opportunity to take effective nature protection measures in a timely manner.

Experience accumulated during the establishment of ANKOS-VG system on the Don, Kura, Moscow, Tom' and other rivers proved that is it necessary to consider the establishment of each new system as an independent development. The system must be able to adjust to the peculiarities of water bodies, climatic and geographical characteristics of their basins, nature of pollution and incoming waste waters, differences in existing structures of the regional economy and Roshydromet's territorial departments involved in the monitoring of water bodies. It is impermissible to treat the establishment of new systems as a simple replication of a complex of technical means since such an approach does not provide an economically sound and functional system. In order to make reasonable decisions on the development of different components of the ANKOS-VG system, to organize preproject works and to introduce new systems, the Hydrochemical Institute has worked out the "Temporary Practical Guidelines for Deployment of Technical Means of ANKOS-VG Surface Water Pollution Control System," which made it possible to efficiently substantiate the selection of a structure for a new system while also accounting for its conditions of deployment on water bodies.

The first stage of any developed automated system must include all the components comprising the structure of ANKOS-VG for the purpose of ensuring its functioning as a system operating on a limited territory. Further development of the systems should be considered a growth in the quantitative composition of its components.

Currently, the developed methods for the establishment and application of ANKOS-VG automated systems limit their use for operational control, assessment and short-term forecast of water pollution to rivers (watercourses). The necessity of obtaining operational hydrochemical information on accidental pollution of water reservoirs, lakes and river mouths requires methods for the establishment and application of automated systems different from ANKOS-VG. These types of water bodies need specific technical means, practical guidelines and mathematical provision. Operational control of waste waters from major pollution sources is also obligatory in this case.

Table 7.2. Approximate list of parameters measured by components of the ANKOS-VG system

No.	Parameter	ACS	MHCL	SHCL	Portable HCL
1	Sampling depth	○			○
2	Temperature	○	○		○
3	Turbidity	○	○		○
4	Specific el. conductivity	○	○	○	○
5	pH	○	○		○
6	Redox potential	○	○		○
7	Dissolved oxygen	○	○		○
8	Chloride	○	○	○	○
9	Sulfates			○	
10	Nitrates	○	○	○	
11	Nitrites	○	○	○	
12	Phosphates	○	○	○	
13	Hydrocarbonates		○	○	
14	Carbonate		○	○	
15	Fluorides	○	○	○	
16	UV light transmittance	○			
17	Sulfides				
18	Cyanide		○	○	
19	Sodium	○	○	○	
20	Potassium	●			
21	Calcium	●			
22	Magnesium				
23	Ammonium	●	○	○	●
24	Iron (total)	○	○	○	○
25	Copper (II)	○	○	○	○
26	Zinc (II)		○	○	●
27	Chromium (III, VI)	○	○	○	
28	Strontium				
29	Cadmium (II)	●			○
30	Mercury (I, II)	●			●
31	Lead (II)	●			●
32	Manganese				
33	Arsenic (III,V)				
34	Silicate	○			

Table 7.2. (Continued)

No.	Parameter	ACS	MHCL	SHCL	Portable HCL
35	COD		○	○	
36	Carbon (organic)	○	○	○	
37	Nitrogen (organic)		○	○	
38	Fats			○	
39	BOD		●		●
40	Synthetic surfactants			○	●
41	Phenol			○	●
42	Oil products		○	○	●
43	$\sum$ of organic acids			○	
44	Pesticides			○	
45	Phosphorus (organic)			○	

Notes:
●—developed methods and means, ○—planned methods and means

The growing intensity of water use and accordingly, the increasing requirements for reliability and timeliness of information obtained by ANKOS-VG systems determine the necessity to broaden capabilities of their methodological, mathematical, software and technical provisions. First and foremost, this applies to the list of controlled parameters extended by the inclusion of new parameters describing the organic constituent of pollutants; to the introduction of a system of different, complex parameters of state; to the assessment of water pollution (current and forecasted) not only in controlled points, but also at the points of water consumption (water use); to the increase of forestalling terms (from three to ten days) of short-term forecasts for points of water consumption (water use); to the improvement of the operating stability of systems and to other issues.

7.2. Automatic Control Stations

Automatic control stations (monitors) are complex, multifunctional devices providing automatic measurement of physical and chemical parameters of water conditions and composition along with the collection, processing, storage and transmission of data. Automatic control stations serve as an initial component of automated systems of observation and control over surface waters and are used as autonomous measuring devices. The distinctive feature of these monitors is their continuous, automatic operation with a low demand for maintenance.

The main problems solved by automatic stations include:

- Detection of sharp, short-term changes in water composition and level of pollution caused by anthropogenic impact and natural factors
- Transmission of EHP signals to corresponding decision-making authorities
- Monitoring of chemical pollution by specified parameters
- Determining trends of chemical parameters within long periods of time, etc.

Monitors may be used to receive data required for mathematical models and different indices (coefficients, parameters) of water quality.

Thus, monitors shall be considered a sort of working tool for controlling pollution and managing water resources, as well as a means for solving practical issues.

Usually, automatic control stations consist of the following main parts:

1) Sampling and measuring module
2) Electric signal forming/transformation module
3) Data registration and output module

The operating principle of an automatic station may be schematically shown as follows. Water from the controlled area of a water body is pumped out into a sample container, where one or more parameters are measured. Here, water comes into contact with sensors intended for the reproduction of electric impulses proportional to the activity or ion concentration of the measured parameters.

The signal generating module receives electric impulses and amplifies (transforms) these signals into values appropriate for detection by the registration module. The last module records the signals on a special paper (punched tape) for a teletype printing device (different combinations are possible), which transmits data through radiotelephone communication to a remote control center (dispatching post, data processing center, directing agency).

If the sensors are installed directly in the water flow, then there is no need to use a sampling pump and container. In this case, however, there is a risk that the sensors may be damaged by debris, floating garbage, etc.

Monitors of the foregoing type may be mounted in a fixed frame and put on a mobile trailer, boat, buoy (beacon).

Semisubmerged monitors used for monitoring pollution of lakes, water reservoirs and estuaries (with a depth of up to 300 m) are of particular interest. Such stations usually consist of a power unit, signal amplifiers and a data recording device in a waterproof container with connected measuring sensors.

As an example of a foreign automatic control station measuring not only standard parameters (pH, Eh, temperature, concentration of suspended matters, specific electric conductivity and dissolved oxygen), but also water contamination parameters, we should mention the SM-1200 monitor, which is produced in six different modifications. It is a component of the automated water quality control system of Otometid Environmental Systems (USA). Out of 28 parameters controlled by the system, 21 pertain to physical properties and chemical composition of water (including bromides, iodides, fluorides, calcium, hardness, nitrates, sulfides, etc.). The connection of an additional module allows for the determination of ions of ammonium, chromium (VI), copper, iron, orthophosphates and residual chlorine. Some parameters are measured if their concentrations increase to the level of waste waters.

Horiba (Japan), along with the monitor for measuring standard water quality parameters (WARA-22 model), produces automatic monitors for determining COD, continuous measurement of carbon content, etc., while Yanagimoto manufactures WPM-2000, an automatic station providing continuous measurement of 12 parameters.

The Dutch company, Nismegen Holland, has developed and manufactured an automatic control station for determining MPC of heavy metals based on voltamperometry.

The Nayada-Standard automatic station (Czech Republic) is a modular, unified set of equipment for the continuous measurement of physical properties and chemical composition of surface, ground and drinking waters. An emergency program for water bodies allows, along with automatic measurement with time intervals of 5, 15, 30 and 60 minutes, samples to be taken for further detailed analysis.

In Russia, the development of measuring converters for automatic stations was mainly based on the use of the following physical and chemical methods: potentiometry, conductometry, photometry, polarography, inverse voltamperometry, etc.

The first domestic automatic station of the ASCSW type was intended for operation both as a part of an experimental automated system for the monitoring and control of water pollution (in particular for the basin of the Moscow River) and in an autonomous mode.

The station (Figure 7.2) was comprised of:

- Pump and hydraulics (PH) for taking samples of water in the western point of the water body and transferring them to measuring devices
- Measuring devices (MD) transforming physical and chemical parameters into electric signals
- Measurement and transformation equipment providing analog and digital transformation of the signal and delivering it to a registering and storing device
- Communication equipment for the transmission of data to an information processing center (IPC)

The ASCSW has made it possible to obtain information on the physical properties and chemical composition of surface waters, relatively clean waters after treatment and industrial/drinking waters. The system measured 7-9 controlled parameters (with an interval of 0.5, 1, 2, 3, 6 and 12 hours) and automatically took samples of water if the maximum allowable concentrations were detected (or upon command from the dispatching post) (Table 7.3).

In addition, as opposed to earlier manufactured stations, the ASCSW was capable of controlling two levels of a water body (mounting levels of water intake devices) and was equipped with an emergency power supply and device for the discharge of water from measurement cells in case of a prolonged power cutoff.

Table 7.3. Parameters, range and error of ASCSW measurements

Parameter	Range of measurements	Measurement error
Water temperature, °C	0—40	±0.5%
Water level, m	0—12	±1%
pH	4—10	±0.1%
Eh, mV	-600...600	±5
Specific electric conductivity, S/m	10^{-4}—10^{-2}	±5%
Suspended matter, mg/l	0—2000	±8%
Dissolved oxygen, mg/l	0—25	±1%
Sulfide, mg/l	pS = 15—19	0.5
Copper, mg/l	0.02—1	

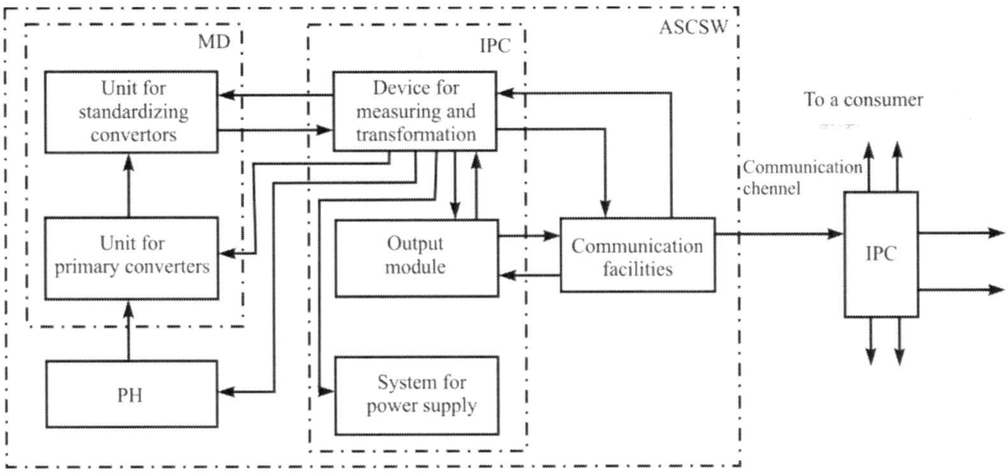

Figure 7.2. Structural scheme of the ASCSW automatic control station.

Table 7.4. Parameters, range and error of ASKZV-G measurements

Parameter	Range of measurements	Measurement error	Measurement method
pH	4—10	±2.2	Potentiometric
Eh, mV	-700...700	±1.9	
Temperature, °C	0—40	±1	Thermistor thermometer
Specific electric conductivity, S/m	$1\cdot 10^{-5}$	±3	Conductometric
Concentration of dissolved oxygen, mg/l	0—20 0—50	±5.5	Polarographic
Turbidity (BB)	0—250 0—1,000	±10.5	Nephelometric
Water level, m	0—3 0—6 0—12	±1.5	Manometric
pCl^-	1—3.5	±4.5	Potentiometric (ISE)
PNO_3^-	1—4	±3.8	same
pNa^+	1—5	±3	»
pF^-	1—4	±3.8	»
$[Cu^{2+}]$, mg/l	0—0.1	±20.5	Photocolorimetric
$[Fe^{3+}]$, mg/l	0—1	±20.5	same
$[Cz^{6+}]$, mg/l	0—1	±20.5	»
$[PO_4^-]$, mg/l	0—2	±5.5	»
$[NO_2^-]$, mg/l	0—0.5	±5.5	»
Absorption of UV light by natural water, %	0—100%	±5.5	Photometric

Later on, an improved modification of the ASKZV-G automatic station was made, which was used as an initial component of automated stations for the monitoring and control of surface waters of the ANKOS-VG type in a number of priority areas of the country (in particular, for the Don River).

The structure of the ASKZV-G station featured measurement facilities, discharge devices and information processing and registration equipment, including data transmission equipment.

Ion selective, photoabsorptiometric and other converters were added to the base complex of the measurement blocks of the automatic station, which extended the functional capabilities of ASKZV-G to 12-15 parameters (including ions of chlorine, fluorine, phosphate, nitrate, nitrite, etc.) (Table 7.4).

ASKZV-G provides real time measurement with intervals of 0.5, 1, 1.5, 2, 3 and 4 hours (in accordance with the internal program or under command from the information processing center). It also takes care of averaging the measurements, controlling excessive MPC, registration, storage and output of data, and, if necessary, taking samples of water and their storage. The main complex of technical means for automatic collection, processing and transmission of information is provided on Figure 7.3.

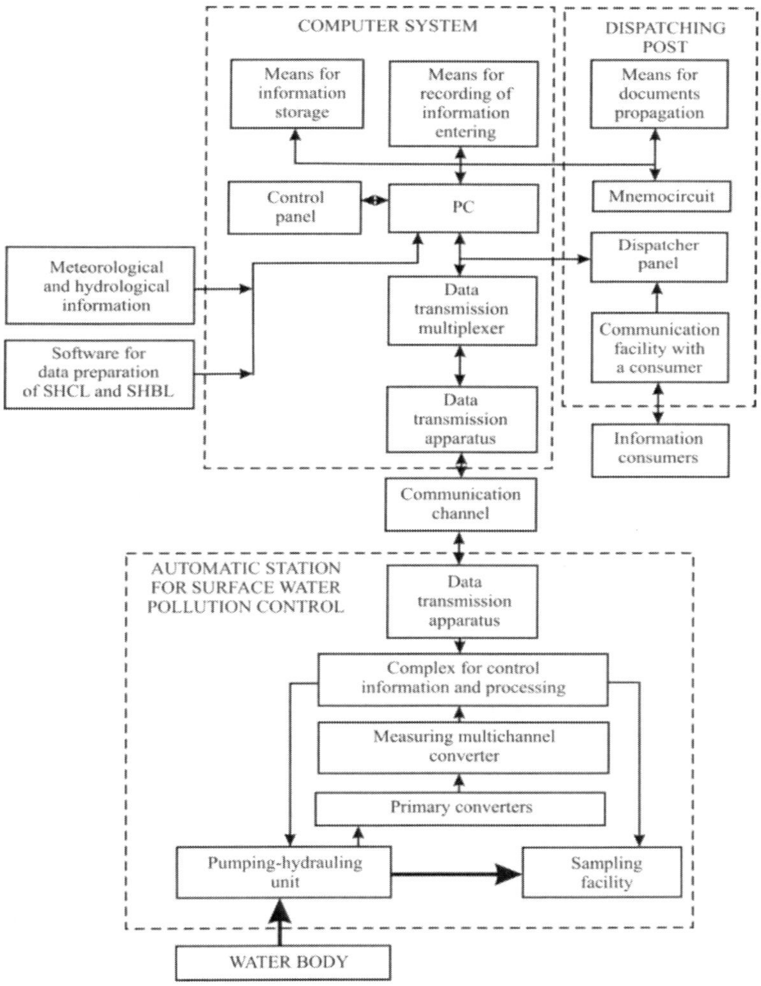

Figure 7.3. The main complex of technical means for automatic collection, processing and transmission of information.

An automatic station of the next generation, AMA-201M, was based on the automatic multichannel analyzer of the same name. It was also designed for an ANKOS-VG automated system and for measurement of 12-17 parameters with an interval of 1, 2, 3 and 4 hours within 14 days (Table 7.5).

Table 7.5. Parameters and range of measurements by AMA-201M

Parameter	Range of measurements	Measurement method
pH	4—10	Potentiometric
Eh, мВ	-50...700	»
Temperature, °C	0—40	»
Specific electric conductivity, S/cm	10^{-4}–10^{-3}	Conductometric
Concentration of dissolved oxygen, mg/l	0—20	Potentiometric
Water level, m	0—3 0—12 0—16	Manometric
Turbidity, mg/l	0—50 0—1,000	Nephelometric
Activity of ions		
chloride, pCl^-	1.0—3.5	Potentiometric
nitrate, pNO_3^-	1.0—4.0	»
fluoride, pF^-	1.0—5.0	»
sodium, pNa^+	1.0—4.0	»
Concentration on ions, mg/l		
copper, Cu^{2+}	0—0.1	Photocolorimetric
iron, Fe_{tot}	0—1.0	»
chromium, Cr (VI)	0—1.0	»
phosphate, P_{min}	0—2.0	»
nitrite, NO_2^-	0—0.5	»
Absorption of UV light, %	0—100	Photometric
Parameters to be introduced		
Activity of ions		
calcium, pCa	1.0—4.0	Potentiometric
potassium, pK	1.0—4.0	»
ammonium, pNH_4		»
Concentration		
silicon Si, mg/l	0—15	Photocolorimetric
lead Pb, mg/l	0—30	Inverse voltamperometric
copper Cu, mg/l	0—30	»
cadmium Cd, mkg/l	0—5	»
zinc Zn, mkg/l	0—50	»
arsenic As, mkg/l	0—50	»
phenols 0—100, mkg/l	0—5	»

The AMA-203 automatic multichannel analyzer, with multipurpose software and configurable measuring blocks, is the basis for a new generation of stations. The ASKZV-2 is designed to take measures in a discrete, uninterrupted mode with an interval of 1, 2, 3 and 4 hours, to collect, process and control information. The MPC is designed to store (from several hours to three days) the obtained information and to transmit it automatically via dial-up lines or radio communications. AMA-203 is also able to accept and execute commands from a dispatching post, to take samples and store them in case of exceeding MPCs, to output information and to provide continuous operation of all systems with maintenance intervals of at least 15 days.

The use of new metrologically certified methods and engineering solutions allows for the reliable control (in the range 0-3 MPC, 0-30 MPC) of 12-18 parameters from the list of all developed methods. Future researches in the field of automatic control of surface waters pollution may increase the number of parameters measured by automatic stations up to 25-30 or more.

This type of automatic control station (AMA-203) is the most advanced model as compared with the earlier developed stations (ASCSW, ASKZV-G, AMA-201, AMA-201M).

We should note that the latest domestic automatic stations are as good as the best foreign products, which along with their low price, has increased their export demand (in particular, Germany imports the AMA-201M station).

The analogs of automatic stations and analyzers for water quality control manufactured in the USA (Beckman, Honeywell, etc.), Germany, France, England, Holland, Hungary and the Czech Republic are designed for simultaneous control of 10-12 parameters from the total list of 15-20 parameters. Foreign automatic stations cannot adjust their measurement range for a certain parameter by 10 times upon command from a dispatching post, which in its own turn does not register the high level of pollution and necessitates the use other means (for example, sending for a mobile laboratory). Stations like Nayada-80, SM 1200-5, etc., require maintenance once per 4-7 days, while the latest Russian models may be serviced once per 7-10 or less frequently. This fact was confirmed by the experimental operation of an automated system on the Don River. Unfortunately, in the 1990s, the works related to the automatic control of water pollution were suspended for well-known reasons despite their unquestionable urgency.

7.3. Mobile Hydrochemical Laboratories

An important component of the operational monitoring, its corresponding automated systems of monitoring and control over pollution of surface waters (ANKOS-VG) are mobile hydrochemical (hydrobiological) laboratories (MHCL, MHBL), which are designed for the operational survey of a water body, detection of EHP, contouring of high (extremely high) pollution zones, tracking of their movement and solving various other issues.

Mobile laboratories allow us to carry out a great amount of hydrochemical, hydrobiological and hydrometric works, including sampling, on-site measurement of physical properties, chemical and biological parameters, conservation of samples, and their storage and delivery to permanent laboratories for complete analysis.

The on-site, operational instrument measurements of different parameters ensure accurate results without distortions that could be caused by the analysis of samples in permanent laboratories after their prolonged storage and transportation.

The Russian mobile hydrochemical laboratory (MHCL-1) was developed at the behest of the State Committee for Science and Technology operated by the Hydrochemical Institute in cooperation with organizations of the Ministry of Tool Engineering, Automation and Management Systems.

The statement of this problem was defined by the necessity to organize a system of operational monitoring and to provide it with effective means of operational control over national territories (especially those that are exposed to violent anthropogenic impact).

Commencement of works related to the design of the Russian mobile hydrochemical laboratory was preceded by thorough study of the demands made by major consumers of MHCL-1 and the experience of other countries in designing similar technical means, in particular France (for controlling river waters pollution by six parameters measured by a mobile automatic station), Germany (for operational chemical, biological and radiological researches), Poland.

AW-13 (Poland) and WL-2-15 (Germany) were used as analogs (the basis) for the development of MHCL-1. Design deficiencies inherent in these models (unsuitable for long trips lasting more than one day, normal temperature conditions cannot be maintained in the cabin at violent fluctuations of ambient temperature, safe labor conditions and movement of expert group are not provided to the full extent, no places for rest, etc.) were not only overcome, but functional capabilities were also expanded in the process of development and while making technical decisions.

On the basis of express methods (photometry, potentiometry, ionometry, biotesting) and a complex set of portable devices and analyzers (FK-120, KFK-2, I-120M, AKV-106, microscopes MBS-9 and Biolan), the mobile hydrochemical laboratory MHCL-1 is capable of on-site measurement of about 30 physical, chemical and biological parameters (Table 7.6) including products of phyto and bacterial plankton along with two-three pollutants (test methods). MHCL-1 can also take care of sampling and the conservation of samples for further analysis in permanent laboratories (determination of specific pollutants; number, species composition and biomass of bacterial plankton, zooplankton and zoobenthos).

Mobile hydrochemical (hydrobiological) laboratories may be mounted on automobiles, buses, vessels, boats or helicopters. Mobile laboratories mounted on a car are the most popular modification due to the large scale of surveyed territories, great distance from water bodies to permanent laboratories, contrast of climatic factors, road conditions, etc.

In order to provide high cross-country ability, the laboratory was mounted on a four-wheel drive bus. As compared with its foregoing foreign counterparts, MHCL-1 has several advantages. The laboratory may be operated in autonomous mode without refueling for a period of seven days and a range of 200 km. The laboratory received its power (for measuring devices, refrigerator for air-conditioning and heating system, etc.) supply from an onboard electrical network or from two petrol power units—accumulators. The bus cabin was divided into two rooms; one was a dwelling unit for the research group and the other, a work module with a 7 m^2 laboratory.

The following problems were successfully solved:

1) Operational communication with dispatching post or CEM
2) Arrangement of devices and auxiliary equipment with due account of labor safety and convenience
3) Minimization of vibration and oscillation of equipment
4) Maintenance of acceptable onboard temperature conditions (from 10 to 35°C) at an ambient temperature from -30 to 40°C.

The stock of reagents and expendable materials was sufficient for 100 measurements. More ware and reagents would allow additional measurements of several other parameters to be made, for example, anionic synthetic surfactants, permanganate value, etc.

Parameters that are to be measured on-site by MHCL-1 should be selected in accordance with the assigned objective and planned time of field work.

On the basis of the conducted expeditionary tests of MHCL-1 on water bodies in different areas of the country (Bashkiria and the Rostov Region by standby mode and imitation of emergency discharges by injecting florescent tracers into waste waters), we have developed a recommended practice for the effective autonomous use of the laboratory (without an ANKOS-VG system) in order to carry out the following works:

- Operational inspection and detection of the presence (or absence) of volley discharges of contaminated waste waters (upon signal)
- Observation, inspection and detection of EHP zones, forecast and tracking of their movement
- Determination of pollution source (enterprise responsible for volley discharge of waste waters), ascertaining of causes and preliminary assessment of impact

Table 7.6. List of parameters determined by the mobile hydrochemical laboratory (MHCL-1)

Parameter	Unit of measure	Instrumentation	Method	Range of measurement
Chromaticity	degree	KFK-3	Color	0—100
Transparency	cm	Secchi disk	Visually	0—50
Temperature	°C	AKV-106	Amperemeter	0—40
Odor	points	—	Organoleptic	0—4
Electric conductivity	S/m	AKV-106	Conductometer	0.01—1
Turbidity	mg/l	AKV-106	Turdbidimeter	0—5
Ox-redox potential	mV	AKV-106	Potentiometer	-150...700
pH		AKV-106	—	2—12
Dissolved oxygen	mg/l	AKV-106	—	0—20
Oxidizability	mgO_2/l	—	Titrimeter	0—100
Nitrite nitrogen	mg/l	KFK-2	Photometer	0—0.5
Nitrate nitrogen	mg/l	I-120M	Potentiometer	0—10
Ammonia nitrogen	mg/l	KFK-2 I-120M	— —	0—140 0.1—10
Hydrogen carbonates	mg/l	I-120M	—	0—500
Chlorides	pCl un.	AKV-106 I-120M	— —	0—4 1—4
Sulfates	mg/l	—	Titrimeter	0—60
Calcium	mg/l	—	—	0—1,000
Magnesium	mg/l	—	—	0—20
Sodium	pNa un.	AKV-106 I-120M	Potentiometer —	1—4
Potassium	mg/l	I-120M	—	0—20
Fluorides	pF un.	I-120M	—	1—5
Silicates	mg/l	KFK-2	Photometer	0—0.05
Total mineralization	mg/l	—	Rated	0—2,000

Parameter	Unit of measure	Instrumentation	Method	Range of measurement
Copper ions	mkg/l mg/l	KFK-2 PFKD-1000	Photometer »	1—50 1—100
Lead ions	mkg/l mg/l	KFK-2 PFKD -1000	Express »	1—100
Cadmium ions	mkg/l mkg/l	KFK-2 PFKD-1000	Photometer »	2—20
Chromium (VI)	mkg/l	KFK-2	»	1—300
Total chromium	mkg/l	KFK-2	»	1—500
Aluminum	mkg/l	KFK-2	»	1—200
Total iron	mkg/l	KFK-2	»	3—500
Synthetic surfactants	mkg/l	KFK-2	Express photometer	
Phenols	mkg/l	KFK-2	Photometer	1—500
Oil products	mkg/l	NV-1	IR photometer	0.05—5.0
Acute toxicity of water	points		Bilogical	

- Collection of operational information, sampling, operational analysis by MHCL-1 and preparation of samples for transportation to permanent laboratory.

Mobile laboratories as a part of automated systems for monitoring and control (ANKOS-VG) may be used to ascertain and eliminate causes of failures in automatic stations and even to fulfill their program functions with a frequency of 1.5-2 hours. Moreover, MHCL-1 may be used as a quasipermanent hydrochemical laboratory during the operational inspection of a water body by an aerial observation station (helicopter) based on remote sensing methods of control. Finally, it is possible to use MHCL-1 for the purposes of surveillance monitoring related to sampling, preparation and transportation of samples, which may be combined with the direct functions of a mobile laboratory in operational monitoring. Use of MHCL-1 by operational and network divisions involved in both operational and surveillance monitoring of surface waters, increases the effectiveness of such control by several times.

MHCL-1 shows the highest effectiveness while being a part of the automated systems of observation and monitoring (ANKOS-VG), adjusted for the operational inspection of watercourses (with a flow rate of 5-100 m^3/sec). Mobile laboratories are not appropriate for controlling major rivers due to the difficulties in providing safety for hydrochemical (hydrobiological) and other sensitive works.

Tests of MHCL-1 as a mobile means of water control for the purposes of operational and surveillance monitoring have proved its efficiency in cases of major anthropogenic accidents (disasters) with violent and large-scale impact on the natural environment, including surface waters. MHCL-1's methodological support allows it to carry out operational investigations of volley discharges that are generally made by enterprises in heat and power engineering, ferrous metallurgy, utility, agriculture and a few other factories, though the laboratory's capabilities are limited in cases of enterprises involved in petrochemistry, coke chemistry and organic synthesis associated with the contamination of surface waters by highly toxic organic substances, oil products, metal ions, etc.

New domestic developments (including those made by the Hydrochemical Institute) in methodological support, technical means of measurement and auxiliary equipment designed for the operational, express, on-site measurement of high concentrations of pollutants, have made it possible to overcome the foregoing deficiencies and expand the range of parameters controlled by mobile laboratories.

Operational efficiency of mobile laboratories may be raised dramatically by revising and replacing the current measuring devices and equipment with modern, certified, analytical technical means that are widely available on the market and produced by leading instrument engineering companies

Prospective models (the upgraded MHCL-1M and new MHCL-2) of mobile hydrochemical laboratories are designed to use metrologically certified methods developed by the Hydrochemical Institute for determining a number of priority pollutants: ions of copper, lead, cadmium, chromium (VI), total chromium, aluminum, iron, synthetic surfactants, phenols, oil products and acutely toxic water.

Instrumentation of the laboratory includes the following new portable devices:

- Ion meter with a digital indication and a set of ion selective electrodes
- Digital pH-meter
- Oxygen analyzer
- Photocolorimeter-densitometer, PFKD-1000, implementing methods of photometry for the quantitative measurement of pollutants and densitometric measurement of a range of heavy metals and several organic compounds, including express measurement of hydrogen sulfide
- Block for acutely toxic water
- IR photometer (analyzers of oil products) for measuring total concentration of oil products
- Aqua-Test express laboratory
- A set of test papers with URB-1 for pumping water samples under pressure, etc.

The development of important stages of sampling (conservation, storage and transportation of samples to permanent laboratories) improves the quality of the obtained results due to the general decrease in error in the measurement of the fractional analysis of polluted waters.

7.4. Permanent Hydrochemical Laboratories

A permanent hydrochemical laboratory is one of the main components of the ANKOS-VG system and is intended for the detailed (complete) analysis of water samples under an extended program providing identification and determination of concentrations for specific pollutants, the on-site measurement of which is technically impossible or unreasonable for economic or other reasons.

Mobile hydrochemical laboratories are used to take samples of water, bottom sediments or biological material from water bodies and, if necessary, they control and retain samples from automatic control stations, even looking after the transportation thereof to permanent (hydrochemical or hydrobiological) laboratories.

A permanent hydrochemical laboratory, as well other information and measurement components of the ANKOS-VG system, should conform to a higher level of requirements in order to provide topnotch analytical support, which includes a complex of metrologically certified methods for the measurement of overall indices and specific substances, technical means used for these methods, auxiliary devices and equipment (including equipment for storage, initial processing and preparation of samples for analysis), laboratory ware, chemical reagents and other expendables.

The diversity of controlled physical parameters and the chemical composition of waters (inorganic and organic substances) determines the necessity to use practically the whole arsenal of modern analytical instrumentation, in particular, photocolorimeters, spectrophotometers (for UV, visible and IR ranges), spectrofluorimeters, atomic absorption spectrophotometers, ion meters, polarographs, gas chromatographs, different special analyzers based on advanced/combined methods and auxiliary equipment.

The operational monitoring of surface waters within the framework of the ANKOS-VG system is associated with the acquisition and processing of a great amount of information content that is supplied by the simultaneous measurement of a significant number of hydrochemical and other parameters at short time intervals.

The solution to such an intricate problem under conditions of prospective expansion of the system's capacities, the tightening of requirements to express measurements and the increase of their accuracy and reliability is possible only on the basis of further instrumentalization and automation of the whole process of natural water analysis.

The foregoing versatile problem led to the establishment of new and improved existing methods and corresponding instrumentation for automatic and computerized control carried out on-site and in permanent and mobile laboratories.

Parameters controlled by the automatic stations, mobile laboratories and permanent laboratories were determined on the basis of the developed and approved priority list of pollutants and water quality indices (consisting of 131 items) subject to control by divisions of the Roshydromet via the ANKOS-VG systems and in accordance with the analysis of their methodological support.

The assessment of prospects for the development of essentially new methodological support systems and the establishment on the basis thereof of automatic technical means shows that the major part of substances from the priority list is now detected and can be determined in the future in permanent laboratories with the use of complex equipment.

The list of parameters controlled by automatic stations and mobile hydrochemical laboratories may be determined by the customer's requirements and may vary within the ranges of 25-40 and 30-40, respectively.

Thus, the necessity to computerize water analysis demanded the improvement of already existing equipment for the purpose of enhancing their metrological characteristics and operation reliability, and in order to establish essentially new technical means (different types of analyzers, measuring blocks, equipment complexes and auxiliary devices) designed for e use in permanent laboratories and other components of control systems.

The major objectives for automated water analysis include:

- Reduction of analysis costs (labor, time, material)
- Increase of data reliability (measurement accuracy) on account of parallel measurements
- Unification of metrologically certified methods and technical means
- Increase of obtained information, if necessary

As is known to us, the cost of one manual measurement, depending on the applied method, is 3-10 times higher than the cost of an automatic measurement. The cost of information obtained by SHCL or MHCL is 6-8 times higher than that obtained by ACS. With the apparent economic advantage of solely using ACS in the ANKOS-VG system, we should consider the fact that the automatic station is the most expensive choice, while its information capabilities are quite limited at the present stage. The acquisition of necessary (by volume and list of controlled parameters), express and detailed information to solve the operational objectives of ANKOS-VG may only be ensured by considering all system components as a whole.

A vital part of the strategy of automatic water control includes field tests aimed at the establishment of new and improved versions of existing instruments. Meanwhile, the necessity to make an on-site, estimated analysis of water samples has determined the importance of express (rapid) semi-quantitative (test) methods for measuring pollutants with high (> 10 MPCs) concentrations. As a result of conducted research, a complex of express methods was developed. This complex satisfies the requirements of simplicity and quickness of measurements while pairing them with high sensitivity and selectivity in the presence of a considerable amount of accompanying inorganic and organic substances. The foregoing complex includes methods for detecting copper, zinc, iron, cadmium, mercury, lead, chromium, phenols, anionic and synthetic surfactants, oil products and other pollutants, as determined by mobile hydrochemical laboratories, observation points and posts.

Modern physicochemical methods of potentiometry (ionometry), spectrophotometry, inversive voltamperometry, chromatography, etc., are used to automate chemical analysis and establish highly productive analyzers. Methodological support elaborated by the Hydrochemical Institute in cooperation with joint contractors consists of a large list of metrologically certified methods, standard samples and test means for measuring different water parameters. The methods intended for further automation of water chemical analysis must satisfy the following requirements:

- Simple analytical procedures and acquisition of an initial signal proportional to the concentration of the determined component
- High sensitivity, accuracy and selectivity
- Metrological provision
- Economic feasibility of technical means based on these methods

While selecting a method, the following points must be examined in detail: all operations and analytical procedures, their sequence, the assessment of labor intensiveness and opportunities for simplification, improvement, replacement, etc.

All analysis operations may be divided into three conventional groups:

1) Selection and conservation of samples, completion of documents and their transportation to permanent laboratories (consumption of time up to 40%)
2) Preparation of samples for measurement: filtering, dosing, extraction, precipitation, etc.; taking samples for analysis, injection of reagents, mixing, thermostatting; preparation of reagents, standards, etc. (consumes up to 40-50% of the time)
3) Measurement and recording of signals and processing and registering data (up to 10% of the time)

The automatic measurement of different water components along with the application of the aforementioned methods shows that the analysis of water samples may include a considerable number of operations and analytical procedures. For example, measurements involving widely used photocolorimetric methods consist of 28 consecutive operations, 25 of which may be automated. The design of such an analyzer is a complex technical challenge.

The contribution of each operation (in accordance with a manual algorithm) to the total cost, its comparison with the estimated cost of automatic measurement, the opportunities to reduce expenses on operations related to the sampling and transportation of samples on account of automatic samplers and the reasonable planning of a work program and route for a MHCL allows us to draw a conclusion on the opportunity to automate photocolorimetric measurements made in the laboratories. The most reasonable method was to use a two-three automatic, complimentary devices, which automated the process of analysis by up to 95-98% of the total scope of work.

The experimental model of such an analyzer, with automatic operation from 3rd to 26th position and a performance of 60 measurements per hour, was designed by the Hydrochemical Institute. A prototypes of a discrete autoanalyzer of surface waters, APV-102, with an automatic operation from 5th to 26th position and a performance of 100-120 measurements per hour, was developed at the end of the 1970s in cooperation with Analitpribor, Trade Scientific and Production Association. In comparison with Tekhnikoy analyzers, APV-102 provided higher performance and measurement accuracy.

Another important aspect of automatic water pollution control was the design of auxiliary equipment. The state of methodological and technical support in the preparation and auxiliary procedures for chemical analysis affect the operational control and reliability of obtained data to a great extent. Patented and scientific research has shown that difficulties in the preparation of samples for analysis stemmed from filtering large volumes of water samples from areas that are very near a water body. Filtering 1.0-5.0 l of water takes from 2 to 6 hours. In cases of high or extremely high pollution, filtering time for express analysis in permanent or mobile laboratories of an ANKOS-VG system is slowed down by 5-10 minutes with a total duration of all analysis operations of 25-30 minutes on the high end.

The time loss of laboratory staff associated with auxiliary operations equals approximately 25-40% from the total labor hours, which explains the interest in minimizing such time losses. In Russia, devices for the initial preparation of water samples were not given proper attention. The Hydrochemical Institute suggested a complex solution with some urgent points directly addressing the foregoing challenge. On the basis of pilot studies and analysis of methodological and technical solutions, it was possible to design auxiliary equipment (including methodological support and design documentation) for the following purposes:

1) Filtering samples
2) Photochemical destruction of dissolved organic substances
3) Gas extraction of highly volatile and volatile organic substances dissolved in water
4) Punching water samples through reactive test papers

The developed complex of auxiliary equipment intended for use in permanent and mobile laboratories reduces labor and time costs and decreases distortion of the chemical composition of water and its measurements (heavy metals, total nitrogen, sulfur, phosphorus, etc.) and solves the issue of detecting highly volatile and toxic organic substances (for example, chlorohydrocarbon).

The design of automatic analyzers is a leading trend in the strategy of computerized control of surface waters pollution in the framework of the ANKOS-VG system.

Due to the fact that we do not have an opportunity either to examine or name all of the developed technical means for automatic control of water as applied to ANKOS-VG, we shall describe only the most remarkable products.

Among analyzers and devices designed for permanent and mobile laboratories, we should note a portable complex analyzer of the KAP-105 type with a submerged (up to 15 m) sensor for measuring up to nine parameters. Additionally, there is a special complex of equipment, KAV-1, designed for sampling, initial processing and on-site storage of samples, and for preserving water composition during the transportation of the samples to a permanent laboratory. The laboratory potentiometric analyzer, LPA-1, comes with computer control and data processing that includes the preliminary automatic preparation of samples, allowing it to detect any of four parameters from the following list in a mode of discrete, continuous control through the use of the ion selective electrodes: pH, Eh, pCl^-, pNO_3^-, pF^-, pK^+, pNH_4^+ and pCa^{2+}.

The inversive voltamperometric analyzer, IVA-1, is designed to measure microconcentrations of copper, lead, cadmium and zinc ions in surface waters through an integrated graphitized electrode without a preliminary concentration or removal of dissolved oxygen. The U-III device for measuring organic carbon is based on the photochemical decomposition of organic substances to carbon dioxide with chemical oxidation and further coulometric detection.

The portable combined water control autoanalyzer, AKB-106, is an essentially new model, which excels over some foreign models (U-7 Khariba, Japan) in its number of detected components and their metrological characteristics. The analyzer uses potentiometric, voltamperometric, turbidimetric, tensometric, conductometric and thermistor methods of analysis. The device is designed to determine the physical properties and activity of ions. Together with its primary functions (automatic measurement in different modes), the autoanalyzer is capable of carrying out additional operations, for example, calibration, control, etc. The main characteristics of the AKV-106 analyzer are compared with those of KAP-105 in Table 7.7.

Specifications and capabilities of AKV-106 are higher compared with KAP-105. Use of AKV-106 increases operational control by 20-30 times.

The absence of Russian analogs for foreign autoanalyzers, like Tekhnikoy, and flow-injection autoanalyzers, like FIA STAR, along with the high demand for their use in operational monitoring have, initiated development in this direction.

A breadboard model of the laboratory flow-injection autoanalyzer, PIA-Start, was designed by the Hydrochemical Institute. The analyzer is based on a modular principle and includes an automatic sampler, pump and dosing device, reactor and a block for photometric, ionometric and voltamperometric detection.

The application of flow-injection autoanalyzers reduces the number of samples required for analysis, time of sample processing and analysis and decreases the consumption of reagents (by ten or even hundred times), providing herewith increased measurement accuracy. Connection of such a high-performance autoanalyzer with a computer provides a modern technology for the acquisition, processing, storage and output of information in an automatic mode.

The portable photocolorimeter-densitometer, PFKD-1000, is based on photometric and densitometric methods of pollutant detection in permanent and mobile hydrochemical laboratories. The device is designed for reactive test papers and plates with an attached layer of an absorbent like Siluphol for thin-layer chromatography. Methodological support developed for PKFD-1000, including metrologically certified methods, makes quantitative measurements of iron, copper, zinc, phosphates, nitrites, silicates, phenols, ASS, oil products, etc. PKFD-1000 is equipped with a digital processing device that outputs a value of concentration with due account of preset coefficients from a calibration chart.

Aqua-Test Express Laboratory is a new modification of the Autonomous Hydrochemical Laboratory for Water Express Analysis developed by the Institute of Water Chemistry in 1990. The laboratory is designed to monitor surface water pollution and may be used in field conditions directly near a water body and in a permanent/mobile laboratory.

Aqua-Test Laboratory consists of portable devices, a set of auxiliary equipment, a set of laboratory ware, expendables and containers.

Methodological support of the foregoing laboratory includes over two dozen ionometric, photometric, densitometric and titrimetric methods for detecting priority pollutants with a measurement error. Table 7.8 contains the list of measured parameters and the ranges of their concentration.

Table 7.7. Comparison of main specifications of AKV-106 and KAP-105

Parameter	Measurement range		Limit of absolute measurement error	
	AKV-106	KAP-105	AKV-106	KAP-105
pH	0—14	from 4 to 10 un.	±0.1 from 4 to 10 ±0.3 from 0 to 4from 10 to 14	±0.1 un. pH
pNa^+	0—4	from 0 to 3 un. at (pH-pNa) 3	±0.1 from 0 to 3 un. ±0.3 from 3 to 4 un.	±0.1 un. pNa
pCl^-	0—4	from 0 to 3.5 un.	±0.1 from 0 to 3.5 un., ±0.3 from 3.5 to 4 un.	±0.1 un. pC1
Eh, mV	-150...700	from -700 to +700	±10 mV	±15
Turbidity	from 0 to 50 mg/l from 0 to 500 mg/l	from 0 to 250 mg/l		±12.5 mg/l
Spec. elect. conductivity	from 1 to 0.01 S/m from 0.1 to 0.001 S/m	from 0 to 1 S/m	±10 %	±0.4 cm/m
O_2	from 0 to 20 mg/l	from 0 to 15 mg/l	±0.5 mg/l	±1 mg/l
T	from 0 to 40°C	from 0 to 40°C	±0.5°C	±5°C
Depth	from 0 to 15 m	from 0 to 15 m	±0.5 m	±0.5 m

Photometric and densitometric methods are implemented by the portable photocolorimeter, PKFD-1000, and IR-photometer (for oil products). Amperometric measures of BOD_5 are carried out by a BOD tester. Ionometric methods with application of ion-selective electrodes equip some devices with a self-contained power supply, either a pH-meter (millivoltmeter pH-150) or Ecotest-01 analyzer to measure specific electric conductivity, pH and water temperature.

The Hydrochemical Institute has developed a multilevel scheme and conditions for the measurement of total content of oil products in water with chromatographic separation from coextractive substances of hydrocarbons and their further measurement analyzers of oil products, which may be used in permanent and mobile laboratories.

In the long-term, after modification and complement of the device with a special peripheral unit, the range of its application may be extended to the detection of total content of organic substances (analogous to TOC (Total Organic Carbon) in national standards of several countries). Aqua-Test Express Laboratory provides representativeness of measurement data, speedy data acquisition and improved labor conditions.

In those same years, the Hydrochemical Institute has also developed a permanent, self-contained hydrochemical laboratory intended for adoption by Roshydromet's division due to the development and certification of new networks of SSMEC laboratories. For the first time, a modular approach was used for the laboratory's equipment, with modern analytical, technical means; auxiliary laboratory equipment; ware; reagents and expendables for the implementation of certified methods for detecting priority pollutants in accordance with the requirements of Roshydromet and the Russian Federal Standard. This list for Roshydromet's network of laboratories includes more than 100 Methodic Papers in order to ensure accurate monitoring of surface water pollution in the Russian Federation.

Table 7.8. List of parameters controlled by Aqua-Test Express Laboratory

Parameter	Method of analysis	Certified range of measurement, mg/l
Ammonium	Ionometry	0.40—18.0
ASS	Photometry	0.010—0.40
BOD	Titrimetry	1.0 (> mgO_2)
Hydrocarbonates	»	5—500
Total iron	Densitometry	0.010—0.10
Hardness	Titrimetry	0.4—8.0 mmol/l
Potassium	Ionometry	2.0—390
Calcium	»	4.0—4,000
Copper	Densitometry	0.001—0.010
Sodium	Ionometry	2.30—2,300
Nitrates	»	0.620—6,200
Nitrites	Photometry	0.005—0.30
Oil products	IR-photometry	0.040—2.0
pH, æ	Ionometry	2—12
Transparency	Visually	
Silicate	Photometry	0.10-1.60
Sulfates	Turbidimetry	0.50—50
Phenols	Photometry	0.002—0.25
Phosphates	»	0.005—0.20
Fluoride	Ionometry	0.190—1,900
Chloride	»	11.0—3,500
Chromium (VI)	Photometry	0.001—0.030 0.010—0.10
Chromaticity	»	0—70 degrees
Zinc	Densitometry	0.010—0.10

7.5. DATA PROCESSING CENTER

The data processing center, as an information and management component of ANKOS-VG, includes a dispatching post, computer unit, laboratory of metrological support and services for all technical means comprising the system.

The technical means of ANKOS-VG's foregoing center simultaneously functions as an operational component of CEM by executing the following primary objectives:

- Organization of automatic acquisition of information from automatic stations in accordance with a preset program (on dispatcher's or programmer's command) on a real time scale
- Organization of automatic transmission of commands related to the management of separate components of automatic stations and reception of messages on the operation of their main parts
- Operational decoding and consistency check of information received from ACS, MHCL, SHCL (SHBL)
- Initial automatic processing of information and assessment of water pollution at the places where automatic control stations are installed
- Detection of trends in pollution level of areas controlled by ACS (CEM), along with the detection of high and extremely high pollution in the controlled areas and between them
- Automatic signaling to ANKOS-VG (CEM) if high or extremely high pollution is detected or forecasted and in case of station failure (fire, shutdown, etc.).

A dispatching post equipped with a computer requests measurement data from automatic stations via telephone network and from mobile hydrochemical laboratories via telephone radio channel. If the connection is established, data communication equipment is used for reception and buffering the storage of obtained results. Processing of an incoming message includes: checking and recovering (if required) the format, conversion a message from transmission format to processing format, conversion of measurement units, consistency check and control of service parameters describing the state of the system's technical means.

Data from mobile and permanent hydrochemical laboratories is received by a computer system as frequently as once a day. If necessary, the system is capable of receiving hydrological and meteorological information, as well as data related to pollution control of other environments.

In accordance with software settings, the obtained information undergoes reprocessing, which serves as a source for detection of changes in a water body, volley discharges of waste waters, assessment and operational, short-term forecasting of water pollution. Parameters of a controlled water body are displayed on a mnemocircuit. All obtained information is stored on a magnetic medium for the required period of time. Documentation may be further printed and transmitted to consumers via telecommunications.

References

[1] Stradomsky V.B. Issues of Automatic Control of Surface Water Composition // Hydrochemical Materials. Vol. LXIII. — St. Petersburg: *Hydrometeoizdat,* 1976. – P. 3—12.

[2] Environment Protection Equipment: Nomenclature Reference. — M.: *CNIITEIL Publishing House,* 1978.

[3] Stradomsky V.V., Zavesa M.P. Development of Instrumentation for Computerized Analysis of Surface Waters // Hydrochemical Materials. Vol. LXXXIV. — St. Petersburg: *Hydrometeoizdat,* 1983. – P. 3—25.

[4] Belousov A.P., Stradomsky V.B., Mukhin V.M., Tsirkunov V.A. Complex Portable Devices for Control of Water Pollution // Hydrochemical Materials. Vol. XXXIV. — St. Petersburg: *Hydrometeoizdat,* 1983. – P. 77—85.

[5] Stradomsky V.B. State and Prospects of Automatic Control of Surface Water Chemical Composition // *Hydrochemical Materials.* Vol. XCII. — St. Petersburg: *Hydrometeoizdat,* 1984. – P. 26—36.

[6] Operations Manual for Mobile Hydrochemical Laboratory MHCL-1, PU 1.570.004 RE. — St. Petersburg: *Hydrometeoizdat,* 1985. —35 p.

[7] Belousov A.P., Nikanorov A.M., Zavesa M.P., Gekov V.F. Current State and Prospects of Ground Operational Automatic Control of Surface Water Pollution // *Modern Issues of Regional and Applied Hydrochemistry.* — St. Petersburg: *Hydrometeoizdat,* 1987. – P. 206—216.

[8] Kimstach V.A., Vinnikov Y.Y., Stradomskaya A.G. State and Prospects of Analytical Control Methods of Surface Waters // *Hydrochemical Materials.* Vol. CV. — St. Petersburg: *Hydrometeoizdat*, 1989. – P. 3—13.

[9] Kimstach V.A., Nikanorov A.M. Methodological Principles of Surface Water Operational Control // *Hydrochemical Materials.* Vol. C. — St. Petersburg: *Hydrometeoizdat,* 1991. – P. 3—13.

[10] Khoruzhaya T.A., Bryzgalo V.A., Predeina L.M., Korotkova E.I. Methods of Bioindication and Biotesting in Studies of Freshwater Ecosystems under Accidental and Extremely High Pollution // *Hydrochemical Materials.* Vol. C. — St. Petersburg: *Hydrometeoizdat,* 1991. – P. 47—68.

[11] Belousov A.P., Zavesa M.P., Muzykov A.G. Use of Automated Systems for Control of Surface Water Pollution (ANKOS—VG) for Operational Monitoring. // Hydrochemical Materials. Vol. C. — St. Petersburg: *Hydrometeoizdat,* 1991. – P. 69—78.

[12] Belousov A.P., Avdeev V.N., Rybnikov V.G., Muzykov V.G., Bystrova L.F., Alafyev S.V. Combined Freshwater Analyzer AKV-106 // *Hydrochemical Materials.* Vol. C. — St. Petersburg: *Hydrometeoizdat,* 1991. – P. 136—146.

[13] Stradomskata A.G., Lyapkina N.S. Express Detection of Oil Products in Water Bodies // *Hydrochemical Materials.* Vol. C. — St. Petersburg: *Hydrometeoizdat,* 1991. – P. 147—151.

[14] Vinnikov Y.Y., Anokhin S.V., Boeva L.V. Chromotomass-spectrometric Measurement of Qualitative and Quantitative Composition of Organic Substances in case of Accidental Discharges intro Water Bodies // *Hydrochemical Materials.* Vol. C. — St. Petersburg: *Hydrometeoizdat,* 1991. – P. 172—183.

[15] Muzykov A.G., Belousov A.P., Rybnikov A.G., Tsirkunov V.A., Zavesa M.P. Experience of Application of Mobile Hydrochemical Laboratory (MHCL-1) in Operational Monitoring of Waters // *Hydrochemical Materials.* Vol. C. — St. Petersburg: *Hydrometeoizdat,* 1991. – P. 79—84.

[16] Regulatory Document 52.24.464—95 Practical Guidelines. Densitometric Measurements of Zinc in Water with Reactive Test Papers. Introduced on 01.01.95. — Rostov-on-Don, 1995.

[17] Regulatory Document 52.24.465—95 Practical Guidelines. Densitometric Measurements of Copper in Water with Reactive Test Papers. Introduced on 01.01.95. — Rostov-on-Don, 1995.

[18] Regulatory Document 52.24.465—95 Practical Guidelines. Densitometric Measurements of Iron in Water with Reactive Test Papers. Introduced on 01.01.95. — Rostov-on-Don, 1995.

[19] Federal List of Measuring Methods Authorized for Use in the Field of Monitoring of Environmental Pollution. — Moscow, 1996. – P. 16—36.

[20] Nikanorov A.M. Hydrochemistry — St. Petersburg: *Hydrometeoizdat,* 2001 — 444 p.

Chapter 8

SPECIAL KINDS OF OBSERVATIONS IN INLAND SURFACE WATER MONITORING

As previously stated, the inland surface water monitoring system includes so-called *special observations* arranged to resolve water quality problems besides those that relate to surveillance and operative observations. Moreover, the tasks of the special observations minimize their peculiarities for each specific case. These observations can be carried out with both a limited and unlimited validity term. The observations held in 1980s, aimed at the research concerning self-purification of water bodies located at the different regions of the country, may serve as an example of the former type. The special observations currently held within the framework of the State Monitoring of Russian Waters Quality have no limited validity term. The following observations may be referred hereto as observations of the base line state of water bodies, observations aimed at the definition of evacuation of hazardous substances into the seas, observations aimed at the quality evaluation of transboundary surface waters and transfer of polluting substances across the frontier, general reference observations, reference observations aimed at the definition of organochlorine pesticides, observations of Lake Baikal's state and observations in line with the points provided by the Global Environment Monitoring System (GEMS). Each kind of special observation is based on the network of points currently attached to the modal observation points [1].

This comparison of points applies the information received during modal observations with a higher degree of efficiency. Nevertheless, the indices pertaining to space, time and components of the modal point's network do not always satisfy the requirements of the special observation, and are in need of improvement.

Activities within each kind of observation are regulated by special documents. The following may be taken as examples: Guidelines to administrations of Hydrometheorological Service as of arrangement and carrying out the observations of surface inland waters pollution [2] (investigation of self-purification processes), Methodological guidelines for arrangement and carrying out of observations over river pollution with discharges from agricultural lands and evaluation of evacuation of salts, fertilizers and pesticides therewith [3], Methodical guidelines for calculation of evacuation of organic, biogenic matters, pesticides and trace elements with river flow [4], Guidelines for rationale of monitoring system and calculation of river flow evacuation of oil products [5]; Transboundary surface waters state monitoring subsystem [6] and Working schedules of monitoring of Lake Baikal state. Observations made at the points listed from the Global Environment Monitoring System are held in accordance with the norms of RD 52.24.309—92 [1] under a possibly more precise observance of the GEMS Guidelines [7]. The documents presented refer to the activities of the special observations held by the network both in the past and currently. There are also a range of new documents concerning the arrangement of observations in the oil, gas and natural gas liquid fields development [8, 9] and on observations in the coal mining districts and concurrent industries. Several other documents are continuing to be developed.

This chapter further explores the special observations that serve as the basis of the subsystem for monitoring the state of transboundary surface waters (SMSTSW), the National Surface waters quality monitoring Subsystem within GEMS, the subsystem for anthropogenic eutrophication monitoring of fresh waters and the subsystems for local monitoring of water contamination in impact regions.

8.1. Transboundary Surface Waters Monitoring

The human community has since long understood that the problem of environmental pollution has no limits. To a great extent, this refers to transboundary waters. The "Convention on the protection and use of transboundary water sources and international lakes (Helsinki, 1992)," [10] underlines that the protection and use of transboundary water bodies are important and urgent tasks, the effective accomplishment of which can only be ensured by means of enhanced international cooperation. Moreover, the Convention:

- Defines transboundary waters as those surface or ground waters that mark, cross or are located on boundaries between two or more states
- Gives special attention to river basin management
- Identifies the principal purpose of monitoring, which is to ensure the retrieval of sufficient and reliable information to manage river basin resources
- Mentions the necessity of all riparian countries' participation in the definition of aims, monitoring programs and measures of water resources management

In Europe, the transboundary waters and the numerous tributary flows thereof are used intensively for navigation, industrial needs and agriculture, power production, recreation and sewage discharge. Untreated or badly treated domestic sewage, chemical pollution caused by the activity of industrial and agricultural enterprises, seepage from old and new waste deposits, accidental pollution and incoming pollutants with precipitation constitute a threat to the state of transboundary waters and their corresponding water ecosystems.

Given the importance of this issue, the ECE target group for monitoring and assessment has, in its first step, developed the guiding principles of water monitoring and quality assessment for transboundary rivers. This became the basis for an *international instrument*, known as the "Protection of Transboundary Waters [11]."

These guiding principles are based on the following approaches:

- All riparian countries are to take part in monitoring measures
- Monitoring is an element of river basin water resources management

- A river basin shall be considered part of a complex with tidal firths
- Observations are to be held in their complexity (qualitative and quantitative parameters of water, flora and fauna, pollution sources and consequent circulation of polluting matters)
- Information can be retrieved from primary sources, and the main sources shall be the results from the national monitoring systems
- The monitoring process shall be viewed as a consequence of interconnected operations, starting from the definition of information needs up to the utilization of information product
- The monitoring process is improved with respect to the information retrieved and appearance of new information needs

The following structure of the guiding principles has been introduced.

Definition of problems. Taking the water management activities into account, the specific problems of river basins are defined.

Information needs. Based on the problems defined, the information needs to solve specific problems and the achievement of specific monitoring aims is identified. The monitoring aims may be the following: assessment of an actual river basin state, control of compliance with the conditions foreseen by the discharge permits, control of pollution-fighting measures efficiency, provision of timely signalization for accidental pollutions, profound study of separate problems, etc. This specialization in aims has to first and foremost identify what the collected information is needed for (e.g., for decision-making process).

Strategy of monitoring and state assessment for water bodies. This strategy defines the monitoring approach and the programs thereof. It foresees a holding of three kinds of monitoring (water bodies monitoring, early warning monitoring and pollution sources monitoring), a complex assessment by means of a three-component approach (physical and chemical analysis of water, suspended particulate matter, sediments, and organisms; ecotoxicological assessment with application of bioassay and biological methods of early warning and biological research) and the integration of monitoring activities for the purpose of economic efficiency.

Monitoring programs. Taking the chosen strategy into account, the following is then identified: a sampling site; definition parameters, including hydrological parameters; recurrence and terms of observation; methods of sampling and analysis of samples; methods of sample transportation and storage and after that, the samples are taken, delivered and analyzed.

Data management. The data retrieved within the framework of the monitoring programs shall be certified, archived and made available. This data is processed in order to be transformed into materials satisfying the needs of specific users. Data exchange is done in a developed exchange format. Data communication is the final stage of data management and links the collection of information with its communication to the users. The materials for distribution shall be prepared regularly.

Monitoring quality management. This shall render efficiency to the system (an extent to which the materials retrieved from the system can satisfy the need for information) and effectiveness (retrieval of information at minimal financial and personal costs), as well as the traceability of processes and events (registration and awareness of the processes and events allows measures to be taken for the improvement thereof). Monitoring quality management is based on the overall policy developed by a collegiate body established by the riparian parties; this body determines the necessary quality level. Within the framework of a quality provision system, it is necessary to show in the documents agreements on processes and products of each element of the monitoring cycle; on requirements for information materials; on the utilization of standards regarding methods and means of sampling, transportation, storage and analysis of samples; on data certification and storage; on ways of calculation and on statistical methods. The implementation of international standards is highly desirable. The implementation of national standards is permissible, but these should provide for the retrieval of compatible data by riparian countries.

In the Russian Federation, transboundary surface water monitoring is effected in two ways: monitoring of separate river basins (with the assistance of the neighboring states) and monitoring of frontier area water sources (within the system of the Russian Federal Service for Hydrometeorology and Environmental Monitoring). In both cases, significant attention is paid to the observance of the fundamental propositions of the international document. In the former, the observations are held according to the propositions set forth by the agreements between the parties. In the latter case, the observations are regulated by departmental document RD 52.24.508—96 [6] which contains the following provisions:

- The subsystem for monitoring of state of transboundary surface waters (SMSTSW) is included in the System of the Surface Water Monitoring (SSWM) as a special unit.
- Within the SMSTSW, the water bodies and watercourses crossing the borders of neighboring states (e.g. Irtysh river) or determining such borders (e.g., Narva), and those located on the border (e.g., Lake Chudsko-Pskovskoye) are all monitored.
- The tasks of the SMSTSW are to support the observations over the state of the transboundary surface waters, assessment of the state thereof, assessment of the pollutants and suspended sediments transported across the state border by the river waters, timely announcement in case of high pollution levels indication, determination of tendencies for alterations in aqueous run-off volume and content of chemical elements, suspended sediments and the time travel thereof and provision of information to the users.
- The subsystem is based on a network of regime observation points and hydrological stations of the Russian Federal Service for Hydrometeorology and Environmental Monitoring, SSWM; these stations are located directly on the border or in its proximity.
- The Hydrologic Service and the Russian Federal Service for Hydrometeorology and Environmental Monitoring , SSWM conduct observations, gather, process and store data according to the principal normative documents of these services [1, 12—14], as well as pursuant to other documents which help to fulfill the provisions mentioned therein.
- The peculiarities of the SMSTSW as compared to SSWM may reveal themselves in the observation programs (the calculations, the number of sampling sites and the frequency of sampling may be higher/more frequent than those made in the course of regime observations) and are reflected in the generalization of the retrieved data (in addition, the assessment of the state in terms of water mass transport, chemical matter transport and suspended sediment transport is calculated, as well as any discrepancy of transport and the anthropogenic element of transport).

- The across-the-border transport of chemical elements is calculated with the following formula.

$$G = \sum_{i=1}^{m} w_i \bar{c}_i, \tag{1}$$

where G is matter yield, m is the number of base periods, w_i is the aqueous run-off volume, i is the base period and c_i is the arithmetic mean of concentration within the base period i. The calculations are made for one year or for a number of years if the annual mean is sought. The annual means for the matters that have no relation between the concentration and water flow rate are taken as the base periods, and the annual periods with regard to the phases of a hydrological regime are accepted as base periods provided the aforementioned relation has been established.

Currently, there are transboundary surface water monitoring subsystems that include over 60 observation points. Approximately one half thereof foresee a possibility of chemical substances transport calculation [15].

The current data processing is affected automatically. Based on the information from the previous year, the sections for the Russian Federation Annals of surface waters quality and the Reviews of Environmental pollution in the Russian Federation, the information is prepared and the data on specific regions is processed for a multiannual period [15, 16].

8.2. National Program of Monitoring and Water Quality Assessment in the Russian Federation within the Global Environmental Monitoring System

The idea of establishment of a Global Environmental Monitoring System (GEMS) was expressed during the UN Stockholm Conference in 1972. The actual principles thereof were formulated in 1974 during a special assembly in Nairobi (Kenya) where the roles of the agency and the UN member countries were specified.

Beginning in 1977, the international program GEMS/Water has been implemented within the GEMS framework. This program is being carried out in cooperation with the World Health Organization (WHO); World Meteorological Organization (WMO); the United Nations Educational, Scientific and Cultural Organization (UNESCO) and the United Nations Environmental Program (UNEP). The global data center for GEMS/Water is located in Burlington, Canada.

The program's first stage objectives were:

- Cooperation with participating countries in the establishment of new systems for water monitoring and reinforcement of the existing systems
- Improvement of reliability and compatibility of data concerning quality of water within the participation countries and in the areas between them
- Assessment of pollution proliferation and long-term based forecasts of water pollution with specific persistent and hazardous chemical substances

In 1990, the decision to transition to the *second stage* of the program was made; at this stage, the stress was shifted from monitoring to data interpretation. The program objectives set for the future aimed to accomplish the following [7]:

- Distribution of data concerning the global assessment of fresh water quality as it is relevant to the issues of health care, the state of fresh water ecological systems and other crucial global problems connected to the environment, governments, scientists and society
- Distribution of data to governments, scientists and society concerning the transport and, if necessary, the assessment of nutrient flows and toxic chemical and other pollutants from the basins of the largest rivers on the boundary line between the ocean and continent
- Consolidation of the international monitoring networks for water quality in developing countries, including improvement of analytical possibilities and guarantees of data quality

A new network structure has also been offered, which includes 40-50 backing stations, 300-400 stations for trend detection and 60- 70 stations for examination of global river flows. Taking the new objectives into account, the third edition of the program guidelines have been prepared and published [7].

The GEMS/Water monitoring objectives for the third stage are as follows:

- Determination of natural, fresh water quality provided the absence of significant direct human impact
- Detection of long-term trends under the critical conditions of water quality
- Assessment of chemical substance flows from the basins of the largest rivers on the boundary line between the ocean and continent

A selection of the GEMS/Water stations is made taking into consideration a range of the principle criteria (Table 8.1.) A list of indices for the definition and frequency of sampling pursuant to recommendations are shown in Tables 8.2 and 8.3.

Table 8.1. Dislocation criteria for GEMS/Water stations

Station type	Main selection criteria	Water body type
Backing station	Minor intact basins; absence of pollution sources; absence of immediate human activity (including roads); absence of basins with widely represented metallic grounds; distance of at least 100 km from the main pollution sources (e.g., cities, industrial enterprises, etc.)	Upstream lakes (with tide changes within 0.5 to 2 years); upstream areas of the rivers
Trends detection station	Middle-sized basins; average response rate to pollution and alterations in land utilization; presence of a number of environmental pollution factors, (or) essentially with one leading kind of activity (e.g., industrial, housing, agricultural, mining pollution, etc.)	Lakes/water research (with tide changes within 1—3 years); rivers

Station type	Main selection criteria	
Global river flow examination station	The basin priority is defined by the importance of the river synodic, population, main kinds of human activities and the importance of the river under consideration for the adjacent water bodies; the station closest to the estuary shall not be caught in the tides; a station shall be specific for the river with respect to the water section; the drain measurement data shall be available at the location of a water quality monitoring station	Rivers

Table 8.2. List of parameters defined within the GEMS/Water Program

No.	Parameter	No.	Parameter
	General water quality		Dissolved salts
	indices	1	* Calcium
1	* Water drain/level	2	* Magnesium
2	* Suspended solid particles	3	* Sodium
	(R)	4	* Potassium
3	* Temperature	5	* Chlorides
4	* pH	6	* Fluorides (PV)
5	* Electrical conductivity	7	* Sulfates
6	* Dissolved oxygen	8	* Alkalinity
7	* Transparence (V)		Biogenic Matters
	Ion Balance	1	* Nitrates and nitrities
1	Cations, total	2	* Ammonia
2	Anions, total	3	Dissolved organic
3	Sodium adsorption coefficient		Nitrogen
	Organic Matter	4	Particulate organic nitrogen
1	Dissolved organic carbon	5	* Dissolved phosphor, total (R,V)
2	Particulate organic carbon	6	* Particulate phosphor, total
3	BOD	7	* Non-filtered phosphor, total
4	COD		(R,V)
5	* Chlorophyll "a" (R,V)	8	* Chemically active silicon earth (R,V)
	Non-organic Pollutants		Solid Particles
1	Dissolved aluminum	1	Particulate aluminum (G)
2	Aluminum, total	2	Particulate arsenic (G)
3	Dissolved arsenic	3	Particulate cadmium (G)
4	Arsenic, total	4	Particulate chrome (G)
5	Dissolved boron	5	Particulate copper (G)
6	Boron, total	6	Particulate iron (G)
7	Dissolved cadmium	7	Particulate lead (G)
8	Cadmium, total	8	Particulate manganese (G)
9	Dissolved chrome	9	Particulate mercury (G)
10	Chrome, total	10	Particulate selenium (G)
11	Dissolved copper	11	Particulate zinc (G)
12	Copper, total		Organic Pollutants
13	Dissolved iron	1	Aldicarb
14	Iron, total	2	Aldrin
15	Dissolved lead	3	Atrazine
16	Lead, total	4	Benzene

Table 8.2. (Continued)

No.	Parameter	No.	Parameter
17	Dissolved manganese	5	2,4-D
18	Manganese, total	6	Dieldrin
19	Dissolved mercury	7	Benzene hexachloride
20	Mercury, total	8	Carbohydrates, total
21	Dissolved nickel	9	Polynuclear carbohydrates
22	Nickel, total	10	PCB
23	Dissolved selenium	11	Phenols
24	Selenium, total		Microbial contamination
25	Dissolved zinc	1	Fecal coliform bacteria
26	Zinc, total	2	Coliform bacteria, total

Notes: The principal parameters subject to definition at all GEMS/Water stations are marked with an asterisk; (R)—basic parameters for river stations only; (V)—basic parameters for the lake/water research stations only; (PV)—basic parameters for ground water research stations; (G)—parameters of significant importance for global river flow research.

An important element of the monitoring system is solid particles monitoring (monitoring of suspended matters, tractional alluvia and bottom sediments). Nevertheless, this kind of monitoring is not spread wide enough mainly due to difficulty and high costs of methods of water phase and solid particles separation. Taking the economic possibilities into account, the GEMS/Water Program recommends that first the bottom sediment analysis be carried out. The GEMS principles in the former USSR were established by academician Y. A. Izrael [17]. Russia has officially had the GEMS/Water since 1992. The Hydrochemical Institute of the Russian Federal Service for Hydrometeorology and Environmental Monitoring (located in Rostov-on-Don) is the national center for the program's implementation. Twenty-seven sites out of the regime of the observation point's network within the surface water monitoring system were selected to be included in the RF National GEMS/Water subsystem within the GEMS/Water network. These points serve for assessment of:

- Anthropogenic impact (eight points)
- Evacuation of matters into the seas (eleven points)
- Background state of water bodies (eight points)

Table 8.3. Recommended sampling frequency for GEMS/Water stations

Station	Water Body	
	Rivers, streams	Lakes/Reservoirs
Backing station	*Min.*: 4 including high and low water levels *Optimal*: 24, i.e., the sampling shall be carried out daily every fortnight to determine the total amount of suspended particles	*Min.*: 1 under thermal turbulence (at drain points from the lake) *Optimal.*: 1 under thermal turbulence, and one vertical profile at the end of the stratification period
Trends detection station	*Min..*: 12 for large drainage basins (ca. 100,000 km^2) *Max.*: 24 for smaller drainage basins (ca. 10,000 km^2)	*Eutrophication research*: 12 incl. twice every month during summer *Other, Min..*: 1 under thermal turbulence *Max.*: 2, with 1 under turbulence and 1 under maximal thermal stratification

Station	Water Body	
	Rivers, streams	Lakes/Reservoirs
Global river flow research station	Large basins (over 200,000 km^2) (1); 6—for specific particulate metals (2); 12—for other indices. Smaller basins (less than 200,000 km^2) (1); 24—for main monitoring indices (3); 12—for biogenic matters, organic pollutants and specific metals (4); 6—for particles analysis (2)	

Notes:

(1) For global river flow research stations, keeping a continuous record of water flow is recommended, together with a weekly sampling in order to identify the total amount of suspended particles.

(2) For particulate arsenic, cadmium, chrome, copper, mercury, lead, selenium and zinc.

(3) For temperature, pH, electric conductivity, dissolved oxygen, calcium, magnesium, sodium, potassium, chlorides, sulfates, alkalinity, nitrates, nitrites, total filtered and non-filtered phosphorus, silicon, chlorophyll "a," dissolved organic carbon and particulate carbon, and dissolved organic nitrogen and particulate nitrogen.

(4) For dissolved fractions and particles of aluminum, iron, manganese, dissolved arsenic, cadmium, chrome, copper, lead, mercury, selenium and zinc.

The national subsystem shall provide for:

- Observations in the points under a coordinated program
- Implementation of control over the quality of analytical definitions
- Operation and dispatch of data retrieved in the points of observations in an approved form to the GEMS/Water headquarters (in Burlington, Canada).

Within the framework of the Russian Federation national subsystem, the GEMS/Water program foresees the definition of the following water composition elements and properties: flow, temperature, electric conductivity, hydrogen index, suspended matters, biochemical oxygen consumption (acc. to BOC_5), ionic sodium, potassium, calcium, magnesium, carbohydrates, chlorides, sulfates, ammonia nitrogen, nitrite and nitrate nitrogen and total phosphorus, silicon, copper, zinc, iron, lead and mercury.

The recurrence in the terms of observations comply with the ones set forth by RD 52.24.309—92 for annual observations [1].

The observations at the points, analysis of sampled water, control over the quality of analytical definitions and the distribution of the data to the Hydrochemical Institute is done via the network branches of the Russian Federal Service for Hydrometeorology and Environmental Monitoring.

The Hydrochemical Institute bears the Russian function of a methodical center, exercising control over the analysis quality, control and correction of data retrieved [18] and the revaluation and dispatch of data to the GEMS/Water headquarters.

8.3. Monitoring of Anthropogenic Eutrophication of Land Surface Waters

The processes of development and transformation of the water ecosystems run significantly faster under modern conditions when compared to the past because they are no longer as dependent on the natural factors valid on the scale of geological time, but instead are affected to a larger extent by anthropogenic factors. The increase of the anthropogenic impact is accompanied by chemicals being drained from the continents into the seas and oceans, the spreading of surface waters acidulation across vast territories of the country due to the transboundary transport of sulfur and nitrogen and the increased concentration of these matters in the atmosphere. Among the global processes, the speed of such events has increased notably within the last decades; one can mention the process of anthropogenic eutrophication that expanded across many water bodies around the world causing the deterioration of land surface waters quality [19—21].

Under the term of *anthropogenic euthrophication*, an increase of biological productivity within the water bodies is understood to exist; it has been caused by the enrichment thereof with nutrients coming from human activities [22]. According to J. Odum, it is the eutrophication that determines the transition of a water ecosystem from one level of a dynamic homeostasis onto another that is initially connected with the state of the drainage basin and economic activities in the drainage basin area [23—24].

In general, *eutrophication is a process through which the water ecosystem transforms as a result of incoming biogenic and organic matters to a water body; the main system index of this process is the reorganization of structure and metastasis of a biotic community.*

8.3.1. Causes, Factors and Processes of Anthropogenic Eutrophication

At the present stage, the causes of eutrophication may be represented in the following scheme (Figure 8.1). The main pollution sources introduce organic and mineral compounds and thermal pollution into the water bodies. The major part of the pollutants forms the most unusual compounds, the cumulative effect of which is hardly predictable even with modern, state of the art technology. With this, matter with cumulative properties may either depress or stimulate the metastasis of water ecosystems depending on the strength of the “dose” and the duration of impact.

One of the main causes of increased eutrophication is the accumulation of biogenic matters in an aquatic environment, therefore, the regime of biogenic elements is viewed as a source of potential eutrophication, and the nature of a water biotic community is deemed to be an effect of the former.

The eutrophication of water bodies takes place not only due to incoming mineral compounds of phosphorus and nitrogen, but also due to incoming nitrogenous and phosphorous organic matters with industrial and domestic pollutants; these matters are directly consumed by bacteria [25].

Separate elements of the eutrophication mechanism can be united into a general scheme shown in Figure 8.2, which is based on currently developed ideas [26].

As a rule, the eutrophication process covers all the links of the trophic chains within the freshwater ecosystem. The alterations taking place in the biotic community are to the changes occurring in the habitat under the influence of physical and chemical factors. A diversity of methods of adaptation, in its turn, determines the significant flexibility of the biotic communities.

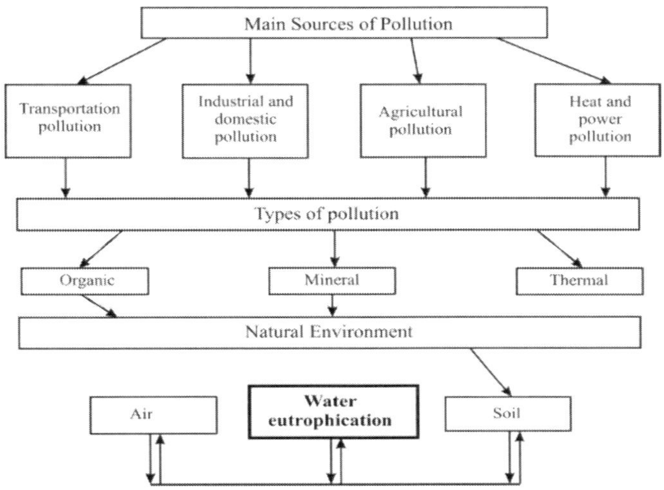

Figure 8.1. Scheme of anthropogenic eutrophication of surface waters.

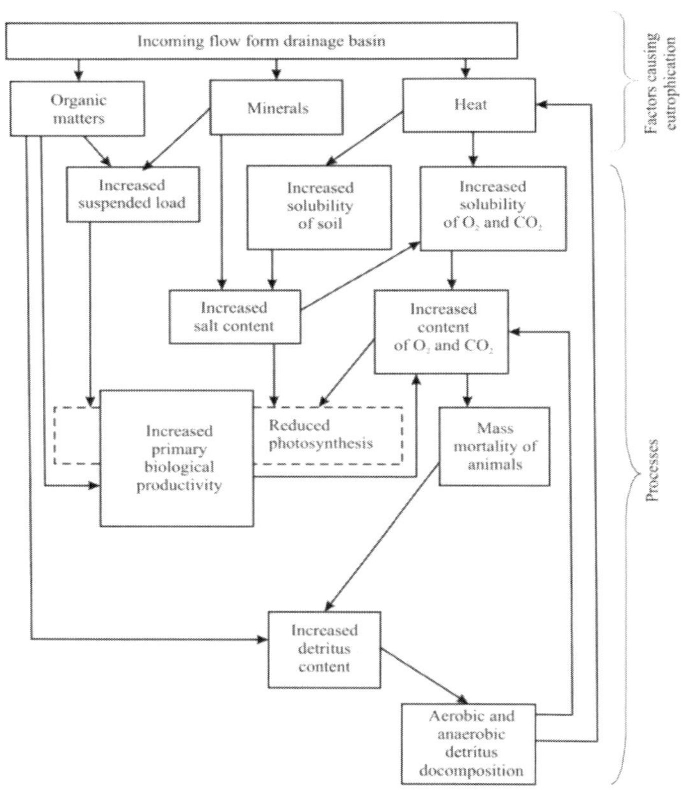

Figure 8.2. Factors and processes determining and accompanying anthropogenic eutrophication.

Research concerning the main ways to transform the structural and functional arrangement of ecosystems for the detection of the limits of permissible anthropogenic load on the ecosystems is of primary importance when it comes to finding solutions for the problems connected with water resources protection and management. One of the considerable difficulties in the assessment of the consequences of anthropogenic eutrophication of water bodies and drainages is the alteration of the ecosystem's functions; in the first stage, this alteration is almost imperceptible due to homeostatic mechanisms. In most cases, such dysfunction is hardly discernible from the natural variations in development of water ecosystems. The latter may include the inter-season and interannual deviations of hydrochemical, hydrobiological and hydrological processes, etc.

The assessment of anthropogenic eutrophication consequences in water bodies and the detection of the eutrophication scales on the background of a natural variability can only be identified on the basis of results from multiannual regime observations over hydrobiological, hydrochemical and hydrodynamic parameters held within the framework of the state monitoring system. A network of hydrobiological observations provides for the monitoring of biotic status (Figure 8.3) and is meant to trace the functioning peculiarities of specific aquatic organism communities on every trophic level, as well as to observe the alterations of status in separate water biocenoses under the influence of anthropogenic impact factors [19]. During the first 15-20 years, the hydrobiological network has gathered actual materials concerning the intra- and interannual spatiotemporal variability of qualitative and quantitative indices of planktonic and benthal communities of water bodies [19, 27—29]; these materials distinctly reflect the external traits of increased anthropogenic eutrophication. The water bodies accounted for include reservoirs (Invankovskoye, Cheboksarskoye, Gorky, Kuibyshev, Saratov, Tsimlyamsk, etc.), rivers (Lower Volga, Lower Don, Seversky Donets, Moskva, Klyazma, Oka, Pakhra, Samara, Yt, Syt, Kotorosk, Sanikhta, etc.) and lakes (Ladoga, Chudsko-Pskovskoye, etc.).

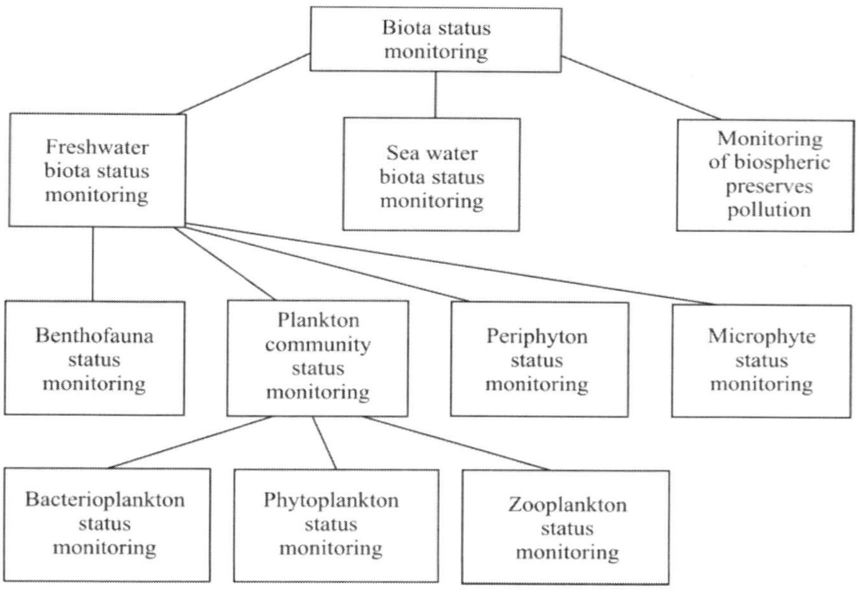

Figure 8.3. Structure of biota status monitoring in surface waters.

As is known, the level of organization within separate species and the communities thereof depends on the state of environment. In polluted environments, the nature of alterations in the biocenotic structural organization reflects the core of this correlation.

The enrichment of a water environment with biogenic matter (nitrogen and phosphor compounds) is one of the main causes of the following structural transformations in specifically researched communities of aquatic organisms:

a) Phytoplanktonic changes:

- Increase of the overall phytoplankton population due to the increase of the cyanophyta number, up to “flowering”
- Depauperization of species composition with a trend for domination of one or two species
- Domination of cyanellae, even in springtime
- Decrease of relative population of green algae and dysfunction of seasonal development of an entire community
- Increase of a-saprobic population gen. *Oscillatoria* and *Hitzschia,* which can vegetate even under conditions of oxygen regime dysfunction

b) Zooplanktonic changes:

- Enhanced quantitative development of a community with transformations of group and species compositions
- Increased frequency of recurrent, abnormally high overall population
- Decrease in the number of species due to dropout of oligo- and mesosaprobes
- Increase of highly tolerant

c) Zoobenthal changes:

- Increase of community biomass, uneven ecological structure of the community with implicit domination of separate species
- Decrease of species diversity due to the mortality of clear water organisms
- Constant or periodic increase of oligochaetes population, which is characteristic for bottom silting

d) Vphytoperiphytonic changes:

- Depauperization of species composition and increase of cyanellae population development
- Increase of a-saprobic population occurrence frequency, beginning with species such as *Hitzschiapalea* and *Oscillatoria subtilissima*
- Depressed development of green filamentous population development

The profound structural alterations in the aquatic organism communities occur due to an increase and acceleration of anthropogenic eutrophication and the toxic impact of separate pollutants on biota; the incoming level of these matters into water bodies remains quite high.

At that, the toxic impact effect may be exhibited at the direct influx of toxic matters and during their generation on the definite stages of the anthropogenic eutrophication process.

The obvious difference between the planktonic and benthal communities is presently characteristic for the majority of aquatic ecosystems; this is defined not merely by hydrological peculiarities of a water body and the nature of a long-term anthropogenic impact, but to a greater extent, by the decreased stability of the ecosystem during its transition from one trophic state into another under the conditions of increased eutrophication in a water column. As a rule, the eutrophication process of water bodies begins in the phytoplanktonic type.

As is known, the phytoplankton plays a leading role in water quality indication in water bodies where human impact may have both eutrophic and regressive effects. A witness thereof is consolidated information from a multiannual period concerning phytoplanktonic communities in water regions, lakes and watercourses in Russia; the consolidated work was made by a group of people from the Hydrochemical Institute: V.A. Bryzgalo, L.S. Kosmenko, L.P. Sokolova, A. M. Korshun, et al. Based on this consolidated information, water ecosystems were ranked based on transformations in phytoplanktonic communities [19] (Figure 8.4).

Characteristic structural alterations in water ecosystems are defined by the following traits:

- Permanently positive biotic balance in the upper part of trophogenic stratum during summer
- Noticeable intra-season changes of photosynthesis, oxygen content, free carbon dioxide content, carbonates content and active aqueous response during summer

The following natural modifications of phytoplankton development common for all water bodies are revealed in the early stages of anthropogenic eutrophication:

- Alteration of species composition in the dominating complex
- Alteration in importance of separate species within the dominating complex and of the entire systematic groups within the overall population and planktonic biomass
- Increase of absolute quantitative values of populations in all principal groups
- Decrease of the relative population of diatoms within the overall population of phytoplankton
- Shift of balance between mangocellular and parvocellular species in the direction of domination of the latter, especially in an early stage of eutrophication

Among the most evident manifestations of consequences of phytoplanktonic eutrophiucation is the water bloom that occurs as a result of the self-regulation process being interrupted within the ecosystem and the domination of one or several of the most adapted species of algae [27—30]. The most acute competition takes place in the summer period. An oppressive majority of summertime dominant diatoms in the oligotrophic water bodies are displaced into summer and autumn plankton, while during the summer, the cyanellae, green algae, xanthophytes and pyrrophyta dominate.

The bloom is an indirect indicator of the deteriorated sanitary state of a water body. On the other side, the blooming is in itself a source of biological pollution for the water bodies, with all the negative hygienic and sometimes epidemiological consequences that it implies. During the blooming endemics, the taste of water is deteriorated, the filters of treatment facilities are clogged and they demand a large volume of pure water for rinsing. Such breakdowns are especially dangerous when they occur in the major research reservoirs where an intensive bloom consequent to anthropogenic eutrophication tends to spread rapidly, and the masses of cyanophyta floating on the surface may be driven by winds towards the water intakes. Thus, excessive growth of cyanellae and discharge thereof into the reach of the Tsimlyansk waterworks facility caused a terminal breakdown of its water intake filters [31].

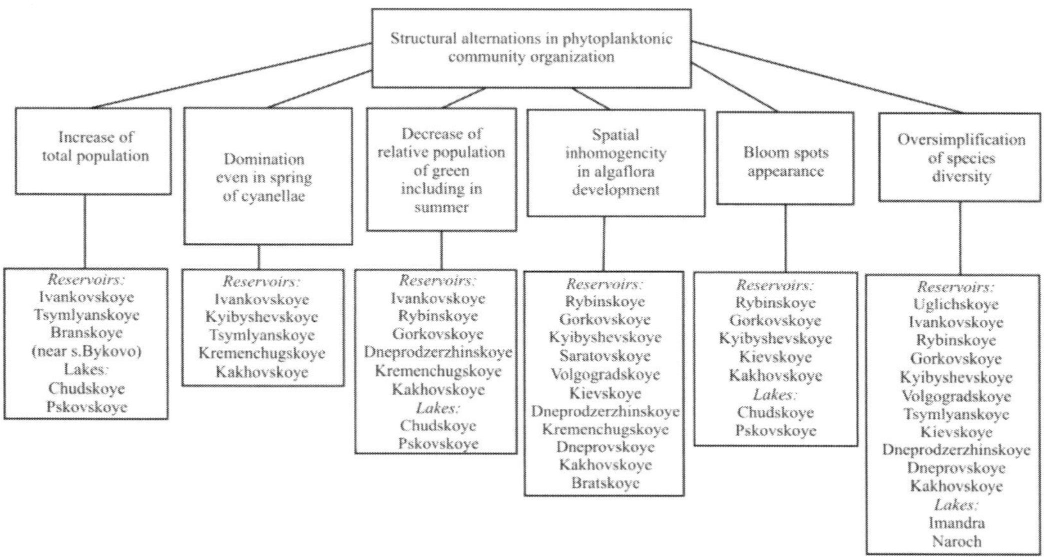

Figure 8.4. Ranking of eutrophicated aquatic ecosystems according to characteristic features of structural alternations in phytoplanktonic community organization.

One fact is worth mentioned separately: the blooming occurs as a result of anthropogenic impact on the ecosystem and is an adaptive response of the ecosystem; it can be deemed a new stage of its existence under the conditions of an altered environment. As the long-term research shows [20], an abundant blooming of cyanellae is facilitated by low current, high transparence, high content of motile dissolved matter and influx of biogenic elements. Another important role in the stimulation of algae development comes from the increased revitalization of the environment and decreased oxygenic saturation.

An intensively increasing process of anthropogenic eutrophication makes it necessary to monitor the arrangement of the anthropogenic eutrophication of surface waters and to support work this with methodic and methodological support.

8.3.2. Arrangement Principles of an Anthropogenic Eutrophication Monitoring Subsystem

A subsystem of freshwater ecosystems anthropogenic eutrophication (SFEAE) is listed within the monitoring system for land surface waters of the Public Service for Environmental Monitoring [32].

The main objectives of this subsystem are as follows:

- Observation over anthropogenic eutrophication processes development
- Assessment of changes in the structural organization of aquatic organism communities under the conditions of short-term or chronic external impact
- Detection of long-term trends in anthropogenic eutrophication for the purpose of forecasting their development and elaboration of measures for water management

These objectives have been largely fulfilled (during the last 15-20 years) through the existing networks of the Public Service for Environmental Monitoring for hydrochemistry and hydrobiology, and with the help of scientific institutions from the Russian Federal Service for Hydrometeorology and Environmental Monitoring pursuant to a uniform program overseeing the utilization of standardized observation methods, as well as taking responsibility for the complex observations of water bodies (tracing the hydrological, hydrochemical and hydrobiological parameters).

The first stage of SFEAE was established on the basis of the points of the existing network backed up by a vast range of observations over a complex of parameters. Accumulated research results assess the consequences of anthropogenic eutrophication in the following ways:

- By means of comparison of points and terms of priority hydrochemical and hydrobiological parameters
- By means of investigation of spatiotemporal variability thereof
- By means of detection of peculiarities in development of aquatic organic communities under the increased anthropogenic impact
- By means of ranking water body districts according to their pollution level accounting for the eutrophicating and degenerative effects of human impact

The formation of a stations network for monitoring the anthropogenic eutrophication process is carried out based on the analysis of the location of the routine monitoring stations existing within the global monitoring network for surface land waters [32].

At that, the following provisions are being observed:

- The most favorable conditions for anthropogenic eutrophication reinforcement are created in shallow areas of lakes and reservoirs, in ponds and in regulated watercourses in which the hydrological conditions of an ecosystem functioning have been changed drastically
- Irregularity of human impact and, consequently, of physical and chemical conditions of the aquatic environment may cause the occurrence of a range of microsystems within one ecosystem

A network of points for observations is selected on the basis of the characteristic ecological status of a water body, pursuant to the results of a reconnaissance survey, for the purpose of detecting spatial irregularities in the allocation of values that identify the functioning of an ecosystem of the entire water body and of separate biotypes within the water body.

During the reconnaissance survey of a water body, the sampling is made in the entire water area for the period that is significantly less than the spatial and temporal variability of the water body with a space between the sampling points that helps to avoid an omission of abnormal areas of the surveyed water body. These conditions are maximally supported through the utilization of remote observation methods [33] that were covered in detail in Chapter 6.

Depending on the peculiarities of a water body, the survey may be singular within the period of maximal phytoplankton growth, or repeated at the different stages of the ecosystem's development, but shall in any case last no longer than one year. We may refer *exempli gratia* to the results of an experimental reconnaissance survey made at the Tsimlyansk Reservoir (June-August 1986) with the help of a rapid recording from a helicopter. This method of data collection:

- Reduce the time intervals between samplings
- Covers the water face in a short period (6 hrs) with a high sampling rate (Figure 8.5)
- Processes the fresh biological material that is of special importance in the assessment of productivity processes
- Lays a water area of a reservoir out in biotopes where the regime observations are necessary, in accordance with the given parameters

The sampling was made in 160 helicopter hoverings, and analysis was done for 30 integral samples (see Figure 8.5). The analysis of phytoplanktonic community samples revealed a significant discrepancy in phytoplanktonic population allocation (Figure 8.6), with a general tendency to increase thereof in shallow areas and areas in the proximity of dams. The reconnaissance survey results facilitated an optimization of the regime network for the reservoir via its extension through operative observation points (Figure 8.7).

Thus, during the selection of regime observation points, the preference lies with the following areas of water bodies:

- The areas that already have operating control points, provided that these do not contradict the results of the reconnaissance survey
- The areas affected by both existing and planned sources of pollutants influx, which makes the potential reinforcement of the eutrophication process in freshwater ecosystems possible
- The areas distinguished by the most obvious manifestations of discrepancies in the structural and functional organization of aquatic organism communities consequent to human impact

A program of supervision over anthropogenic eutpohication shall provide the integral assessment of ecological state of a freshwater ecosystems (for water, sediments, suspended matter, hydrobionts) according to a range of priority parameters (hydrological, hydrochemical and hydrobiological) [32].

In carrying out systematic control over the anthropogenic eutropication of the ecosystems, it is necessary to include the following parameters in the list of especially important ones: hydrochemical parameters—dissolved oxygen, biochemical oxygen demand (under BOD_5), ammoniac nitrogen, nitrite nitrogen, nitrate nitrogen, overall nitrogen content, phosphate phosphorus, overall phosphorus content; hydrobiological parameters—quantitative and qualitative indices of phytoplankton growth, total bacterial plankton population, primary production and destruction of organic matter and qualitative and quantitative indices of macrozoobenthos development.

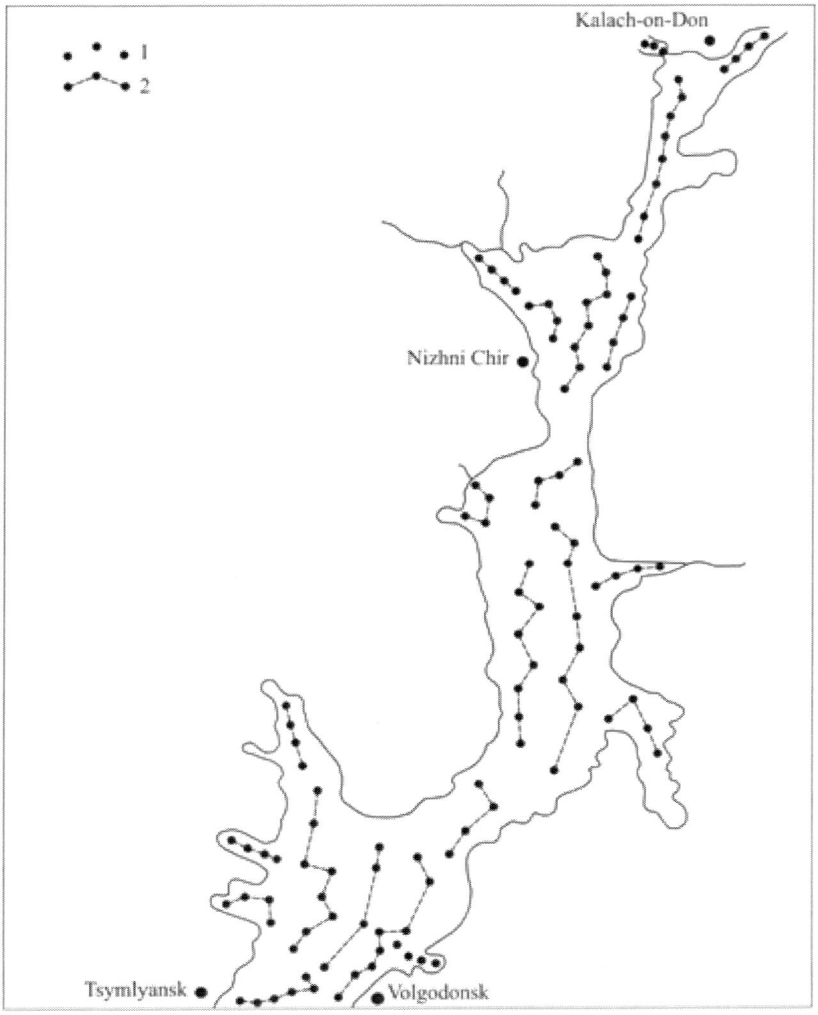

1–sampling points, *2*–integral sampling.

Figure 8.5. Scheme of sampling points located at the Tsimlyansk Reservoir.

The following must be taken into consideration in order to define the frequency and terms of sampling:

- A monitoring system must be "biologically equivalent," i.e., it has to consider the natural rate of recurrence within the aquatic communities that is determined by regional peculiarities of a water body; with this, the terms of monitoring shall coincide with the periods of the maximal quantitative development of biota

- A degree of ecosystem stability that is determined based upon the data of multiannual research
- The most rigid conditions (terms) of influx of the matters stimulating the eutrophication processes within an aqueous environment

It is expedient to carry out the observations of anthropogenic eutrophication at least once a month during the vegetative period covering at that point the peak period of spring floods and summer high water. Provided that a range of regime observations is being carried out in order to allow the determination of the repeatability of various system components, it is possible to reduce the frequency of control by means carrying it out within the most indicative terms of the eutrophication process display.

We may take as an example for consideration the observations of anthropogenic eutrophication held at the Kuibyshevskoye reservoir. Here, the following factors come into play:

- Generic character of a water body subject to frequent cases of phytoplanktonic eutrophication
- Availability of a rather long range of observations
- Wide dynamic spread of hydrochemical and hydrobiolocical indices alterations

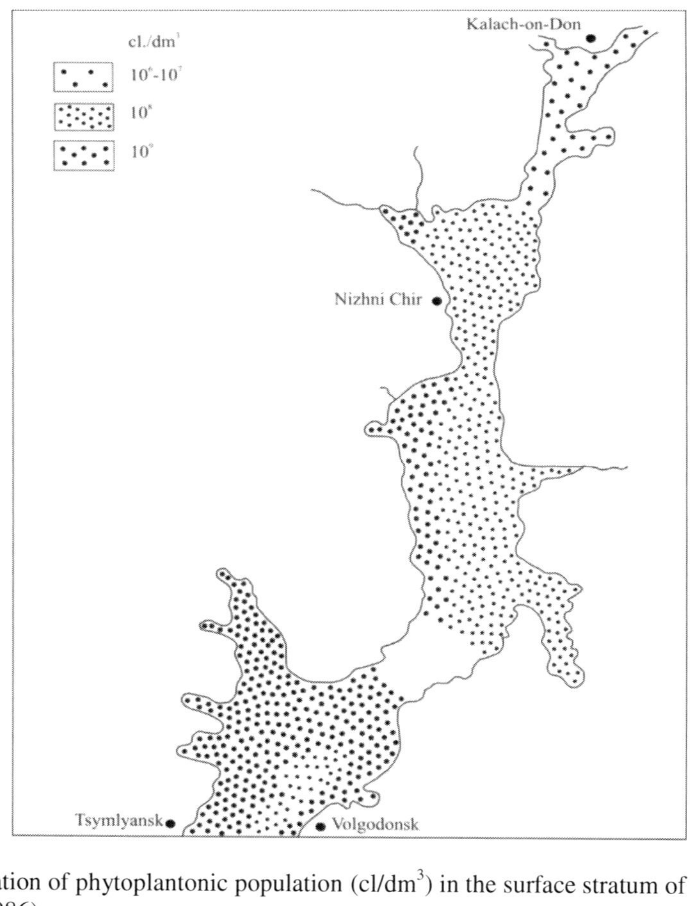

Figure 8.6. Allocation of phytoplantonic population (cl/dm^3) in the surface stratum of the Tsimlyansk Reservoir (July 1986).

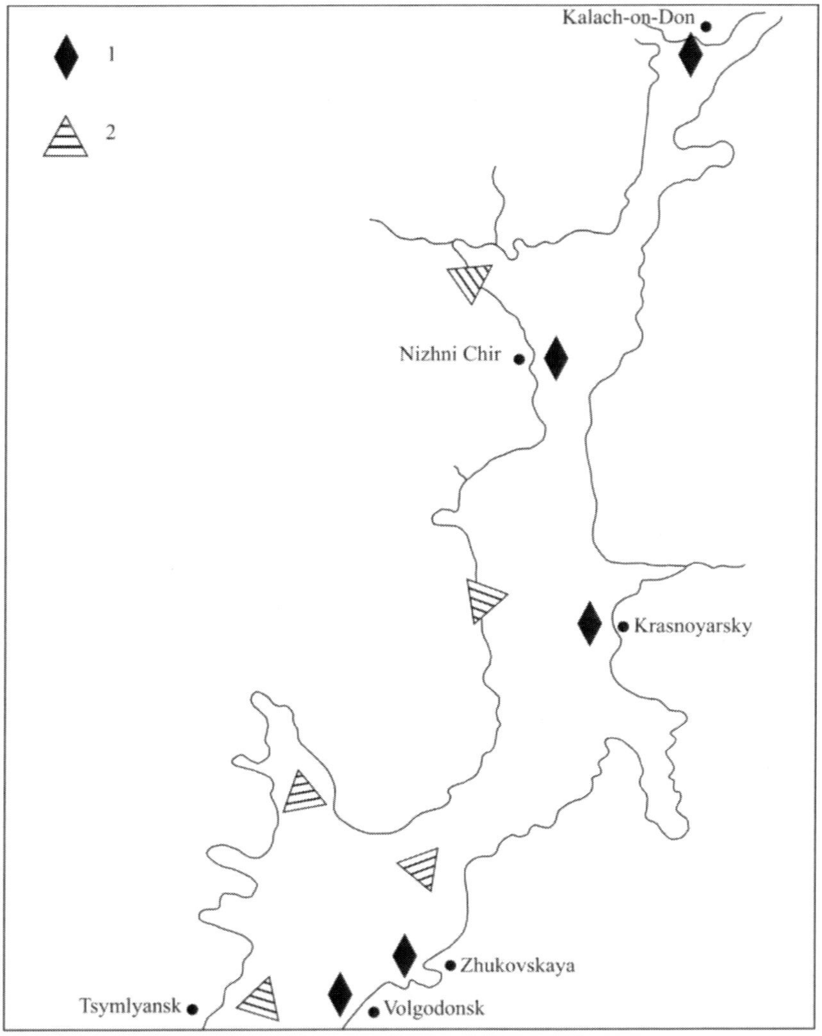

Figure 8.7. Index map of regime (backing) (1) and operative (2) hydrobiological observation points at the Tsimlyansk Reservoir.

As reference material, the multiannual data of chemical and biological status observations of the Kuibyshevskoye Reservoir were made from 1980 until 1990 at 13 hydrochemical and 9 hydrobiological points within the global monitoring (Figure 8.8). The spatial discrepancy was evaluated based on the value variations of phytoplankton population and its main groups as well as the most significant forms of nitrogen and phosphorus to observe the vital functions of this community of aquatic organisms (nitrite, nitrate, ammoniac nitrogen, phosphate phosphorus, total nitrogen and total phosphorus).

The generalized results (see Table 8.4) and the comparative evaluation thereof allowed us to make a conclusion that the water research was heterogenic in its content of nitrate, ammoniac forms of nitrogen and in the total phosphorus content.

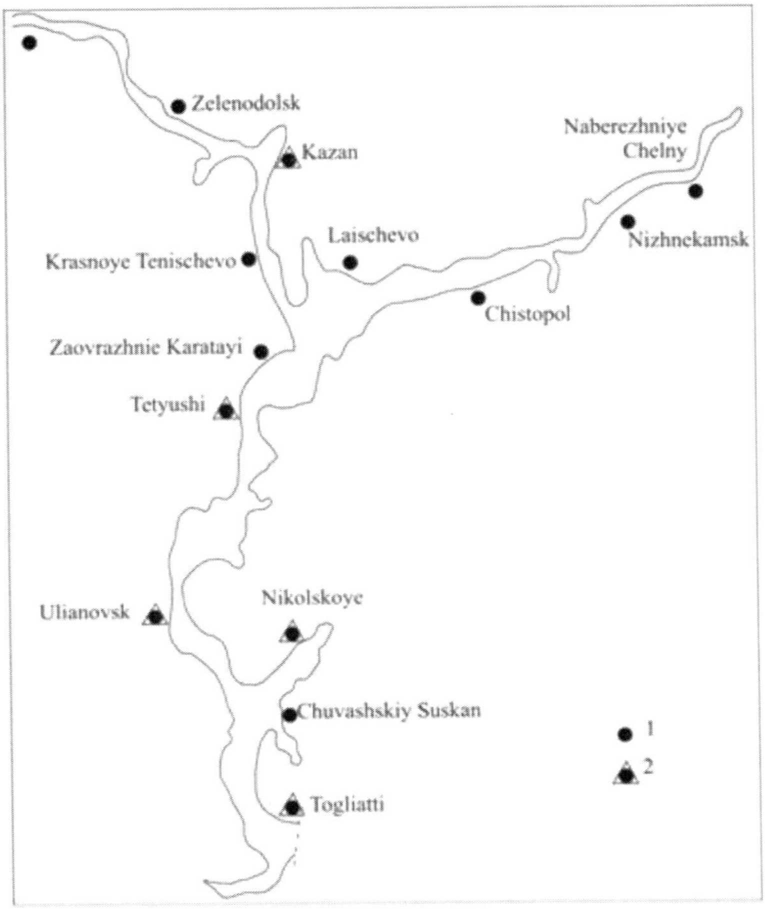

1–GMS points, *2*–anthropogenic eutrophication observation points.

Figure 8.8. Location scheme of observation points at the Kuibyshevskoye Reservoir.

Abnormally high discrepancies in concentrations are characteristic of the areas near the cities of Volzhsk, Togliatti, Ulianovsk, Chistopol and Naberezniye Chelny. Variations of ammoniac ions were noticed near the cities of Kazan and Zelenodolsk. The following areas are distinguished by their level of phytoplankton development (Figure 8.9):

- Areas with intensive growth of algal flora, especially in summer (up to "blooming") near Togliatti and the settlement of Nikolskoye
- Areas with weak growth of phytoplankton during the entire period, near the city of Naberezhniye Chelny

Given the discrepancies in an ecosystem with respect to biogenic elements content and the quantitative values of phytoplankton growth, in selecting the points for observations of anthropogenic eutrophication, preference was given to the water body areas where the above indicators deviated to be more or less than the average values for the water region. Besides, the hydrologic and morphometric peculiarities of the water body were also taken into consideration during selection of the observation points.

Table 8.4. Concentration of nitrogen- and phosphorus-containing compounds, and total population of phytoplanktonic communities at the Kuibyshevskoye Reservoir for the period 1980-1990

Observation point	Concentration, mg/l					Total phytoplankton population, kcl/ml
	nitrogen			phosphorus		
	nitrite	nitrate	ammoniac	phosphate	total	
Volzhsk	Not detected[1]—0.09	0.02—2.06	0.02—1.29	Not detected—0.15	0.01—0.42	—
Togliatti	Not detected —0.25	Not detected —0.90	Not detected —3.31	Not detected —0.20	0.01—0.80	0.4—343
Chuvashskiy Suskan	Not detected —0.19	Not detected.—1.00	Not detected —1.41	Not detected—0.07	0.01—0.20	—
Nikolskoye	Not detected —0.14	Not detected—0.77	Not detected —1.51	Not detected—0.22	0.01—0.22	0.3—104[2]
Ulianovsk	Not detected —0.38	Not detected—0.78	Not detected —2.89	Not detected—0.31	0.01—0.83	0.5—90.3
Tetyushi	Not detected —0.20	Not detected—0.88	Not detected —3.70	Not detected—0.09	0.01—0.32	0.8—82.5
Zaovrazhnie Karatayi	Not detected —0.16	Not detected—0.94	Not detected —0.95	Not detected—0.07	0.01—0.17	—
Laishevo	Not detected —0.18	Not detected—0.88	Not detected —0.95	Not detected—0.06	0.01—0.22	—
Chistopol	Not detected.—0.15	Not detected—0.79	Not detected —5.47	Not detected—0.13	0.01—0.19	0.6—23.9
Naberezhniye Chelny	Not detected —0.20	Not detected—1.56	Not detected —2.62	Not detected—0.35	0.01—1.29	0.07—14.3
Krasnoye Tanyschevo	Not detected —0.18	Not detected—0.66	0.12—1.55	Not detected—0.10	0.01—0.13	0.8—30.0
Kazan	Not detected —0.26	Not detected—0.56	Not detected —6.31	Not detected—0.19	0.01—0.31	0.6—90.4
Zelenodolsk	Not detected —0.26	Not detected—0.83	Not detected —2.82	Not detected—0.31	0.01—0.89	1.0—29.8

Notes:

[1] Beyond detection limit.

[2] Phytoplankton research conducted near the settlement of Krestovo Gorodische.

Thus, for the solution of tasks connected with the studies of anthropogenic eutrophication, and its consequent impact on the development of an ecosystem, the areas near the cities of Kazan, Tetyushi, Ulianovsk, Togliatti and the settlement Nikolskoye were selected as points for obligatory observations (see Figure 8.8).

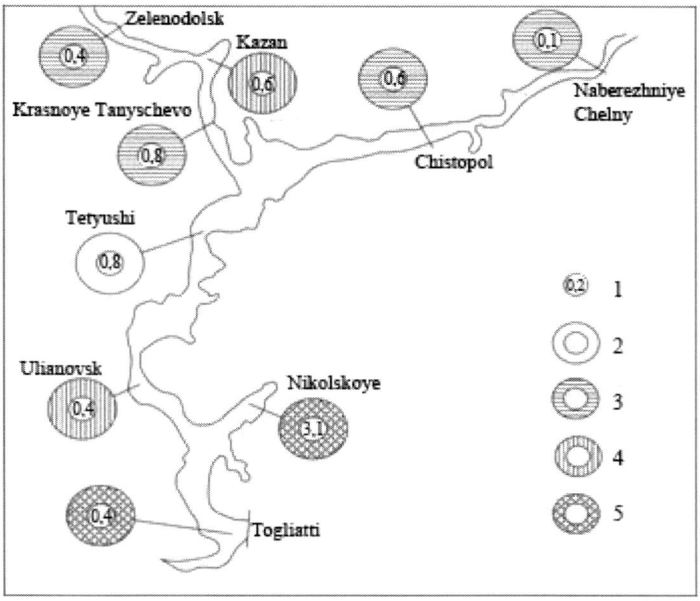

1—minimal value, *2*—*5*—maximal values: not more than 10^3 (*2*), 11^3—50^3 (*3*), 51^3—100^3 (*4*) and 101^3—500^3 (*5*).

Figure 8.9. Index map of total phytoplankton population allocation (cl./ml) in the water area of the Kuibyshevskoye Reservoir.

Special attention was paid to an area near the city of Naberezhniye Chelny, where a high level of biogenic load created the potential possibility of anthropogenic eutrophication process enhancement.

Special observations for the season were held at Suskanskiy Bay within the Kuibyshevskoye Reservoir (daily sampling was done at the biological midday for 8-14 days) in order to detect the short-term alterations of the ecosystem and to level out any apparent mistakes due to short-term variances of the indices observed. The following considerations were taken into account in choosing the terms for the daily surveys (May, August and October):

- The water region becomes free from ice on average from April 20 to April 28, and at the latest, within the first ten days of May (the beginning of the vegetation period)
- The initial ice formation in the forms of fastland ice and slush appear between November 7–November 12 (end of vegetation period)
- The content of inorganic nitrogen and phosphorus varies from the maximal value to the minimal one in spring (May), and these are determined during the intensive weed growth in summer (July to August); in autumn, the content tends to increase
- The season development process of phytoplankton development is characterized by an outburst of diatom vegetation in May, peak life activities of phytoplankton in August due to increased development of cyanellae and the repeated domination of diatom weed in October

The results of investigations held at the Kuibyshevskoye Reservoir lead to the following conclusions:

- The widest variance range of a total population of planktonic communities occurred in August; in this month, the total phytoplanktonic population was, as a rule, much higher than in May or October (see Table 8.5)
- Within the summer period, the cyanellae steadily dominated
- Among the biogenic elements, the nitrate and ammoniac forms of nitrogen were the most variable; the concentration thereof is subject to significant intra- and interannual variances with a general tendency to increase in spring (Table 8.6)

Thus, from the viewpoint of the anthropogenic eutrophication process of development, the Kuibyshevskoye Reservoir has two characteristic periods: the spring period (where there is an accumulation of a potential reserve of biogenic elements that define the later development level of a biotic water community) and the summer period (during which the ecosystem displays the most obvious response to the settled mode of biogenic elements).

8.3.3. Assessment of the Anthropogenic Eutrophication Level in Freshwater Ecosystems

The method developed by the Hydrochemical Institute for freshwater ecosystems pollution evaluation on the basis of phytoplanktonic communities' development was used in the development of the principles of anthropogenic eutrophication assessment [34].

The statistic results from the multiannual surveillance information concerning the quantitative indices of phytoplanktonic communities at more than 30 Russian water bodies and water courses confirmed the viability of freshwater ecosystems eutrophication assessment based on the priority factors of phytoplankton community system reformation under the increase of anthropogenic eutrophication. Assessment of the anthropogenic eutrophication effect is based on the research concerning the spatiotemporal allocation of the phytoplanktonic population according to the variance character thereof. The nature of total population values variance is determined on a relative static series condensation (Π_r) and on the thickest value (M_r) within the modal interval [32]. The calculation is made with the following formulas:

$$\Pi_r = \frac{W}{K} \tag{2}$$

in which W is the frequency (%), i.e., the frequency prevalence of one or another variant (or interval) within the sum of frequencies and

$$M_r = X_{M_r min} = K \frac{-W_{M_r-1}}{(W_{M_r} - W_{M_r-1}) + (W_{M_r} - W_{M_r+1})} \tag{3}$$

where $X_{M_r min}$ is the lower boundary of the modal interval, K is the interval difference or the interval magnitude, W_{M_r} is the frequency prevalence of the modal interval, W_{M_r-1} is the frequency prevalence in a previous interval to the modal interval and W_{M_r+1} is the frequency prevalence in a consequent interval from the modal interval.

Table 8.5. Total population variance in planktonic communities at Suskanskiy Bay in the Kuibyshevskoye Reservoir

Planktonic community							
Phytoplankton	May	thnd. cl./ml	—	1.8—10.3	1.7—3.4	2.2—11	1.5—7.0
	August		8.8—706	8.9—432	19.7—300	8.5—1,150	32—107
	October		—	3.5—14.3	1.4—5.8	4.0—11	—
Bacterial plankton	May	mln. cl./ml	—	1.2—2.3	2.1—4.5	2.1—2.5	2.5—2.7
	August		1.1—4.6	2.4—5.1	3.0—5.7	2.2—2.7	2.6—2.7
	October		—	—	2.1—4.7	2.3—2.6	—
Zooplankton	May	thnd. spc.m^3	—	12 — 79	14—84	5.2—75	0.4—145
	August		53 — 337	66 — 788	12—1,100	103—758	4.0—412

Note: п—number of daily samplings

Table 8.6. Concentration (mg/l) variance of specific nitrogen, phosphorus and carbon forms at Suskanskiy Bay in the Kuibyshevskoye Reservoir

Index		Years				
	Sampling month	1982 (n = 8)	1983 (n = 10)	1984 (n= 10)	1985 (n = 14)	1986 (n = 11)
Nitrate nitrogen	May	—	0.03—0.68	n.d.—0.03	0.19—0.45	0.39—0.63
	August	0.10—0.72	0.04—0.14	n.d.—0.10	n.d.—0.29	n.d.—0.07
	October	—	0.19—0.21	0.09—0.24	0.12—0.31	—
Ammoniac nitrogen	May	—	0.28—0.41	0.03—0.22	—	0.02—0.10
	August	0.18—0.70	0.33—0.98	0.01—0.20	n.d.—0.13	n.d.—0.16
	October	—	0.04—0.12	0.01—0.23	n.d.—0.07	—
Inorganic phosphor	May	—	0.01—0.03	n.o—0.02	n.d.—0.05	0.01—0.02
	August	0.01—0.05	0.02—0.07	0.08—0.11	0.02—0.08	0.02—0.06
	October	—	0.02—0.04	0.03—0.06	0.02—0.08	—
Organic carbon	May	—	6.7—14.2	7.7—11.5	7.5—13.9	3.8—5.7
	August	8.3—22.6	9.0—18.3	7.7—16.9	4.0—7.5	4.2—8.5
	October	—	11.4	10.1—14.3	5.0—7.0	—

Notes: п—number of daily samplings; n. d —below detection limit

The effects of anthropogenic impact on the aquatic ecosystems are defined based on the relative static series condensation, Π_r, and the thickest value of a total phytoplanktonic population within the modal interval, M_{tp} (Table 8.7).

The combination of low Π_r and high M_{tp} values is characteristic of ecosystems with a high eutrophication level.

The assessment of anthropogenic eutrophication is based on the calculation of the following statistical measures:

- Thickest value M_{tp} of a modal interval within the static series of a total phytoplanktonic population
- Detection frequency of high phytoplanktonic population values α

$$\alpha = \frac{n_j'}{n_j} \quad (4)$$

in which n_j' is the number of analyses results in which the values of a factor under definition were above or below the most or the least frequent variance range and n_j is the total number of values.

β is the multiplicity of excess or (or decrease) in a total population as compared to the average value based on the most frequent β values:

$$\beta = \frac{high\ value}{M_{tp}} \ or\ \beta = \frac{M_{tp}}{low\ value} \quad (5)$$

- Thickest modal interval value, M_{ts}, of species diversity within a phytoplanktonic community
- Thickest modal interval value, M_{rd}, of the relative population of a dominating species
- Thickest modal interval value, M_{rc}, of the relative cyanellae population in phytoplankton samples taken in May

An assessment of the anthropogenic eutrophication level within the freshwater ecosystems is made on the basis of the aggregate of the statistical measures (Table 8.8).

As an example, an interpretation of data results for quantitative phytoplankton development in separate researches and stream flows can be considered (Table 8.9).

Table 8.7. Classification of effects of anthropogenic impact on aquatic ecosystems

Anthropogenic impact effect	Statistical measures of total phytoplanktonic population variance	
	Thickest value, M_{tp}, thnd.cl./ml	Relative condensation, Π_r, %
Ecological devolution	0.5	100—300
Ecological devolution elements	< 1.5	50—100
Anthropogenic strain with the elements of ecological devolution	< 2.0	20—60
Anthropogenic strain	< 5.0	50—100

Anthropogenic eutrophication	> 10.0	< 30

Table 8.8. Classification of levels of a freshwater ecosystem's anthropogenic eutrophication

Statistical measures of total phytoplanktonic population development	Eutrophication level		
	low	moderate	high
M_{tp}, thnd.cl./ml	2–10	2–20	1–10
α_1, %	10–40	10–40	20–50
α_2, %			5–50
β	5–50	50–200	10–300
M_{ts}	20–30	10–20	10–25
M_{rd}, %	30–50	50–100	20–70
M_{rc}, %	< 5	30–50	40–60

Notes:
α_1 and α_2 stand for frequencies of high and low values (respectively) of total population values detected

The following measures are to be defined:

a) Thickest value of the most frequent values:

- Total population, M_{tp} (thnd.cl./ml)
- Total amount of species, M_{ts}
- Relative dominating population, M_{rd} (%)
- Relative cyanophyta population, M_{rc} (%)

b) Multiplicity of excess in relation to the thickest value of total population high values, β
c) Static series condensation concerning the total phytoplanktonic population, Π_r

The value of the relative condensation of a static series, Π_r, and the thickest value of the modal interval, M_{tp}, are used for the preliminary classification of an ecosystem under examination in accordance to its response to anthropogenic impact (according to Table 8.10). An assessment of a freshwater ecosystem's status, judging based on the eutrophication level thereof, is possible through the utilization of aggregated statistic measures of structural arrangement of a phytoplanktonic community—such as the modal interval thickest values of the total population static series, total amount of species, relative population of dominating species within a vegetation period, relative population of cyanphyta in May samples of phytoplankton, multiplicity of total population high values in excess as compared to the thickest value and a system of ecological status evaluations developed by V. A. Abakumov [35] (Table 8.11).

Table 8.9. Initial values of priority parameters in the development of phytoplanktonic communities

Water body	Observation point	Observation year	Observation duration	Total amount of definitions
Lakes				
Kenon	Chita, within the city borders, and at the area of thermal discharge	1989—1993	May-September	40
Water regions				
Ivankovskoye	Bezborodovo, Konakovo	1989—1992	May, July, August	36
Kuibyshevskoye	Zelenodolsk, Kazan, Krasnoye Tenishevo, Nabereyniye Chelny, Chistopol, Laishevo, Ulianovsk, Zaovrayhniye Karatai, Tetyushi, Nikolskoye, Chuvashskiy Suskan, Togliatti		May, July, October	653
Saratovskoye	Zolnoye, Syzran, Yekaterinovka, Balakovo	1985—1990	May, July, October	382
Tsimlyanskoye	Krasnoyarskiy, Zhukovskiy, Volgodonsk	1984—1987	May-September	61
Veselovskoye	Budenovskiy, Proletarsk, Novoselovka	1981, 1988—1991	May, June, August, October	51
Rivers				
Don	Volgodonsk, Rostov-on-Don, Koluzaevo, Azov, Peschanniy and Perevoloka branches	1981—1994	April-October	287
Seversky Donets	Belaya Kalitva, Kamensk-Shekhtinskiy, Krasnodonetskiy	1981—1984	April-October	59
Lower Volga	Kamenniy Yar, Liebyazhye, Astrakhan	1984—1990	May, July, August, October	79
Sanikhta	Chkalovsk	1988—1990	May-October	18
Moskva	Barsuki, Moscow, Korzunovo	1981—1989	May-October	52
Pechenga	Pechenga	1984—1992	May, June, August	69
Ushakovka	Irkutsk	1987—1992	May, July, September	36
Kolosyoki	Above the firth	1984—1992	May, July, August	115

Table 8.10. Evaluation of the effects of anthropogenic impact on freshwater ecosystems regarding statistical measures of phytoplankton total population

Water body, observation point	Statistical measure		Ecological status acc. to V.A. Abakumov [28]	Anthropogenic impact effect
	M_{tp}	$П_o$		
Kuibyshevskoye Reservoir, upper part	61.2	0.4	Elements of ecological devolution	Anthropogenic eutrophication
Saratovskoye Reservoir	22.3	1.5	Anthropogenic strain	id.
Volga River, Astrakhan	10.8	4	id.	»
Sanikhta River, Chkalovsk	9.0	8	»	»
Kenon Lake, Chita	1.2	45	Elements of ecological devolution	Ecological devolution
Moskva River, village of Barsuki	0.9	84	id.	id.
Pechenga River, station Pechenga	0.5	86	»	»
Ushakovka River, Irkutsk	0.26	169	Ecological devolution	»
Kolosyoki River, above the firth	0.10	172	id.	»

8.4. Local Monitoring of Surface Land Waters Pollution at Impacted Urbanized Areas

A wide, human-induced introduction of powerful technical means for increased efficiency in mining, processing and utilizing natural resources engenders the increased negative effects of anthropogenic impact on the natural environment.

The fuel and energy complex (FEC), which includes the coal mining, oil, gas and thermal power industries, is one of the largest pollution sources of natural objects in these vast territories. The multi-industrial nature of the FEC, efficiency of main and related production facilities, multiplicity of pollution sources, diversity of raw materials and chemical agents, peculiarities of implemented technologies and waste treatment methods, which contain a wide range of pollutants and toxic compounds, are the origin of the complex and manifold industrial impacts on natural objects and their effects.

High pollution levels alters the status of natural ecosystems up to the degradation thereof, and high morbidity rate among population, reduced life duration, etc., are all characteristic of many regions with developed FEC production facilities. This allows us to consider and classify such territories (natural objects) as the areas of ecological crisis and/or ecological disaster (Federal Law of the Russian Federation "Of the Environmental Protection," 2002; "Criteria of ecological status of the areas used to detect the ecological emergency zones and ecological disaster zones," 1992 [36, 37]). Decisions concerning problems with the evaluation and prediction of the natural environmental state in such zones are impossible without a special complex for monitoring the pollution of natural objects. Such a system should be adapted to the specific features of FEC industrial factories and their technogenic effects. The existing system of routine observations over the air, soil, land and [1, 2, 38] subsurface water pollution and its spatiotemporal definition does not comprehensively provide the information necessary for an adequate evaluation of the ecological adversity level for natural objects in the intense anthropogenic impact zones and for the development of measures to remediate these impacts.

Table 8.11. Eutrophication level assessment on statistical measures of phytoplanktonic community structural organization at separate aquatic ecosystems

Water body, object	Statistical measure						Eutrophication level	Ecological status acc. to V.A. Abakumov [28]
	M_{tp}	$M_{ов}$	$M_{од}$	$M_{ос-з}$	β	$П_о$		
Water reservoirs								
Ivankovskoye	0.9	11	83	50	134	30	Moderate	Ecological devolution of benthic biotic communities
Kuibyshevskoye, lower part	61.2	18	23	0	6	0.4	High	Anthropogenic strain
Saratovskoye	22.3	—	55	0	—	2	High	id.
Tsimlyanskoye	4.0	18	38	—	16	12	Moderate	»
Veselovskoye	2.5	11	—	—	14	13	Moderate	Ecological devolution of benthic biotic communities
Rivers								
Don, Volgodonsk town	3.7	20	20	—	13	14	Low	Anthropogenic strain
Seversky Donets, Belaya Kalitva	2.2	22	20	—	7	20	Low	id.
Seversky Donets, Ust-Donetskiy	2.0	23	19	—	13	23	Low	Anthropogenic strain with elements of ecological evolution
Volga, Verkhneye Lebyazhye settlement	6.6	31	44	—	4	9	Low	Anthropogenic strain

The aforementioned reasons, alongside the high probability of emergencies and consequences thereof at different stages of FEC production facilities (for excavation of energy raw materials, transportation and processing thereof, for the production of heat and electric power), predefined for the specialists of the Hydrochemistry Institute the development of methodical and methodological principles and approaches for arranging and carrying out complex observations within a framework of local natural objects pollution monitoring for impacted urbanized areas. It also helped specialists in this field prepare scientific generalizations and normative and technical documents [8, 9, 39—41].

The main principles of natural environmental state monitoring arrangement at areas of chronic pollution caused by FEC objects exemplifies the special observations held in the regions subject to man-induced consequences from the coal mining industry, which is one of the main branches of the fuel industry.

8.4.1. Main Pollution Sources for Natural Objects in the Areas of Coal mining Industry Development

The coal mining industry includes the excavation and processing (cleaning and briquetting) of mineral coal, which is mainly used in the production of electric power at thermal electric power stations, in ferrous and nonferrous metal industries in the form of coke, as a raw material in the chemical industry (manufacturing of plastics, spirits, explosives, etc.) and in civil engineering (manufacture of cement, glass, etc.). Russia's territories hold around 30 coal mining fields, and the total coal stock is estimated to amass to 5.3 trillion tons [39]. The coal resources are allocated unevenly on the country's territory with around 90% of it being concentrated in the eastern regions (Kuznetskiy coal basin is the main coaling station).

The coal mined in the different fields varies significantly in its metamorphization grade, location, occurrence depth and composition, all of which determines the scales, the methods of mining and the peculiarities of the industrial complexes developing on the basis of such fields. Coal is comprised of an organic (combustible) mass that is formed from the compounds of carbon, oxygen, hydrogen, sulfur, sometimes phosphorus and inorganic infusions. The inorganic component (ash) left after coal combustion is mostly (95-98%) comprised of silicates (generally, aluminum silicates), sulfides (mainly iron sulfides) and carbonates (calcium, magnesium and iron carbonates).

The most harmful component of inorganic coal infusions is sulfur (main forms: sulfate, sulfide and organic sulfurs, of which, the latter is firmly included in the organic mass), which is the precondition of acid mine drainage formations and acid discharges into the atmosphere.

Alongside the mentioned inorganic components, coal is comprised of the following elements: arsenic, manganese, lead, cadmium, copper, beryllium, nickel, vanadium, etc. Mining and processing coil entail an influx of a significant amount of mine (quarry) waters and refuse, drain waters, and atmospheric discharges from the refineries, community boiler houses, coke factories and other related industrial objects. All of this waste goes into the natural environment (Tables 8.12 and 8.13).

Mine waters are created during coal mining and are mainly comprised by underground waters drained through excavations, atmospheric sediments partially filtered through them, and run-off from the nearest water bodies.

Table 8.12. Main pollution sources for natural objects in the areas of the coal mining industry and related industry developments [39]

Pollution sources and their impacts on natural objects	Impact recurrence	Environmental objects		
		Natural waters	Atmospheric air	Soils
Mining and cleaning of coal				
Discharge of mine (quarry) waters into water bodies	Constantly	DI	AI	AI
Discharge of mine (quarry) waters into surficial lows	Constantly	II	AI	AI
Discharge of pollutants through fan units	Constantly	II	DI	II
Refineries				

Discharge of used waste waters				

Table 8.12. (Continued)

Pollution sources and their impacts on natural objects	Impact recurrence	Environmental objects		
		Natural waters	Atmospheric air	Soils
into water bodies	Periodically (during exchange of circulating water with fresh water)	DI	AI	AI
into surficial lows	Periodically	II	AI	DI
Waste water overflow from sludge dumps	Occasionally	II	AI	DI
Dust discharge into atmosphere	Constantly	II	DI	II
Refuse from coal mines, coal strip mines, refineries				
Aeolian travel of coal decomposition products	Constantly	II	DI	DI
Offset of coal and refuse decomposition products via infiltrating water's surface prechannel flow	Periodically (during rain, flood and snow melting periods)	II	AI	DI
Offset of pollutants with smoke from burning of refuse	During burning of refuse	II	DI	II
Industrial and community boiler houses				
Waste waters discharge	Periodically (during operation of boiler houses)	DI	AI	AI
Discharge into atmosphere	Periodically	II	DI	II
Sites of coal storage and transportation				
Dissipation of coal dust with wind, raw material losses in the adjacent territories	Constantly	II	DI	DI
Objects using coal as raw material				
Coke and by-product manufacturing facilities				
Waste waters discharge	Constantly	DI	AI	AI
Discharges into atmosphere	Constantly and periodically	II	DI	II
Allocated drain from adjacent territories	Periodically (during rain, flood and snow melting periods)	DI	AI	DI
Related infrastructure objects				
Discharge of household waste waters	Constantly	DI	AI	DI
Allocated discharge from the territories meant for development	Periodically (during rain, flood and snow melting periods)	DI	AI	DI
Parking and traffic areas, storage of gases and lubricants				
Allocated discharge from adjacent territories	Periodically (during rain, flood and snow melting periods)	DI	AI	DI
Discharge of exhaust gases by transportation vehicles	Constantly	II	DI	II

Notes:

II—indirect impact (via other environmental objects), DI—direct impact, AI—absence of impact

Table 8.13. Number of atmospheric discharge sources for pollutants (from the example of Gukovougol JSC, 1993)

Sources of pollutants	Number of sources		Sources equipped with waste treatment facilities	
	Total	Including organized sources	Total number	Share from the total number (%)
TOTAL	535	506	125	23.4
Including			1	
industrial boiler houses	148	148	30	20.3
community boiler houses	300	300	37	12.3
dehydration plants	5	5	5	1,001
aspiration bunkers	53	53	53	100
refuse	29	—	—	1

Note: A dash stands for the absence of this parameter.

The chemical composition of coal mine waters is defined mainly by the composition of the underground waters from which they are formed. Coal mine waters are distinguished by an increased content of suspended matter, a higher total ionic concentration as compared to the same underground waters (mostly due to sulfates) and, moreover, by an alteration in ionic composition (transition of hydrocarbonate class into a sulfate class, and in separate cases—into a chloride class).

The coal mine waters are subdivided into alkaline, neutral and acid waters. The latter occur directly in the excavations consequent to interaction with brassil (pyrite) contained in coals. When snow melts in the spring and in summer and autumn rains, the acidity of coal mine waters may increase by 1.3 to 2 times as compared to the winter time [42].

Low pH values of coal mine waters (highly acidic waters may have a pH value of 2...3) define the increased concentrations of manganese, aluminum, iron (up to dozens, hundreds and even thousands of milligrams per liter, correspondingly) and other heavy metals including mercury, cadmium, lead, beryllium, etc.

The ecological danger is also stipulated by the fact that the coal mine waters are saturated with oil products and organic aromatic compounds including polynuclear aromatic hydrocarbons (PAH) and their derivatives.

Coal mine waters discharge (after mechanical cleaning in retention ponds) into water bodies and surficial lows (draws, gullies) is accompanied by the contamination of natural objects with inorganic and organic pollutants.

Refuse dumps are formed on the surface during coal mining (in the Donetskiy coal basin, each ton of coal mined releases an average 300 to 320 tons of refuse [43]) and due to waste from coal cleaning.

The refuse dumps are a potential source of pollution with inorganic compounds—sulfates, chlorides, heavy metals—for natural environment objects due to the transport thereof by air currents and washing away of atmospheric sediments.

Burning refuse dumps are a special ecological danger, because they contaminate natural objects (mostly, the atmospheric air) with sulfur dioxides, nitrogen oxides and polynuclear aromatic hydrocarbons.

Coal cleaning (either dry cleaning or washing), made at refineries and coke and by-product manufacturing facilities (located at the mines or outside the mine territories), is accompanied by the formation of solid (refuse, coal and dust, etc.) and liquid wastes contaminating natural environmental objects. Regardless of the method or cleaning, the toxic and strictly limited organic compounds are used as oil wetting agents, collectors, frothers and dense media (Table 8.14).

A considerable part of waste waters contaminated with the aforementioned reagents comes into the water bodies and surficial lows even when the water supply is recycled and causes contamination of water and soil.

Among the stationary pollution sources, *industrial and residential boiler houses* are the most impactful, emitting at least 40% of harmful exhausts into the atmosphere. Under the adverse meteorological conditions, the pollutants may be spread 1-2 km away from the source in concentrations above the maximum permissible concentration. The waste waters from boiler houses comprise about 35% of the total water consumption volume and contaminate the water bodies with oil products and suspended matter.

Table 8.14. Agents used most frequently in the coal cleaning process [9]

Utilization	Matter
Oil wetting agents	Surface acting agents, desulfated kerosene, alkaline products of oil processing
Dense suspensions	Carbon tetrachloride with the addition of benzene and bromophenol, zinc and calcium chlorides
Weighting of the suspension	Minerals—lodestone, pyrite, sand, chalk and clay
Suspension thinning agents	Liquid silica glass, sodium hexametaphosphate, amines, wood glue, starch, etc.
Collector agents	Kerosenes with high content of aromatic and unsaturated compounds (lightened, desulfated, tractor kerosenes), wastes from oil refineries occurring from manufacturing of isooctylene, floating reagents
Frothing reagents	Higher alcohols and distillery slops of aliphatic, terpene and aromatic alcohols, and high boiling resins, frothing agents, mixture of light and middle oils, camphorated oils, sodium, potassium, etc., chlorides.

Coal coking at the coke manufacturing facilities (main product—coke with a carbon content more than 96.5%), which often accompanies the mining fields, is the most dangerous environmental pollution source among the coal mining industry objects.

Coking products are comprised of coking gas— hydrogen, methane and its homologs, carbon oxides and toxic compounds (aromatic carbohydrates, including naphthalenes and benzenes; hydrogen sulfide with a minor content of carbon bisulfate and thiophen; cyanide compounds and pyridine bases); coal tar components—phenolic and aromatic compounds, including polynuclear aromatic carbohydrates and heteronuclear compounds containing sulfur and nitrogen; raw benzene components—benzol, toluene, acids, trimethylbenzenes with sulfur compound impurities, etc.

The matters that arise from coal coking come into atmosphere with smoke and damp water, and then as atmospheric sediments, they go into soil and water bodies, including underground waters, via waste waters.

Heat and power plants (HPP) powered with coal take first place among the power manufacturing facilities. HPP discharges include sulfur dioxide (45% from the total volume coming into the atmosphere from other sources), carbon and nitrogen oxides, flue cinder containing heavy metals and polynuclear aromatic hydrocarbons including benzapyrene.

Waste waters (flushing water, recycling water, etc.) tend to show increased mineralization and contain oil products, phenyls, fluorides, arsenic, vanadium and flushing solution elements, such as, sulfuric acid, hydrochloric acid, fluoride acid, acetic acid, magnesium, iron hydroxides, etc.

8.4.2. Basic Types of Technogenic Impact

The influence of coal mining industry branches and associated industries on the state of the environment is characterized by the following types of impacts:

- Geophysical (transformation of terrain relief, disruption, mechanical damage to soil mantle)
- Hydrogeological (change of ground water level, ways of running and offloading points, exhaustion of water-bearing horizons, change of water status in soils)
- Hydrological (change of water content, water status of surface water bodies)
- Physical and chemical (contamination of ambient air, soils, surface and ground waters with different substances, including acute toxic matter)
- Biological (change of trophic status of a water body, rearrangement of biocoenoses, decrease of productivity, species diversity, etc.)

The scale, level and nature of environmental contamination in the area of technogenic impact of the coal mining industry production facilities are determined by:

- Original composition of produced coal
- Method of coal production, processing technology, its usages as a raw material for power fuel
- Sources of contamination, methods used and level of production waste treatment and disposal

The aforementioned types of technogenic impact condition are:

- Mosaicity of natural bodies contamination and the level in the territory of impact
- Complex composition of hard, gaseous and liquid production waste as well as ways the contaminants migrate into atmospheric air, soil, surface and ground waters
- Variety of effects and responses of natural bodies to technogenic impact (salinization, acidulation, acute or chronic waters toxicity, etc.)

8.4.3. Surface and Ground Waters Contamination

Main sources of technogenic surface and ground waters contamination in the area of mineral coal production, processing and usage are mine waters, waste waters of industrial and municipal boiler houses, washing plants, coke and by-product production and the overall fuel and energy complex.

Filtered from tailing ponds waste waters and waste waters from slag collection points, the contaminants have considerable influence on ground waters at the alluvial sediments horizon where they are then offloaded to nearby water bodies.

Contaminants from main contamination sources are delivered to the nearest water bodies, to depressions of terrain relief both with arranged waste waters discharge and with non-arranged, distributed (surface and ground) run-off, as well as during the process of a mass transfer.

According to the results from research conducted by a range of authors [42, 44, 45], a tenfold increase of mineralization, sulfates and chlorides concentration (Table 8.15) as well as fluorides, oil products and heavy metals content in surface waters has been observed alongside a hundredfold increase of aluminum, arsenic, mercury and manganese content.

The content of suspended materials slagging the riverbeds has sharply risen. The high sorption capacity of most of the contaminants, including heavy metals, has conditioned the accumulation thereof in benthal deposits and hydrobionts.

According to information provided by special research, [46] from 20 to 30 PAH, including priority ones (benzo(b)fluoranthene, indenopyrene, benzo(a)pyrene and others) are identified in benthal deposits of water bodies in the areas of waste waters from coke and by-products production discharge.

Similar changes have been observed in the chemical composition of ground waters.

Table 8.15. Chemical composition of mine and river waters during a low water period (Rivers of Priazovye are provided as an example) [44]

Water	Sum of ions, g/dm	Ion composition, %-eq[1]					
		Cl^-	SO_4^{2-}	HCO_3^-	Ca^{2+}	Mg^{2+}	$Na^+ + K^+$
the Krepenkaya River							
Mine	2.53	14.7	53.1	32.2	21.0	18.8	60.2
River	2.30	12.8	49.9	37.3	31.2	13.4	55.4
the Nagolnaya River							
Mine	2.00	16.5	58.2	25.3	17.8	27.2	55.0
River	1.86	19.0	59.0	22.1	33.8	10.2	56.0
the Miusik River							
Mine	1.94	10.1	71.6	18.3	19.9	30.0	50.1
River	1.71	10.0	69.8	20.2	26.4	27.4	46.2
the Mius River (taking into account its feeders)							
Mine	2.13	14.4	55.4	30.2	17.9	25.6	56.2
River	1.57	18.1	52.2	29.7	24.2	26.1	49.7
the Krynka River (taking into account its feeders)							
Mine	2.19	18.8	52.8	28.4	13.1	16.9	70.0
River	1.75	14.8	54.1	31.1	33.6	18.1	48.0
the Kalmius River (taking into account the Gruzskaya River)							
Mine	2.78	27.6	56.5	15.9	17.8	19.5	62.7
River	2.42	24.2	59.6	16.2	28.0	19.2	59.8

[1] Amount of equivalent material.

8.4.4. General Principles of Special Surveys for the Local Monitoring Arrangement

The main purpose of special surveys for the local monitoring arrangement and environmental plans in coal mining industry areas and associated industry developments is the evaluation and forecast of changes in the condition of natural objects, revelation of impact zones of intensive technogenic contamination and pollution sources and assessment of the territories' (zones, bodies) ecological status.

Basic principles and provisions used in the system of natural bodies' contamination security monitoring form the basis of special surveys of local monitoring being carried out. [9, 41, 47].

The special surveys (local monitoring) are based on the following principles:

1) Complexity of the survey over the contamination of objects of all natural environments (atmospheric air, soil, surface and ground waters), which is provided for by indices that are subjected to supervision in all natural environments. The evaluation of aquatic ecosystems conditions includes surveys in water, hydrobionts and benthal deposits and is carried out by means of hydrochemical, hydrobiological, toxicological and hydrological indices [1, 48]. For anthropogenic changes, the evaluation and identification of information concerning natural (background) conditions of bodies (water, benthal sediments for a water body) is required as well.
2) Systematic (regular) character of surveys.
3) Representativeness of sampling points at natural bodies.
4) Obligatoriness (reasonability) of evaluation of a natural body's state (water body or part thereof) during the revelation of impact zones with elevated chronic contamination levels pursuant to "Criteria" [37] with the usage of additional parameters and indices (with regard to ground and aquatic vegetation, health of population, etc.). A final conclusion is made on the basis of the results from 2-3 years of research.

During special surveys arrangement, the following shall be taken into consideration:

- Contaminants disposal from main and associated industries (to adjacent territories, to the nearest water bodies, terrain depressions, etc.) may be performed both with arranged run-off (with waste waters) and with non-arranged, distributed run-off (ground and surface-down slope) as well as with gaseous emissions to the atmosphere with its subsequent transfer by air masses.
- Influence of non-arranged contaminants run-off appears mostly during periods of melting snow and rain showers. In areas where permafrost rocks are widespread, intensive carry-over may be observed during the whole warm period. Surveys during periods of spring floods and during rainfall floods are the most informative.
- Flow of substances (including contaminants) running from production facilities to adjacent territories migrates from the watershed on the slope of relief to the nearest water body (or depression). For this reason, the selection of a sampling point at the

water body shall be executed while taking into consideration the place of contamination, source location at the water shed and terrain relief determining hydraulic slope.

- Presence of hydraulic relation between surface and ground waters may shape the contaminating impact of the latter on water bodies.
- During technogenic impact, special attention shall be paid to small water bodies (water flows) with a low dilution capacity potential and low buffer capacity, as well as to local natural anomalies and the probability of its emergence in the survey area.
- For the integral evaluation of technogenic impact by production facilities on conditions of surface water, a comparative method should be used, i.e. comparison of survey in control and background (conventional background).
- For the quantitative evaluation of impact, it is reasonable to draw up a hydrochemical balance for the water body (or part thereof) receiving the flow of contaminants with concentrated or distributed run-off [49].

The arrangement of special surveys for local monitoring calls for a three-stage work performance.

The first stage. Collection of initial data, including large-scale maps and layouts of deposits with production facilities located within the territory of the deposit; data concerning the mineral coal composition, volume and composition of waste waters, used technology, and used chemical reagents; data concerning arranged and non-arranged run-offs (emissions) of contaminants into the atmosphere and water bodies; information on types of treatment, capacities of treatment facilities, etc. and data concerning natural bodies contamination and the presence of other sources of contamination.

The second stage. On the basis of collected information, analysis zones of the highest technogenic load are selected, a structure of the surveys network and approximate list of indices subjected to supervision is formed and a reconnaissance survey of natural bodies (water bodies) is carried out for the revelation of zones with the most intensive technogenic load, as well as its main contaminants (first of all, acute toxic).

The third stage. The survey program is specified, taking into account the revealed intensive technogenic load and specificity of contamination sources; optimization of the analysis layout is performed and particular methods for specific contaminant determination and identification are selected.

This specified sequence of work shall be observed with regard to all natural bodies (atmospheric air, soil, surface and ground waters).

8.4.5. Development of a Survey Network and Program

The density of the survey network for local monitoring in the coal mining industry area and associated industry developments is determined by peculiarities of the hydrographic network, the location of production facilities (sources of contamination), waste run-off, established tasks and economic capacities. The designed network shall include both points of monitoring observations from the Roshydromet network and additional points taking into consideration the locations of contaminant delivery with an arranged run-off and distributed run-off from a territory with lateral afflux from transfer by water masses, etc. The selection of stations, including a background station, shall be carried out in accordance with the specified RD [1]. Surveys on surface waters shall be carried out jointly with ground waters sampling at points within the geological network of the surveys.

Sampling periodicity and terms are established taking into account the specificity of production facilities operation, hydrological regime of water bodies and the regime of wastewaters discharge (winter and summer-autumn low water, spring floods and rainfall floods).

Points of bottom sediment sampling shall coincide with the water sampling network and correspond to zones of the highest levels of water contamination and maximal sedimentation (waste waters run-offs, zones of lateral inflows backing, riffles) as well as take into account the diversity of its types in terms of morphological character [47].

Bottom sediments sampling shall be carried out at least two to three times a year (flood time, rainfall flood, before water freezes over).

For the evaluation of the water body's ecological status, the episodical hydrobionts capture and analysis thereof for toxic (acute toxic) content is necessary as well. Bottom sediments and hydrobionts, sampled beyond the limits of a contamination source's influence, are examined as part of the background.

A recommended list of substances subject to supervision in natural bodies was determined taking into consideration the examined coal mining branch of the FEC, the nature and composition of analyzed raw materials, the technology of its production, its processing and its practical use (Table. 8.16).

Besides, the properties and typical peculiarities of substances, representing the biggest hazard to the environment and human beings, were examined with the application of the following criteria: cancerogenecity, mutagenecity, stability, occurrence, contamination level and others.

Table 8.16. List of supervised contaminants during special surveys on natural bodies' contamination in the coal mining industry areas and associated industry developments

Body	Substances
Atmospheric air	Sulfur dioxide, carbon oxide, nitrogen oxide, ammonia, polycyclic aromatic hydrocarbons including benzo(a)pyrene (selectively), light aromatic, alkane-naphthene hydrocarbons, hydrogen sulfide and sulfides, aluminum, vanadium, cadmium, copper, arsenic, nickel, tin, mercury, lead, zinc
Water	Polycyclic aromatic hydrocarbons including benzo(a)pyrene (selectively), oil components (volatile and non-volatile alkane-naphthene hydrocarbons, gums, asphaltenes), hydrogen sulfide and sulfides, fluorides, iron, manganese, aluminum, vanadium, cadmium, copper, arsenic, nickel, tin, mercury, lead, zinc
Soils, bottom sediments	Polycyclic aromatic hydrocarbons including benzo(a)pyrene (selectively), oil components (hydrocarbons, gums, asphaltenes), coal tars, sulfides, fluorides, iron, manganese, aluminum, vanadium, cadmium, copper, arsenic, nickel, tin, mercury, lead, zinc

Survey data processing and generalization includes the evaluation of water bodies contamination in terms of MPC exceedance (or background) for separate components, a formalized total (for all substances with concentrations exceeding the MPC) index of chemical contamination of the water (ICC), a bottom sediments accumulation factor (SAF) and an accumulation factor (AF) for hydrobionts.

REFERENCES

[1] RD 52.24.309—92 Methodological directions. Nature preservation. Hydrosphere. Arrangement and carrying out of monitoring observations over surface waters contamination on Roskomgidromet network. — SPb.: *Gidrometeoizdat,* 1992. — 67 p.

[2] Methodological indications to directions of Gidrometsluzhba. Methodological directions on arrangement and carrying out of surveys and supervisions over environmental media contamination (OGSNK). — L.: *Gidrometeoizdat*, 1977. — 59 p.

[3] Temporary methodological directions on arrangement and carrying out of monitoring of rivers pollution with run-offs disposed from agricultural lands and salts, fertilizers and pesticides carried over by the run-offs. — L.: *Gidrometeoizdat,* 1981. — 41 p.

[4] Temporary methodological directions on calculation of organic, biogenic substances, pesticides and microelements carried over by river run-off. — M.: *Gidrometeoizdat,* 1983. — 32 p.

[5] Methodological recommendations on substantiation of observation system and calculation of oil products carried over by the river run-off. — L.: *Gidrometeoizdat,* 1990. — 19 p.

[6] RD 52.24.508—96 Methodological directions. Arrangement and functioning of subsystem of transboundary surface waters monitoring. — M.: *Rosgidromed,* 1999. — 44 p.

[7] GSMOS/Water. Practical guide. The 3rd edit. — UNEP / WHO / UNESCO / WMO. Berlington, Ontario, 1992. — 313 p.

[8] RD 52.24.354—94 Methodological directions. Arrangement and functioning of system of special surveys on surface water conditions in the areas of oil, gas and gas condensate deposits development. — L.: *Gidrometeoizdat,* 1994.

[9] RD 52.24.581—97 Recommendations. Arrangement and functioning of system of special surveys on conditions of the environment in the area of coal mining industries and associated industries development. — L.: *Gidrometeoizdat,* 1999. — 38 p.

[10] Convention on transboundary waterways and international lakes protection and usage, Helsinki, March 17,1992. — UNO, 1992. — 23 p.

[11] Protection of Transboundary Waters. Guidance for policy and decision-making. — Water Series № 3 — Economic Comission for Europe. United Nations, New York and Geneva, 1996. — 36 p.

[12] RD 52.04.107—86 Instructions for hydrometeorological stations and centres. Ex. 1. — L.: *Gidrometeoizdat,* 1987. — 183 p.

[13] Instructions for hydrometeorological stations and centres. Ex.7. P. 1. — L.: *Gidrometeoizdat,* 1973. — 476 p.

[14] Instructions for hydrometeorological stations and centres. Ex. 6. P. 1. — L.: *Gidrometeoizdat,* 1978. — 384 p.

[15] Shlychkova V. V., Matveeva N. P., Khomenko A. N., Nikanorov A. M. On monitoring of transboundary surface waters condition // Scientific conference on results of researches in the sphere of hydrometeorology and monitoring of environment pollution in CIS countries devoted to the 10th anniversary of Intergovernmental

Hydrometeorology Council, / Theses of the reports. Section 5. — SPb.: *Gidrometeoizdat*, 2002. — P. 103—106.

[16] Shlychkova V. V., Matveeva N. P., Khomenko A. N., Chestnokova I. A., Musaeva T. S. Conditions of transboundary surface waters // Yearbook of the Russian Federation surface waters quality, 1993 — Obnisk: Ed. VNIIGMI—MTsD, 1994. — P. 319—350.

[17] Izrael Y. A. Ecology and natural environment supervision. — M.: *Gidrometeoizdat,* 1984. — 560 p.

[18] RD 52.24.509—96. Methodological directions. Order of works performance on control of hydrochemical data quality. — SPb.: *Gidrometeoizdat,* 1999. — 55 p.

[19] Nikanorov A. M., Bryzgalo V. A. Freshwater ecosystems in conditions of anthropogenic eutrophication // *Hydrochemical materials.* — T. 114. — 1999. — 266 p.

[20] Hendersson-Sellers B., Marklend H. R. Dying lakes. Reasons and control of anthropogenic eutrophication. — L.: *Gidrometeoizdat,* 1990. — 279 p.

[21] Ecological modifications and criteria of ecological regulation // Papers of international symphosium. Nalchik, 1—20 June, 1990. — Л.: *Gidrometeoizdat,* 1991. — 384 p.

[22] Reymens N. F., Yablokov A. V. Dictionary of terms and notions, related with wild-life protection. — M.: *Science,* 1982. — 144 p.

[23] Odum Y. Ecology. V. 1. / Under ed. of V. E. Sokolova. — M.: *Mir,* 1986. — 328 p.

[24] Odum Y. Ecology. V. 2. / Under ed. of V. E. Sokolova. — M.: *Mir,* 1986. — 366 p.

[25] Mordukhay-Boltovsky F. D., Rivyer I. K. Invertebrates as indicator of water bodies eutrophication // Scientific bases of control of surface waters quality on hydrobiological indexes: Papers of Soviet-English seminar. — L.: *Gidrometeoizdat,* 1977. — P. 28—33.

[26] Plotnikov V. V. On crossroads of ecology. — M.: *Mysl,* 1985. — 208 p.

[27] Methods of natural waters bioindication and biotesting. Ex. 2 / Under ed. of V. A. Bryzgalo and T. A. Khoruzhaya. — L.: *Gidrometeoizdat*, 1989. — 276 p.

[28] Bryzgalo V. A., Nikanorov A. M., Sokolova L. P., Korshun A. M. River of Far East in conditions of modern anthropogenic impact. — *Water resources.* — V. 27, No 2. — P. 245—253.

[29] Bryzgalo V. A., Korshun A. M., Nikanorov A. M., Sokolova L. P. Hydrobiological characteristics of lower areas of the Don river in conditions of sustained anthropogenic impact // *Water resources.* — 2000. — V. 27, No 3. — P. 357—363.

[30] Sirenko L. A. Eutrophication of continental water bodies and some task on its control // Scientific bases of control of surface waters quality on hydrobiological indexes. — L.: *Gidrometeoizdat,* 1981. — P. 137—154.

[31] Datsko V. G., Vasilieva L. G. Yearly phytoplankton production in Tsymlyanskoye water storage basin // *Hydrochemical materials.* — 1964. — V. 36. — P. 75—83.

[32] RD 52.24.620—2000 Methodological directions. Nature preservation. Hydrosphere. Arrangement and functioning of subsystem of freshwater ecosystem anthropogenic eutrophication monitoring. — SPb.: *Gidrometeoizdat,* 2001. — 40 p.

[33] Methodological recommendations on remote-sensing methods of surface waters quality control. Ex. 1 of 3. — L.: *Gidrometeoizdat,* 1987 and 1989. — 32 and 47 p.

[34] RD 52.24.564—96 Methodological directions. Nature preservation. Hydrosphere. Method of freshwater ecosystems contamination evaluation on indexes of phytoplankton associations development. — SPb.: *Gidrometeoizdat,* 1999. —32 p.

[35] Guidance on hydrobiological monitoring of freshwaters ecosystems / Under ed. of V. A. Abakumov.— SPb.: *Gidrometeoizdat,* 1992. — 318 p.

[36] Federal Law "On environment protection" dated 10.01.02 No 7-ФЗ // *Rossiyskaya gazeta.* Jan, 15 2002. (14).

[37] Criteria of ecological situation of the territories evaluation for revelation of environmental emergency situation and zones of ecological catastrophes. — M.: *Minprirody,* 1992. — 58 p.

[38] Methodological directions on principles of system of water bodies water quality observations and monitoring arrangement on Goskomgidromet network within OGSNK framework. — L.: *Gidrometeoizdat,* 1984. — 39 p.

[39] Nikanorov A. M., Strdomskaya A. G., Ivanik V. M. Local monitoring of water complexes contamination in area of high technogenic impacts of fuel and energy complex.— SPb.: *Gidrometeoizdat,* 2002. — 225 p.

[40] Ivanik V. M., Fedorova E. A. Recommendations. Evaluation of surface waters contamination status in the areas of oil and gas deposits development and influence of anthropogenic impact on it. — L.: *Gidrometeoizdat,* 1997. — 48 p.

[41] RD 52.44.2—94 Nature preservation. Hydrosphere. Complex inspection of contamination of nature environment with intensive anthropogenic loading. — SPb.: *Gidrometeoizdat,* 1996.

[42] Posokhov E. V. Sulfate waters in the nature // *Hydrochemical materials.* — 1972. V. 58. — 167 p.

[43] Tatyachenko G. M. Perfection of mines waste dumps extinguishing techniques and technologies. — *Nedra,* 1994. — P. 211—215.

[44] Peltikhin A. S. Mine waters influence on formation of water content of Priazovye rivers // *Hydrochemical materials.* — 1985. — V. 93. — P. 114—126.

[45] Peltikhin A. S. Classification and chemical content of mine waters of central and north-western parts of Donbass // *Hydrochemical materials.* — 1988. — V. 104. — P. 114—126.

[46] Deynichenko N. V. Role of polycyclycal aromatic hydrocarbons in ecology of water bodies of Azov basin // *Author's abstract. dis... cand . of ch. scs.* — Krasnodar, 2000. — 21 p.

[47] RD 52.24.609—99 Methodological directions. Arrangement and carrying out of observations over contaminants content in benthal deposits. — SPb.: *Gidrometeoizdat,* 2001. — 23 p.

[48] Nikanorov A. M., Khoryzhaya T. A., Brazhnikova L. V., Zhulidov A. V. Monitoring of waters quality: toxicity evaluation. — SPb.: *Gidrometeoizdat,* 2000. — 156 p.

[49] Nikanorov A. M. Hydrochemistry. — SPb.: *Gidrometeoizdat,* 2001. — 444 p.

SECTION III.
CERTAIN THEORETIC ASPECTS OF ECOLOGICAL MODELING AND STANDARDIZATION

Chapter 9

WATER QUALITY MONITORING AND INTRA-WATER BASIN PROCESSES

9.1. STANDARDIZATION OF SURFACE WATER QUALITY

Freshwater ecosystems are complex natural bodies in which closely interrelated physical, chemical and hydrobiological processes take place simultaneously. Unfortunately, during their studies, hydrochemists, hydrobiologists and hydrodynamicists do not often work together. Meanwhile, only complex research makes it possible to obtain an objective picture of changes taking place in the ecosystems under the influence of natural and anthropogenic factors.

Nowadays, the unique regulatory criteria of water quality evaluation are maximum permissible concentrations (MPC) of contaminants, which are used in state monitoring system (SMS) of Roshydromet and other monitoring systems.

The list of MPC includes more than one thousand sanitary and hygienic items and more than 1,200 fishery standards for separate components and chemical compounds in the water[2]. Despite the constant increase in the number of regulated indices, the MPC system is not comprehensive and does not update standards for newly created substances and compounds in time. These substances annually increase by the thousands and according to some information, increase even more rapidly than that. In this respect, water quality control with the revelation and evaluation of the role of all contaminants in the water is almost unrealistic both from a technical and economic point of view.

A similar situation is also observed with regard to hydrobiological monitoring, which requires a lot of time and considerable costs to determine the interrelation and role of separate components in the ecosystems of surface waters. According to G.G. Vinberg [1], this causes a rapidly increasing stock of results in biological descriptive research that even nowadays is practically unavailable for surveying, extracting (for the most significant data), generalizing and applying. In other words, there is a *paradox of lack of knowledge in the case where an excess of information* is observed.

---

[2] List of fishery standards of maximum permissible concentrations (MPC) and approximately safe impact levels (ASIL) of contaminants. — M., 1999. — 303 p.

At the time of its implementation into the regime of monitoring surveys in Russia, the MPC system played a definite, positive role in the evaluation of water bodies' contamination and served as a basis during the mutual comparison of separate ecological systems and the degree of contamination across different regions of the country.

In recent years, however, with the newfound knowledge about water systems as functioning mechanisms and the increase of anthropogenic pressure on water basins and water courses, hydrochemists, ecologists and other specialists have expressed opinions have about the system of standards do being satisfactory.

There are some obvious drawbacks of a system of water bodies' contamination evaluation on the basis of MPC, which, according to our opinion, may be outlined as the following:

a. MPC ideology is based on the evaluation of the contaminants' actions on the organismic level, after which those (evaluations) are transferred to the level of population, which is methodologically incorrect. It should be added that test organisms do not always represent all hydrobiological associations of water ecosystems. According to O. M. Kozhova and B. K. Pavlov [2], the concentration of the contaminants that do not lead to considerable disturbance of vital activity of separate zooids in the experiment (in the case of MPC evaluation) may lead to considerable changes of species populations.
b. Criteria system on the basis of MPC does not take into account synergism and antagonism of different contaminants. Such processes as contaminant accumulation by different water organisms, for example, algae, with the subsequent release thereof during mass (seasonal) dying off are beyond the vision of these criteria as well. This drawback is fully related to the failure of taking into account the role of biota and benthal deposits that are actually participating in the processes of contaminant migration and transformation in water ecosystems. These facts were proved by special experiments conducted by us and are described in the upcoming sections of the book.

2) Examined system does not make it possible to evaluate the degree of MPC exceeded and the duration of the high contaminant concentrations' influence on ecological conditions of water bodies. Meanwhile, specified factors are of great significance for the correct interpretation of ecotoxicological consequences of such impact. It shall be added to above mentioned point that the problem of continuous existence (months, years) of not only high but also low contaminant concentrations, as well as its influence on zooids, populations and biocoenoses of freshwater ecosystems, requires scrutiny and special examination.
3) Another drawback deals with the fact that the same values of maximum permissible contaminant concentrations are used for water basins of different physical and geographical zones. For this reason, subsequent comparisons of water body contamination levels cannot be considered accurate in view of sometimes sharp differences between natural environments.
4) Many properties of the contaminants, including those important for surface water evaluation, such as toxicity, depend on a specific hydrochemical situation before they become apparent. Even from the simplest ecotoxicological point of view of contaminants, such as heavy metals, it is very difficult to forecast and take into

account the effects of this impact on water organisms. The same could be related to the temperature, which as it rises, increases the toxicity of most chemical substances as well. According to data provided by some American researches [3], however, for DDT, dimethrine and other compounds, acute toxicity decreases with increased temperature. In particular, the high toxicity of aluminum in waters with reduced pH has been revisited and established in recent years.

5) MPC system does not take into account processes through which contaminants transform after delivery thereof to a water body. Meanwhile, these processes are very complex and take several stages (Nikanorov, Trunov, 1999) [4]. Herewith, intermediate materials are often more toxic than the initial contaminants. Frequently, they are different in terms of toxicity from the initial substances not only by their intensiveness, but also by the mechanism of action [5, 6]. Transformations of phenols to more toxic hydroquinoles or the formation of mutagenic nitrose- or asoaromatic substances may be named as examples.

The specified examples are sufficient to conclude that looking at the MPC as the standard (for example, for sanitary-hygienic reasons) aimed at mankind's heath protection, does not mean that it protects the ecosystem itself. Some toxic components affect different levels of the biota organization in water ecosystems, as shown in Table 9.1 [7].

From the ecological deadlock to which the existing system inevitably leads due to the impossibility of dealing with thousands of new substances, there lies the foundation for a fundamentally different concept of a regulatory and legal framework arrangement for the evaluation of anthropogenic impact on water ecosystems. It shall be based on knowledge of the intra-water basins processes, the type and effects of contaminant transformed and their influence on biota and bottom sediments. In this case, the general reaction of the ecosystem on external disturbances will be studied instead of the reaction of test organisms on separate contaminants.

9.2. Study of Intra-Water Basin Processes in Water Quality Monitoring

In our judgment, the new method for a surface water quality monitoring system development shall be based on large-scale physical, chemical and hydrobiological research of intra-water basin processes in different geographical zones of the country. Water bodies shall be ranged according to types of water use and ecological maximum permissible concentrations; therefore, protecting ecosystems from destruction and preserving them within the framework of water use requirements shall become the ultimate objective.

The aforementioned ecological MPC (or maximum permissible ecological loads MPEL) shall serve as criteria for the ecological well-being of a water system's preservation in accordance with the selected priorities.[3] The full transition of the supervision system from MPC to MPEL is associated with some consequences; in cases when a small amount of acute

[3] In the context of examined issue, the notion of ecological well-being preserves elements of indeterminancy and is ambiguously interpreted by different researchers. It requires special substantiation, which shall be accepted by the scientific community.

toxic (poisonous) chemical substances and compounds gets into the water reaction of the water ecosystem, these measurements may not be adequate to address the hazard that threatens people who are at risk of using contaminated water as potable water or for domestic needs. A reasonable combination of standards from MPC and MPEL employed in protecting both human beings and the ecosystem may be optimal solution.

In other words, an *integrated approach* provides full information not only about the quality and properties of water, but also accounts for the processes running in the water body and creating environment for life forms, is necessary for the full characterization of natural waters. Hence, it follows that preventive measures as a consequence of efficient monitoring for water ecosystems protection will provide the expected result only in cases where all *basic mechanisms of water quality formation in the water ecosystem* are known. It is well known that water ecosystems, as objects of study are extremely complex. Their functioning is subject to a large number of interrelated factors. It is very difficult to model these objects in a laboratory environment. In many cases, the same result may be obtained through different ways or with different combinations of initial parameters. This fact determines the probability of mistakes in cases of transferring research results obtained by looking first at models and then at real water bodies.

Table 9.1. Some effects of toxic contaminants on different levels of biota organization

Level	Direct effect	Indirect effect	Adaptation mechanism
Organism	Physiological disturbances, changes in conduct and fertility, disturbance of development	Disturbance of vital functions because of living environment deterioration, improvement the organism's state due to a decrease of tenseness in the biotic relationship	Physiological, behavioral reactions and modification morphogenetic reactions
Population	Changes of structure, dynamics and speed of biomass turnover	Change of equilibrium populations density, competitive displacement	Adaptive rearrangement of structure, natural selection, imitation
Group of competing populations	Changes of species composition and species diversity (structure of the group) and dynamic structure	Change of total density of specimens in the group, displacement of the group from biocoenoses	Displacement of labile and immigration of resistant species
Biocoenosis	Change of fertility, simplification of vertical structure	Rearrangement of biocoenoses owing to change of abiogenous conditions	Adaptive rearrangement of biocoenoses structure
Biogeocenosis (ecosystem)	Disturbance of circulation of elements, change of hydrochemical indices	Is determined by landscape changes	Adaptive rearrangement of ecosystem structure

The usage of laboratory models (aquariums or other artificially created ecosystems) in water ecology is theoretically possible, but the range of limitations taken into account has two basic peculiarities concerning the models:

- They do not represent an identical copy of any water body, but only, according to the figurative expression of Y. Odum, “living working models” reproducing certain mechanisms of a water ecosystem’s functioning.
- They, to some extent, correctly reflect the modeled situation but herewith the real correlation of true and false is not known exactly at each specific conditions.

At the same time, this approach has some positive features: in laboratory conditions certain mechanisms of a freshwater ecosystem’s functioning may be studied at considerably lower expenses when the measured disturbing influence is known. Besides, it is possible to obtain approximate information concerning the impact of any influence on natural bodies.

There are two ways of overcoming the above mentioned hardships. The first one is to considerably roughen the natural processes described in the models by practically excluding the physical, chemical and biological character of phenomena taking place in water and bottom sediments and replacing the description of these processes with systems of mathematical equations with appropriate factors. This way has certain prospectives. Due to the insufficiency of our knowledge about processes, however, running in depth water ecosystems could result in mathematic modeling that is erroneous in practical application because of forced simplifications.

The second approach, which in our opinion is considerably more realistic, is to include a physical model, making it possible to reproduce basic processes in a miniature model by preserving all the main parameters of modeled ecosystem as an intermediate step between abstract and mathematical models and natural ecosystems in the modeling chain.

Herewith, full-scale physical modeling is understood to be a specially organized procedure of arranging and processing field research when changes on the phenomena scale are possible but nature thereof is preserved.

In terms of water ecology, we tested using high efficiency methods of full-scale modeling; in this way, the model is represented by certain parts of the examined water body itself (experimental subsystem-mesocosm). It is extracted from the maternal water ecosystem in such a way that in terms of examined properties, both systems—modeling and modeled—preserve a high level of similarity during the research process.

The detailed substantiation of techniques and methodology of full-scale water ecosystems ecological modeling is examined in a monograph by T. Nikanorov (1999) [4]. In this paper, only the most important results of the research conducted by the author jointly with N.M. Trunov and employees of the Hydrochemical Institute will be specified.

9.2.1. Technical Means and the Technology of Full-Scale Ecological Modeling

Experience in the practical application of mesocosms at different water bodies shows that in order to conduct short-term model experiments, such as studying the processes of self-purification from easily oxidable substances, studying toxicity dynamics during forecasts of emergency contaminants discharge, etc., the approximate volume of the mesocosms should be equal to 2-5 m^3.

For mesocosms of a specified volume, it is quite easy to select an appropriate area in the littoral zone of practically each water body; these mesocosms are easy to establish and cater to. We used two types of mesocosms that were different both in terms of volumes and in terms of structures of insulating containers [8] (Figure 9.1).

Insulating containers for mesocosms of the first type were rigid columns made of organic glass with volume of 200-250 dm^3 (see Figure 9.1*a*). Mesocosms of this type are used in cases when long-term expositions are necessary during which the process of the experiment shall be changed frequently (for example, in cases of selection of content for acute and subacute toxicological experiments). Containers of this type may be easily installed and removed from the water body by one person. One small water body area that is able to accommodate a sufficient number of the containers for conducting of experiments with repeatability, etc., may be located.

The containers are manufactured from 5 mm. plexiglas. The lower edge cuts into bottom sediments and is made from thicker plexiglas or is bordered from the outside with a stainless steel band and is sharpened. The structure is strengthened with metal angles fixed from outside at the corners.

The water mixture, owing to natural turbulence taking place in real water bodies, is almost absent in mesocosms of this type. The electric mixer fed by an accumulator storage battery is housed in the structure of the mesocosm with due acknowledgement of this fact.

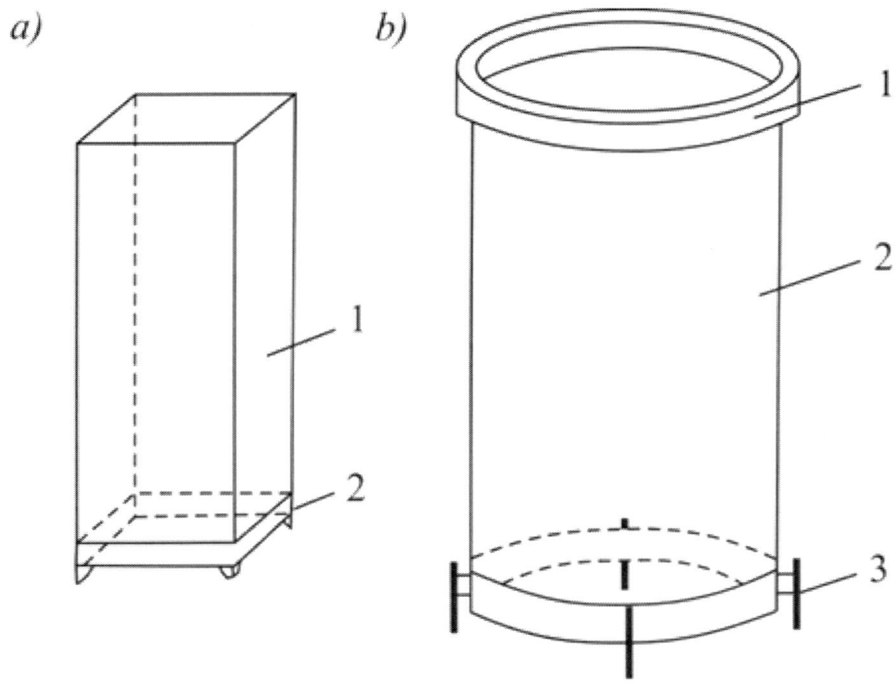

a) Rigid container of plexiglas: *1*—case of plexiglas, *2*—metal angles; *b)* Film-type container: *1*—float of foam plastic, *2*—polyethylene film, *3*—metal base

Figure 9.1. Basic scheme of insulating containers.

To conduct sustained experiments, mesocosms of greater size are necessary. In this case, the microecosystem is insulated with flexible polymeric film fixed in the water body in such way as to secure the microecosystem's insulation on one hand and provide the least distorted picture of turbulent pressure pulses along the entire insulating surface on the other hand. It is necessary to bring together hydrodynamic conditions of modeling and modeled systems, which are extremely important for a long-term experiment.

In the structure of film insulating container, shown on Figures 9.1*b* and 9.2, thin polymeric film made from common polyethylene or polyethylene reinforced with glass fiber makes up the insulating surface. The lower end of the film and its lateral edges, which are joined in a sleeve form, are fixed to a metal base in such a manner as to exclude contact of the internal content of mesocosm to the metal surface. The metal base with fixed its film is hammered into the soil 10-20 cm deep, thus providing not only insulation for the experimental ecosystem but also mechanical stability of construction in case of heavy roughness on the water surface.

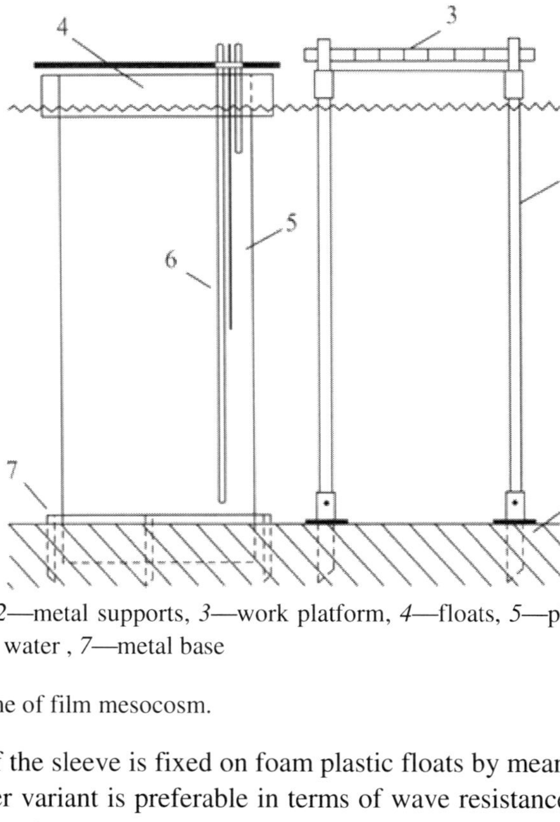

1—bottom sediments, *2*—metal supports, *3*—work platform, *4*—floats, *5*—plastic lateral surface, *6*—pipes for sampling water , *7*—metal base

Figure 9.2. Basic scheme of film mesocosm.

The upper end of the sleeve is fixed on foam plastic floats by means of rigging wires (see Figure 9.2). The latter variant is preferable in terms of wave resistance; besides, in this case, the film is fixated in such a way that the insulated volume is less shaded.

The cross dimensions and height of mesocosms may differ depending on the depth at the area of modeling. Usually, mesocosms with a volume of 2- 5 m^3 are used for one or two-long week experiments.

For long-term experiments, as well as in cases where it is necessary to include representatives of higher trophic links to the experimental ecosystems mesocosms, greater dimensions are required. Herewith, the film enclosures should have a natural bank line with the shallow area as one of its borders (Figure 9.3). In this case, however, it is usually difficult to create numerous identical ecosystems sufficient for the experiment. The reason for the difficulties is the spatial heterogeneity of morphometric indices of the riverbed (channel), properties of bottom sediments, aquatic vegetation, etc. It is worth bearing in mind that such mesocosms are more difficult and expensive to maintain and that experiments conducted therein are harder to reproduce.

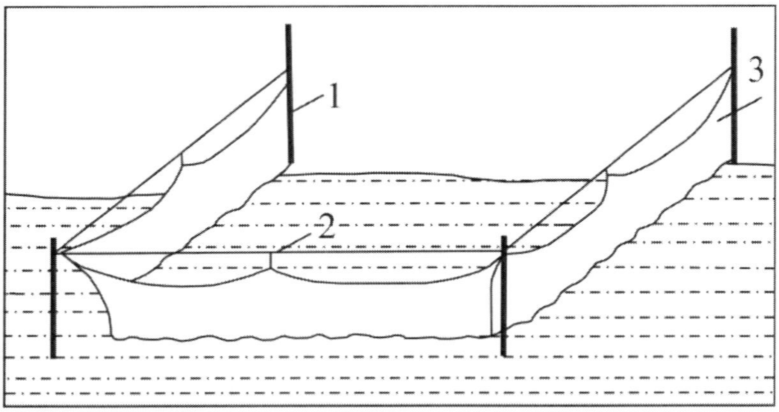

1—metal bars, *2*—rigging wires, *3*—film of mesocosm

Figure 9.3. Film mesocosm with bank line area.

Selection of modeling area and installation of mesocosms. In order to create series of homogenous microecosystems, it is necessary to provide the same basic conditions as that were these ecosystems are formed:

1) Part of the water body in which the mesocosms are installed shall be quite homogenous in a morphometric sense and shall have identical composition of bottom sediments and aquatic population.
2) Mesocosms installed in the water body are included in the experiment only after the adaptation period expires, lasting as a rule for some days during which cross water exchange between all mesocosms takes place.

Herewith, approximately 10% of the volume shall be refreshed in each mesocosm. This is done in order to equalize the initial chemical and biological indices of the experimental ecosystems before the experiment is started.

For easier installation, removal and maintenance of mesocosms in a water body, demountable, wooden ladders and a work platform are also installed (Figure 9.4). The presence of the platform makes it possible to work with several mesocosms at once and not only reduces the time of work performance with regard to all mesocosms, but also sharply increases the quality of executed operations.

On the bank, close to the place where the mesocosms are installed, the first field laboratory is equipped in which water samples not admitted for fixation are processed. Since field experiments are always limited by time, it is essential to have an opportunity to record changes of any indices as soon and as correctly as possible. Mobile chemical and biological laboratories are the most convenient for usage in this context. They are mounted on the base of automobile trailers and have a self-contained power-supply from onboard power generator [8].

Special attention shall be paid to such indices as content of dissolved oxygen, pH and Eh. Especially for work with mesocosms, we built flow-through cells for the measurement of the aforementioned indices (Figure 9.5*a*).

The cell shown in Figure 9.5*b* makes it possible to measure immediately at the water body not only the content of dissolved oxygen but also BOD as well as perform production-destruction analyses.

Application of examined additives, water sampling and assays performance. The scheme of the examined additives application and subsequent water sampling for chemical and biological assays, as well as the programs for chemical and biological research, are assigned by time on the basis of conditions and tasks of the experiment.

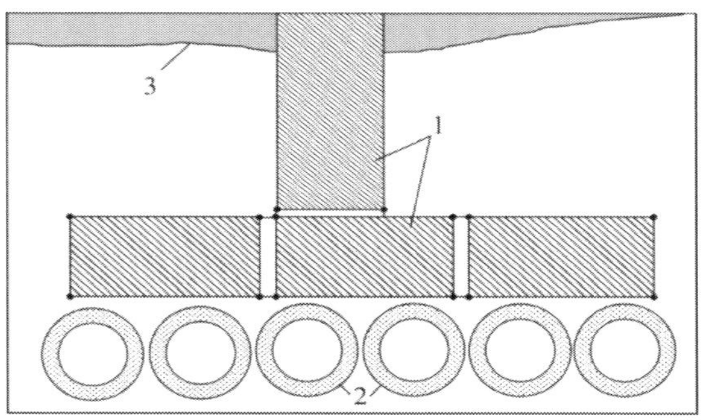

1—demountable ladders, *2*—floats of mesocosms, *3*—bank line

Figure 9.4. Small mesocosms in a water body (top view).

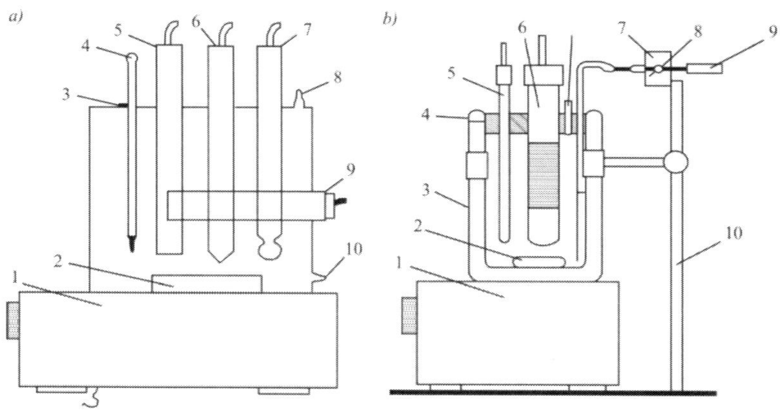

a) 1—magnetic mixer, *2*—magnetic stick, *3*—hole for buffer solutions introduction, *4*—thermometer, *5*—measurement electrode for NO_3, *6*—auxiliary electrode, 7—measurement electrode for pH, *8*—hole for vacuum creation, *9*—measurement electrode for CO_2, *10*—hole for water sample delivery

b) 1—magnetic mixer, *2*—magnetic stick, *3*—glass balloon, *4*—rubber plug, *5*—thermometer, *6*—dissolved oxygen content meter, *7*—hole for vacuum and excessive pressure creation, *8*—hole for water sample delivery and discharge, *9*—valve on intake duct, *10*—support

Figure 9.5. Cell for potentiometric measurement of pH, Eh and the content of NO_3 and CO_2 (*a*) and for oxygen content measurement during the production-destruction processes investigation (*b*).

During additives application to experimental mesocosms, the uniform distribution of the applied substance throughout the entire volume of the mesocosm shall be provided for. Usually, the application of additives is accompanied with water mixing into the mesocosm. Herewith, two important points shall be observed:

1) Mixing shall be quite correct in order to not to disturb the balance in the water system—bottom sediments.
2) The same mixing procedure shall be executed (but without the addition of chemical substances) with the reference mesocosm as well.

Water shall be sampled not earlier than 2-3 hr. after the additives are applied. An exception is made in cases of investigation of swift-flowing processes of an examined substance's interaction with the water ecosystem, for example, the dynamics of toxicity or transformation of easily oxidizable substances.

The strict necessity to reduce measurement error requires the observance of the following important conditions of sampling and measurements *in situ*.

1) Representativeness of sample. Taking into account the special heterogeneity of physical, chemical and biological indices in the mesocosm as well as in the water body itself, water samples for assays shall be taken at different horizons and shall be analyzed separately or with the application of a methodological technique of integrated sample preparation. In cases of instrumental recording of such indices *in situ*, such as the content of dissolved oxygen and temperature, measurements are carried out at different horizons as well. As a rule, the subsurface and bottom horizons are marked out.
2) Completeness of comparison criteria checking. Pursuant to this condition during the experiment, water samples are taken simultaneously in all mesocosms and in the reference area of the maternal ecosystem. The appropriate assays on all physical, chemical and biological indices included in the list of comparison criteria are conducted.

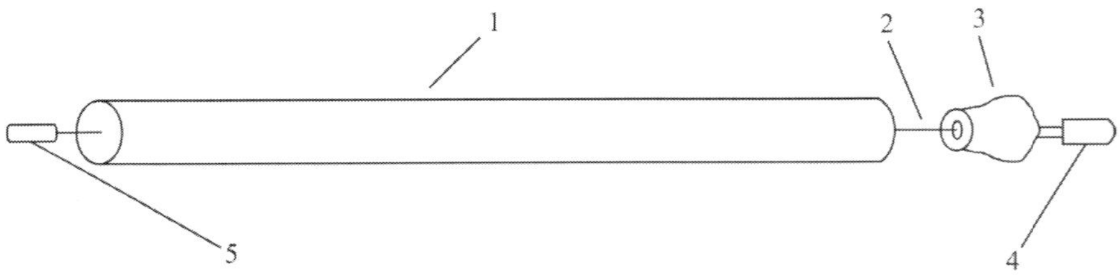

1—plastic pipe, *2*—rod of the lock, *3*—plug, *4*—valve for pouring out the, *5*—handle.

Figure 9.6. Rod-type sampling tool for water sampling at specified horizons.

Herewith, the frequency of sampling is determined by characteristics of the most variable indices. It is known from the experiments that during investigation of most of mentioned processes, the comparison criteria shall be measured at least every 2-3 hr.

For quick sampling of an integrated sample from small mesocosms, taking into account specified requirements, we have developed and used a special rod-type sampling tool in the form of a plastic pipe, the lower end of which is equipped with a plug-type lock (Figure 9.6).

During water sampling, the sampling tool with its lock closed is lowered to a specified depth, then the lock is opened and water fills the sampling tool. After the water has filled the tool, the lock is closed and the sampling tool is hoisted up; the sample is poured out into bottles through a valve in the lower end of the sampling tool.

9.2.2. The Problem of Similarity and Adequacy of Models and the Reproducibility of the Intra-Water Basin Processes Modeling Results

Among the problems, which nowadays restrain the wide usage of physical modeling in natural systems study, the basic one may be singled out—the substantiation of the degree of similarity (resemblance) of the model to the examined natural body (original).

It is known that the problem of similarity is one of the most important and at the same time, one of the most challenging problems of physical modeling. I. Newton laid the foundation of this problem's solution and formulated conditions concerning mechanical phenomena resemblance.

In the sphere of hydrodynamics, the research of V. Frud and O. Reynolds were fundamental. Foundations of general theory, explaining conditions of similarity in any physical nature, are laid out in the papers by N.E. Zhukovsky, D. Relay and F. Buckingham.

At the later stages of scientific development, great contribution to the creation of the modern similarity theory was made by papers from researchers such as A.N. Kolmagorov, V.A. Venikov, L.I. Gutenmakher, I.M. Tetelbaum, A. M. Suchilin, P. M. Alabuzhev and others.

Under physical modeling, such research is understood to support the fact that real processes and objects may be reproduced in an experimental way by preserving their physical nature or using other similar physical phenomenon. A necessary condition of physical modeling is the observation of the principle of object-original and model similarity. Under physical similarity, such correspondence between two phenomena with a common nature is understood to exist so that data obtained on the running processes during the investigation of one phenomena may be applied to many other similar processes [9—14].

The attainment of full similarity between two systems (the examined natural body and its physical model) is in most cases very difficult and very often an unrealizable task. Using a precisely full-scale model when the model is represented by part of the real natural system (object of investigation) preserves its maximal similarity to the original with regard to the most important indices. Only in these cases can the researcher find simple similarity criteria and carry out the model experiment correctly.

Herewith are the most important *provisions of similarity theory*:

1) During the modeling of such complex systems as intra-water basins processes, only the most important details in specific research are permitted to be extracted and taken into account.

2) During this research, it is permissible to use approximate similarity and factors; those that have an insignificant influence on natural processes may be modeled roughly or neglected.

Indeed, any system, any natural phenomena, may be related with an immense number of all kinds of processes and trials of cognition thereof, so it cannot be effective without the creation of a model, simplified in comparison with the original. This simplification, however, shall preserve these aspects, which are significant during the examination of the complex system.

In other words, pursuant to the first provision, understanding the unattainability of absolute similarity, *it is important to arrange the model investigation in such a way as to provide reflection of the investigated object by the model only with respect to those aspects that are significant in this investigation to the specified task.*

The second provision states that, *similarity theory admits approximate similarity at which point factors having insignificant influence on natural processes may be modeled roughly or neglected.* In other words, during approximate modeling there are deliberately admitted and assessed errors related to some degree with the simplification of the physical representations concerning the system [10—12].

As was previously mentioned, natural ecosystems include a large number of interacting components; these systems do not have clear boundaries and the role of casual factors is significant, etc. Even the composition of natural water is almost impossible to model in laboratory conditions [15].

Successfully overcoming these difficulties in the practice of natural modeling is possible in cases of the application of the notion of a *model's adequacy* besides the notion of ecosystems similarity. A model system, by means of which the established task may be successfully attained, is called "adequate" for this purpose.

It shall be emphasized then, that the notion of adequacy does not coincide with notions of completeness, exactness and correctness (validity); adequacy means that these requirements are observed not generally but only to the extent that is sufficient to attain the purpose. It is necessary to examine a model's adequacy as one of basic principles of creation thereof because any model of a water body is fundamentally poorer than the original; it is the fundamental property of the model.

The problem of a physical model's adequacy regarding its original (modeled natural system) can be solved only through its maximal convergence with regard to the most important characteristics of modeling and modeled systems [16—22].

In actual practice, it is implemented through the creation of spatially insulated subsystems (experimental microecosystems) described above. Herewith, the definite volume of the examined water, jointly with the populating communities of autotrophic and heterotrophic organisms, appropriate area of bottom sediments and sometimes part of bank line, are included in the experimental microsystem.

Such artificially insulated microecosystems have sufficient similarity with the examined maternal ecosystem with regard to many of its basic properties. They are also available for thorough internal and external inspection. For these reasons, such microecosystems are one of the most important instruments in making it possible to study chemical and biological processes in water ecosystems. Herewith, any kinds of ponds, plastic and metal enclosures, artificial pools, trays and other structures are used as experimental ecosystems.

These experimental ecosystems are obtained by means of insulation of specific volumes of water and bottom sediments in natural conditions for modeling natural water ecosystems and are defined by the term, "mesocosms."

For a long time, this term was used synonymously with the term, "microcosm." In 1992, in Potsdam, however, a European working group (European Workshop on Freshwater Field Tests) decided to delimit the terms "mesocosm" and "microcosm" depending on the dimensions of the experimental microsystem.

Research on mesocosms is conducted at a transient stage from laboratory experiments in aquariums when compared to the natural water body [19, 23, 24]. The closer the basic parameters of the experimental and examined ecosystem are, the greater is the extent of correspondence of the results obtained on the model to the characteristics of the examined natural ecosystem. At the same time, large mesocosms require much more material and financial means for their creation and maintenance thereof. Besides, it is more difficult to reproduce any experiment carried out at such mesocosms.

Most commonly, dimensions of mesocosms are determined by the specific conditions of the planned experiments and first of all, by the number of trophic levels within the experimental ecosystem.

In a review, V.V. Alekseev [25] diagrams the correlation between the volume of a modeled ecosystem and the number of trophic levels it has. It is clear from the diagram that each new trophic level corresponds with a considerably greater volume. Mesocosms containing side by side producers and consumers of the first and the second order and intended to last for a period of several months shall have a volume of at least 1,000 m^3. This volume is necessary to maintain populations at a sufficient level for the alimony of the highest trophic link (i.e. baby fishes) as well as for sampling.

For modeling metabolism of a planktonic community with phytoplankton, bacterial plankton, protozoa and rotifers, a sufficient volume for the experimental microecosystem should be approximately 0.005 m^3. At the same time, if predators are included in the experiment, volumes of the insulated microecosystems shall exceed 102-103 m^3 [22, 23].

In certain papers, it is specified that there exists a high similarity of hydrochemical indices in mesocosms and a water body, where the identity of plankton contained therein [26], certain increase of animal plankton biomass due to absence of its consumption by fish [27] and the preservation of qualitative composition of the plankton community, are the same outside and inside the mesocosm for four weeks [28]. Mesocosms with a volume of 50 dm^3 are recommended for usage during a period of time not exceeding 4 days [27], 300 dm^3—not exceeding 10 days, 600 dm^3—not exceeding 18 days and 1.5 m^3—not exceeding one month [29]. During comparison of the experiments, results in the 1.5 and 30 m^3 mesocosms did not reveal considerable differences in the dynamics of biogenic substances of phytoplankton, zooplankton and bacterial plankton. These results as well as pure economic considerations make some researchers recommend 1–2 m^3 capacities for ecotoxiclogical experiments with planktonic communities without fish in eutrophic water objects and 10 m^3 capacities in oligotrophic water objects [30, 31].

Most mesocosms include only planktonic communities, although model ecosystems including benthal organisms and fish have been created as well [32, 33].

The range of experiments in mesocosms conducted by us from 1989-1990 for the evaluation of the contaminant's transformation speed makes it possible to judge the approximate degree of similarity of modeled ecosystems in the period of monitoring.

Mesocosms examined in the experiments isolated square columns of water (together with benthal deposits) with sides of 2 x 2 m and a height of 1.8 m. Its volume amounted to 6.7-6.8 m^3. In 1989, during experiments at Isakovskoe Reservoir (Luganskaya region, Ukraine), where the objects of investigation were mercury and nitrates, a running range of hydrochemical and hydrobiological parameters in reference to the mesocosm and water basin was obtained (Figures 9.7 and 9.8).

In 1990, for the examination of copper inflow to ecosystem components, 4.5–4.8 m^3 mesocosms were installed at the small Lake Krivoe (Rostovskaya region). The mesocosms were represented by cylinders with diameters of 2 m with soft walls of polyethylene film, the edges of which were submerged in bottom sediments by means of a special foundation. On the basis of the analysis of indices, characterizing the state of the model biocoenosis (see Figure 9.8) leads to the conclusion that quite a good reproduction of chemical and physical processes dynamics may be made during the monitoring period.

The mathematical treatment of the results of experiments at Lake Krivoe (Figure 9.8*a, b*) made it possible to deduce the appropriate descriptive equations on the basis of a graphic representation of species compositions and phytoplankton biomass dependence for the duration of observations (Figure 9.9).

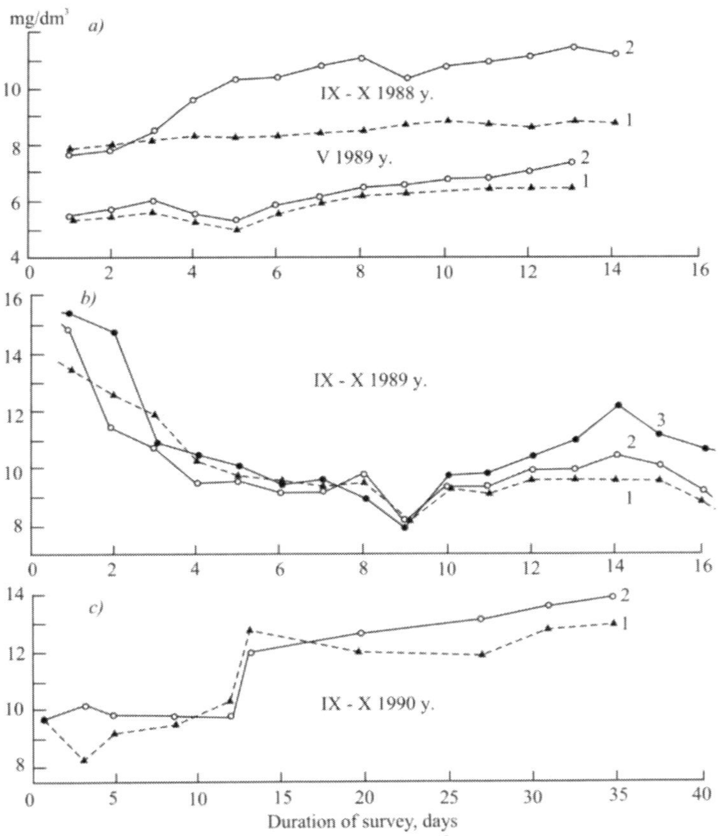

a) The Seversky Donets River: *1*—water body, *2*—1.1 m^3 reference mesocosm; *b)* the Isakovskoye Reservoir: *1*—water body, *2* and *3*—1.1 and 6.8 m^3 reference mesocosms; *c)* Lake Krivoe: *1*—water body, *2*—4.5-4.8 m^3 reference mesocosms

Figure 9.7. Dynamics of dissolved oxygen concentration (mg/dm^3) in mesocosms and water bodies.

Specified dependencies had the following general view:

$$y(t) = A\exp(-bt)\cos(ct) + \alpha \quad (1)$$

where A is the initial height of the diagram, b is the curve decay rate, c is the rate of curve growth after passing minimal values, t is time and α is the shift of the diagram along the Y-axis.

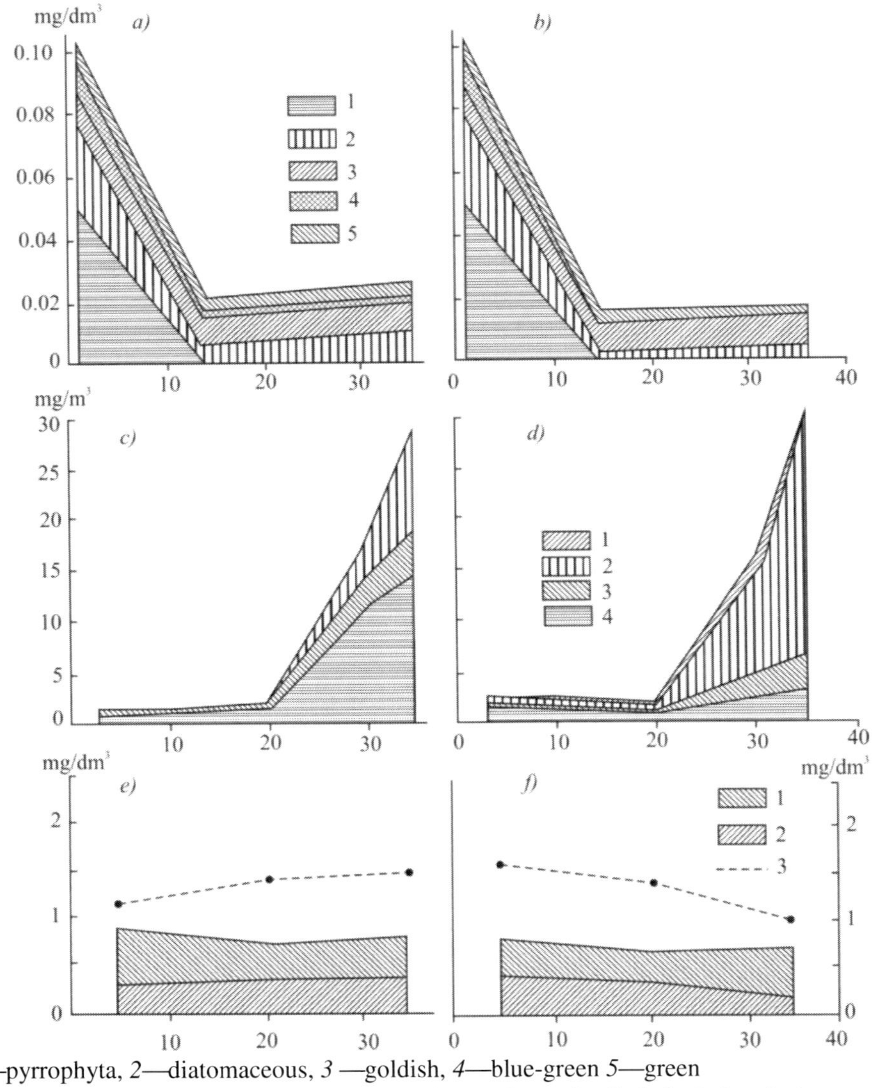

a, b: *1*—pyrrophyta, *2*—diatomaceous, *3* —goldish, *4*—blue-green *5*—green
c, d: *1*—Cyclopoidae a, *2*—Calanoidae a, *3*—Chydoruacphaer, *4*—Ceriodaphniaretical
e, f: *1*—bacillus, *2*—coccus, *3*—number of bacteria

Figure 9.8. Dynamics of species composition and plankton biomass (*a, b*), zooplankton (*c, d*) and bacterial plankton (*e, f*) in Lake Krivoe (*a, c, f*) and in a 4.5 m³ reference mesocosm (*b, d, e*).

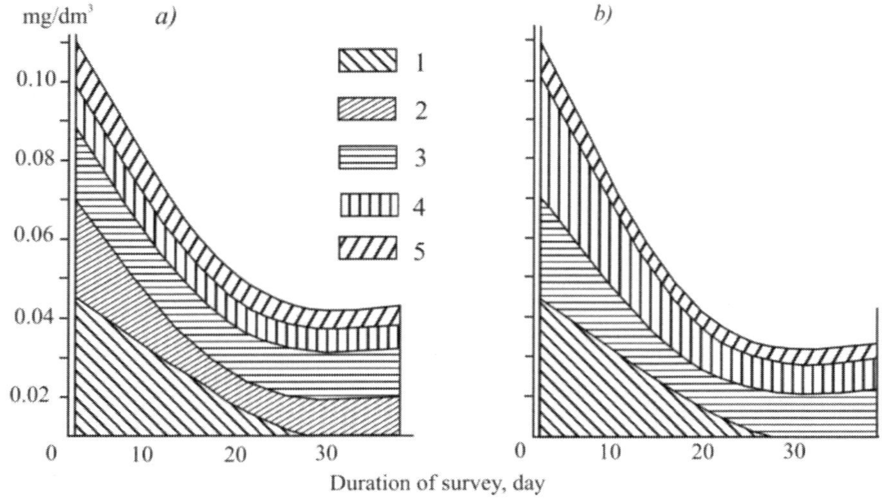

1—pyrrophyta, *2*—diatomaceous, *3*—goldish, *4*—blue-green, *5*—green

Figure 9.9. Dependence of plankton biomass species composition on duration of observations.

During full-scale experiments conducted at Lake Balaton (Hungary) [34], similar (in terms of structure) mesocosms with an insulated volume of 17 m^3 were used to make forecasts of the eutrophication processes. Satisfactory similarity during 30-35 days with the preservation of biocoenosis structural composition and with a certain increase of biomass of the communities was noted. The usage of opaque, white film reinforced with cloth in the structure created a certain darkening inside the mesocosms, reducing the influence of the mesocosm's increased diameter with respect to its depth.

Therefore, it is possible to judge the degree of the model approximation to the original by comparing the range of parameters describing the structure and properties of the modeling and modeled objects.

A great variety of water mesocosms used by different authors for the investigation of different chemical substances' interaction became the main reason for the development of peculiar standards regarding the mesocosm structure in order to conduct the experiments and process the obtained results. For this purpose, meetings of working groups of specialists regarding the theoretical and applied issues of mesocosm applications for the natural environment's protection were held in the USA and Europe [23].

The observed expansion of mesocosm usage in theoretical and applied works may be explained by the fact that this research makes it possible to reveal general regularities of an ecosystem's reaction to unfavorable factors of the environment as well as to forecast structural and functional transformations of natural ecosystems under the influence of some contaminants. In some cases, experiments conducted with the usage of mesocosms become elements of compulsory programs for chemical substances and surface waters interaction. Thus, in the procedure of state ecological registration of new pesticides in some countries (the USA, Canada, Germany and others), the examination of new preparations in experimental conditions on mesocosms is provided for [23, 35, 36].

Despite the great variety of schemes of experiments with the application of mesocosms (microcosms), the dimensions and the structural peculiarities thereof, two main groups of used microecosystems may be selected nowadays:

1) Experimental microecosystems (microcosms and mesocosms) created outside beyond the examined water body (small ponds, pools, plastic or metal capacities, etc.)
2) Microecosystems placed immediately in an examined water body

The study of experimental microecosystems is similar to the study of water bodies. As a rule, all basic hydrochemical indices, pH, Eh values, content of dissolved oxygen, nitrate, nitrite, ammonium nitrogen, mineral and organic phosphorus and silicon, etc., are monitored [38, 39]. From the list of hydrobiological indices, preference is given to those that have a relatively short response period, among which are chlorophyll content, number and biomass of microorganisms; protozoa and aquatic plants, as well as species composition of the latter; intensity of photosynthesis; production and destruction of organic matter, etc. [39].

Despite the visual identity of the processes running in experimental and maternal ecosystems, it is necessary to provide for functional similarity of the examined natural ecosystem and its model—experimental microecosystem—in order to transfer the natural ecosystem's data, which is obtained as the result of the experimental ecosystem, or the model's, examination. Herewith, the accuracy with which any object of a natural ecosystem (or the whole ecosystem) is modeled in the experimental ecosystem, to a large extent, depends on the task of survey. If there are no similarity criteria, it is practically impossible to assess theoretically the degree of modeling and modeled system similarity.

Therefore, the main problem nowadays is in restraining the wide usage of full-scale chemical and biological modeling, by means of experimental microsystem mesocosms, for the investigation of basic inter-water basins processes in the absence of quantitative similarity criteria of modeling and modeled ecosystems.

Taking into account the specificity of used physical models (experimental microecosystems created at the water body itself), we will consider that for the transfer of data, obtained as a result of experimental ecosystems investigation, to natural ecosystems, the functional similarity of the model and modeled object shall first be provided for. Herewith, as it was already mentioned above, the accuracy with which the experimental ecosystem models this or that element of a natural ecosystem (or of the whole ecosystem), to a large extent, is determined by the tasks of investigation.

It shall also be taken into account that *strictly speaking, it is possible to experiment with natural ecosystems only by taking into consideration the fact that in cases of a repeated attempt, it is often impossible to precisely provide even basic conditions of the experiment: brightness, cloudiness, wind direction and force, air and water temperature, inflow of allochtonous organic matters, etc.*

The *reproducibility of the results* obtained in parallel experiments, however, is one of the indispensable conditions of any modeling. The selection of a number of replications directly related with the costs for the experiment depends on this condition. Works with 4-5 m^3 and larger mesocosms are quite labor-intensive and the number of chemical and biological definitions is limited in terms of possibilities. Unfortunately, in practice, one has to be restricted by two, rarely three, replications [40, 41].

Figure 9.10 shows the results of research in mesocosms with two replications with conditions at a river and a lake provided as the example. The experiments were conducted for the evaluation of methanol transformation rate and the rate of copper ($CuSO_4$) inflow to components of the water ecosystem. The initial and calculated values of content are always to some degree different in parallel experiments since volumes of mesocosms with soft walls cannot be calculated exactly. During the test with copper at Lake Krivoe, the experiments were conducted in mesocosms of different volumes. Preliminary conclusions can be made that in the case of experiments with a duration of 20-30 days and a mesocosm volume equal to 4-5 m^3, quite sufficient and examined processes are reproducible mesocosms installed in homogenous conditions with due checking of their water resistance.

Mathematical processing of experimental data on Figure 9.10 makes it possible to describe dependencies of changes in methanol and copper concentrations over time through the following systems of equations:

1) For Figure 9.10*a*:

$$y_1(t) = 240\exp(-0.30t) \quad (2)$$

$$y_2(t) = 240\exp(-0.36t) \quad (3)$$

In general, these types of dependencies are representable by the function:

$$y(t) = A\exp(-bt^m) \quad (4)$$

where *A* is the initial ordinate (height) of the diagram, *b* is the degree of the curve decay and *m* is the curvature of the curve; these dependencies are related to the category of generalized exponentials.

2) For Figure 9.10*b*:

$$f_1(t) = 40K_2(t) \quad (5)$$

$$f_2(t) = 40K_3(t) \quad (6)$$

In a general view:

$$f(t) = 40K_v(bt^m) \quad (7)$$

where $K_v(z)$ is the cylinder function of imaginary argument (Macdonald's function), *v* is the curve removal from the origin of coordinates and the other designations are the same.

3) For Figure 9.10*c*:

$$q_1(t) = 150\exp(-0.61t) \quad (8)$$

$$q_2(t) = 150\exp(-0.70t) \quad (9)$$

These dependencies represent simple exponential functions.

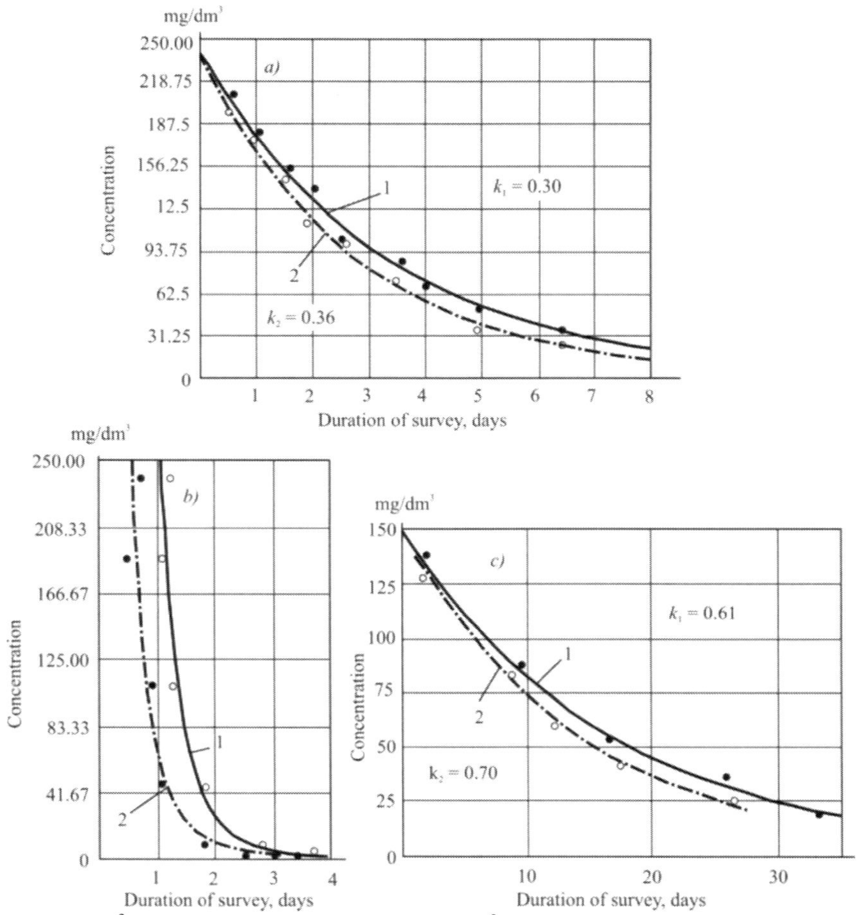

a) 1 and *2*—1.07 m^3 mesocosms, *b) 1* and *2*—1.15 m^3 mesocosms, *c) 1* and *2*—4.8 and 12.7 m^3 mesocosms

Figure 9.10. Process of methanol (*a, b*) and copper (*c*) concentrations in mesocosms and self-purification rate factors for the Seversky Donets River in October of 1988 (*a*) and May of 1989 (*b*), and for Krivoe Lake in September and October of 1990 (*c*).

The experience of full-scale modeling states that the response of a model ecosystem to toxic impact, to a greater extent, corresponds to the response of natural bodies more than the response of laboratory models. Jointly with the increased of accuracy in reproduced natural conditions of model ecosystems, technical and economic difficulties of its creation will increase as well. Therefore, requirements for the models are contradictory; the completeness of the model contradicts its simplicity and the accuracy of the model—its dimensionality, efficiency of the models—impacts its costs implementation. The success of such approach largely depends on the type of compromise selected between these contradictory criteria. In general, modeling, according to a neat expression from R. Shennon, is an "inseparable combination, [an] alloy of science and art" [42].

Understanding of this fact compels researchers to decline the paradigm of classic experimentation, making it possible to divide any complex phenomenon observed during the experiment into utterly small and insulated parts—separate pairs of indicants, separate pairs of reactions, etc., in such a way that any controlled indicant or parameter might be quantitatively examined while maintaining of all other variables at the same level. This means that in the examined approach, there are no reasons to consider the investigated indicants to be bounded so feebly that they are uncontrolled in the experiment levels and could not considerably influence the results of the investigation. Therefore, in this case, it is impossible to accept the classic scheme of the experiment formed in earlier times under the influence of R. Fisher's statistical representations . His findings might have worked perfectly in cases of relatively simple systems when the task of the experimenter was to have minimum influence on the process of the experiment.

During examination of such complex objects as the natural ecosystem, a great number of uncontrolled factors in the experiment are sources of considerable variability. These factors lead to a total dispersion so large that it is extremely difficult to single out any regular tendency on its background. For these purposes, it is necessary to examine quite a wide, varied range of analyzed variables. In other words, this means that in practice, the frequently neglected, great, aggregate dispersion heavily disguises the regular tendency [43].

The physical modeling of natural ecosystems at full-scale conditions does not make it possible to study separate factors independently and simultaneously observe all relations and parameters of the controlled experimental process. In the real life environment of a full-scale experiment, it is not possible to exclude the influence of weather, seasonal and other conditions, when deeply modeling all the processes in examined ecosystem. This means that the ideology of the experiment conducted with regard to such a complex, multicomponent system shall be based on the simultaneous variation of the biggest number of factors that have an influence on the system, if possible.

A single-factor experiment provides results that are fair only for those conditions at which it was conducted. At the same time, at the early stages of a multifactor experimentation's methodology development, when our knowledge about examined ecosystems is insufficient and there is the hazard that factors having a heavy influence on investigated system may get to the number of uncontrolled factors, *it is much simpler and safer to use the total variability of the system by naturally changing the environmental factors.* The justifiability of such approach becomes even more evident when taking into consideration the fact that the purpose of experimentation consists, as a rule, in studying of natural ecosystems at conditions of their natural (or close to their natural) functioning.

9.2.3. Similarity Criteria for Water Ecosystems

There are different ways of evaluating model similarity. Some of them will be examined below. The most acceptable method for physical models is to look at the state variables of the object (continuous and discrete surveys) and state variables of the model evaluation within the examined period of time, during which the similarity is preserved. The definition of the principle of ecological similarity formulated by us in 1989 complies with the aforementioned conditions: *"ecological systems are similar if at similar environmental conditions, at similar dimensional conditions and at similar periods of time values (parameters), the characterizing state of the system is proportional to the values (parameters) of another system regardless of the dimensions of the compared systems."* With regard to modeling in mesocosms, this means that processes in these mesocosms shall run in the same directions even though they may be different in terms of intensity and scale. In case of long-term experiments, researchers determine these differences during their observations. The degree of the differences' acceptability shall be taken into consideration, but it shall be correlated with the *purpose for modeling*. For instance, a mesocosm volume and structure may comply with the requirements of a weeklong experiment for the calculation of a self-purification rate factor for a one-time contaminant application and may not completely correspond to the investigative purposes of consequences in the case of continuous contaminants flowing into an ecosystem with contents close to background ones.

Taking into consideration the complexity of chemical and biological similarity criteria development, we proceeded from the fact that an *experimental ecosystem maximally approximated to the modeled one in terms of chemical and biological indices is used as a physical model of the studied ecosystem.* This condition results from the analysis of basic provisions of similarity theory conducted above and pursuant to which physical modeling of such complex systems as water ecosystems may only be approximate. *The simpler and safer the criterial treatment of such modeling results, the less the original and the model differ from an examined aspect.*

Taking into account the initially approximate type of modeling and stochastic nature of factors influencing the ecosystems, let us analyze the *principal components method* as the basis of similarity criteria development. The theoretical possibility of such an approach's application is based on the fact that the statistical distribution of a multidimensional, random variable subjected to certain laws is invariant with respect to different coordinate systems. Analytically, this means that the same multidimensional statistical object described by a correlation matrix in different coordinate systems corresponds to different matrices. These correlation matrices are transformed during the transition from one coordinate system to another by means of similarity transformation.

The essence of the final transformation of the principal components method consists of the linear transformation of the initial data matrix into a new matrix, possessing a range of important properties for the purposes of the examination. First of all, there is no correlation between the new variables and the preservation of the dispersion value in case of coordinate axes rotation. In this case, the dispersions are redistributed in such a manner that the first principal component (the first proper vector) represents the biggest dispersion. Subsequent proper vectors represent gradually decreasing dispersion. Thus, the basic part of dispersion of observed indicants correlates with the number of proper vectors and each vector represents a known share of the total dispersion [44, 45]. Herewith, the fact that *all initial statistic information collected during the experiment and contained in the initial data matrix is preserved in the correlation matrix and has obtained its factor patterns* is completely evident.

The factor pattern is represented by a system of linear equations:

$$Z_1 = a_{11}F_1 + a_{12}F_2 + ... + a_{1m}F_m$$
$$Z_2 = a_{21}F + a_{22}F_2 + ... + a_{2n}F_m$$
$$\cdots\cdots\cdots\cdots\cdots\cdots\cdots\cdots$$
$$Z_m = a_{m1}F_1 + a_{m2}F_2 + ... + a_{mm}F_m \quad (10)$$

where $Z_1Z_2,..., Z_m$ is observed during the experiment's chemical and biological variables; F_1, $F_2,..., F_m$ are new variables (components) and a_{11}, $a_{12},..., a_{mm}$ are the elements of the factor pattern, named as component (factor) loads.

Such a system has only one solution. This means that factor patterns of this type may be used in the identification of similarity between two systems close in terms of properties.

The system of equations (10) may be easily solved with respect to the variables F_1, F_2,..., F_m, which makes it possible to rewrite it in a more convenient form for interpretation:

$$F_k = (a_{j1}Z_{j1} + a_{j2}Z_{j2} + ... + a_{jm}Z_{jm})/\lambda_k \tag{11}$$

where j = 1, 2,..., m; k = 1, 2, p; a_{jt} is the elements of the factor solution, i = 1, 2,..., n; Z_{jm} represents the initial variables; λ_k is the kth own value of the initial data matrix and p is the number of preserved principal components.

Knowing the main components, it is possible to interpret the behavior of the examined system in terms of two to three principal components that explain practically all of the observed changes within the controlled indices. Herewith, *such an interpretation is frequently much more succinct and impartial than an interpretation using a bigger number of isolated indices* [1, 46].

Taking into account the aforementioned points, let us formulate *the first similarity principle: two ecosystems are similar in space of compared indicants if they have factor patterns that are indistinguishable in statistical sense.*

This obvious idea is difficult to implement in practice because each specific factor pattern not only factor loads, but also the components itself have a certain distinctiveness. In order to establish similarity relations between two compared systems, the following technique is used. On the basis of the measured indices during the experiment, the appropriate factor patterns for both systems are calculated in which the necessary number of principal components is preserved by taking into account the total dispersion described by these components. Usually, as is accepted in most of research using principal components methods, the number p of principal components interprets 70-80% of the indicants of the total dispersion that is selected [44, 47]. By examining the remaining principal components as new variables, the new principal components, generally for two compared systems, are obtained through the second factor of the transformation.

Having a common comparison basis, *the second similarity principle* may be stated: *two ecosystems are similar in the space of compared indicants, if the regression equations of each indicant is built on common principal components from the system of mutually compatible equations.* Accordingly, there is no similarity if the regression equations built on common principal components for the compared ecosystems are different in terms of at least one of the selected (significant) indices.

Selection of space for the comparison criteria as well as the determination of significance values during all statistical calculations is fully under controlled by the researcher. In cases where very strict (restrictive) conditions are selected, the similarity may not be confirmed when they actually take place and vice versa, when selecting more permissive comparison conditions, the correct conclusion may be reached faster, although the probability of erroneous result is greater. In this case, it should be reasonable to endeavor for full coverage. The mechanical inclusion of the widest range of variables for the list of controlled indices does not make sense. It is essential to select a space of comparison criteria to get the interaction of basic structures and functions of the examined system reflected in it.

The analysis of the results of numerous experiments conducted in accordance with the specified methodology at which all examined chemical and biological indices are determined in experimental and reference microecosystems as well as in maternal ecosystem indicates the presence of two main reasons for the absence (disturbance) of similarity between modeled and modeling ecosystems.

The first reason deals with the presence of inevitable artifacts conditioned by the imperfect technique of experiment conducting. Here, the following sources of errors are included:

- Incorrect (suboptimal) selection of place of experiment
- Insufficient volume of experimental ecosystems
- Reduction of brightness due to light absorption and reflection by walls of the mesocosm
- Shuttering and accounting of resources and metabolism products inflow and outflow
- Limitation of organisms migration in closed space which causes the effect of community "shocking"
- Encrustation of the mesocosms walls during long-term exposure in water bodies
- Too long a period of experiment
- Difference of hydrodynamical conditions in modeled and modeling ecosystems

The influence of the hydrodynamical factor, which is quite significant in inter-water basins processes, was taken into account at some special experiments through the method of gypsum structures, giving good results in case of modeling in small volumes and with little water movement [48]. This method makes it possible to examine water movement at a specific point of the volume. The index reflects the intensity of gypsum dissolution per unit of sphere mass (μ_w). Figure 9.11 shows the results of the experiment at Lake Krivoe with bunches of gypsum spheres installed vertically in mesocosms of different volumes and in the maternal ecosystem itself (lake).

Dependencies between the intensity of the gypsum balls' dissolution and the depth of the lake are described by the following equations:

1) For Figure 9.11*a*:

$$y(h) = 15.252 - 6.338h \tag{12}$$

2) For Figure 9.11*b*: line *1*

$$y_1(h) = 14.723 - 6.736h \tag{13}$$

Line *2*

$$y_2(h) = 14.879 - 6.164h \tag{14}$$

Inspection of the experiment's purity in terms of preservation of this type of similarity, which we named similarity of the first type (Nikanorov, Trunov, 1999) [4], is carried out by conducting parallel measurements of all observed experiment indices with reference to the experimental and natural ecosystems along with the calculation of statistical similarity criteria.

Figure 9.12 specifies the time dependencies of two indices (pH and concentration of O_2) for the background area of the Isakovskoe Reservoir and for reference the mesocosm were obtained during the experiment.

The dynamics of both indices have minimal divergence within the limits of measurement errors in the experimental and reference mesocosms within the time period, T_{sim}, during which the processes are practically identical. To increase this time, it is necessary to augment the similarity of the experimental microecosystem with the maternal one, for example, by means of an increase in the mesocosm's volume.

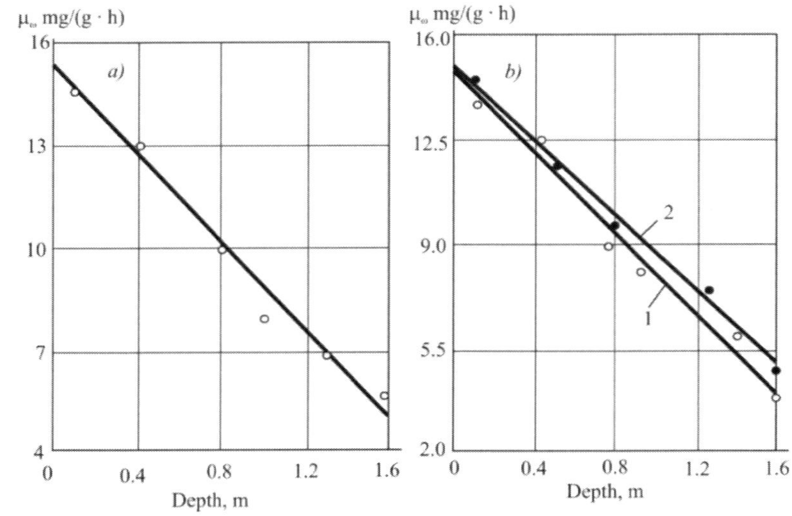

Figure 9.11. Intensity of gypsum balls dissolution (μ_w) acc. to the results of seven repeated experiments at the lake (*a*) and in the mesocosms (*b*) with a volume of 4.5 m^3 (*1*) and 12.7 m^3 (*2*).

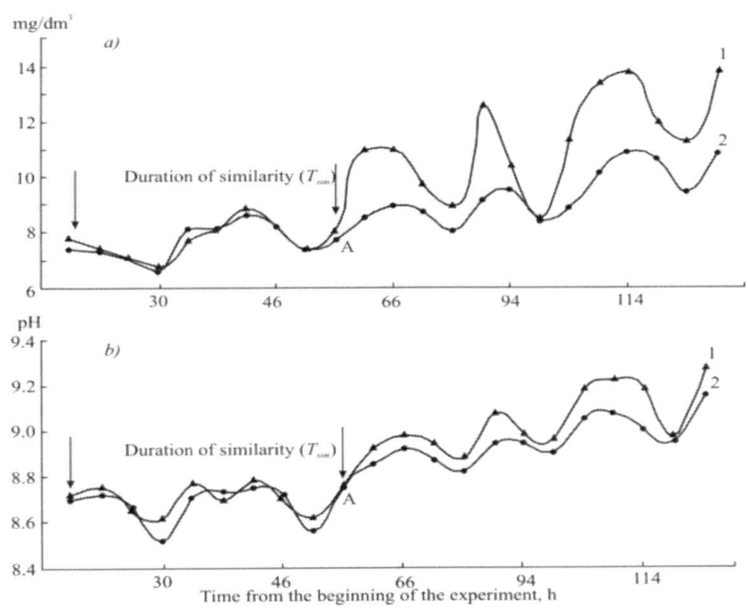

Figure 9.12. Dynamics of concentration (mg/dm^3) of oxygen dissolved in water (*a*) and pH (*b*) in the Isakovskoye Reservoir (*1*) and in the reference mesocosm (*2*).

The second reason for similarity disturbance is related to the impact of the examined factors on the experimental microecosystems, as a result of which the similarity between the experimental and reference experimental ecosystems is broken.

Analysis of the experiment results shows that the dynamics of oxygen concentration dissolved in water is characterized by the ramification of point *A* (see Figure 9.12*a*), which separates identical processes running in the Isakovskoe Reservoir and in the reference mesocosm from the processes in which the divergence of values measured in the water-storage basin and mesocosm volume takes place. The reason for such a divergence is clear; the examined processes run in a different volume of water. Two processes, however, the change of oxygen concentration dissolved in water in the natural environment and in the mesocosm, is similar to each other but have different similarity ratios before and after the ramification at point *A*. Indeed, the two processes $y_1 = f_1(x)$ and $y_2 = f_2(x)$ are similar when there is the constant value, *k*, at which the following condition is fulfilled:

$$f_1(x) = kf_2(x) \quad (15)$$

The proportionality factor changes after passing the ramification at point *A*; it is the same between ramification points of the processes.

With regard to changes in the oxygen dissolved in the water concentration in the natural environment and in the mesocosm during the time period before (see equation (15)), the point *A* similarity ratio for the processes in the natural water basin and in the mesocosm is equal to one. This means that the processes in the natural environment and in the mesocosm are identical.

After passing the ramification point *A* (see Figure 9.12*a*), oscillating processes in the change of oxygen dissolved in the water concentration of a water body and of a mesocosm take place simultaneously but with different functions modulating the processes. In this case, modulating functions are straight lines passing through the crossing point of the lines drawn through single-type extremums of the curves specified on the figure after the ramification point. These straight lines are placed at different angles to the x-axis. Their equations in the coordinate system with the origin at the specified crossing point have the following view:

$$Y_1 = k_1x, Y_2 = k_2x \quad (16)$$

Therefore, the similarity ratio for processes of change in oxygen dissolved in the water concentration in the natural environment and in the mesocosm after passing ramification point A is equal to:

$$k = k_1/k_2 \quad (17)$$

Simple graphical and mathematical calculations show that the similarity ratio for the process at full-scale conditions (see Figure 9.12*a*, curve 1) and in the mesocosm (see Figure 9.12*a*, curve 2) amounts to 5.83.

As an example of such processes, in addition to Figure 9.12, we will show a general view if the processes analogous to dynamics of oxygen dissolved in the water concentration in the natural environment and in the mesocosm after passing ramification point A (Figure 9.13).

Modeling functions passing through the maximal values of the curves are described by the following equations:

$$y_1(t) = 5t + 2t\sin(4t) \quad (18)$$

$$y_2(t) = 2(5t + 2t\sin(4t)) \quad (19)$$

Therefore, the similarity ratio of the processes, shown in Figure 9.13, is equal ½.

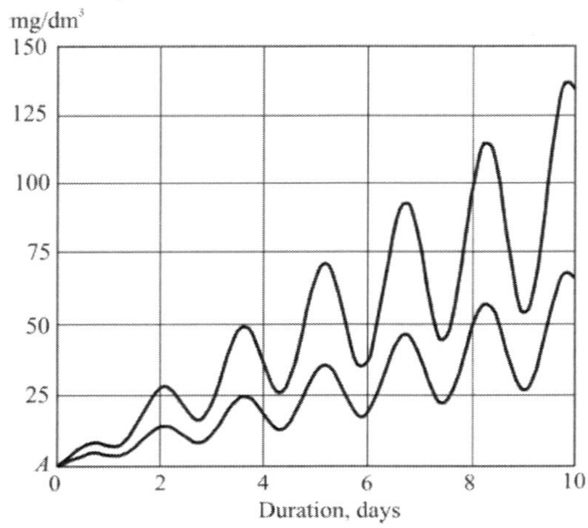

Figure 9.13. Dynamics of oxygen dissolved in the water concentration (mg/dm^3) after ramification point *A* (scheme).

In general, the similarity ratio of the processes is determined after processing the observation results through mathematical statistics and determining the processes' ramification points. Similarity ratios are determined either by using the dimension theory or the shooting method.

The aforementioned applies to pH dynamics in the water basin as well as the mesocosm (see Figure 9.12*b*).

9.2.4. Methodology of Full-Scale Ecological Modeling

Experience in conducting research with the usage of different mesocosms, as well as obtaining the above principles of criterial data for such experiments, made it possible to formulate basic requirements for technique and technical means of modeling. As was noted before, the speed and nature of many process to a considerable extent depends on the range of factors directly or indirectly functioning in a water body (hydrodynamical processes, hydrochemical and hydrobiological regime, etc.). *The influence of many of these factors is of an individual nature characteristic to the specific water body (or part thereof).*

It is natural that these factors shall be taken into account during the physical modeling of physical and chemical processes. Besides, as was specified above, the results of full-scale physical modeling are easier to analyze in view of the similarity theory if the maximal approximation of modeling conditions to conditions of real water body are taken into account.

From this position, the main reason to replace laboratory modeling with full-scale modeling is because of the impossibility of selecting a similarity regime for such vastly different systems as the laboratory aquarium and the real water body.

In our work, special attention was paid to mesocosms installed immediately in the examined water body. *In this case, the models are spatially jointly insulated volumes of water and benthal deposits with populating aquatic organisms.*

As a rule, the experimental volumes are insulated with transparent polymeric materials: plexiglas, polysterene, polyethylene, etc. The following properties of the insulating materials are important:

- It shall not have folds and sharp bends on its surface, which would worsen the correlation of the surface to the volume.
- Its adsorption capacity shall be minimal with regard to examined contaminants.
- It shall not be toxic for biota of mesocosm.
- Adhesional characteristics of the material shall make it possible for peryphitic communities to develop on the insulating surface.
- The insulating surface shall not considerably weaken the brightness of the insulated space to the full exclusion of a mass transfer between the insulated space and the surrounding water medium.

Shape of the isolated water volume shall those that minimize the correlation of the lateral surface of the mesocosm area, S_m, to its volume, S_m/V_m. Insulating surfaces shall be installed in the water body in such manner that "cutout" water and bottom sediments' volume will be subjected to minimal disturbances.

The resulting microecosystem (mesocosm) will be maximally approximated to the initial maternal macrosystem in terms of its chemical and biological parameters.

Mesocosms have the same structure of a water layer and the same (undisturbed) structure of benthal deposits as the initial ecosystem as well as the same initial chemical composition water and water population. Common or quite similar factors are found in the brightness, temperature, pressure and the gradients of the time variations thereof. The limited dimensions of microecosystems make it possible to conduct practically all possible experiments without fear concerning the fate of the water body.

Jointly with the great similarity in terms of the range of physical-chemical and chemical-biological indices, the experimental ecosystems still have some significant differences from the modeled water bodies. These differences, first of all, deal with the inevitable distortions of the space-temporal scales of microecosystems.

The consequence of these distortions has been previously mentioned in terms of the differences of hydrodynamical properties of the model and the original as well as the presence of the developed lateral surface in mesocosms.

The correlation of the volume, V, occupied by the ecosystem, and the area, S, of its natural boundaries (V/S) describes the specific provision of the ecosystem with external boundaries through which different substances come in and escape.

For open (in terms of water exchange parameters) ecosystems, such an ecologically important parameter as the *water exchange index* is indicant of "openness":

$$a = \Delta V/V_o \tag{20}$$

where ΔV is the volume of water coming into the water basin and carrying chemical substances and V_o is the volume occupied by the ecosystem.

It is worth noting that at conditions of modeling by means of mesocosms, it is only possible to provide equality between named parameters for the modeled and modeling ecosystems by comparing two artificial microecosystems. For the case where at least one of compared ecosystems is natural, it is almost impossible to provide equality between these parameters since the *S/V* value for mesocosms is always higher than corresponding value for natural ecosystems.

Therefore, during model experiments conducted through mesocosms, this parameter should be minimized. As a rule, this goal is achieved by increasing the mesocosm's volume. Herewith, it should be taken into account that the increase of the mesocosms dimensions is associated with certain complications and requires appropriate technical means and techniques of experiment conducting.

As was previously mentioned, *for large mesocosms, it is more difficult to select the area of the water body as well as to carry out water sampling for assay and applied additives, etc.* The same notion applies with regard to the inclusion of representatives of higher trophic links into the composition of biotic components in experimental ecosystems, etc.

In this respect, during physical modeling at full-scale conditions, there is the necessity to make determinations with minimal excess of those basic conditions of the experiment so that the basic results of this field research, performed on the experimental ecosystems, may be transferred to the level of the modeled objects.

Before the beginning of the experiment, this can only be done on the basis of past experience and the intuition of the researcher.

Therefore, the procedure is as follows. The parameters of the experimental ecosystems as well as the space of the controlled indicants are selected on the basis of research tasks while taking into consideration some of the excess extent.

During the experiment, the controlled indices in the examined ecosystems are measured with equal accuracy. This means that all the measurements in the mesocosms and at the areas of comparison are carried out using the same methods and instruments under the same conditions.

During the experiment or after it is over, the statistical similarity criteria are calculated using the scheme set forth above for the reference mesocosm and modeled ecosystem.

The experiment is deemed to be conducted correctly if the appropriate similarity criteria are identical during the whole period of research for the reference mesocosm and modeled ecosystem.

Let us examine the determination of characteristics for the mass disintegration rate of a thoroughly studied contaminant, phenol, in order to provide a specific example of factor analysis usage during full-scale modeling of the self-purification process.

The world production of phenols amounts to approximately 3 million tons a year but phenols also have a natural origin.

Moreover, the amount of "natural" phenol by far exceeds the volume of its industrial production. In these conditions, it is no wonder that phenols are not included in the list of priority contaminants for all services monitoring water quality.

The behavior of phenols in natural waters and its properties have been quite thoroughly studied and are being investigated nowadays [51, 52].

The problems associated with phenol's existence in natural waters, however, is still urgent since there are no adequate models in the practice of water protection and design organizations that describe self-purification in water bodies with regard to this substance. The calculation error here reaches 300-400% [53].

The described experiment was conducted against the relatively clean background area of the Seversky Donets River upstream from the town, Rubezhny, in July of 1987.

Cylindrical mesocosms with volume $V \approx 1000$ dm^3 were used (see Figure 9.1a and Figure 9.2). The mesocosms were installed in the river on July 12th at 10 a.m., and on July 13th at 10 a.m. with a one-time additive of phenol (carbolic acid), which was applied in such a manner as to obtain an initial concentration of $C_0 \sim 50$ mg/dm^3. The reference mesocosm remained intact.

The water was sampled in the morning and in the evening from the surface and benthic horizons by means of a rod-type sampler (see Figure 9.6).

The sampled water was treated immediately after sampling at a field laboratory deployed on the bank of the river.

The following general indices were determined in the experimental and reference mesocosms: the content of oxygen dissolved in the water, pH, Eh, water temperature, content of phenol, forms of nitrogen, phosphorus, etc.

Figure 9.14*a* presents the dynamics of the oxygen content dissolved in the water for the experimental and reference mesocosms. It is clear from the figure that approximately 55 hr. after the beginning of the experiment, the dissolved oxygen concentration in the reference mesocosm sharply dropped to values between 0.1-1.5 mg/dm^3, and the new regime remained for quite a long period of time.

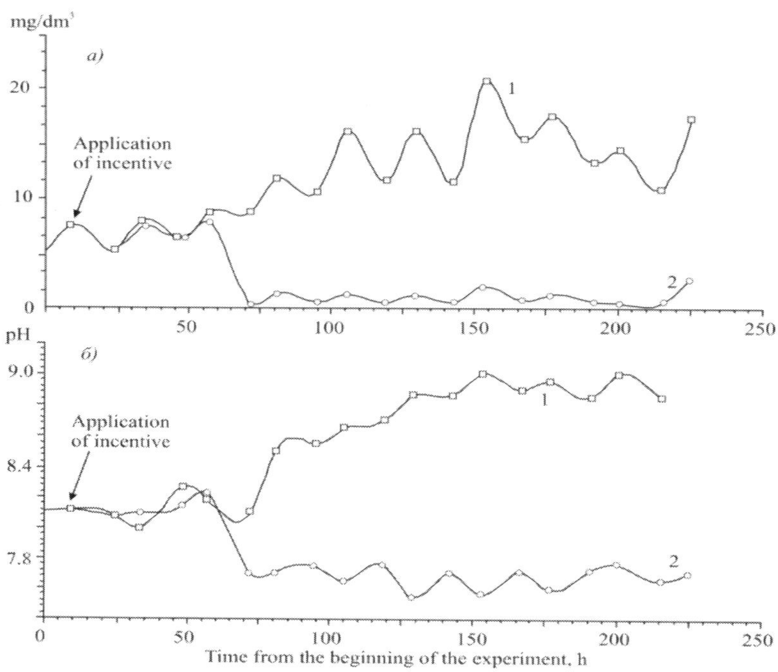

Figure 9.14. Dynamics of concentrations (mg/dm^3) of dissolved oxygen (*a*) and pH (*b*) in the reference water (*1*) and experimental (*2*) mesocosms.

An analogous process was observed for pH (Figure 9.14*b*) as well. It is interesting to note that the addition of carbolic acid does not have any immediate influence on the pH of the mesocosm water. Moreover, soon after the phenol addition, a slight increase in pH values was observed.

Two stages are revealed in the behavior of the experimental ecosystem:

1) From the beginning of the experiment to the sharp transition to a new state
2) After the transition to a new state until the end of experiment

The analysis of these two stages carried out through the principal components method revealed the following.

The first three principal components, corresponding to the first stage of chemical-biological processes described 99.4% of all observed dispersion. Herewith, 73.3% of the dispersion fits into the share of the first component, 18.4% fits the share of the second component and 7.7% fits the share of the third component. The factor loads corresponding to these stages are specified in Table 9.2.

Table 9.2. Factor loads of main components for chemical-biological processes in the experimental mesocosm (numerator—surface, denominator—bottom)

Index	Principal component		
	1st	2nd	3rd
The first stage			
Oxygen content	0.97/0.85	-0.04/0.26	0.23/0.44
pH	0.66/0.56	-0.73/-0.78	-0.16/0.26
Eh	-0.91/-0.94	-0.05/-0.04	0.40/0.33
Water temperature	0.92/0.94	0.37/0.32	0.04/0.01
The second stage			
Oxygen content	0.89/0.76	0.20/0.02	0.34/0.62
pH	-0.74/-0.77	-0.51/-0.59	0.41/0.15
Eh	0.70/0.83	-0.52/-0.14	-0.46/-0.12
Water temperature	0.86/0.95	-0.38/-0.23	0.12/-0.04

It is not difficult to describe this stage of chemical-biological processes in the experimental mesocosm. The following indices have the greatest influence on the first principal component:

- Oxygen content (factor load 0.97) in surface and benthic (0.85) layers
- Eh in benthic (0.94) and surface (0.91) layers
- Water temperature in benthic (0.94) and surface (0.94) layers

Therefore, processes of the first stage may be characterized as reductive-oxidative with temperature participation.

Similar analysis of the results characterizing the second stage shows that the first three components of this phase of chemical-biological processes describe 92.2% of all observed dispersion. Herewith, 66.5% of dispersion fits the share of the first component, 14.2% fits the share of the second component and 11.6% fits the share of the third component. Factor loads corresponding to these stages are specified in Table 9.2 as well.

At the second stage, oxygen does not have such a strong influence on the first principal component as it did at the first stage. The following indices have the greatest influence on this principal component:

- Temperature in the benthic layer (0.95)
- Oxygen content in the surface layer (0.89)
- Temperature in the surface layer (0.86)
- Eh in the benthic layer (0.83)

For this stage, the increase in the role of the temperature factor is typical.

In general, the observed changes and considerable differences of factor patterns shall be marked. Some components of factor patterns, such as the changed signs, bear evidence to the considerable differences within the observed stages.

This means that we cannot describe the observed decay of phenol through only one model since the analysis revealed two stages which differ quite heavily. In the first stages, the experimental microecosystem still preserves similarity to the reference system. The similarity dramatically decreases, however, during the transition process to the second stage when specific microflora using phenol as substratum develops. Herewith, the highest index of phenol decay in the water may be naturally forecasted for the second stage.

Here, attention shall be paid to the following circumstance. In contradiction to the processes discussed during the analysis of Figure 9.12, the processes described by Figure 9.14 are markedly dissimilar. The reason for this lies in the fact that the dynamics of oxygen and pH in the reference mesocosm curvature of the function modulating the oscillation processes takes place; in the experimental mesocosm, such curvature is absent. Figure 9.15 schematically specifies branches of the function modeling oscillation analogous to pH dynamics in the reference and experimental mesocosms. Hence, the conclusion concerning the absence of the general time-independent ratio linking ordinates of oxygen content and pH in the reference and experimental mesocosms follows.

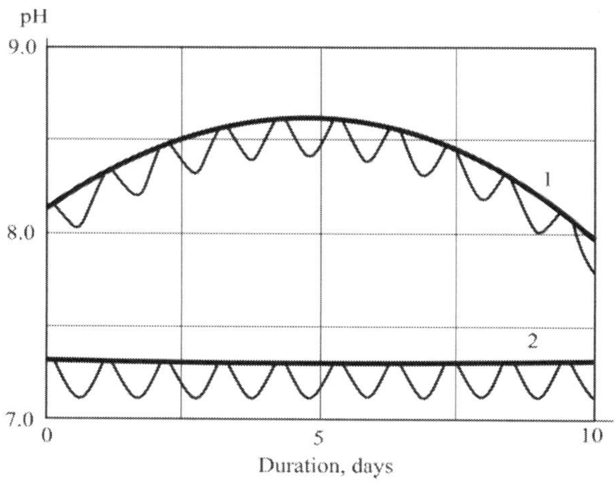

Figure 9.15. Schematic representation of the processes analogous to the processes in the reference (1) and experimental (2) mesocosms.

The model of the phenol decay process (not taking into account two different stages) is specified on Figure 9.16. The dependence of the phenol concentration over time is described by the equation:

$$C(t) = 40\exp(-0.009t^{3/2})+5 \tag{21}$$

The general view of such dependencies is represented by the extended exponential:

$$C(t) = A\exp(-bt^{m})+c \tag{22}$$

where A is the initial height of diagram, b is the degree of the curve descent, m is the curvature nature and c describes the shift of the diagram along the y-axis.

In this case, the phenol decay speed ratio (not taking into account the two stages of the process) changes over time. The value of this ratio in the process of a quick drop of concentration (on the area from 2 to 6 hr.) amounts to approximately $K \approx 0.33$ day^{-1}.

The time dependence of the phenol concentration in the water of experimental mesocosm, taking into account the two stages of the process, is specified on Figure 9.17.

At the initial stage of the process (Figure 9.17*a*), this dependence is exponential and is described by equation:

$$C(t) = 50\exp(-0.1t) \tag{23}$$

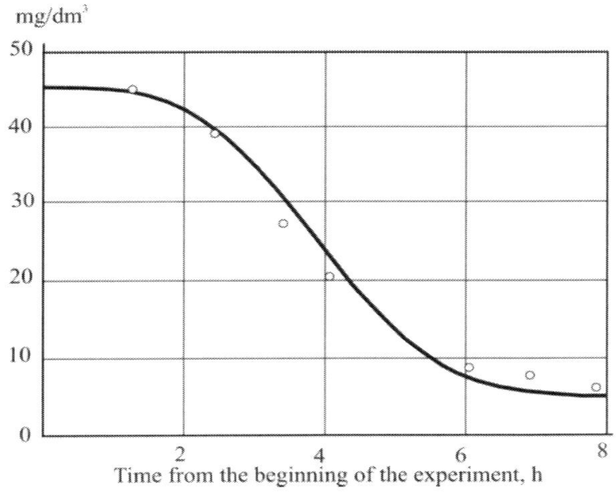

Figure 9.16. Model of phenol decay process without taking into account similarity.

At the final stage of the process (Figure 9.17*b*), the dependence represents the algebraic function:

$$C(t) = \frac{40}{1+\sqrt{t}+(0.1t)^{1.9}} \tag{24}$$

The phenol decay speed ratio observes the similarity conditions amount of $K \approx 0.1$ day^{-1} (see Figure 9.17*a*). At the stage of adapted microflora development, when similarity conditions are not observed (see Figure 9.17*b*), the value, *K*, is calculated by this approximate formula:

$$K = \frac{1}{T}\int_0^T C(t)dt \qquad (25)$$

where *T* is the time of observation (*T*= 5) and *C* is the examined concentration (see formula (24)).

The calculated phenol decay speed ratio for the specified stage amounts to $K = 0.69$ day^{-1}.

9.2.5. Similarity of Modeled and Maternal Ecosystems in the Data for Remote Surface Water Sensing

Along with the aforementioned research concerning the problem of similarity during the modeling of physical-chemical and biological processes in the same models, studies of a contaminant's influence on the optic properties of the waters was simultaneously conducted through remote methods. Detailed results of this research are specified in the section devoted to special types of water quality monitoring (Chapter 6). One of the tasks set for the group of experimenters, headed by B.L. Sukhorukov, consisted in finding remote sensing indices that could be used along with chemical-biological ones for the assessment of similarity and, therefore, in substantiation to the appropriateness of field experiments conducted on the mesocosms, which results in transition to real water bodies.[4]

The ***spectrum of brightness ratio (BR)*** was selected as the index on which comparison of states for the artificial and natural ecosystems was performed [54]. In a previous section, the similarity of modeled and natural ecosystems (mesocosm and water body) was substantiated by the set of physical-chemical indices [4, 55, 56]. The selection of brightness ratio as criterion is grounded by the following reasons. First, this index is integral and describes the state of ecosystem as a whole [57, 58].[5] According to this definition [59], the brightness of radiation ascending from the water surface to the direction, Ω, is described by the dependence:

$$\rho(\Omega) = \frac{B(\Omega)}{B_0} \qquad (26)$$

where $B(\Omega)$ is the intensity of the light reflected to the direction Ω and B_0 is the intensity of the light dispersed by the absolutely white orthotropic platform (ideal lambert diffuser) located on the upper boundary of the scattering layer. If formula (26) is rewritten in the form of:

$$B(\Omega) = \rho(\Omega)B_0 \qquad (27)$$

[4] Remote research was conducted jointly with chemical-biological work on the same mesocosms and was included in the general program of complex field observations conducted under the supervision of the author of this monograph.

[5] It shall be particularly underlined that BR reflects the state of only optically active components in the ecosystem.

then, in physical sense, the brightness ratio may be interpreted as the value characterizing the medium's ability to *transform the incoming radiation*. In our case, it is the water ecosystem's ability to reradiate (disperse, absorb) the falling radiation, B_0.

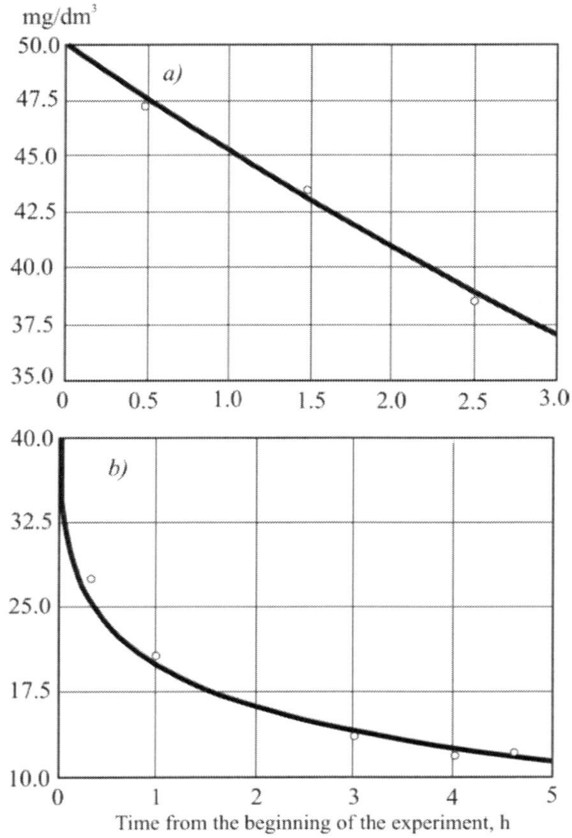

Figure 9.17. Dynamics of phenol concentration (mg/dm^3) in the water of the experimental mesocosm at the initial (*a*) and final (*b*) stages of the process.

All optically active components of the ecosystem take part in the re-radiation process. Components of different genesis, associated with biotic and abiotic content, are optically active: mineral suspended substances, dissolved organic substances and different kinds of plankton—phytoplankton (in general), bacteria and zooplankton. The contribution of the latter two is minimal, though still theoretically possible. Optically inactive components to a greater or lesser extent influence the spectral travel of ascending from water radiation only through optically active components.

Thus, the brightness ratio is the "optical image" of the water ecosystem and is an integral optical characteristic of its optically active components. Phytoplankton are one of the basic indicated components of a reradiating complex. Moreover, phytoplankton are also one of the indicators of the ecological state of a water system (concerning changes in species diversity, concentration, color combination of phytoplankton of different divisions) and are useful in remote spectrometric methods that implement the well-known technology of biomonitoring or, specifically, biotesting [60].

It shall be noted, as well, that the method for obtaining the described data is remote, contact-free. In cases where this method is used, there is no interaction between the instrument and object, i.e., there are no distortions related to a sample's non-representativeness during water sampling and the natural functioning of the modeled ecosystem area is not fundamentally disturbed. The volume of water, which contributes to the intensity of the ascending radiation in the described experiment conducted on the Don river, amounts to about 2 m^3. In other words, during the remote spectrometric survey, the ideal averaging of collected data both in terms of volume and in terms of depth and area is observed.

The maximal openness of mesocosms is obtained through the separation of the water column from the film made of material that lets light through within the spectral area essential for plankton organisms and macrophytes; this amounts to between 400-700 nm. The film is brought into the bottom sediments, but it is not connected to the bottom formed by the natural channel bodies. Although in this case, water from the maternal ecosystem may pass through the bottom sediments, still, in terms of many of examined indices for the ecosystem, it is maintained as quite insulated for intra-water basins processes modeling [61, 62]. The upper edge of the film is attached to the floats, which makes it possible to preserve the ecosystem's isolation during wind-induced or other hydrological phenomena, herewith, not creating additional windage by raising the walls of the insulating volume, which would be an additional obstacle to light's passage through the volume.

Dimensions of the insulated volume to solve the described task were determined on the basis of the following considerations. The basic radiation registered by the remote control device was formed in a cylinder with the height, Z_b (depth of visibility of white Secchi disk). The diameter of the cylinder was approximately determined using the formula $D = (h + Z_б)\varphi$, where h is the height of the remote control device's location over the water surface and φ is the angle of lens. Usually, the pickup angle of the optical device used for remote control (and installed onboard an aerial vehicle), does not exceed 10°, therefore in calculating the formula, $tg\varphi \approx \varphi$. It was noted in some papers, that in "turbid" waters (most ecosystems are related to these kinds of waters), additional lighting owing to multiple dispersions in the volume surrounding the brightness body may be observed. We determined the thickness of such a "surrounding" layer as $Z_b/2$.

Taking into account these specified requirements, the author and B.L. Sukhorukov carried out a multipurpose experiment on one of the branches of the Don River at its mouth—the Mertvy Donets River. The duration of the experiment was 24 days. Four mesocosms of uniform structure, with the possibility to separate the volume of water between the surfaces to the bottom, were installed. Polyethylene film with an approximate thickness of 10 mcm served as the insulating material. The foundation of stainless steel was deepened to the bottom of water body and on the water surface, the film was held in place with a ring fixed on foam-plastic floats. Taking into account the fact that the value Z_b for the Don River in the lower stream does not exceed 1 m, and the height, h, of the spectrometer objective lens installation amounted to 1.5 m, the parameters of the mesocosms were equal to D = 2.2 m, $H \approx 1.7$ m.

For the arrangement of processes of the accelerated mesocosms' ecosystem development, both the one-time application of metal at the beginning of experiment (Cu to mesocosm No. 1 and Hg to mesocosm No. 3), and the daily increase of copper concentration in the water (in mesocosm No. 2), was used. The copper was applied in the form of $CuSO_4$ and the mercury was applied in the form of $Hg(NO_3)_2$ [57, 62]. Mesocosm No. 4 was the reference mesocosm; no substances were additionally applied into it. Besides, the whole set of measurements was carried out immediately in the river—the maternal ecosystem—close to the mesocosms.

Let us adduce the BR spectra describing the mesocosms as artificial ecosystems. Figure 9.18 specifies BR spectra that are normalized for the sum of the BR for the whole spectrum [57, 62], which was obtained immediately in the reference mesocosm and in the river at close proximity to mesocosms. All spectrum pairs that we had at our disposal, proving the non-randomness of all conclusions presented below, are specified.

It is clear from the figure that the structures of the spectra are absolutely identical. Therefore, the development of basic biotic components takes place only in one direction and the insulation of ecosystem, at least in cases of mesocosm usage with such dimensions, practically does not have an impact on the development of the main representatives of the phytoplanktonic community; these are what determine to a great extent the form of the measured spectra. It shall be noted that weather conditions and the hydrological situation remained relatively stable during the experiment.

The stationary, undisturbed ecological process was investigated. There were also no changes in the ecological situation in upper reaches of the river, which might inevitably have had an influence on the spectra of ascending radiation of the river. Therefore, in the second part of the experiment, in accordance with changes in weather conditions (the wind had risen and the water had become more turbid than in the microcosms), to some extent, a more notable difference between the microcosms and rivers spectra is observed. Besides, small chaotic oscillations were additionally noted in the spectrum of the river on the 24th day of the experiment, which is typical for spectra obtained from the water surface where ripples are caused by the wind. Returning to the purpose of this section—studying the similarity of maternal (natural) and model (artificial) ecosystems over a long time period reminds us of basic provisions of similarity theory. Phenomena, processes or systems are similar if at similar moments in time and at the same points of space, the values of the variable indices describing the state of one system are proportional to the values of the appropriate indices of other system.

In cases of physical similarity, the fields of the appropriate physical parameters of two systems shall be similar in terms of time and space. Dimensionless combinations made up of definite parameters of investigated phenomena (systems), as was noted in Section 9.2.3, are named "similarity criteria [63]."

This index was also examined in the papers of academician, A.N. Kolmagorov. In order to prove the systems' (phenomena) similarity, according to Kolmagorov, it is necessary and sufficient to provide that the values of any dimensionless combination, $K = X_{\beta_1}^{\alpha_1} X_{\beta_2}^{\alpha_2} \dots X_{\beta_m}^{\alpha_m}$, where the parameters of phenomena $S^{(1)}$ and $S^{(2)}$ are equal to $k^{(1\ )} = k^{(2)}$. Each dimensionless expression, k, was named by Kolmagorov as *similarity criterion*. In cases when determining parameters are dimensionless, and in cases when $k = X_{\beta_1}/X_{\beta_2}$, where X_{β_1} and X_{β_2}, the determining parameters of the same dimensionality of the so-called "trivial" similarity criteria are formed [63]. Similarity theory examines non-trivial criteria, however, making it is possible to obtain unique equality among all physical-chemical and biological indices of compared ecosystems,[6] but it also substantiates the presence of integral measurable values, which take into account the influence of all measured indices on these values but do not converge on them.

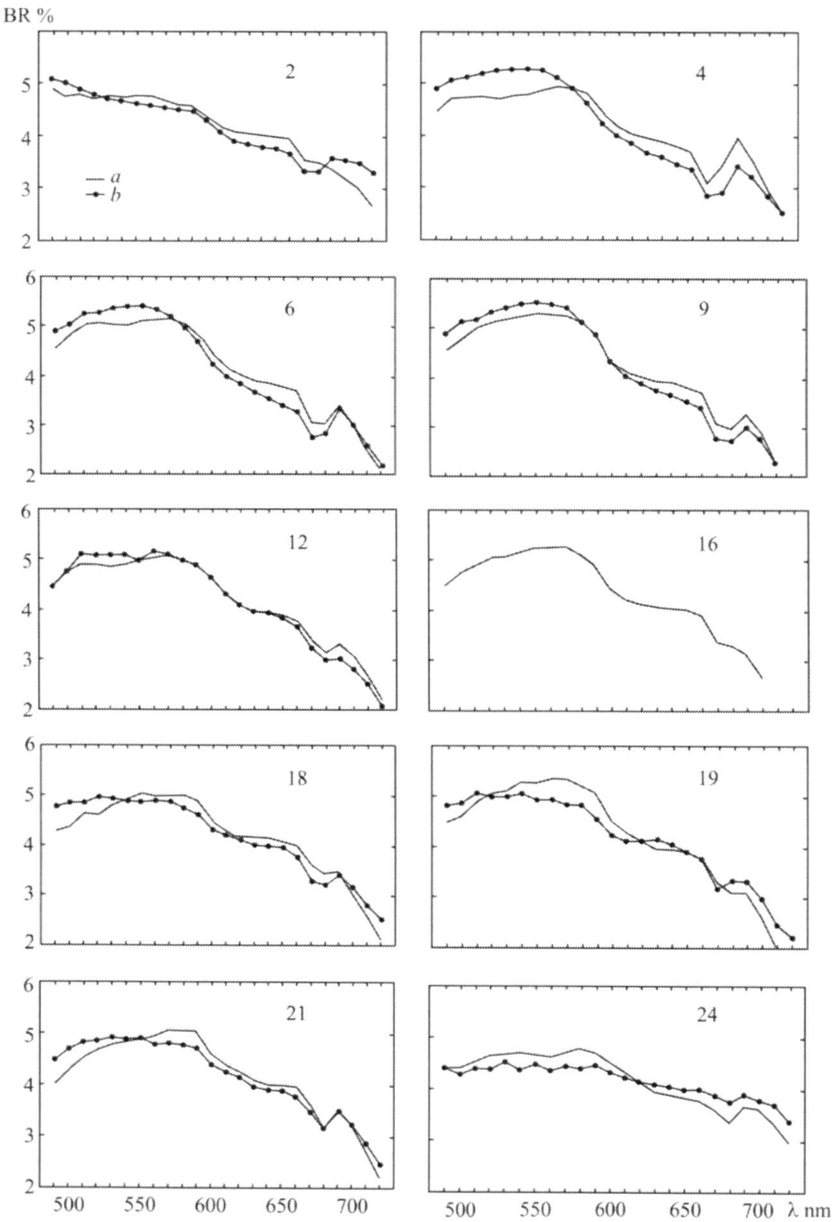

a—river, maternal water ecosystem, *b*—model water ecosystem. *2–24*—days of the experiment

Figure 9.18. Normalized spectra of brightness ratios (BR) for radiation ascending from the water of the maternal and model ecosystems.

[6]This cannot be performed at the actual level of knowledge about water ecosystems.

Taking into consideration that the brightness ratio is the "optical image" of optically active components (OAC) within a water ecosystem, it should be considered a similarity criterion for ecosystems compared by remote methods. The degree of reliability of the criterion increases when choosing the correlation of brightness ratios across different waves. Relations of brightness ratio measured on one of waves to the sum of brightness ratios measured on all waves of the visible spectrum may be selected as trivial similarity criteria. Taking into account the fact that brightness ratio and the brightness of radiation ascending from water are different in terms of their multiplying constant:

$$R(i, j) = B(i, j)/(B(1, j)+\ldots+B(n, j)) \tag{28}$$

where $B(i, j)$ is the Ith count of brightness ratio value for the spectrum j; $i = 1,\ldots, n$ and n is the number of discrete counts of the measured brightness spectrum. We will obtain these normalized brightness ratios through $R(i, j)$ and they will be practically transformed to *similarity criteria* of water ecosystems in terms of the aforementioned components. A *necessary* condition of ecosystems similarity is the coincidence of its optical reflection—the spectra of its brightness ratio. Herewith, it shall be taken into account that the general occurrence of spectra is not a *sufficient* condition of ecosystem similarity since the spectra of brightness ratios ambiguously reflects physical and biological parameters of the water ecosystem. Spectra of brightness ratios are characteristic of *only* optically active components and may be used only in this sense. Alternatively, in other words, *if normalized spectra of brightness ratios of water ecosystems coincide, ecosystems are similar in terms of their optically active components and conditions of functioning during the compared time period.*

"Ambiguity" shall be understood in a well-defined sense. It is impossible to determine the absolute values of OAC concentrations in water bodies with an unknown "matrix" with a preliminarily set accuracy. In the same water body, however, during the same hydrological season, it is possible to determine definitively the orientation of the processes in which OAC participate relative to the concentrations thereof without the examination of its qualitative composition.

We can suppose that if optically inactive components suddenly appear in the ecosystem, their influence will be revealed by optically active components with time delay *a priori* unknown to us. It is evident that this component and the time ambiguity is practically absent in the experiment where the natural development of one and the same ecosystem takes place,[7] even though they are temporarily divided in parts. As was noted above, here we examined the issue of the influence of closure of some part of the maternal ecosystem only on its optical properties. Therefore, the suggested criterion for this specific case may be considered a sufficient condition of ecosystem similarity as well.

---

[7] Experimental proof of the presented considerations can be found in a paper by A.M. Nikanorov and others [61], where data, corroborating the similarity of maternal and model ecosystem in terms of hydrochemical and hydrobiological indices, is specified. The data was obtained at several different water bodies including during the described experiment. It is impossible to speak about "dechphering" the optical spectra or even about the correspondence of hydrochemical and optical indices since the indices used for the evidence of the ecosystem similarity in the paper [61] cannot be related to optically active components. The data, collected during the described experiment, is obtained through fundamentally different methods characterizing the ecosystem from different perspectives and complement each other through remote, contact-free spectrometry and hydrochemistry.

Using this similarity criterion as evidence of the maternal and model ecosystems' similarity in the aforementioned sense, let us execute the procedure of hierarchical classification on basis of spectra array obtained during the experiment [64].

First of all, let us make an important remark. The theoretically definite, accurate similarity criteria equality $k^{(1)} = k^{(2)}$ is not observed in practice. The reason for such a divergence lies in the fact that all experimentally measured chemical, biological and physical values are determined with errors. Therefore, the similarity is really assessed as the inequality $|k^{(1)} - k^{(2)}| < \Delta_U^{12}$, where Δ_U^{12} is some small dimensionless value that has a sense of the level of criteria difference.

Each spectrum can be presented as a vector, R_i, with the dimension $1 < i < k$, where k is the number of discrete counts on which the spectrum is built. It is convenient to use Euclidean distance as the measure of two vectors' (spectra) proximity:

$$d(R_m, R_i) = \sqrt{\sum_{i=1}^{n}(R_{i,j} - R_{i,m})^2} \quad (29)$$

where R_m and R_j are normalized spectra for observations m and j. The distance between spectra groups is defined by the principle of "average connection" as an arithmetic average of all pairwise distances between representatives of examined groups:

$$d(S_p, S_q) = \frac{1}{n_p n_q}\sum_{R_m \in S_p}\sum_{R_j \in S_q} d(R_m, R_j) \quad (30)$$

where S_p and S_q are the pth and the qth spectra groups (classes) and n_p and n_q are the number of spectra in groups p and q, accordingly.

The dendrogram, obtained as a result of such classification, is presented on Figure 9.19. On the x-axis, the number of mesocosms (numerator) and the day of the experiment (denominator) are specified. Letters P and K mark the river (maternal ecosystem) and reference mesocosm (artificial ecosystem not subjected to load). Mesocosms No. 1-3 were contaminated by salts of Cu and Hg [62]. Without detailed analysis of the graph structure, let us mark its peculiarities necessary to answer the question. At the 6th, 9th, 12th, 18th, and the 21st days of experiment, the normalized spectra of the river and reference mesocosm are united at quite a low level of generality A, corresponding to value Δ_A^{PK}, i. e. they belong to the same classes or coincide with a good degree of accuracy.

Herewith, it is significant that the independent existence of maternal and model ecosystems lasts for more than three weeks. The state of the river ecosystem has repeatedly changed, although insignificantly, during this time, which is observed as belonging to measured indices of all classes marked at level A (except for class I, which is formed by the normalized spectra of ecosystem and subjected to a considerable load [62]).

The mismatch of criteria at the 4th and 19th days, and to a lesser extent at the 2nd day, of the experiment shall be noted as well. The following explanation of such behavior of spectra may be given. During the first days of experiment, due to the installation of artificial ecosystems (mesocosms) at the water body, disturbance of the hydrological regime of the ecosystems—both maternal and model—took place. Therefore, the difference could be expected at least by the 2nd day. It was a period of relaxation within the system. The difference observed on the 19th day does not have a satisfactory explanation, especially in view of the fact that by visually comparing the spectra shown in Figure 9.18, it is clear that on the 18th and 21st days, a good coincidence of similarity criteria took place. Most likely, this difference is related to the error of spectra obtainment. As far as all the measurements were carried out *in situ*, uncontrolled changes of the water body surface (ripple, glare), which would be different for the maternal and model ecosystem, could influence the process of measurement.

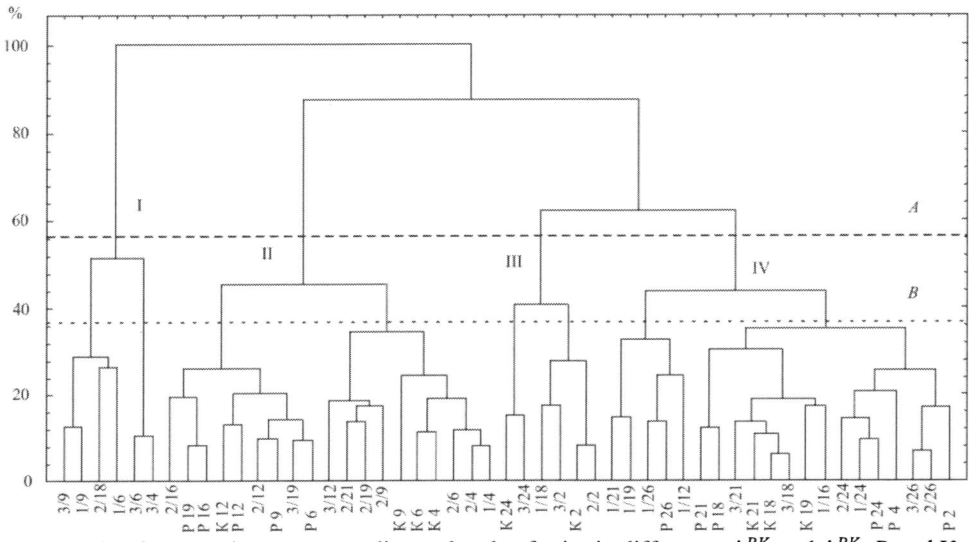

A and *B*—levels of connection corresponding to levels of criteria difference Δ_A^{PK} and Δ_B^{PK}; P and K—the river and reference mesocosm, not subjected to loading; 3/9, 1/9, etc.—numbers of the mesocosm (numerator) and day of the experiment (denominator), I-IV—classes formed by the similar coefficients spectra.

Figure 9.19. Dendrogram for the 51st experimentally measured normalized spectrum of brightness ratio of ascending from water surface radiation (July and August of 1990).

Thus, it is clear from the comparison of normalized BR spectra, that processes related to the change of optically active components concentration in isolated ecosystems of quite a large volume are identical to the processes running in the basic maternal ecosystem. The ecosystem of the reference mesocosm turned out to be quite representative in terms of its optically active components when compared to the maternal ecosystem. The development of the river ecosystem and ecosystem of the reference mesocosm, not subjected to the influence of additional anthropogenic load, takes place almost identically during a similar time period.

The change of ecological state of mesocosms that were subjected to anthropogenic contamination became apparent in terms of changes of its optical state and, therefore, of spectra of ascending radiation.

Let us examine the process of change of such mesocosms ecological state through the example of mesocosm No 1. A very heavy load (approximately 150 MPC) led to clearly expressed changes of spectra forms. The beginning of the ecosystem's optical change had been already marked at the second day after the load application; it reached its maximum between the 6th-9th days (Figure 9.20). By the 20th day, however, the ecosystem's capability to self-purify made it possible to judge its return to a state close to its initial by optical parameters.

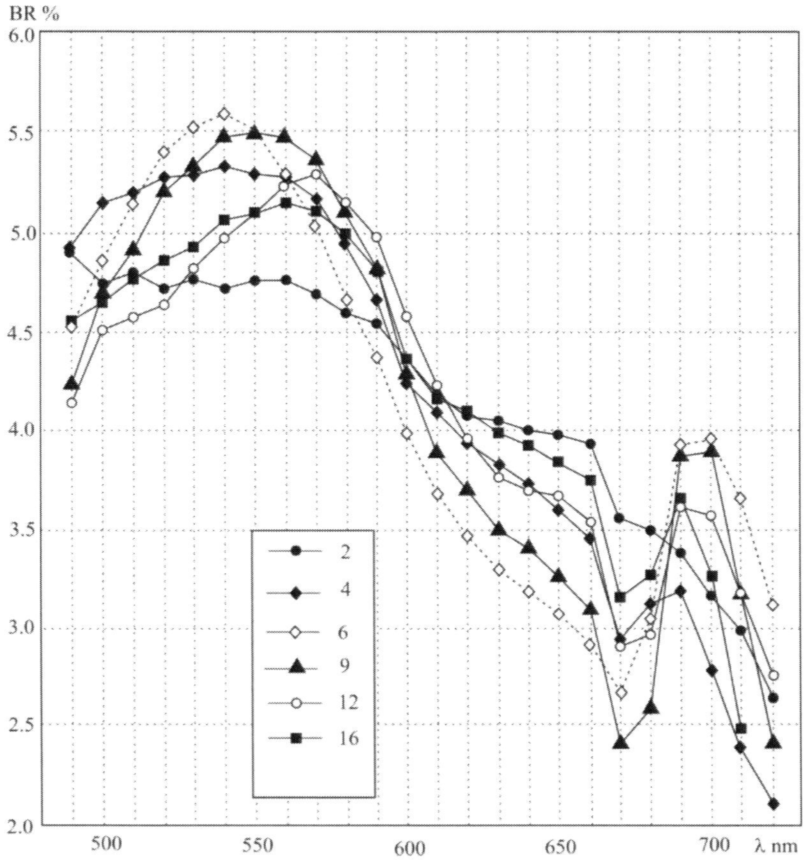

2—river spectrum from the maternal ecosystem, *4-16*—days of the experiment, accordingly

Figure 9.20. Normalized spectra of brightness ratios for contaminated mesocosm No. 1 on different days of experiment.

Thus, it was shown by direct experiment at a natural body that during 24 days of modeling a mesocosm's optical state within a model ecosystem remained similar to the optical state of a maternal ecosystem. Therefore, ecosystems of artificial water formations (mesocosms) with volumes not exceeding 2 m^3, installed at such natural bodies as the Don River, in terms of optically active parameters, remain similar to the maternal river ecosystem. In our opinion, this research brought us one step closer to finding the solution to the key task in modeling theory— appropriateness of transfer of the model experiment results to natural water bodies.

References

[1] Vinberg G. G. Variety and unity of vital phenomena and quantitative methods in biology // *Magazine of general biology.* — 1985. — V. XLII. No 1. — P. 5—18.

[2] Kozhova O. M., Pavlov V. K. Population approach to evaluation of hydrobionts stability // In col.: *Long-term forecasting of ecosystems state.* — Novosibirsk, 1988. — P. 63—69.

[3] Mauer F. L., Ellersieck M. R. Manual of acute toxicity interpretation and database for 410 chemicals and 66 species of freshwater animals // *US Dep. Inter. Fish and Wildlife Sew. Resour. Publ.* — 1986. — N 160, IV. — P. 1—32.

[4] Nikanorov A. M., Trunov N. M. Intra-water basin processes and natural waters quality control. Ser. Water quality. Ex. 2. — SPb.: *Gidrometeoizdat,* 1999. — 155 p.

[5] Kaplin V. T. Actual conditions and main direction of chemical substances transformation processes in natural waters // *Materials of VI all-USSR symposium on actual problems of water basins self-purification and water quality regulation.* — Tallin., 1979. — P. 3—17.

[6] Lesnikov L. A. Accounting of contaminants transformation during commercial fishing water basins protection // *Thesis of rep. of I all-USSR conference on commercial fishing toxicology,* December, 1988.P. 2. — Riga, 1989. — P. 5—6.

[7] Bagotsky S. V., Sanin M. V. To issue of ecologically substantiated regulation of contaminants inflow to water basins // In col.: Ecological standardization and modeling of anthropogenic impact on water ecosystems. Edit. 2. — SPb.: *Gidrometeoizdat,* 2001. — P. 111—117.

[8] Trunov N. M., Teplyakov Y. V., Tavlinova T. I. and oth. Set of technical means for full-scale ecological modeling // In col.: Ecological standardization and modeling of anthropogenic impact on water ecosystems. Edit. 1. — L.: *Gidrometeoizdat*, 1988. — P. 17—24.

[9] Alabuzhev P. M., Geronimus V. B., Minkevich L. M. and oth. Similarity and dimensionality theory. Modeling. — M., *Vysshaya shkola,* 1968. — 206 p.

[10] Venikov V. A., Venikov G. V. Similarity and modeling theory. — M.: *Vysshaya shkola,* 1984. — 478 p.

[11] Guhman A. A. Introduction to similarity theory. — M.: *Vysshaya shkola,* 1973. — 295 p.

[12] Klain S. J. Similarity and approximate methods. — M.: *Mir*, 1968. — 267 p.

[13] Kutateldze S. S. Analysis of similarity and physical models. — Novosibirsk: *Science,* 1978. — 146 p.

[14] Sedov L. I. Methods of similarity and dimensionality in mechanics. — M.: *Science,* 1987. — 430 p.

[15] Stumm W., Forsberd M., Gherini S. Water chemistry // In: *Encyclopedia of Environmental science.* V. 3. — New York e. a., 1983. — P. 1142—1161.

[16] Braginskiy L. P., Velichko I. M., Stcherban E. P. Freshwater plankton in toxic medium. — Kyiv.: *Naukova dumka,* 1987. — 179 p.

[17] Burdin K. S. Usage of artificial Microsystems for solution of theoretical and applied tasks in ecology // In col.: *Man and biosphere.* Edit.2. Stability of ecological systems. — M.: ed. MSU, 1978. — P. 24—144.

[18] Zilov E. A., Stom D. I. Microcosms and its application in experimental hydrobiology// *Successes of modern biology.* — 1989. — V. 107, edit. 1. — P. 150—157.

[19] Pogozhev P. I. Gerasimova T.N. Study of lake zooplankton in mesocosms // *Water resources.* — 1994. — V. 21, No 1. — P. 36—40.

[20] Heath R. T. Are microcosms useful for ecosystems analysis? // Microcosms in ecological research / Ed. J. P. Giesy. — Technical Information Center, U. S., Department of Energy. Washington, D. C, 1980. — P. 333—347.

[21] Ludgren Alf. Model ecosistems as a tool in freshwater and marine research // *Arch. Hidrobiol. Suppl.* — 1985. — 70, N 2. — P. 157—196.
[22] Uhlman Dietrich. Scaling of microcosms and the Dimensional Analysis of Lakes // *Int. Revue ges. Hydrobiol.* — 1985. — 70, N 1. — P. 47—82.
[23] Guidance Document on Testing Procedures for Pesticides in Freshwater Static Mesocosms. — Setae-Europe, 1991. — 46 p.
[24] Odum E. P. The mesocosm // *Bioscience.* — 1984. — 34. — P. 558— 562.
[25] Alekseev V. V. Theoretical and experimental ecological systems researches // *Outcomes of science and technics. Ser. General ecology, biocoenology, hydrobiology.* — 1980. — Edit. 5. — P. 58—104.
[26] Takahasi M., Witney F. A. Temperature, salinity and light penetration structures: controlled pollution experiment // *Bull. Mar. Science.* — 1977. — V. 27, N 1. — P. 8—16.
[27] Bulyon V. V. Technique of experimental assessment of purified run-offs of pulp plant influence on plankton of Lake Baikal. // In col.: *Hydrobiological researches of water basins self-purification.* — L., 1976. — P. 58 — 65.
[28] Brockmann U. N.. Eberlein K., Hentzschel G. et al. Parallel plastic tank experiments with cultures of marine diatoms // Helgoland, wiss. *Meeresuhters.* — 1977. — V. 30, N 1/4. — P. 201—216.
[29] Andrushaytis A. G. Experience of ecotoxicological experiment conducting in Rizhsky bay. Technical procedure // *Experimental water toxicology.* — 1984. — Edit. 9. — P. 78—88.
[30] Kuiper J. Model ecosystemen in der aquatische ecotoxicologie // *Vakbl biol.* — 1985. — BS. 65. — P. 13—54.
[31] Kuiper J. Brochmann U. H., Van den Groenewoud H. et al. Influence of bag dimensions of the development of enclosed plankton communities during POSER // *Mar. Ecol. Prog. Sar.* — 1983. — V. 14. — P. 9—17.
[32] Bernardi R., Giussani G., Pedretti E. L. Selective feeding of zooplankton with special reference to blue-green algae in enclosure experiments // *Mem. 1-st Ital Hydrobiol.* — 1982. — V. 40. — P. 113—128.
[33] Gamble D. S., MPChachlan R., Beaton C. D. Comparative growth and development of autumn and spring spawned Atlantic herring, larvae reared in large enclosed ecosystems // *Mar. Ecol. Prog. Ser.* — 1985. — V. 26. — P. 19—33.
[34] Istvanovics V., Voros L. et al. Changes of phosphorus and nitrogen concentration of phytoplankton in enriched lake enclosures // *Limnol. Oceanogr.* — 1986. — N 31(4). — P. 798—811.
[35] SETAC-Europe. Guidance of a Workshop on Aquatic Microcosms for Ecological Assessment of Pesticides. Wintergreen, Virginia, USA, October 1991. — SETAC Foundation for Environmental Education and the RESOLVE Program of the World Wildlife Fund, 1992. — 56 p.
[36] Touart L. W. Aquatic mesocosm test to support pesticide registrations // Hazard Evaluation Division Technical Guidance Document. U. S. Environmental Protection Agency, Washington D. C, Report No US-EPA/540/09-88-035. — 35 p.
[37] Manual on surface waters chemical analysis / Under edit. of A. D. Semenov. — L.: *Gidrometeoizdat,* 1977. — 542 p.
[38] Uniform methods of waters analysis. Edit. 1. — L.: *Gidrometeoizdat,* 1978. — 145 p.

[39] Manual on methods of surface waters and bottom sediments hydrobiological analysis / Under edit. of V. A. Abakumov. — L.: *Gidrometeoizdat,* 1983. — 239 p.

[40] Kartasheva N. V., Lebedeva L. I. Methodological peculiarities of experiment with natural zooplankton community in insulated volumes // *Experimental water toxicology.* — 1984. — Вып. 9. — P. 162—170.

[41] Kuiper I. Marine ecotoxicological tests: multispecies and model ecosystems experiments // *Ecotoxicological testing for the marine environment.* — 1984. — V. 1. — P. 527—588.

[42] Shannon R. Imitating systems modeling — art and science. — M.: Mir, 1978. — 176 p.

[43] Trunov N. M., Nikanorov A. M., Askalepov B. N. Application of experimental ecosystems for researching natural waters: the problem of similarity // Hydrological, Chemical and Biological Process of Transformation and Transport of Contaminants in Aquatic Environments. Proc. of the Rostov-on-Don Symp., May. 1993—1994, LANS Publication N 219. — P. 429—432.

[44] Iberla K. Factor analysis. — M.: *Statistics,* 1980. — 395 p.

[45] Khaylov K. M. Are ecological principles of aquaculture possible? // *Biological bases of aquaculture in seas of European USSR.* — M., 1985. — P. 40—54.

[46] Khailov K. M. Issues of methodology and technology of hydrobiological researches // *Hydrobiological magazine.* — 1979. — V. 15, №6. — P. 3—12.

[47] Dubrov A. M. Statistical data processing by principal components method. — M.: *Statistics,* 1978. —185 p.

[48] Khaylov K. M. and oth. Production and application of gypsum structures for recording of physicl-chemical interaction of body with moving water in small-scale space // *Sea ecology.* —1988. — Ed. 30. — P. 83—90.

[49] Nikanorov A. M., Trunov N. M., Shakunova N. N. and oth. Full-scale modeling of basic processes of contaminants and water ecosystems interaction // In col.: *Ecological standardization and modeling of anthropogenic impact on water ecosystems.* Edit. 1. — L., 1988. — P. 9—18.

[50] Trunov N. M., Shakunova N. N., Svetovskaya R. V. and oth. Full-scale modeling in water ecology (chemical-biological aspect) // Studying of the processes of chemical composition of natural waters in conditions of anthropogenic impact / Materials of the 28th all-Union hydroch. meet., May, 1984 — L., 1987. — P. 169—170.

[51] Mexhevikina Y. V., Gribovskaya I. V., Ilyashevich T. I. Kinetics of phenol decay in water of the Enisei river // *Water resources.* —1989. — No 5. — P. 183—186.

[52] Dorner Wolf G. Phenole in wasser: ein ekologisches und analytisches problem. *Techn. Rdsch.* — 1986. — 78, N 39. — P. 60—61.

[53] Popov A. N. Forecast of water quality taking into account self-purification and secondry contamination processes // VIII Symp. On issues of water basins water quality. — Tallinn, 1990. — P. 67—68.

[54] Nikanorov A. M., Sukhorukov B. L. Similarity of model and maternal ecosystems on data of surface waters remote monitoring // *Water resources,* 2003. – T. 2.

[55] Nikanorov A. M. Principle of standardization of anthropogenic impact on freshwater ecosystems and ecological similarity // *Hydrochemical materials.* — 1990. — V. 108. — P. 181—188.

[56] Nikanorov A. M. Usage of ecological modeling for determination of limits of sustained freshwater ecosystems development and ecological standardization // *Water resources.* — 1995. — V. 22, No 2. — P. 214—217.
[57] Sukhorukov B. L., Garbuzov G. P., Nikanorov A. M. Assessment of water bodies conditions by spectra of brightness ratio // *Water resources.* — 2000. — V. 27, No 5. — P. 579—588.
[58] Sukhorukov B. L., Garbuzov G. P., Akavets A. A., Nikanorov A. M. Integral indexes of water bodies state during interpretation of remote spectrometric data // *Rep. of RAS.* — 1998. — V. 363, № 2. — P. 278—281.
[59] Optics of ocean. V. 1. Physical optics of ocean. — M.: *Science,* 1983. — 372 p.
[60] Khoruzhaya T. A., Bryzgalo V. A., Predeina L. M., Korotova E. I. Methods of bioindication and biotesting during studying of freshwater ecosystems at emergency and extremely high contamination // *Hydrochemical materials.* — 1991. — V. 100. — P. 47—68.
[61] Nikanorov A. M., Teplyakov Y. V., Urazaev K. K. Similarity of models of water bodies ecosystems: similarity criteria and ways of assessment thereof // *Water resources.* — 1996. — V. 23, № 5. — P. 578—582.
[62] Sukhorukov B. L., Garbuzov G. P., Nikanorov A. M. Optical properties of contaminated water ecosystems // *Meteorology and hydrology.* — 1997, No 11. — P. 5—19.
[63] Physical encyclopaedic vocabulary. — M.: SE, 1984. — 944 p.
[64] Ayvazyan S. A., Buchstaber V. M., Enuykov I. S, Meshalkin L. D. Applied statistics classification and reduction of dimensionality. — M.: *Finance and statistics,* 1989. — 607 p.

Chapter 10

BUFFER CAPACITY OF FRESHWATER ECOSYSTEMS

One of fundamental properties of water ecosystems is its buffering or buffer capacity (ecological capacity, ecological reserve), which becomes apparent in the ability of these natural bodies to neutralize or considerably alleviate the negative impact of anthropogenic loads and natural factors by means of internal chemical-biological and physical processes.

It is evident, that this mentioned ability has its limits, the overrunning of which leads to functional and structural reconstructions of the systems and degradation and destruction thereof in the future.

In essence, buffering determines the limits of permissible influence on water ecosystems and is for this reason, a key parameter in the program of ecotoxicological research and ecological monitoring. The issue of buffering has become especially urgent in the last decades of XX century and is likely to preserve its urgency in XXI century. Anthropogenic load on water ecosystems is increasing continuously; the list of ingredients contaminating water basins and waterways becomes longer as well.

During the last 20 years the author, jointly with A.V. Zhulidov, I.A. Lapin, N.M. Trunov, M.G. Tarasov and oth. [1—16], has been studying the issue of freshwater ecosystems buffering.

Special model experiments on mesocosms, results of field work at real water bodies and theoretical and applied development on the determination of a water ecosystem's buffering capacity have served as material for the research. Some of obtained data concerning the issue is provided in this present chapter.

Buffer capacity is an index describing the share of an anthropogenic load (dose of pollutant), the inflow of which does not disturb the natural functioning of the whole ecosystem.

The buffer capacities of water ecosystems with regard to different contaminants may be unequal may even differ greatly. It shall be noted that this notion cannot be applied to all substances. There are substances that shall not get into water ecosystems at all, like acute toxic pollutants (from groups of xenobiotics) and poisonous substances. These are compounds obtained artificially, which cannot be found in the natural environment even in the smallest quantities. These compounds shall be marked out as a single group and calculation of buffer capacity for such compounds shall not be performed. The presence of such substances at any concentration is not permissible in water ecosystems.

It was noted in many papers, where the issue of a water ecosystem's buffering and its tolerance to the impact of anthropogenic and other factors was studied [17—36], that buffer capacity, within the meaning of specified statement above, is related to the stability (homeostasis) of ecosystems, in particular, with species diversity (alpha and beta), trophic and dimensional structures of populations. In studying the ecosystems, however, we are often not provided with all of the information concerning hydrobiological characteristics of the water bodies.

V. D. Polischuk [37], when studying dynamical characteristics of planktonic communities, came to the conclusion that the lake of a temperate zone contains approximately thousands of living organisms species, and ecosystems of wetlands exceed this by two orders more. Relatively complete information may be provided only concerning the species diversity of some communities or trophic levels. Within separate trophic levels, the stability thereof, according to Murdoch, S. Reice, J. Thorp and oth. [38—40], is the function of the number and intensity of competitive relation between species.

Along with the named biological factors, the following factors, mentioned in some published works, determine buffering of ecosystems as well:

- Trophic levels as well as communities may attain equilibrium with habitats existing independently of each other [21, 42].
- There is a relation between the area of freshwater systems and species diversity of hydrobionts (alpha and beta); besides, structures of macrophytes coenoses influence the diversity and number of hydrobionts [43].
- In some cases, the leading species complex of mass species, included in holocoenotic core, play an important role in the definition of community structure and stability [44].
- Very often, a change of ecosystem state takes place unevenly with the orientation changing many of the most important biological processes [45, 46]. Discontinuity of an ecosystem's state changing under the influence of anthropogenic, and sometimes natural factors, results from theoretical analysis as well when buffering is disturbed due to exceeding the threshold values of such impacts [1, 5, 7, 9, 12—14].

Thus, in analyzing buffer properties of freshwater ecosystems not only specified peculiarities of biocoenoses functioning shall be taken into account, but also the processes influencing the dynamics of water population, such as the interaction of the components of the communities (competitiveness, carnivorism, etc.), difficulties in determination of taxonomical belonging of a range of groups of invertebrates and inferior plants and the insufficient knowledge about homeostasis mechanisms with regard to main groups of pollutants. Nowadays, all these factors do not make it possible to determine buffer capacity through species diversity, although this approach regarding biological factors does seem to be a priority and we give high rating to it [5]. We find it difficult to examine the notion of "buffer capacity" in comparison to the notion of *"stability of ecosystems"* as well.

For the development of substantial models of an ecosystem's tolerance to anthropogenic impact, it is necessary to describe the state of the standardized system through a small number of variables. The revelation of the latter from groups of ten monitored parameters very often becomes one of the main methodological problems. Suffice to say that searching for variables by describing only mechanical movement— position in space and speed—took mankind approximately two thousand years.

The most attractive solution for ecology is searching for a one-dimensional description of the ecosystem state by means of data compression (description of the state by one parameter). The latter makes it possible to arrange and visually compare the state of the controlled ecosystems as more–less, better–worse, closer–farther, etc. The theoretical possibility of such a one-dimensional description results from the well-known Canton theorem, stating that *any n-dimensional cube may include the number of points equal to the number of points in a segment, i.e., any n-parameter state may be replaced with one-parameter.* Herewith, however, with regard to water ecosystems, the key problem of data compression and the constructive definition of such a one-dimensional scale or integral state parameter rests unsolved. A one-dimensional scale of the optical states of large water bodies with the characteristics of brightness of radiation ascending from the water column in the red area of the visible spectrum may serve as example from a closed field of knowledge. The change of brightness depends on changes of an optically active substance's concentration (chlorophyll "a," seston). With accuracy between 90-95%, however, these changes are described by means of one parameter:

$$\alpha = \Delta C_C/\overline{C}_C - \Delta C_{chl}/\overline{C}_{chl} \qquad (1)$$

where $\overline{C}_C$, $\overline{C}_{chl}$ are the average values of seston and chlorophyll "a" concentrations in the water body, respectively and $\Delta C_C = C_C - \overline{C}_C$ and $\Delta C_{chl} = C_{chl} - \overline{C}_{chl}$ are deviations from the average values in this part of the water area [47].[8]

Of course, there is no full analogy between the specified example of optical characteristics usage for the assessment of a phytoplankton community state and the data compression for the purposes of definition an ecosystem's state. In practice, the one-dimensional scale at present is determined on the basis of the researchers' experience and intuition. This fact explains the variety of approaches and integral parameters that are used in the ecological standardization of ecosystems states.

It shall be noted, that even the adequate definition of a one-dimensional scale may not provide a complete idea of the ecosystem's development, just as knowledge of part of the material point of a motion path does not make it possible to forecast an object's further trajectory without knowledge of motion law. Therefore, it is important to know not only and not so much the order of states but also the ways (laws) of transition from one state to another, i.e., to have a model of the ecosystem that demonstrates its development (motion—by analogy, with mechanics).

With regard to dynamic systems, *tolerance may be defined as the property of moving a system expressed with little deviation from some trajectory in cases of small-scale disturbances.* Concretizing the notion of "small-scale disturbances," different notions of tolerance may be attained. If the disturbances are understood as changes of the initial state of the system, and deviations are understood as measure of the disturbed motion (functioning) difference from the undisturbed portion, we arrive at Lyapunov's notion of tolerance, even, orbital. If in case of disturbances, the laws of trajectory motion preserve some properties of undisturbed trajectories, we arrive at the Lagrange or Poisson stability.

[8] For more details concerning remote methods ofmonitoring, see Chapter 6.

Thus, in trying to formulate the notion of "tolerance," it shall be precisely identified, which disturbances are accepted or considered, which insensibility is of the most interest, or which responses to disturbances are equivalent (unequivalent).

A.D. Armand [48] published a review of definitions of tolerance with regard to geographic systems. These definitions suit water systems as well. Generalizing the large number of works concerning this issue, he presents the following meanings of specified notion:

- Probability of preservation of the object during certain time period
- Stability of state in time, inertness or constancy of qualitative characteristics
- Ability of former state restoration after disturbance
- Ability of adaptation to changed conditions, transition to a new equilibrium state and flexibility
- Ability to preserve some vital parameters at definite levels by means of other parameters of flexibility or homeostasis
- Ability to "suppress" outer signal by repeated transfer thereof from one element to another
- Ability not to react by means of "shielding"
- Ability to continuously (but not eternally) accumulate harmful impacts (for example, contaminants) without visible damage to system
- Ability to easily let the contaminants through the system in such a manner that those do not have time to harmfully impact the system
- Ability to preserve the production function in a social and economic environment
- Absence or quick decay of oscillations in the system
- Ability to maintain the trajectory of development or the orientation of trend

P. Kindlmann and J. Leps [49], examining the tolerance of ecosystems from mathematical and ecological points of view, came to the conclusion that the notion of "tolerance" developed in the physical-mathematical fields of science and may be theoretically applied to ecosystems. From now on, however, it shall be taken into account that when *selecting parameters and variables, with regard to which the tolerance concept is examined, it is essential that extracted herewith, the subsystem of the specific ecosystem would be relatively autonomous, making it possible to neglect the influence of the discarded variables. Besides, the time and spatial scales within the framework of which the tolerance concept is examined shall be logically coherent with selected types of equivalence responses to disturbances.* It shall be added to the aforementioned that examining the notion of tolerance to specific impacts, but not general tolerance, is the matter of discussion. It is a crucial factor since *depending on the impact nature (type of contaminants) within the same system may be more tolerant, less tolerant or completely intolerant.*

Failure to take into account these factors makes the notion of the "tolerance of ecosystems" free in any substantial sense. It shall be especially emphasized that the acceptance of unconditioned assumptions is impermissible. It may cause the tolerance concept to be inapplicable to real ecosystems, particularly for the following reasons [48, 50]:

- Stability of system in quite little ambit of equivalence without specification of dimensions of the area

- The fact that disturbances with a permanent nature can be found in ecosystems is neglected.

In virtue of the abovementioned and other, more frequent reasons, the notion of tolerance in ecology and geography in most cases is of a nonformalized nature.

The most important property of any ecosystems, including freshwater and wetlands, is that during the process of its evolution, they pass *successions*—consecutive, irreversible and directed changes of communities in the same place of location thereof. Successions appear either under the influence of allogenic (exogenous) processes (for instance, by means of biogenic elements inflow from outside, as in the case of anthropogenic eutrophication) or as a result of natural autogenous (endogenous) processes [51, 52]. There are many classifications for successions [53]. Following V.N. Sukachev [54, 55], however, as a rule, three basic types of succession are singled out:

- *Syngenetic*—dynamics of coenoses in young ecotopes (for example, on benthic deposits of young water storage basins)
- *Endogenetic*—dynamics of coenoses under the influence of the processes running in communities itself as a result of its vital activity
- *Exogenous*—dynamics of coenoses under the influence of outside factors

Sometimes, another succession category is singled out—hologenetic, under which the process of communities changing is intended to result from changes at the level higher than the communities level, for example, at landscape level. Changes of biocoenoses under the influence of the climate of the region, or under the influence of changes caused by a river valley's development, etc., may serve as an example.

Each such stage in the succession series may be actually be considered a new ecosystem at least by virtue of the fact, that only some species from the preceding community pass into its next stage. Therefore, the *ecological standardization of anthropogenic impact on freshwater ecosystems shall take into account the time stage of the ecosystem development within the series of its succession for which the developed standardization principle or specific calculations of the ecosystem buffer capacity are intended.*[9] In other words, there shall be a peculiar complex of guidelines regulating ecological standardization, which would seamlessly correspond to the succession row of the ecosystem within the limits of its series. Besides it shall be taken into consideration that even climax communities, representing final (and therefore, predictable) stages of succession,[10] theoretically remain invariable up to the moment of considerable climatic (or geological) changes and may be of at least two types:

- *Climatic*, if the ecosystem consists of communities typical for this physical and climatic natural zone

---

[9] It is specified in the section devoted to standardization that it shall be based on knowledge of inter-water basins from a chemical-biological standpoint, including successional processes, and take into account the nature of water consumption in a water body. A trial to quantitatively assess tolerance (through buffer capacity) with regard to anthropogenic acidulation was made by the author, jointly with I.A. Lapin and M.G. Tarasov.

[10] It shall be noted that climax communities of freshwater systems in some cases visually are not considerably different from intermediate stages (ecosystems) in a succession row within its series.

- *Edaphic*, if specific peculiarities are inherent to the ecosystem, as a rule not resulting directly from natural peculiarities of the physical and climatic zone (for example, as a result of action from a pyrogenic factor or anthropogenic eutrophication)

In speaking about standardization of pollutants and their inflow into freshwater ecosystems, special attention shall be paid to the *existence of two types of optimums for organisms:* physiological and ecological [52]. The *physiological optimum* almost always is relatively stable and changes only during ontogenesis; it is always wider than the ecological and its boundaries may be defined in case of the exclusion of competitive interrelated actions.

The *ecological optimum* fully depends on and is determined by competitive interrelations in communities[11] and just by virtue of this fact, it can never be stable, but instead is continuously changing under the influence of outside ecological factors and ontogenesis. The phenomenon of physiological and ecological optimums is fair not only for separate species of plants and animals, but, what is especially important, for communities and ecosystems, including freshwater and wetlands.

Even in terms of species diversity, the disturbed freshwater ecosystems and wetlands without a complex structure have unevenness in living organisms' distribution. When looking for simplicity, the only vegetational component of these ecosystems (in particular, superior littoral-aquatic vegetation) in most cases is its heterogeneity, which is determined by the location and quantity of synusia composing the phytocoenosis of wetlands. In this case, as it was noted by S.V. Viktorov and other researchers earlier [56] with regard to phytocoenosis, the examined ecosystems as unitary whole are composed of spatially limited and morphologically different elementary parts, each of which may to some extent have independent significance. This fact makes it possible to theoretically dismember the set of environmental conditions at the present area of an ecosystem to its elementary components of habitat. Such an idea makes it possible in some cases to reveal relations of different components of freshwater ecosystems and wetlands within the environment. This approach, however, as shown in our research,[12] requires a great deal of prudence during when interpreting results. This is the case of the uncritical transition of the concept regarding heterogeneity of living organisms' communities to the heterogeneity of the ecotope; here, notions of the *physical (geochemical) heterogeneity of habitat and biological heterogeneity of biocoenosis* are mixed. As far as it can be judged from the literature, this problem was first established and investigated in indicative geobotany with regard to ground ecosystems (see [56]).

The essence of the problem may be illustrated through the following example. During research concerning the monitoring of a wetland's anthropogenic contamination, the heterogeneity of different areas of the ecosystems in terms of physical and geochemical indices shall be taken into account and recorded by technical means of control. Studying biocoenoses within the limits of separate wetlands areas (freshwater ecosystems), however,

---

[11] Interspecies relations (competition, carnivorism), as a rule, narrows the ecological niche of each species in comparison with the expectations based on physiological and morphological parameters of organisms; intraspecies competition usually leads to the widening of the ecological niche of the species [57].

[12] See monographs: Nikanorov A.M., Zhulidov A.V., Pokarzevsky A.D. Biomonitoring of heavey metals in freshwater ecosystems. – L., 1985; Nikanorov A.M., Zhulidov A.V. Biomonitoring of metals in freshwater ecosystems. – L., 1991; Nikanorov A.M., Zhulidov A.V., Emets V.M. Heavy metals in the organisms of Russia wetlands. – L., 1993.

reveals that they are heterogeneous in terms of physical and chemical parameters. Its "homogeneity" comes in terms of biological parameters (for example, absence of synusia and phytocoenosis change). In other cases, the change of synusias and even phytocoenoses reflects not the physical and geochemical heterogeneity of habitat, but its biological heterogeneity. The latter is determined by the habitat forming a role for phytocoenoses or zoocoenoses itself, as is clearly seen in the example of beaver habitats in the Voronezhsky biospheric reserve. In some cases, biological heterogeneity of wetlands (freshwater ecosystems) becomes definite even within the boundaries of separate local areas whose conditions necessitate its purposeful study during the ecological survey as part of the arrangement of works on biomonitoring of anthropogenic factors for an ecosystem's contamination and ecological standardization. Therefore, "differentiation of these two types of heterogeneity and a clear idea of which heterogeneity (physical or biological) we determine in each specific case is strictly compulsory" [56].

In view of the fact that "ecosystem" does not have clear boundaries (unlike biocoenosis), an immense number of ecosystems can be singled out and observed within the wetlands (freshwater ecosystems). From all diversity, *from our point of view, for full scale research on an ecosystem's buffer capacity, it is important to only consider ecosystems forming a succession row within its series.* In this case, the investigated ecosystems show a genetic affinity to each other. Herewith, the optimal case is one where the structure of such a row, its relation with definite natural process and sequence of different stages during the latter are revealed beforehand by means of long fixed observations. In this case, such a succession row may be considered an indicator of a definite process. Often, however, in the practice of field research, the situation arises when having revealed this or that row within the wetland (freshwater ecosystem), it is necessary to decide whether this row is genetic (succession) or there is no genetic affinity between its links (ecosystems). Besides, if a genetic series is revealed, it is necessary to determine the process or group of factors with which it is associated and use this genetic series for the determination of "areal" of the process within the wetland (freshwater ecosystem).

In order to distinguish a genetic (succession) series from other series, members of which are not allied in ground geobotany [56], the following indicants are used: nature of boundaries between the members and the grade of specific (taxinomical) similarity of neighboring members in any link (ecosystem) of the relict fragments of neighboring links (ecosystems). In most cases, these principles are applicable to wetlands and freshwater ecosystems in a narrow sense of the word.

In any undisturbed biocoenoses, the composition of the populating species is not casual but reveals regular features, the most important of which is the *ecological equilibrium* both between the organisms and with the environment. As was shown by G. Walter [52], this equilibrium is conditioned by:

- Interspecies competitiveness
- Ecological dependence of living organisms of different species coexisting in biocoenosis, which, actually creates the possibility of its existence in this community
- Presence of complementary species (separated in time of their development or spatially), which in appropriate habitats complement each other (in time and/or space) or at least, do not hamper the existence of other species

In competitive interrelations, the so-called "habitat factors" have crucial meanings [52]. Depending on the nature of integrade influence of these factors and the intensity thereof, these factors favor the increase in quantity of different living organisms' species. Herewith, the natural change of habitat factors rarely causes the immediate death of any species; usually it may take place through an indirect nature—as a result of the reduction of competitive vital capacity of one species in comparison with another [52]. Thus, abiogenous ecological factors in different manners influence the living organisms since the degree of its intensity to a considerable extent depends on the competitive interrelations between organisms.

The phenomenon of species complementarity in biocoenosis of wetlands (freshwater systems) becomes apparent in three forms:

- Temporary change of aspects (for vegetational communities): in zoocenosis temporary arrangement, becoming apparent in changes of lohoses[13] and phalanges [58—61]
- Spatial isolation: this type of complementarity becomes apparent in the form of such phenomenon as layerage in phytocoenosis and chorologic distribution of animals on strata and in areas in zoocenosis
- Trophic isolation: this type of complementarity takes place only with regard to animals and needs no comments

The phenomenon of complementarity being expressed in isolation of living organisms in biocoenoses favors the formation of freshwater systems with utmost species saturation. In its turn, this leads to the fact that *biocoenoses in equilibrium with ecotope increases its productivity, the extent of which is limited and corresponds to the resources of the ecotope.* The latter provision may be elevated to the category of regularity.

The aforementioned makes it possible to explain why it is so difficult for an adventive species to penetrate into biocoenoses: undisturbed communities, maximally using all material resources of the habitat practically do not leave free any ecological niches for species that are alien to the community. One should remember, however, that biocoenoses of wetlands (freshwater ecosystems) reaches maximal species saturation during the period of time that may last for decades.

Speaking about *ecological equilibrium* of wetlands and freshwater ecosystems, it shall be especially underlined that the process is *dynamic*; within each ecosystem, different directed processes take place that are determined (with regard to coenoses), in particular, both by endogenous factors (change of species competitive ability depending on the nature of the exogenous processes) and by exogenous factors. Under certain conditions, ecological equilibrium may be disturbed first, and then successions will take place after.

It shall be especially noted that the evolution of freshwater ecosystems and wetlands under conditions of anthropogenic impact may take place in different directions. As was shown by V.A. Abakumov [60, 61, 62, 63] for water ecosystems, at initial stages of contamination, the increase of biocoenoses metabolism intensity takes place (i. e. metabolic progress is observed) and is accompanied by complex of phenomena attributed to the state of

---

[13] Lohos—set of specimen in population that are at the same stage of development. Lohoses, due to their ecological functions in biocoenoses, form jointly with the appropriate geophysical and geochemical conditions with interdependent complexes—phalanges. The latter may differ not only by sets of forming lohoses, but also by the biological state of separate lohoses [61].

ecological progress. An increase of biocoenosis species diversity and spatial heterogeneity leads to complication of interspecies relations and time structure. With the further increase of anthropogenic contamination, the intensity of the biocoenosis metabolism keeps growing up to a certain critical level, after which the direction of the processes characterizing the ecological state reverses. The biocoenosis species diversity decreases, interspecies relations become simpler, spatial heterogeneity decreases and the time structure becomes simpler (i. e. phenomena attributed to the state of ecological regress).

Jointly with these two directions of metabolic progress and related with two alternative ways of biocoenoses ecological structure changes, there is another direction or method of ecological structure change. This way is *ecological modulation*—such biocoenosis reconstruction, which does not change the general level of the ecological system arrangement (for example, change of dominant species, change of composition of principal complexes, etc.). Ecological modulations are the most widespread directions of metabolic progress both in typical water ecosystems [60, 61] and in wetlands.

Mechanisms of ecological modulations are not fully clear yet. Most likely, the most important processes determining modulations in freshwater ecosystems will be the processes at the phalanges level.

10.1. Calculation of Water Ecosystems Buffer Capacities (Natural and Model)

It was shown in a previous section that a lot of factors influence the hydrobiological parameters, buffer capacity, stability, ecological equilibrium and ecological tolerance of freshwater ecosystems. It is completely evident that the role of all these factors both taken together and separately is very difficult to account for at this level of knowledge during buffer capacity calculation. It is impossible to develop a technique of buffering assessment with regard to all the groups of priority contaminants.

On basis of these specified reasons, it was decided to concentrate efforts on phytoplankton—producers, frequently determining the range of the most important hydrochemical and hydrobiological indices. Copper was selected from the chemical elements on basis of the following considerations:

- It is included in the list of heavy metals related to priority contaminants
- Is a biofriendly element and has considerable biological activity
- Is characterized by clearly its expressed ability to get completely dissolved organic matter in natural waters

The Usman River within the boundaries of the Voronezhsky Reservoir was taken as the model biotope where long-term observations of phytoplankton biomass and dissolved copper were conducted.

10.1.1. Hydrobiological Factors

As is clear from Figures 10.1 and 10.2, the following regularity may be tracked in the seasonal dynamics of plankton biomass in the Usman River: biomass increases from winter to summer (autumn) and drops in the winter. Seasonal copper concentration in the water of the Usman River bears an inverse dependence; the copper concentration is maximal during the winter period and is minimal in the summer (autumn).

Quantitatively, the relation between concentrations of dissolved copper and phytoplankton biomass is established by processing the results of full-scale research through mathematical statistics. For an adequate description of the studied phenomena, however, it is necessary to determine the preliminary form of relation between the examined variables, which depends on the nature of the studied processes and is determined by the investigation of its physical-chemical and biological regularities. After that, parameters of the obtained equation and its constraints are quantitatively defined by statistical processing of full-scale results.

A priori it can be considered, that copper (M) coming into the waterway is distributed into the following components:

M_v—copper, remaining in its dissolved form

M_b—copper, adsorbed and accumulated by phytoplankton (all hydrobionts in a more common case)

M_d—copper, migrated to benthal deposits during the process of suspended and mineral material sedimentation

M_{ad}—dissolved copper adsorbed by the surface of benthic deposits directly from water

M_{av}—copper adsorbed by mineral suspension and contained in it

In this case, the most common relation between the concentration of dissolved copper and phytoplankton biomass is examined. Therefore, more frequent processes determining the behavior of copper in the river may be neglected. In general, the equation of the material balance of copper can be presented in the following form:

$$M = M_v + M_{ad} + M_{av} + M_b + M_d \quad (2)$$

With reference to the quantity of copper adsorbed by benthic deposits, in the most general view, it is proportional to concentration of metal in the water of the river, C_{m_v}:

$$M_{ad} = \alpha' \; C_{m_v} = \alpha' \; M_v/V \quad (3)$$

and

$$M_{ad}/V = (\alpha' \; V)(M_v V) = \alpha' \; M_v V$$

Hence, finally we obtain:

$$M_{ad} = \alpha' \; M_v \quad (4)$$

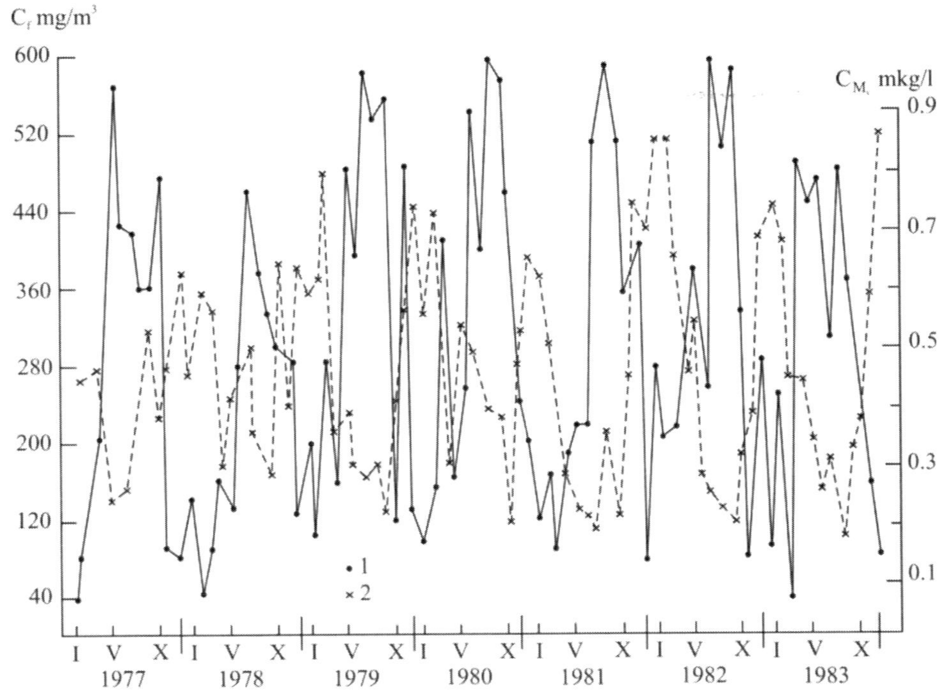

Figure 10.1. Average long-term variability of plankton biomass (1) and dissolved copper (2) content in the Usman River and in the territory of Voronezhsky biospheric reserve [1].

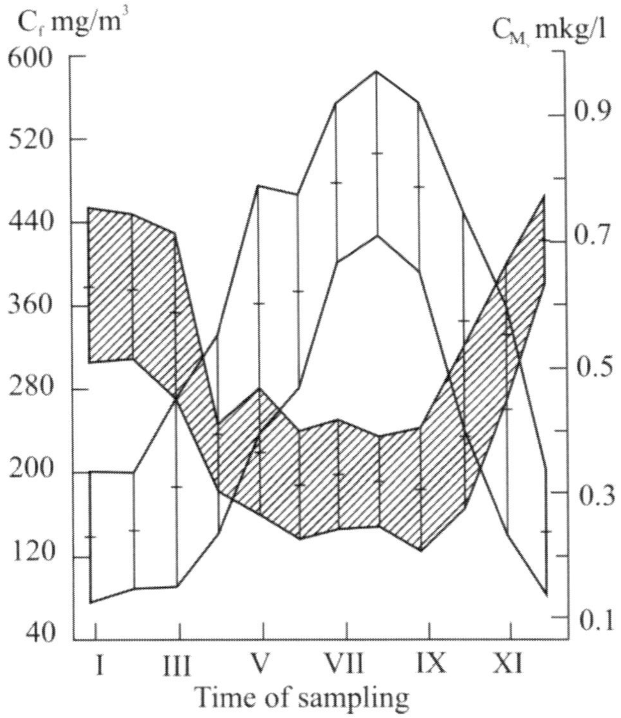

Figure 10.2. Confidence intervals of 95% accuracy for an average, long-term plankton biomass and dissolved copper content (cross-hatched) in the Usman River and the Voronezhsky biospheric reserve; 1977-1981, 1983-1984 [1, 8, 9].

In the first approximation, the amount adsorbed on the suspension metal is proportional to the amount of suspension and concentration of metal in the water, M_v/V. Taking into consideration the physical-geographical and geological peculiarities of the Voronezhsky reserve territory and ecological peculiarities of the Usman River itself, however, the content of the suspension in the Usman River may be taken as a slightly changing value during the year (but during the period of spring floods, the suspension content increases). Taking into account the fact that this work deals with the most common dependencies, the equality determining amount of copper adsorbed on mineral suspension in a volume of water may be presented in the following form:

$$M_{av}/V = \gamma M_v V,\ \text{or}\ M_{av} = \gamma M_v \qquad (5)$$

where γ is the proportionality factor.

The accumulating capacity of hydrobionts with regard to dissolved metals is usually evaluated by the accumulation factors K_n[2], understanding that under these factors, the relation of metal concentration in the body of the specimen of examined species (per unit of dry material mass) to the content of these metals is the same in terms of mass volume of water from the hydrobionts' habitat. Accordingly, the expression for copper content in phytoplankton, taking into account the accumulation factors, may be presented in the form:

$$M_b = K_n M_v C_f \qquad (6)$$

where C_f is the mass of phytoplankton in a unit of water volume.

The amount of copper passed to bottom sediments with fossil phytoplankton may be described by the formula:

$$M_d = K_n M_v \beta C_b \qquad (7)$$

where β is the mortality rate.

Taking into account (4)-(7), expression (2) may be written as follows:

$$M = M_v + \alpha M_v + \gamma M_v + K_n M_v C_f + K_n M_v \beta C_f \qquad (8)$$

or

$$M = M_v[1+\alpha+\gamma+K_n(1+\beta)C_f]$$

Passing to the copper concentration per unit of water volume and introducing symbols:

$$M_m = M/V;\ C_{m_v} = M_v/V;\ a = 1+\alpha+\gamma;\ b = K_n(1+\beta) \qquad (9)$$

we obtain the equation that links the concentration of dissolved copper with plankton biomass in a unit of water volume:

$$C_{m_v} = C_M/(a+bC_f) \qquad (10)$$

or, dividing a and b by C_m for convenience of statistical processing:

$$C_{m_v} = 1/(A+BC_f) \quad (11)$$

where

$A = a/C_m$; $B = b/C_m$.

By introducing the new variable Y = 1/ C_{m_v}, we obtain the equation of line:

$$Y = A+BC_f \quad (12)$$

Taking into account the existing natural daily variability of indices C_{m_v} and C_f (see above), the determination of their statistic relation was performed among the averaged values.

For averaging the entire data, an array for all years of the surveys was divided into 12 groups in terms of months. Average values C_{m_v} and C_f were determined in each group (Table 10.1). Obtained average long-term yearly changes of C_{m_v} and C_f (see Figure 10.2) qualitatively corroborate the form of the constraint equation (11). According to data from Table 10.1, taking into consideration (12) valuations of equation (11), the parameters were defined through the least-squares procedure.

Thus,

$$C_{m_v} = 1/(0.94+0.00477C_f) \quad (13)$$

The correlation factor between the average long-term values in terms of months is equal to 0.91.

Dependence (13) is specified on Figure 10.3. Presented on Figure 10.3, the functional relation of the dissolved copper concentration (C_{m_v}) and phytoplankton concentration (C_f) is described by formula:

$$C_{m_v} = 0{,}3 + \frac{0{,}4}{1+0.025\,C_f^4} \quad (14)$$

The general view of the analogous relation is described by the expression:

$$f(x) = a + \frac{b}{1+cx^{2k}} \quad (15)$$

where the parameter a defines the shift of the curve along the y-axis; the sum $a + b$ defines the vertex of the curve (at $x = 0$); the parameter c determines the "width" of the diagram: at $c > 1$, the function of the curve quickly drops on ordinate $f = a$, at $c < 1$ (as on Figure 10.3) and the descent of the curve ordinates is slow; the parameter k (k = 2, 4, 6,...) describes the speed of the curve approaching asymptote $f = a$.

The deciduous point (is marked with a cross) corresponds to April—the time of the Usman River opening and flooding during which the suspension content and, therefore, amount sorbed on this suspension of copper increases. As far as the share of copper sorbed on the suspension, it may lead to the disturbance of relation (11), which is obtained under the condition of constancy of inorganic suspension concentration during the year.

It shall be noted that if to find relation (12) on the basis of immediate C_{m_v} and C_f determination in samples, taken at the same day (total number of such pairs was 107), the regression relation of these indices turns out to be weak (correlation factor approaches to 0.2). This result reflects the general fact of relation distortion at incorrectly set regressors [64].

In the investigated case, confounding factors during the revelation of the interrelation of the processes show the considerable unevenness of dissolved copper and, especially, plankton distribution in space and time. Related to this fact is the weak representativeness of the separate samples. In all, the changeability of the dissolved copper concentration and plankton biomass under the impact of these factors in samples simultaneously taken from the Usman River reaches accordingly four times (December 29, 1980) and 41 times (August 29, 1984), and in samples taken with a time interval of several hours and days, it was even greater. As a result of this fact, averaging considerably levels the variability of the studied processes. It is carried out in such a manner as to provide efficient noise suppression in combination with the insignificant distortion of basic processes. Taking into account the aforementioned, the absence of correlation between the biomass of aquatic plants and concentration in chemical substances in the environment (in particular, biogenic materials) marked in literature becomes understandable [65, 66].

Table 10.1. Average long-term change of dissolved copper concentrations and phytoplankton biomass in the Usman River water for 1977—1981 and 1983—1984 [35]

Month	C_{m_v} mkg/l	C_f mg/m^3
January	0.637 ± 0.120	139.8 ± 59.7
February	0.636 ± 0.107	143.7 ± 55.1
March	0.593 ± 0.129	184.8 ± 96.3
April	0.536 ± 0.056	238.5 ± 87.3
May	0.386 ± 0.093	357.4 ± 117.7
June	0.319 ± 0.079	371.1 ± 91.2
July	0.332 ± 0.084	471.9 ± 77.0
August	0.316 ± 0.070	504.4 ± 79.1
September	0.307 ± 0.091	472.9 ± 82.4
October	0.401 ± 0.123	341.7 ± 103.4
November	0.545 ± 0.106	252.9 ± 114.4
December	0.706 ± 0.058	144.0 ± 58.2

The kinetics of metals absorption by phytoplankton consists, as a rule, of two stages: light, during which ion binding on cell surfaces at elevated pH during photosynthesis takes place and dark, characterized by the transportation of cells inside at lower pH values due to an increase of local carbonic acid concentration during breathing [67]. At the same time, the speed of metal absorption by phytoplankton, intensity of metal concentration decrease in the water, increase of content thereof in bottom sediments and growth of aquatic plant biomass in a water basin are interrelated [6, 67—72] and vary during the day. It shall be noted that the concentration of metals in cells of phytoplankton to a considerable extent depends not only on the metal content in the natural environment [65, 73—76], but also on brightness, the pH gradient on the border membrane, environment, equilibrium constant, speed of diffusion (accumulation through cell membranes), physiological state and taxonomical appurtenance of aquatic plants[14] [68, 77—81].

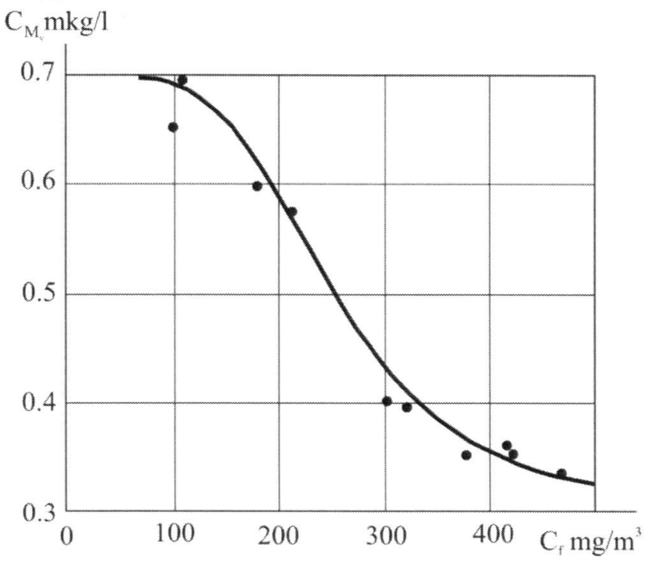

Figure 10.3. Average long-term relation of dissolved copper concentration and phytoplankton biomass in the Usman River part of the Voronezhsky biospheric reserve, from 1977-1981 and 1983-1984.

Taking into account the aforementioned daily variability of metals accumulation by phytoplankton, the daily variability of dissolved metal forms in the water becomes understandable. Concerning the latter hydrophysical peculiarities of water courses and water basins, the unevenness of microecological conditions within the water column (both vertically and horizontally) means that errors related with sampling and analysis of dissolved metal forms and other factors have considerable influence as well. All these factors also automatically determine the daily variability in the distribution of a phytoplankton biomass. Summarizing the aforementioned, it can be concluded that the content *of chemical substances in the waters of freshwater systems is directly related to the total mass of hydrobionts and their seasonal dynamics in these ecosystems* [1, 8, 9].

Additionally it shall be noted that the description for phytoplankton dependence on biomass change in a season of the year is in general fair for zooplankton and periphyton [6, 82—84] and may probably be considered common for all hydrobionts.

Speaking about the seasonal changeability of hydrobionts mass in freshwater ecosystems and its role in the formation of the chemical composition of freshwater ecosystems, it shall be emphasized that with regard to some hydrobionts, in particular, *bivalve mollusks, similar geochemical function is executed by means of filtration, which varies significantly in terms of the seasons of the year.* The most intensive filtration activity and nutrition is observed in the summer months increasing with the growth of temperature to values close to 30°C (Figure 10.4).

---

[14] As was revealed by laboratory research, the speed of phosphorus consumption by phytoplankton is determined by size of its cells and does not depend on its taxonomical position [86]. This provision, however, is disputed [87] and needs additional corroboration.

Further increase of temperature leads to almost full cessation of filtration which is related to the disturbance of normal activity of cilia of ciliate epithelium and adductor muscles . In autumn, winter and early spring, at low temperatures, water filtration by mollusks slows down and at temperatures between 1-3°C; filtration is stopped by the majority of species typical for temperate latitudes. Optimal values of temperature are lower than the top temperature threshold and may differ in terms of origin and habitat species [85].

Considerable differences of temperature influence on the exchange rate become most clearly apparent at 20°C. At these conditions, the biggest rate of exchange by freshwater mollusks is observed, while maximal filtration rate is attained at 30°C. These differences may be explained by the fact that for oxidative processes in mitochondria, a temperature of 20°C is optimal while the most intensive motion activity of a ciliary apparatus is observed at 30°C [85].

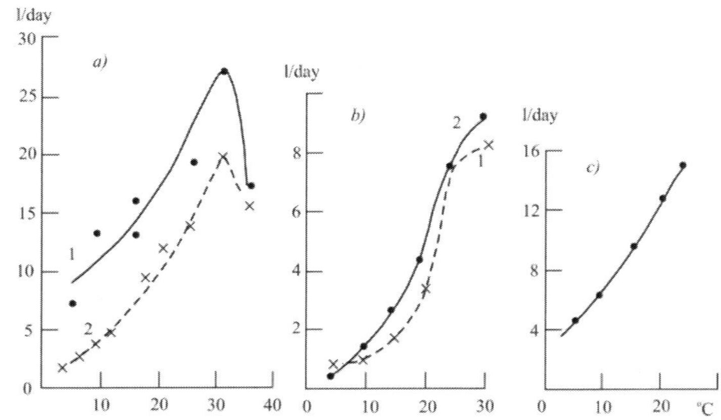

a) 1—Vniopictorim (mass 5 g), 2—U, tumidus (mass 4 g); b) 1—Dreissenapolymorpha, 2—Sphaeriastrumrivicola (mass 0.8 g); c) Hyriopsisschlegelii

Figure 10.4. Speed of filtration by mollusks at different water temperature values [85].

Filtration intensity and exchange rate of bivalve mollusks grow with the increase of its mass and size (Figure 10.5).

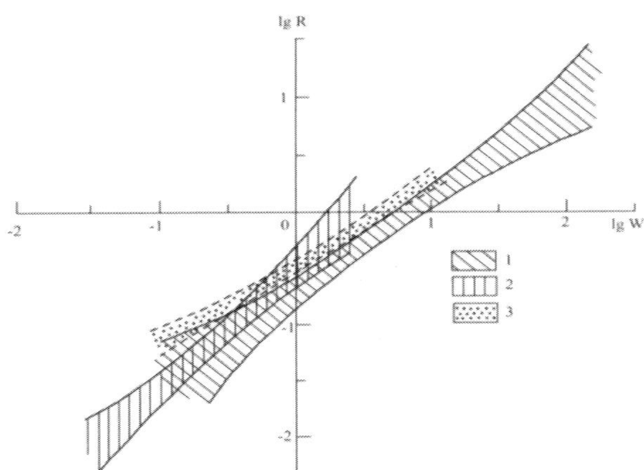

1—Unionidae, 2—Sphaeriidae, 3—Dreissenapolymorpha

Figure 10.5. Relation of exchange rate logarithm (lg$R$) and mollusks mass logarithm (lg$W$) [85].

The concentration of particles suspended in water influences the filtration rate as well. An inverse relationship was revealed between these parameters (Figure 10.6). In the area of low concentrations, the filtration activity is maximal.

The processed data from Figure 10.6, in graphic form, is presented in Figure 10.7.

Rate V of water filtration depending on concentration C of dissolved oxygen is described by the following formula:

$$V(C) = \begin{cases} 27.5C \ (0 \leq C \leq 0.2) \\ 1 + K_{0.9672}(C) \ (0.2 < C \leq 4) \end{cases} \quad (16)$$

where $0 \leq C \leq 0.2$ is the area AB where the filtration rate increases, $0.2 < C \leq 4$ is the area BC where the filtration rate slows down and $K_{0.9672}(C)$ is the Macdonald function of the second genre.

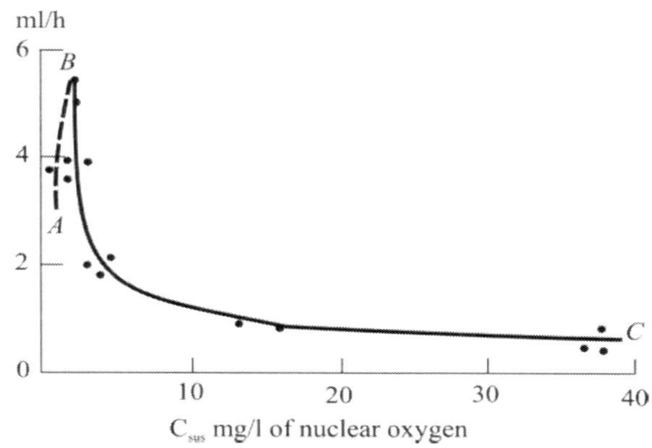

Area *AB*—filtration rate increases; area *BC*—filtration rate slows down.

Figure 10.6. Relation of water filtration rate (ml/h) by mollusks, *Sphaeriumsuecicum*, and concentration of suspended particles C_{vzv} (mg/l) [85].

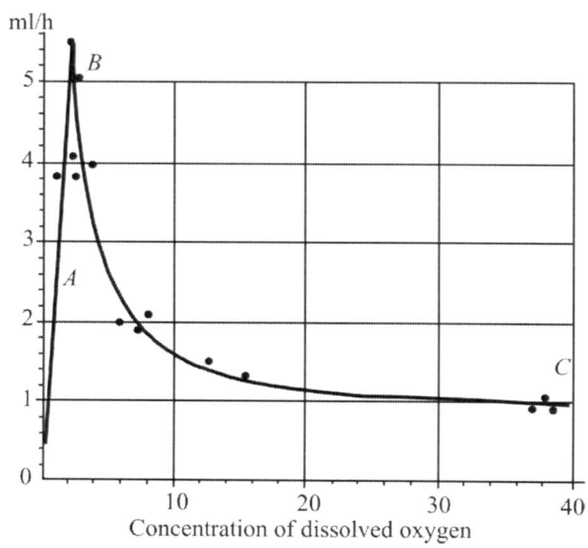

Figure 10.7. Relation of water filtration rate and concentration of dissolved oxygen.

Examining the issue of hydrobionts influence on the chemical content of freshwater ecosystems, it shall be noted that this influence is to a considerable extent related to the intensity of metabolism.

The exchange rate of hydrobionts is determined not only by seasonal variations of ambient temperature but also by peculiarities of life cycles, particularly by the degree of physiological or reproductive activity [85].

Thus, in describing hydrobiological factors that have and influence on freshwater ecosystems buffering with regard to heavy metals, it shall be noted that the most significant consequence of hydrobionts activity is the partial excretion of metals coming into the ecosystem with their subsequent burial in benthal deposits during the process of sedimentogenesis.

Taking into account that heavy metals are extracted from water by means of complex formation and by dissolved organic matter as well, the ecological consequences of the two examined processes for living organisms will depend on the *correlation of a complex formation rate* of heavy metals that have entered into the ecosystem by dissolved organic matter (and metals transition to less toxic forms) on one hand, and metals accumulation rate by hydrobionts from the other.

Comparative data concerning kinetics of these competing processes is absent in the literature. In a paper, L. Walker [73] demonstrates that during cultivation of blue-green algae, *Chroococcusparis*, in water with heavy metals (Cu, Cd, Zn), a 2 ppm concentration up to 90% of the total metals amount has been sorbed by cells during 1 minute, and 10 minutes after the fixation is almost full [88]. Herewith, at pH = 7, the sorption capacity of cells reached for cadmium, copper and zinc is 53, 120, 65 mg/g of the dry mass respectively.

10.1.2. Complex Formation Capacity of Organic Matter Dissolved in Water

From ecotoxicological positions, the form of heavy metals occurrence in water is a crucially significant property. It conditions the degree of the pollutant toxicity and bioavailability for water organisms.

The toxicity of natural water in the most general view is determined, first of all, by the content and concentration of chemical components in the water. Owing to this fact, the direct dependence between these parameters should be expected. Special research on this issue discovered, however, that there is no direct, unambiguous relation between general chemical substances concentration in the water and its toxicity; the toxicity of the natural water by far is more related to the physical-chemical and chemical state of the chemical elements in the water [89—108]. Nowadays, the present provision is a paradigm in the ecology.

All forms of chemical elements (including metals) in natural waters may be provisionally divided into two groups [109]: 1) physical-chemical forms contained in certain fractions of surface waters, which, in its turn, are subdivided either in terms of size or in terms of lability with regard to used technique of analysis and 2) chemical forms, which are characterized either by certain chemical compounds or by valent state. According to this classification, chemical forms of elements are part of the physical-chemical forms which in their aggregate provide the total content of elements in surface waters.

Nowadays, the technique of analysis makes it possible to distinguish the following physical-chemical forms of metals in natural waters [109]: general, suspended, dissolved, free-ionic and distribution of complexes in terms of molecular (ionic) masses within the range of 500-150,000. Due to the complexity and laboriousness of studying such forms, however, the investigation of physical-chemical metals forms in accordance with the simplified classification scheme became widely used in most laboratories. In accordance with the scheme, metals in suspended and filtered forms[15] are determined (by means of filtration through 0.45 mcm filter). They are also determined in the latter form, in ion-exchangeable (free metal ions, simple inorganic and organic complexes) form and in bound (mostly colloidal fraction with molecular mass less than 1,000) form [109, 110—113]. Metals contained in the suspended fraction form bindings with suspended particles on the basis of adsorption on the surface of minerals (like quartz, pyrite, dolomite, titanite, barites), aluminosilicates, metal oxides or by means of complex formation with humic compounds on suspended particles.

For more detailed study of the ecological role of metals in surface waters, and in particular, for the evaluation of its bioavailability, in some cases, another scheme of metals forms extraction is applied that is dependent on the time of contact with resin "Chelex-100" and the usage of stripping voltammetry to determine the relative metals lability in natural waters (labile, of middle lability, inert) [96, 109].

Labile metals fractions, as a rule, leave a small, often negligible, part of its content on the surface of fresh waters. The main part of the metals is bound with organic, colloidal and suspended fractions.

Depending on the applied analytical methods and ways of separation, several schemes are suggested for chemical metals forms classification in surface waters. The following metals forms division is considered the most acceptable: simple hydrated ions ($M^{2+}aq$), simple compounds and complexes with inorganic (MCl_4^{2-}, MCO_3) and organic (M-amino acids and nitrates) ligands, stable compounds and complexes with inorganic (MS, $MSiO_3$) and organic (M-huminates) ligands and metals adsorbed (absorbed) by inorganic (M^{2+}—$Fe_2O_3 \cdot nH_2O$, M^{2+}—$MnO \cdot nH_2O$, M^{2+}—aluminosilicates) and organic (M^{2+}—humic acids, M^{2+}—organical detrita) colloidal particles [114].[16]

All surface fresh waters contain dissolved organic matters up to 80% of which consist of highly oxidizing polymers like humic substances coming into freshwater ecosystems from soils. The remaining part of dissolved organic matter represents a mixture of polypeptides, polysaccharides, fatty acids and amino acids, hydrocarbons, vitamins and other substances actively subjected to (unlike humic substances) biotransformation and disintegration in fresh waters.

All humic substances contained in surface fresh waters are usually divided (often provisionally) into three groups [111, 114]: humic acids (molecular mass 5,000-100,000), fulvic acids (molecular mass 300-2,000) and undissolved humuses. These are humic substances that bind metals ions, participate in cation exchange, contain free radicals, adsorb metals hydroxides, polypeptides, hydrocarbons, etc. [115]. Special research has revealed (see reviews [114, 109]) that metals ions binding with humic and fulvic acids are carried out through (and due to) functional groups continuously forming ions, and especially, as a result

---

[15] Besides these two phase forms, the film form (surface microlayer) is singled out in surface waters containing maximal chemical elements (both in dissolved and suspended states) and organic matter concentrations.

[16] Due to specific chemical properties of mercury, the above mentioned forms of (M2+) MeHg+, Hg22+, Hg° and Me2Hg shall be added.

of chelation. Exactly owing to this fact, surface fresh waters in general possess coordinating unsaturation and actively bind ion forms of heavy metals coming into freshwater ecosystems. For an explanation of this phenomenon and the difficulties (often hard to overcome) in identification and quantitative determination of all dissolved organic ligands, the *concept of the complex forming capacity of surface waters characterizing the degree of its ability to chelate and disguise metals ions* was suggested.

Taking into account that in view of low heavy metals concentrations in surface fresh waters, it is not possible by means of modern technique of assays to determine its chemical forms and therefore to retrace unambiguously its specific toxicity for living organisms, we consider only the fundamental relation of toxicity with the physical-chemical state of metals in dissolved form, the investigation of which is important for the development of the general concept of freshwater ecosystems buffer capacity with regard to metals.

When examining the issue in accordance to the aforementioned points, in the most general form, the evaluation will be as follows: ion forms possess maximal toxicity, and in the case of such ions, the transition to stable, high molecular complexes of metals, the toxicity sharply drops at times to full nontoxicity[17] [91, 93, 94, 96, 97, 99, 100, 105—107, 111, 116]. Therefore, the degree of metals toxicity in natural waters shall be evaluated not on the basis of its general concentration but from the calculation of the amount of metals in ionic forms contained in the water [100]. Besides, according to laboratory studies [97, 111,118—120], metals generally penetrate into the cells as free ions and as some other labile forms, while metals chelated with high molecular dissolved organic matter not only lose their toxicity but frequently become indigestible for fresh water organisms. The latter statement, however, is disputed by some specialists [105], which may probably be considered a misunderstanding— it is more correct to speak about the biological indigestibility of metals chelated in high molecular dissolved organic matter only since low molecular complexes almost freely penetrate cells [119].

Returning to the role of the dissolved organic matter in natural waters, in the change of toxicity of different physical-chemical metals forms, it shall be noted that *dissolved in natural waters, organic matter practically does not suffer seasonal oscillations and the action of this factor in a single mechanism determining the buffer capacity of freshwater ecosystems may be accepted as a constant.*

In general, in accordance with the Irwing-Williams series, humic complexes of different metals are located as follows [121]:

$$Mg < Ca < Cd \approx Mn < Co < Zn \approx Ni < Cu < Hg$$

Without further dwelling on the analysis of this issue, we would like only to illustrate the significance of the complex forming capacity of dissolved organic matter in the evaluation of a freshwater system's buffer capacity with regard to metals through the following example.

Complex forming capacity of 1 l. of water from the Usman River in Voronezhsky biospheric reserve amounted to 70 mcg, i.e. dissolved organic matter in this river makes it possible to chelate up to 70 mcg of copper in one liter of water [6]. Taking into account that

---

[17] Some particular exceptions from this rule are examined in the papers of U.Borgman [117, 93]. Besides, in case of general organic carbon if a concentration in the water exceeds 5 mg/l, it is necessary to take into account the possible toxicity of the dissolved organic matter itself.

the concentration of dissolved forms of copper in the Usman River amounts to between 0.2-0.9 mcg/l, the comparison of the total copper concentration in the Usman River with the obtained value of the complex forming capacity of organic matter shows that the *buffer capacity conditioned by this factor is more than sufficient for copper to be in a low-toxic state.* According to our information, for the Usman River, fulvic acids play a leading role in metals binding, amounting to 57% of the dissolved organic matter in the river (6.8 mg/l of carbon). Being strong natural complex forming agents, fulvic acids determine a high degree of metals chelating in natural waters. In the case of the Usman River, this amounts to 94% of the copper being dissolved. Let us specify the complex forming capacity of waters of the other studied rivers for purposes of comparison: the Shuya River—104 mkg of copper per 1 l of water and for the Lopan River (Kharkov)—12 mcg per 1 l of water.

Table 10.2 specifies the conjugated the values of complex forming capacity of dissolved organic matter in natural waters and concentrations of the complex forming substances in different water bodies of the former USSR.

The results were obtained through the exclusion-chromatogrphic technique on porous glass [11], based on the division of high molecular and low molecular dissolved organic matter on sephadezes and molselectes [122—125]. Determinations of complex forming substances concentrations (organic carbon, humic and fulvic acids, humic substances) were performed through approved techniques [126, 127].

10.1.3. Kinetic Characteristics of Processes of Metals Accumulation by Biotic and Abiotic Components of Freshwater Ecosystems

On the basis of the general provisions of a system approach during ecosystems tolerance to heavy metals influence, it is reasonable to examine major units determining the buffer capacities of freshwater ecosystems.

As was specified above (see also [129]), there is an interrelation of several significant factors determining metals distribution and migration in abiogenous and biogenous water ecosystems forms the basis of freshwater ecosystems buffer capacities with regard to heavy metals: the accumulating capacity of hydrobionts, complex forming capacity of organic matter dissolved in natural waters and depositing capacity of bottom sediments. Thus, the behavior of heavy metals in freshwater ecosystems may be provisionally represented in the form of three simultaneously running and competing, with regard to metals, processes: metal absorption by biota, forming of metals complexes with dissolved organic matter of natural waters and bottom sediments:

$$M + \mathrm{B} \rightarrow M[\mathrm{B}]$$
$$M + \mathrm{DOM} \rightarrow M[\mathrm{DOM}]$$
$$M + \mathrm{BS} \rightarrow M[\mathrm{BS}]$$

where M is the ionic form of the metal; B represents biota; DOM stands for dissolved organic matter of natural waters; BS stands for bottom sediments and M[B], M[DOM], M[BS] are metal, bound with biota, dissolved organic matter and bottom sediments, accordingly.

It is completely evident that heavy metals distribution in freshwater ecosystems depends on the rate of metals binding with different components of ecosystems.

Let us have a closer look on the depositing role of hydrobionts and factors determining the levels of metals accumulation.

Among the metals present in surface waters in ionic and adsorbed on suspensions form, the biologically available fractions are singled out and relatively easily penetrate into the membranes of hydrobionts cells. In their turn, all kinds of hydrobionts are provisionally divided into three groups. The first group (type A) includes the major part of modular organisms (bacteria and all autotrophs). These organisms enrich themselves mostly with metals in ionic or alkylate form in the water. The second group (type B) is hydrobionts that are able to swallow matter suspended in the water and bottom sediments (unitary organisms and some modular heterotrophs). Levels of metals (microelements) in these organisms depend on two factors: concentration of metals and amount thereof in solid phase in suspensions and in the upper layer of bottom sediments. P. Campbell and A. Tessier [130] suppose that the distribution of any metal (*M*) in a competing water system, organic mineral substratum may be represented with this simplified model:

$$M^{2+} + \equiv S(1) \xleftrightarrow{K\,a\,(1)} \equiv S(1) - M$$
$$M^{2+} + \equiv S(2) \xleftrightarrow{K\,a\,(2)} \equiv S(2) - M$$

or in general view:

$$M^{2+} + \equiv S(n) \xleftrightarrow{K\,a\,(n)} \equiv S(n) - M,$$

where ≡ S(1), ≡ S(2), …, ≡ S(*n*) is the number of free spots on the substratum for metal adsorption (clays, humic and fulvic acids, hydroxides of Fe, Mn, Al, Si, etc., that may serve as substrata binding the metals) and ≡ S(*n*) — *M* is the amount of metal *M* adsorbed by the substratum.

K*a*(1), K*a*(2),..., K*a*(*n*) is the common (average) equality constant for the appropriate reactions:

$$Ka(n) = \frac{\equiv S(n)-M}{[M^{2+}]\equiv S(n)} \qquad (17)$$

where $[M^{2+}]$ is the concentration of free ions of metal in the water.

In accordance with the presented model concentration of metal *M*, bound with any substratum, depends on its concentration in the water ($[M^{2+}]$), intensity of reaction (*Ka*(*n*)) and capacity of substrata for metals adsorption (≡ *S*(*n*)).

In real ecosystems, the direct definition is possible only for the first parameter—content of any metal in the water. Two other parameters of the model are difficult to determine in a natural water body.

It shall be emphasized that the concentration of adsorbed (or exchanged) metal itself is insufficient for the determination of free ions $[M^{2+}]$ in the water or for the evaluation of biologically available forms content in it. High concentrations of adsorbed ions are evidence of an abundance of adsorbing surfaces only.

It shall be noted that the model provides for two important assumptions:

- Metals distribution in the system water—substratum is formed at the conditions of equilibrium
- All substrates act independently in the system, as "ideal mixtures of pure substrata"

Both specified assumptions are doubtful since real water ecosystems are in a state of dynamic (unstable) equilibrium and the above mentioned adsorbing substrata, most probably, represent mixtures of clayey (and other suspended) particles covered with bacterial organic matter, humic and fulvic acids and oxides of metals (Fe, Al, Si, etc.).

Table 10.2. Complex formation capacity (CC) of dissolved organic matter in natural waters (with regard to Cu(II)) and concentration of complex forming substances in different water bodies of the former USSR [128]

Water body	Concentration				
	CC, mcmole/l	Organic carbon, mg/l	Humic substances, mcmole/l	Humic acids, mcmole/l	Fulvic acids, mcmole/l
Lakes					
The Plescheevo Lake	0.39	10.3	1.58	0.64	0.94
The Krivoe Lake (Rostov region)	1.27	23.5	0.78	0.18	0.6
The Ladozhskoe Lake	1.78	12.4	1.18	0.34	0.84
Water storage basins					
The Novodnestrovskoe	0.45	6.5	0.82	0.3	0.52
The Kievskoe	0.45	13.5	1.46	0.38	1.08
The Rybinskoe	1.76	13.5	1.84	0.86	0.98
The Tsymlyanskoe	2.14	6.5	0.54	0.24	0.3
Rivers					
The Don River	0.26	6.8	1.24	0.04	1.2
The Kura River	0.79	4.8	0.49	0.29	0.2
The Orsha River	10.49	—	8.95	2.1	6.85
The Northern Dvina River	4.21	—	5.3	0.8	4.5
The Izhma River	3.32	—	4.62	0.82	3.8
The Laya River	3.20	—	4.45	0.45	4.0
The Sheksna River	2.67	14.1	7.48	1.8	5.68
The Burnaya River	2.32	9.7	2.82	1.16	1.66
The Pechora River	1.23	9.7	0.8	0.3	0.5
The Ural River	0.55	4.5	0.47	0.19	0.28
The Lena River	0.48	8.7	1.58	0.42	1.16

With regard to hydrobionts of type A and B, interactions of type metal—metal in the form of competition between ionic forms for spots of metal transition on cell membrane—may influence the process of metal absorption from solutions as well. In diagram form, this mechanism may be represented as follows:

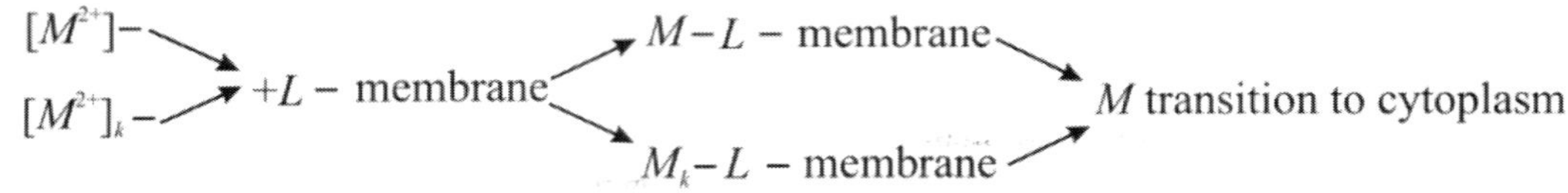

where $[M^{2+}]_k$ is the competing metal L is the spot of metal transition on the cell membrane.

With regard to hydrobionts of type B, in view of metals transportation on food chains jointly with swallowed organic mineral particles, competition may be observed between the spot of transition through the cell membrane from the water solution on one hand and the surface of organic mineral particles (food substrate) on the other hand:

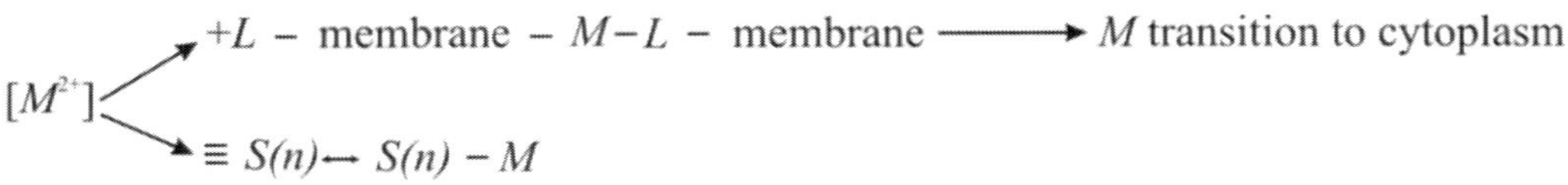

where = $S(n)$ are spots of metal adsorption on organic mineral particles.

In our opinion, the diagrams specified above quite authentically represent mechanisms of metals interaction with organic and inorganic ligands in surface waters. Difficulties, however, do not consist in understanding specified mechanisms, but in the determination of real scales for the competing processes for accounting thereof in theoretical calculations.

Buffer capacity of freshwater ecosystems with regard to heavy metals will be determined not only by potential possibilities of metals ionic forms binding through dissolved organic matter, depositing capacity of benthal deposits and accumulating capacity of biota, but also by *comparative kinetics* of metals absorption by various components of the ecosystems.

Let us introduce the following symbols: C_B is the biomass of hydrobionts, C_{DOM} is the DOM concentration and C_{BS} is the capacity of bottom sediment per unit of water volume. Then, the concentration of linked forms of heavy metals is designated through $C_{M[B]}$, $C_{M[DOM]}$ and $C_{M[BS]}$, and the metal concentration in the solution at the specific time moment is represented through C_M.

In the most general view, the rate of metals content in abiogenous components of freshwater ecosystems decrease owing to metals absorption by hydrobionts and may be provisionally considered proportional to metals concentration in the water and biomass of living organisms populating this biotope [131]. Similar dependence may be considered fair for DOM and bottom sediments [132]. As a result, using introduced designations, we can write rates of three competing simultaneous processes in the form of differential equations systems of the first order:

$$\text{д}C_{M[B]}/\text{д}t = k_1C_BC_M$$
$$\text{д}C_{M[DOM]}/\text{д}t = k_2C_{DOM}C_M$$
$$\text{д}C_{M[BCS]}/\text{д}t = k_3C_{BS}C_M \qquad (18)$$

where k_1, k_2, k_3 are constants of the competing process rates.

Within the framework of the present model for time periods comparable with the time of heavy metals volley coming into the studied water body, concentrations of the examined ecosystems component are accepted as constant. In view of this fact, values k_1C_B, k_2C_{DOM} and k_3C_{BS} are constant and are designated as K_B, K_{DOM} and K_{BS}. For determination of current heavy metals concentration in the solution, C_M, it is necessary to introduce the value $C_{M\Sigma}$, representing concentration, which corresponds to the total content of linked forms of metals. This concentration is equal to the change of bound forms of metals in separate components of the ecosystem:

$$C_{M\Sigma} = \left(C_{M[B]} - C^{(o)}_{M[B]}\right) + \left(C_{M[DOM]} - C^{(o)}_{M[DOM]}\right) + \left(C_{M[BS]} - C^{(o)}_{M[BS]}\right) \quad (19)$$

where $C^{(o)}_{M[B]}$, $C^{(o)}_{M[DOM]}$, $C^{(o)}_{M[BS]}$ are the concentration of bound forms of metals (mkg/l) in appropriate components of the ecosystem per unit of water volume at an initial moment in time. Then, the C_M value may be described by the following equation:

$$C_M = C^{(o)}_M - C_{M\Sigma} \quad (20)$$

where C_M is the initial concentration of metal incoming into the ecosystem.

As a result, the system of equations (18) takes the following form:

$$\partial C_{M[B]}/\partial t = K_B(C^{(o)}_M - C_{M\Sigma})$$
$$\partial C_{M[DOM]}/\partial t = K_{DOM}(C^{(o)}_M - C_{M\Sigma})$$
$$\partial C_{M[BS]}/\partial t = K_{BS}(C^{(o)}_M - C_{M\Sigma}). \quad (21)$$

By summing the equations of system (21) and taking into consideration (19), we will obtain common differential equation for the rate of the total amount of bound metal forms change:

$$\partial C_{M\Sigma}/\partial t = (K_B+K_{DOM}+K_{BS})(C^{(o)}_M - C_{M\Sigma}) \quad (22)$$

By introducing the symbol:

$$K_{\Sigma v}=vK_B+K_{DOM}+K_{BS} \quad (23)$$

We obtain the solution of equation (22) as follows:

$$C_{M\Sigma} = C^{(o)}_M\left(1 - e^{-k\Sigma t}\right) \quad (24)$$

By subsequent mutual dividing the equations of system (21) and through integrating the obtained correlations from the beginning of the process up to actual moment, we will obtain:

$$\left(C_{M[B]} - C^{(o)}_{M[B]}\right)/K_B = \left(C_{M[DOM]} - C^{(o)}_{M[DOM]}\right)/K_{DOM} = \left(C_{M[BS]} - C^{(o)}_{M[BS]}\right)/K_{BS} \quad (25)$$

From equalities (25), let us express $C_{M[DOM]}$ and $C_{M[BS]}$ through $C_{M[B]}$ and substitute it to (19). As a result of this operation, we will obtain correlation of metal, bound by biota, concentration and total concentration of bound forms of metal:

$$C_{M[B]} - C^{(o)}_{M[B]} = C_{M\Sigma} K_B / K_\Sigma$$

By analogy we have:

$$C_{M[DOM]} - C^{(o)}_{M[DOM]} = C_{M\Sigma} K_{DOM} / K_\Sigma$$

$$C_{M[BS]} - C^{(o)}_{M[BS]} = C_{M\Sigma} K_{BS} / K_\Sigma \qquad (26)$$

Obtained correlations taking into account expressions (24) and initial metals concentrations in appropriate components of the ecosystem $C^{(o)}_{M[B]}$, $C^{(o)}_{M[DOM]}$ and $C^{(o)}_{M[BS]}$ give the solution of system (18), i.e., time dependence of bound metals forms concentration:

$$C_{M[B]} = C^{(o)}_{M[B]} + C^{(o)}_{M} K_B(1 - e^{-k_\Sigma t})/K_\Sigma$$

$$C_{M[DOM]} = C^{(o)}_{M[DOM]} + C^{(o)}_{M} K_B(1 - e^{-k_\Sigma t})/K_\Sigma$$

$$C_{M[BS]} = C^{(o)}_{M[BS]} + C^{(o)}_{M} K_B(1 - e^{-k_\Sigma t})/K_\Sigma \qquad (27)$$

The system of equations (27) represents the simplest mathematical model of heavy metals behavior in a water ecosystem.

With the purpose of experimental checking of the suggested model of heavy metals distribution among basic components of water ecosystems, we conducted full-scale modeling of metal behavior in a water ecosystem with a one-time metal introduction. Copper, Cu^{2+}, possessing considerable biological activity and clearly expressed ability to react with the formation of complexes with DOM of natural water was selected as the model heavy metal [133].

Experimental works were conducted in the background area of the Usman River in the territory of Voronezhsky biospheric reserve.

The experiments were conducted in special plexiglas mesocosms with a height of 1.5 m and approximate volume of 120 l with 20 l damper (for more details concerning mesocosms see [134]). Mesocosm I was the reference mesocosm. Mesocosms II and III had ionic copper (II) with a concentration in solution of 100 and 300 mkg/l, respectively, added.[18] In mesocosm IV, bottom sediments, cut out by edges of its foundation, were closed by polyethylene film. The initial concentration of copper (II) in mesocosm VI was 100 mkg/l as well.

After the metal additive was applied, its total concentration reduced in average by 10%, which is accepted by us as an experimental error. Integral samples (0.5 l) for assay were taken every 20-24 hours over three days. Water samples were filtered through a membrane filter

[18] The present experiment was conducted in two variants: with additions of Cu2+ 500 and 1,000 mkg/l, as well as 100 and 300 mkg/l. In the proposed work, the results of only the second variant of the experiment are considered since it was not possible to track the direction of physical-chemical and biological processes in the first variant.

(0.45 mkm) and the following was measured in filtrate: content of dissolved organic carbon, humic substances, general concentration of dissolved copper (II), concentration of ionic forms of copper (II) and forms of copper (II) linked with DOM (humic substances, proteins, polypeptides). Such indices as temperature, pH, Eh and oxygen concentration were measured directly in the mesocosms. Dissolved organic carbon was measured in accordance with the technique [135]. Ionic and bound forms of copper (II) were divided chromatographically with the usage of ionic-exchange celluloses [136]. The concentration of copper (II) in all fractions was determined photometrically [137]. Suspended forms of copper (II) were analyzed by method of neutron activation analysis [2].

Analysis of the results of ionic and bound forms of copper (II) concentration change in mesocosms during the period of the experiment (Figure 10.8) makes it possible to judge the possible behavior of the model metal in a real water ecosystem. Thus, the comparison of changes of the ionic copper (II) concentration in mesocosms II and IV shows the depositing role of bottom sediments in the process of the metal excretion from water. In the conditions of the experiment (see Figure 10.8), a considerable part of the model heavy metal (up to 37%) passed to bottom sediments. A higher copper (II) concentration in mesocosm III made it possible to evaluate the potential complex forming capacity of humic substances (2.2 mgS/l), representing a major part of DOM of studied waters which almost corresponds to 3.65 μM Cu(II)/l. The value of copper (II) content in the suspension (almost by 80% consisting of plankton organisms) was evaluated on the basis of the results of neutron activation analysis.

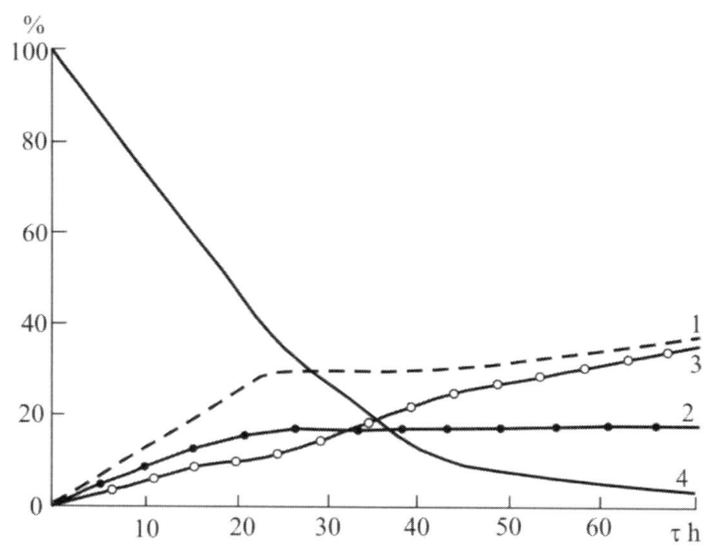

1-3—content of copper (II) in upper layer of bottom sediments (*1*), in DOM (*2*) and biota (*3*), *4*—content of ionic copper (II) in the water

Figure 10.8. Distribution of copper (II) in water ecosystem at conditions of a model experiment in mesocosms.

The distribution of copper (II) between basic studied components of freshwater ecosystem is specified in Table 10.3.

It shall be noted that during the experiment, the increase of pH values from 7.4 to 7.7 and of oxygen concentration on average by 0.4-0.7 mg/l was observed. Besides, the floristic similarity factor (according to Serensen) between the Usman River and mesocosms I-IV amounted accordingly to 61, 36-48, 45-46 and 37-55%. The same factor amounted accordingly to 26-47, 37-39 and 30-38% during comparison of mesocosm I with mesocosms II-IV.

Table 10.3. Distribution of copper (II) (mkg/l) between basic components of a model water ecosystem

Component of ecosystem	Time from the moment of HM additive application				
	Beginning of exposition	16 h	24 h	44 h	70 h
Ionic form of copper in the water	90.0	52.0	33.0	8.0	3.0
Bottom sediments	0	18.5	26.5	27.0	33.0
DOM	0	11.0	15.0	15.0	15.5
Biota	0	7.5	10.0	22.5	31.0

In general, the distribution of copper (II) between basic components of model water ecosystem (see Table 10.3) obtained during combined examination of experimental data for mesocosms II-IV, is presented in Figure 10.8. As it is clear from the figure, the most intensive reduction of ionic copper (II) concentration in the water medium was observed during the first 24 hours. It is conditioned by the fact that during this period, processes of copper (II) depositing by bottom sediments (26.5%) and its binding by DOM (15.0%) and biota (10.0%) run in the most efficient way. During the subsequent period, the intensity of sorption processes by the surface of bottom sediments notably slows down and leads to the increase of copper (II) content by only 6.5% at the end of the experiment. The concentration of model metal in DOM remains practically constant—fluctuations do not exceed 0.5%. This is evidence of the attainment of the maximal complex formation capacity by DOM, which is present in the water at the moment of the experiment. Dependence that is more complicated is typical for copper (II) accumulation by biota; the intensity of metal absorption by aquatic organisms considerably increases 24 hr. after the beginning of the experiment.

It shall be noted that based on the results of the conducted experiments, constants of rates of metal coming into biota and chelating by dissolved organic matter of natural waters turned out to be close in terms of their value, which is evidence of the "commensurability" of these processes in natural freshwater ecosystems. This fact unambiguously indicates that in *natural freshwater ecosystems, in the case of heavy metals coming into dissolved organic matter even at a considerable amount thereof, they cannot fully bind the most toxic ionic forms of metals up to the moment when those begin to be absorbed by biota.* Therefore, specific ecotoxicological consequences of metals coming into freshwater ecosystem will be determined by (among other things) rates of its complexes forming with dissolved organic matters and absorption by biota. Discovered phenomenon shall be taken into account during the solution of the problem of ecological standardization. Although it complicates, in general, the concept of freshwater ecosystems buffer capacities with regard to metals, one of the ways the marginal ecological load on water bodies takes into account the peculiarities of substances' migration and transformation in standardized water bodies has been discovered [2].

Herewith, however, it shall be taken into account that there is a relation between time of emergence of effects of metals influence and the level of biological arrangement of the system; during the transition from cellular to over-organismic level, the time of the effect from metals influence emergence increases (Figure 10.9).

The obtained data (see Table 10.3) was used by us in determining the parameters of the mathematical model. Through the use of mathematical statistics, the following values of parameters were obtained: K_{Σ} = 1.03; K_{B} = 0.30; K_{DOM} = 0.26 and K_{BS} = 0.47 day^{-1}. In general, divergences between experimental and calculated data in most cases do not exceed 20%. For the results reflecting sorptional-accumulating capacity of biota, however, with regard to the model metal at the initial time period (16 hr. after copper (II) was introduced), a discrepancy reaching 60% was observed. Most likely, this is related to the uncontrolled hydrobiological processes running in the mesocosms.

On the basis of the material specified above, quite a simple procedure of water ecosystems buffer capacities compared to one-time heavy metals discharge determination was developed with the mesocosms technique

Defining the expression made it possible to calculate the buffer capacity. We will use relations (25) and (26) in describing the distribution of the model metal that came into the ecosystem between its main components. Using expression (25), let us represent the difference of bound copper (II) forms concentration (DOM and bottom sediments) through changes of its content in biota and substitute it into (19):

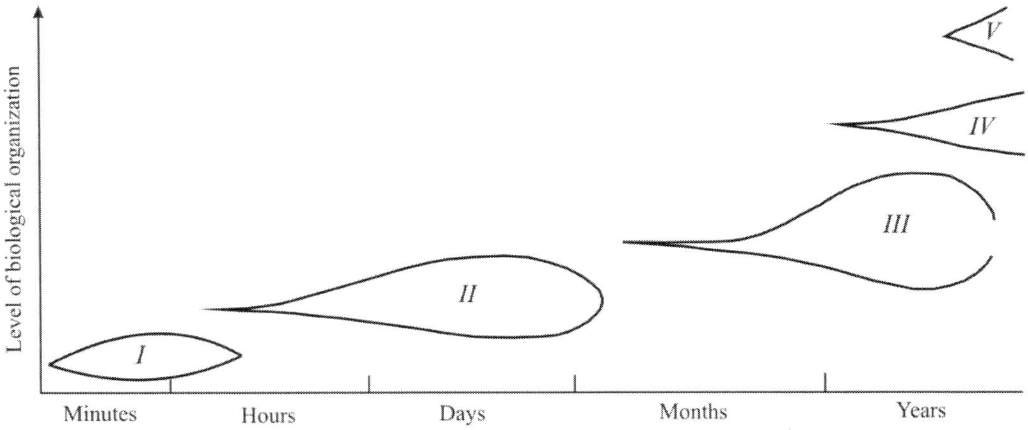

I—cellular level (free-living unicellular eucariotes), *II*—organismic (free-living multicellular organisms), *III*—populational, *IV*—coenotic (sustained change of functional and structural parameters), *V*—ecosystemic (sustained change of functional and structural parameters)

Figure 10.9. Time of emergence of metals' influence effect depending on the level (*I*—*V*) of biological organization of the systems.

$$C_{M\Sigma} = \left(C_{M[B]} - C^{(o)}_{M[B]}\right) + \left(C_{M[B]} - C^{(o)}_{M[B]}\right) K_{DOM} / K_B + \left(C_{M[B]} - C^{(o)}_{M[B]}\right) K_{BS} / K_B \quad (28)$$

Simplifying the obtained relation and taking into account (23), we get:

$$C_{M\Sigma} = \frac{K_{\Sigma}\left(C_{M[B]} - C^{(o)}_{M[B]}\right)}{K_B} \tag{29}$$

The buffer capacity of the water ecosystem (C_{Buf}), as an index, describes the concentration of heavy metal inflow, of which to the water ecosystem will not fundamentally disturb the natural character of its functioning, is defined by the maximal concentration of metal that may be accumulated by biota from a water medium ($C^{crit}_{M[B]}$). By transforming (29), we will obtain:

$$C_{Buf} = \frac{K_{\Sigma}\left(C^{crit}_{M[B]} - C^{(o)}_{M[B]}\right)}{K_B} \tag{30}$$

During calculations, it is more convenient to use a concentration of heavy metal per unit of dry weight (M_B). Therefore, taking into the account errors on the model and introducing the assurance factor, K_z,[19] the expression *for the buffer capacity of the freshwater ecosystem with regard to contamination by heavy metals* may be written in the form:

$$C_{Buf} = \frac{K_z K_{\Sigma} C_B(M_B^{crit} - M_B^{(o)})}{K_B} \tag{31}$$

where C_B is the biomass with a recommended value of $K_z = 0.75$.

10.2. Methods for Water Ecosystems Buffer Capacities Determination

Along with the methods of freshwater ecosystems buffer capacity calculation specified in the previous section, other ways for its determination with the usage of notions of permissible and critical loads are applied.

Existing definitions of buffer capacity (ecological capacity, ecological reserve),[20] both of water and of other ecosystems, include the following basic stages:

- Measurement of values of indices, characterizing physical state of ecosystem, and comparing thereof with critical values
- Determination of permissible limits of change of state of each ecosystem component by particular indices or its state in general by a range of indices (by integral index) and critical limits

---

[19] Introduction of the assurance factor reduces the calculated value of the buffer capacity, thus reducing the maximal permissible ecological load on the ecosystem (increases level of its assured safety).

[20] Izrael Y. A. Ecology and monitoring of natural environment conditions. — M.: Gidrometeoizdat, 1984.

Permissible limits correspond to “normal” physical, chemical and biological states of the ecosystem. Herewith, the permissible degree of deviation from the normal state is deviation, which with the course of time, may be eliminated by the system itself. Critical limits have two sets of values—maximal and minimal, and the attainment of such values of the ecosystem state leads to its destruction or suppression.

The buffer capacity of the system is determined as the difference between maximum permissible (or critical) and actual values of the index, averaged in the examined time interval.

Drawbacks of this way are low accuracy and reliability conditioned by partial accounting of data that describes the ecological load: the absence of a clear quantitative relation between buffer capacity and ecological load.

Specified drawbacks have been, to a considerable extent, overcome in development through the original method of buffer capacity calculation, implementing modern approaches to solve such problems.

The basis of the supposed method is a dosed influence on the ecosystem within the limits corresponding to its stationary regime, measurement of the ecosystem response to the impact and the specific formalization of relation between them, which is used in the practice of standardizing ecological categories of “impact,” “reaction,” “load” and “buffer (ecological) capacity.”

For these purposes, the expression linking the reaction of the ecosystem with the appropriate impact is examined as $E=E(R)$, where $0 \leq R \leq 1$ is the relative value of impact equal its relation to the effective factor of its threshold value and E is the appropriate reaction. The included E-function is positive and turns into zero at R = 0 and R = 1.

As far as the load growth during an increase both of impact and of reaction is evident, the approach proved to be reliable in most various fields of science is used. Elementary load, ΔF, is defined as a product of the reaction value and increment of the impact:

$$\Delta F = E(R)\Delta R \tag{32}$$

And the load, as a definite integral from the reaction on impact:

$$F(R) = \int_0^R E(R)dR \tag{33}$$

The buffer capacity is determined as the load that the ecosystem shall bear before attainment of a sustained stationary regime. In expression (33) impact R^* corresponds to this load.

Then, the buffer (ecological) capacity, S, is described by the expression:

$$S = \int_0^{R^*} E(R)dR \tag{34}$$

The sequence of performance of four operations suggested by the specified approach is displayed with the example of the water basin mesocosm model:

1) Change of background concentration of influencing contaminant and parameters of water ecosystem reaction (for example, frequency of mollusks breathing)

2) Dosed influence on water body by selected contaminant and simultaneous measurement of the mollusks' motion rate
3) Experimental definition of concentrations diapason R^* of the contaminant, within which the rates are in a sustained stationary regime; criterion of this regime are small variations E at considerable changes R
4) Measurement of the buffer capacity of the examined water basin with the usage of the following formula and parameters values:

$$S = \int_{k=0}^{n} E(R_n)\, \Delta R_n \tag{35}$$

R	R_0	R_1	R_2	...	$R^* = R_n$
E (R)	E_0	E_1	E_2	...	$E = E_n$

In this specified way, the ecosystem load is determined exclusively as its internal state (property) and the notion "ecological capacity" for separate living organism is equivalent to the notion of "nonspecific resistance."

In the described way, the buffer capacity of the system is understood as its ecological reserve, the consumption of which does not influence the ecological well-being of the system. The definition of full (utmost) buffer capacity (full ecological reserve), i.e., maximal (critical) load that leads to the destruction (disorganization, suppression) of ecosystem, is of special interest. In this case, expression (34) takes on the form:

$$S_{max} = \int_{0}^{1} |E(R)|\, dR \tag{36}$$

with substituted expression (35) values of parameters E_n and R_n will be: $R_n = 1$ and $E_n = 0$.

The module of reaction values in expression (36) takes into account those cases when the reaction of the system is not positive to any impact and at some loads, it may change the sign.

The examined method makes it possible to establish a quantitative relation between the ecological load and buffer (ecological) capacity of a water ecosystem. Herewith, the ecological load is determined as the product of measured values of the system reaction, and measured values of the impact increment and buffer capacity is determined quantitatively as a load corresponding to the sustained stationary regime of the ecosystem's functioning.

10.3. Method for Buffer Capacity to Anthropogenic Acidulation Evaluation

Water ecosystems acidulation is one of the most progressive forms of anthropogenic impact on the environment [137]. Reduction of pH value in the water body not only leads to disturbance of normal functioning of hydrobionts, but may also favor the aggravation of toxic impact caused by other contaminants or cause the transition thereof from bottom sediments and suspensions to water medium [138—141].

In view of this fact, research aimed at the development of methodological and systematic aspects of freshwater ecosystem buffer capacities (ecological reserve) evaluation becomes topical [14]. The research was conducted within the framework of the concept of anthropogenic impact on the freshwater ecosystem standardized development [7]. The *index of the freshwater ecosystem buffer capacity to acidulation may be determined as the maximal amount of strong acid (per 1 l of water) that does not change initial pH value in cases of inflow into a water basin or water course.*

It may be supposed that a freshwater ecosystem's buffer properties compared to anthropogenic acidulation depend on a range of interrelated factors. The most important factors are natural water or the matrix of natural water,[21] aquatic biota, bottom sediments, hydrodynamical and meteorological factors. Natural water plays a significant role in hydrogen ions concentration reduction in case of anthropogenic acidulation. Many substances included in the matrix of natural water are able to bind hydrogen ions by means of different chemical reactions. Aquatic biota, for example phytoplankton or superior aquatic vegetation, during the process of photosynthesis or during other biochemical processes, are able to increase the value of the water medium pH, thus impeding anthropogenic acidulation [7]. Bottom sediments, as a result of chemical and geochemical processes, running both at the boundary of phases division (water—bottom sediments) and in upper layers of bottom sediments may favor hydrogen ions binding [142].

It shall be noted, that freshwater ecosystem buffer capacity to anthropogenic acidulation will not be a constant value and may vary within a wide range owing to individual peculiarities that are typical for each specific water body or its area. It will also depend on regional and seasonal variability of the factors forming this index.

The purpose of our work consisted in the development of a freshwater ecosystem buffer capacity to anthropogenic acidulation evaluation technique, which was based on the ability of natural water to bind free hydrogen ions in the solution. This property, as it was mentioned above, is conditioned by the qualitative and quantitative composition of water matrix. For example, weak anions of inorganic (carbonates, phosphates, sulfites, sulfides, silicates, etc.) and organic (carboxylic and oxycarboxylic) acids as well as humic acids, fulvic acids and polyphenoles in ionized form or in the form of salts of alkaline or alkaline-earth metals, as a result of their reaction with hydrogen ions, will convert it to a bound state. It shall be noted that the trial to evaluate the buffer capacity of water through general alkalinity has already been made [143]. It may be stated, however, that the calculation of buffer capacity only in terms of alkalinity will be insufficient since, besides the specified inorganic and organic matters that, as a rule, are present in the natural water, organic exometabolites of algae, first of all containing nitrogen-bearing organic matter (amino acids, polypeptides, etc.), are able to interact with hydrogen ions with formation of salts.

The main point of technique, developed by the author jointly with I.A. Lapin, consists in the fact that the filtered (0.45 mkm) 30-40 l of natural water is titrated by 0.1 *M* of HCl. For definition, hydrogen ions equilibrium concentration and pH value is recorded during the titration process. According to experimental data, the equilibrium is attained within 1-2 minutes.

In general, the interaction of the hydrogen ions with different groups of substances may be described by equation:

---

[21] The matrix of natural water is the set of dissolved mineral and organic matter flowing in with surface runoff or atmospheric precipitation into natural waters from rocks or bottom sediments as a result of the life activity of biota or the disintegration of plant remains and animal bodies.

$$H^+ + An^- \leftrightarrow HAn, K = \frac{[HAn]}{[H^+]\cdot[An^-]} \quad (37)$$

Then, taking into account the fact that in natural waters, the total concentration of substances binding hydrogen ions $[An^-]$ considerably exceeds $[H^+]$, it may be supposed that the interaction runs with the correlation of reacting components equals 1:1. As a consequence of this fact, the value, *K*, may be determined from expression (37).

By representing the bound hydrogen ions concentration [HAn] as the difference between the total amount of added hydrogen ions $[H_0^+]$ and equilibrium concentration of hydrogen ions in an unbound state, i.e., in the form $[H_0^+]$ - $[H^+]$, the concentration of ions $[An^-]$ not reacting (at any titration point) may be written as the difference between the value of buffer capacity (C_b) and concentration [HAn]:

(C_b) - [HAn] or (C_b) - $[H_0^+]$ + $[H^+]$

By inserting the obtained expression into (37), we will obtain the value of *K*:

$$K = \frac{[H_0^+]-[H^+]}{[H^+]\{(C_B)-[H_0^+]+[H^+]\}} \quad (38)$$

By representing expression (38) in a more convenient form [143], we will get:

$$\frac{[H^+]}{[H_0^+]+[H^+]} = \frac{1}{C_b}[H^+] + \frac{1}{C_b K} \quad (39)$$

The diagram of $[H^+]/\{[H_0^+] - [H^+]\}$ dependence on $[H^+]$ represents a straight line, the slope factor of which corresponds to $1/C_b$ with an intercept term corresponding to $1/C_bK$. In this case, the value *K* corresponds to the conventional constant of acid-alkali balance (interaction of hydrogen ion with inorganic or organic matter). The higher the value *K*, the more the hydrogen ion is bound.

Figure 10.10 specifies the experimental dependencies obtained for neutral waters in which hydrogen ion binding is carried out mainly by one type of substance. The linear form of dependence, responsible for the correlation of the reacting components equals 1:1 and conforms to this fact. It shall be noted, however, that this situation is not always realized. Figure 10.11 shows examples of when the dependence is not linear. In the last case, hydrogen ions interact with two or more types of organic or inorganic matter, binding it with different strength, i.e., considerable differences of values *K* are observed [144].

Values C_B and *K*, obtained by means of describing the technique for water bodies located in different climatical and physical-geographical zones, are specified in Table 10.4.

Conducted research proved, that this index, based on the matrix ability of natural water to bind to hydrogen ions coming into freshwater ecosystems during the process of acidulation thereof, is the most important factor determining a freshwater ecosystem's buffer capacity to anthropogenic acidulation. In view of the fact that nowadays there are no methods making it possible to evaluate the influence of other factors on buffer properties of freshwater ecosystems, we consider the usage of simple and express techniques for monitoring both separate water bodies and regions influenced by acid rains to be appropriate.

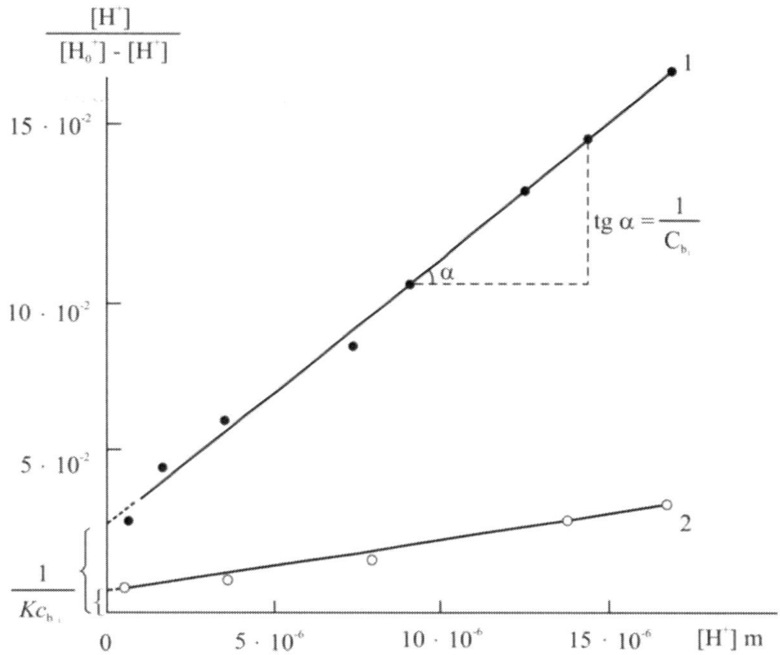

Figure 10.10. Dependence diagrams for calculation of a freshwater ecosystem's buffer capacity for the Chuya River (1) and the Biya River (2).

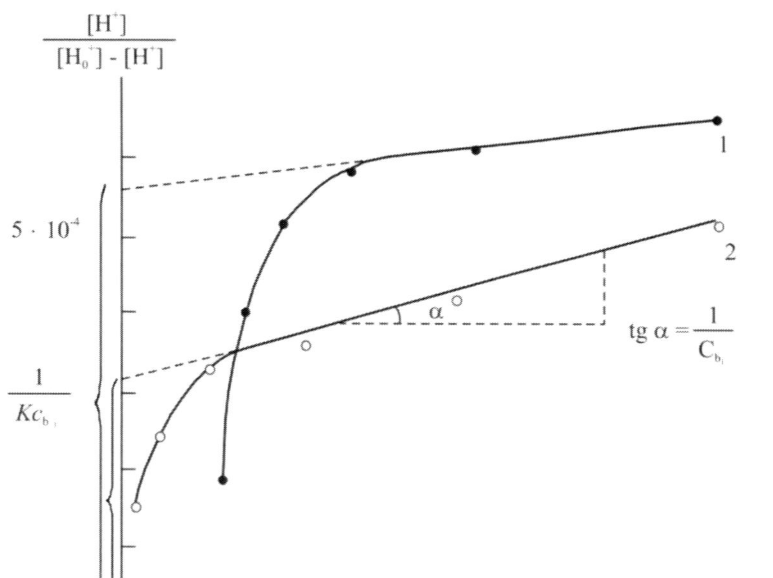

Figure 10.11. Dependence diagrams for natural waters buffer capacity calculation for a commercial fishing pond (*1*) and the Volga River (*2*).

Table 10.4. Values of buffer capacity C_b and constants K of acid alkali balance of several water bodies of the former USSR

Water body	C_b mole/l · 10^4	$K \cdot 10^{-7}$
Commercial fishing pond	62.50	0.03
The Volga River	2.94	1.10
The Didjulis Lake	2.86	2.10
The Don River	2.29	0.90
The Krasnoyarskoe water storage basin	1.20	1.80
The Chulymshan River	1.14	2.20
The Teletskoe Lake	0.77	1.30
The Chuys River	0.63	0.30
The Biya River	0.58	0.28

The basic obstacle in the quantitative definition of the system tolerance consists in the absence of a direct relation between the notion of tolerance and parameters of ecosystem that are providing the tolerance. In the literature, characteristics requiring long-term monitoring of the studied system's functioning (for example, species diversity, trophic and dimensional structure of populations, etc.) and a detailed analysis of the structural relation existing in it are usually suggested for reference.

For a comparative analysis of tolerance to acidulation of different water bodies, J. Turk [144] used titration curves of water and distilled water samples from Island Lake and Marvin Lake (Figure 10.12). These curves made it possible to determine on a qualitative level, the water ecosystem most tolerant to anthropogenic acidulation. From the two lakes, located relatively close to each other, the system of the highland Island Lake is less tolerant.

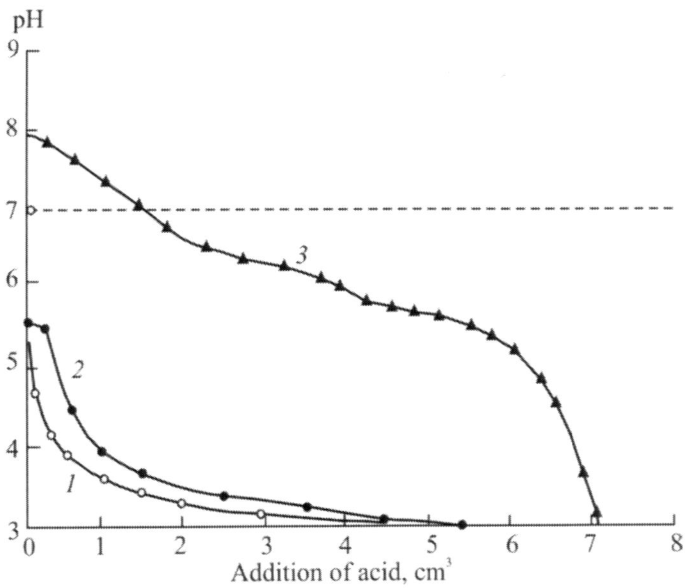

Figure 10.12. Titration curves, displaying the influence of acid additives on distilled water (1) and water samples from Island Lake (2) and Marvin Lake (3).

A similar approach to quantitative evaluation of water systems tolerance (titration method) was suggested by V.V. Batoyan along with his co-authors [145]. Depending on the nature of outside impact and obtained data concerning the response of water ecosystems the authors mark out several kinds of tolerance [145]:

1) Momentary differential tolerance, describing the response of the system to small, instantaneous outside impact, for example, acid rain over the water body. Momentary tolerance may be determined as a relation of the impact value (in this case, the relation of the total amount of acid, coming into the river as a result of short-time acid rain) to the change of the appropriate parameter of the system (in this case, pH of water of the studied water body)
2) Momentary integral tolerance, reflecting the response of system to instantaneous, intensive outside impact (for example, emergency industrial emission, intensive spring snowmelt, etc.). This type of tolerance is determined by the relation of value of impact on the system that is sufficient to bring the system out of the range of typical parameters and to the variation of pH of the water.
3) Long-term consumptive (consumer) tolerance, characterizing the response of the system to small-scale impact during a continuous period (for example, precipitation of sulfur oxides by means of transboundary transition in cases of relatively low acid loads). Quantitative consumptive tolerance is determined in the same way as momentary tolerance.
4) Permanent integral tolerance is the minimal intensity of this type of impact, destroying the system in case of permanent action, i.e., bringing it out of the limits of definite physiological diapason, amounting for most freshwater ecosystems where pH = 4...9 (for example, permanent industrial emission, precipitation of sulfur oxides, exceeding critical load on the ecosystem). It is calculated by analogy with momentary integral tolerance.

On the basis of titration curves, it is not possible to define long-term and permanent consumptive tolerance of ecosystems; only momentary impact is determined, i.e., this is the case when exchange between water and bottom sediments has not taken place yet.

Buffer properties of natural waters are mainly conditioned by free hydrogen dioxide and hydrocarbonate ions contained in it. Besides, the content of humic substances, or in some cases, the elevated content of easily hydrolyzing salts, basic carbonates and hydroxides emerging as a result of carbon dioxide absorption during photosynthesis may influence these properties as well. Therefore, in the determination of water systems tolerance to acidulation, two hydrochemical characteristics are the most informative—oxidability and content of hydrocarbonate ions.

It is evident, that Russian water bodies have a buffer capacity that tends to increase southward, which coincides with geographical and hydrochemical zonality that is related with change of macrocomponent composition of surface waters. Against this background, anomalous areas related with elevated content of organic matter and easily hydrolyzing salts may be singled out. Herewith, the greatest influence is provided by hydrocarbonate ions content in the water, which is conditioned by its high content in natural systems and strong buffer properties in comparison with humic acids.

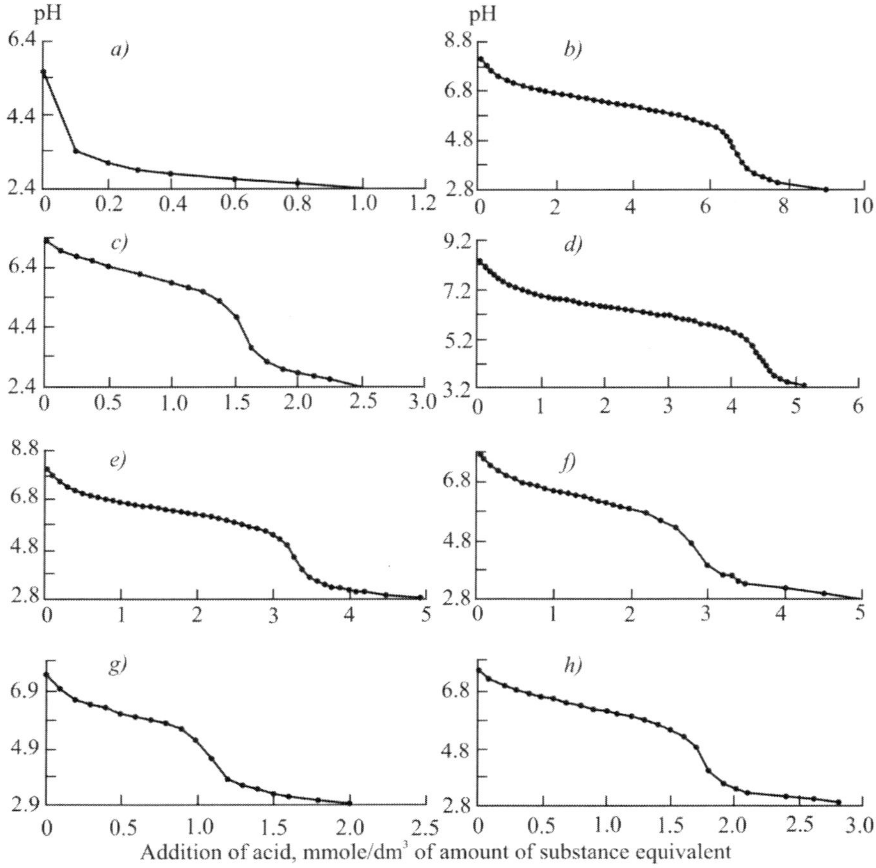

b—the Isakovskoe wsb. in Kommunarsk, *c*—the Bekata River in Ufa, *d*—the Aksay River in Novocherkasskк, *e*—the Don River in Rostov-na-Donu, *f*—the Krivoe Lake in Aksay, *g*—the Angara River in Irkutsk, *h*—the Ushakovka River in Irkutsk

Figure 10.13. Titration curves for distilled water (*a*) and water of different water bodies (*b*—*h*).

Titration by sulfuric acid of filtered and unfiltered water of different natural bodies as well as titration of distilled water was carried out (Figure 10.13). Buffer capacity of unfiltered water as a rule is higher. This difference, however, is insignificant which is explained by buffer properties of suspended matter in freshwater ecosystems containing carbon compounds, organic and other matter. Judging by the results of full-scale experiments on mesocosms in terms of artificial freshwater ecosystems acidulation, the role of suspended matter is considerably less important than the role of bottom sediments.

Distilled water does not possess buffer properties to acidulation (see Figure 10.13). The system of the Isakovskoe water storage basin has the greatest buffer capacity of all the studied bodies (its numerical value is 2.16). The least buffer capacity (0.52) is typical for the Angara River within the city of Irkutsk. Buffer capacity of the other water bodies is specified in Table 10.5.

For comparison of these index values the two aforementioned calculation techniques of buffer capacity for water bodies located in different physical-geographical zones were performed (Figure 10.14, Table 10.5). As it turned out, the series of water bodies ranged in terms of the index "buffer capacity," but were similar when defined by the two independent methods. This is evidence of the fact that an objective index is found based on the ability of the matrix of natural water to bind hydrogen ions coming into the freshwater ecosystems, which to a certain degree characterizes the freshwater ecosystems buffer capacity to anthropogenic acidulation. This makes it possible to range the studied water bodies in terms of tolerance to anthropogenic acidulation.

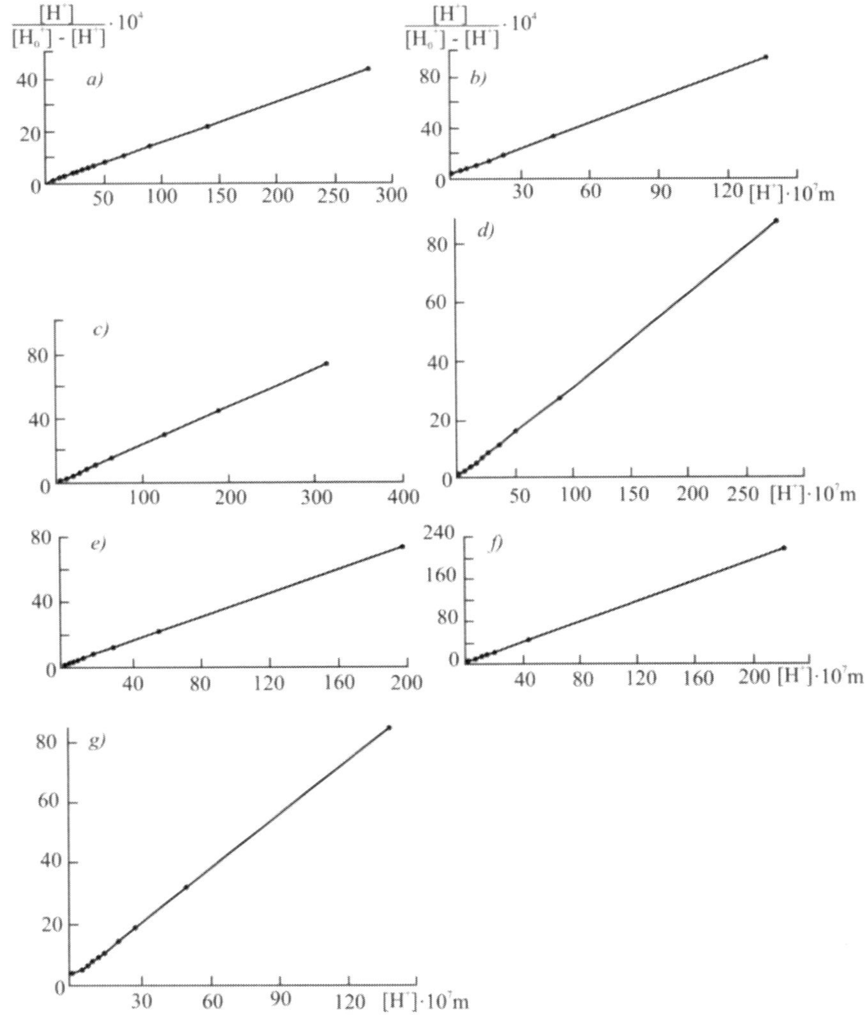

a—the Isakovskoe wsb., *b*—the Belaya River, *c*—the Aksay River, *d*—the Don River, *e*—the Krivoe Lake, *f*—the Angara River, *g*—the Ushakovka River

Figure 10.14. Diagram for calculation of buffer capacity of water from different water bodies by the technique of A.M. Nikanorov and I.A. Lapin [14].

It shall be noted that another advantage of the first method [145] is that it is possible to determine critical acidic load for natural water from the diagram, i.e., that the amount of inflown acid, exceeding of which will lead to a sharp drop of the water pH. Besides, by means of the first method, it is possible to determine the buffer capacity of water and critical acidic load at which pH may be strictly set (for example, in the establishment of boundary conditions of aquatic biota habitation, sensitive to reduction of pH).

Table 10.5. Buffer capacity of natural water from studied water bodies to anthropogenic acidulation

Water body	Buffer capacity (acc. to [98])	$C_B$mole/dm^3·10^3(acc. To [146])
The Isakovskoe wsb.	2.16	6.70
The Aksay River	1.38	4.40
The Don River	1.20	3.30
The Krivoe Lake	1.08	2.90
The Ushakovka River	0.77	1.70
The Belaya River	0.74	1.60
The Angara River	0.52	1.13

The advantage of the second method consists in the fact that by means of this method, it is possible to define the constant of acid-alkali balance, *K* (hydrogen ion interaction with mineral or organic matter). The higher *K* is, the more hydrogen ion are bound [14].

REFERENCES

[1] Nikanorov A.M., Zhulidov A.V., Dubova N.A. and oth. Buffer capacity of freshwater ecosystems for heavy metals and hydrobiological parameters determining it // In col.: *Ecological standardization and modeling of anthropogenic impacts on water ecosystems*. Edit. 1. — SPb., 1988. — P. 58—69.

[2] Nikanorov A.M., Zhulidov A.V., Pokarzhevsky A.D. Biomonitoring of heavy metals in freshwater ecosystems. — L., 1985. — 144 p.

[3] Nikanorov A.M. Principles of standardization of anthropogenic impacts on freshwater ecosystems and ecological similarity // *Hyrdochemical materials*. — 1990. — V. 108. — P. 181 — 188.

[4] Nikanorov A.M., Zhulidov A.V., Dubova N.A. Factors, determining amount of mercury content in hydrobionts of freshwater ecosystems // *Ecology*. — 1988. — No 1. — P. 55—62.

[5] Nikanorov A.M., Zhulidov A.V. Biomonitoring of heavy metals in freshwater ecosystems. — L., 1991. — 311 p.

[6] Izrael Y.A., Nikanorov A.M., Lapin I.A. and oth. Evaluation of buffer capacity of small water basins to heavy metals // *DAN USSR*. — 1985. — V. 283. No 3. — P. 703—706.

[7] Nikanorov A.M., Lapin I.A., Gekov V.F. and oth. Calculation of buffer capacity of freshwater ecosystems for heavy metals // In col.: *Ecological standardization and modeling of anthropogenic impacts on water ecosystems*. Edit. 1. — SPb., 1988. — P. 70—78.

[8] Nikanorov A.M., Zhulidov A.V. Principles and methods of Heavy metals ecological rate-setting in freshwater ecosystems // *River Basin Management*. — Oxford, N.Y., Beijing, Frankfurt, Sao Paulo, Sydney, Tokyo, Toronto: Pergamon Press PLC, 1989. — P. 83—94.

[9] Nikanorov A.M., Zhulidov A.V., Dubova N.A. et al. Buffer capacity of freshwater ecosystems for heavy metals and hydrobiological parameters determing It // *Proc. of*

USA—USSR Symp., Athens, Georgia, October 19—21, 1987. — U. S. Environ. Protection Agency, Athens, GA. Publ. No EPA-600 (9-88)001, 1988. — P. 105—121.

[10] Lapin I.A. Complex formation capacity of natural water in the system of freshwater ecosystems buffer capacity to heavy metals determination // In col.: *Ecological standardization and modeling of anthropogenic impacts on water ecosystems.* Edit. 1. — SPb., 1988. — P. 83—95.

[11] Lapin I.A., Krasuykov V.N., Edigarova I.A. Exclusive-chromatographic method of natural waters complex formation capacity determination // In col.: *Ecological standardization and modeling of anthropogenic impacts on water ecosystems.* Edit. 1. — SPb., 1988. — P. 96—100.

[12] Nikanorov A.M., Lapin I.A., Krasyukov V.N., Sedush V.P. Studying of lead, cadmium and mercury migration and distribution in model freshwater ecosystems // In col.: *Ecological standardization and modeling of anthropogenic impacts on water ecosystems.* Edit. 2. — SPb., 1999. — P. 222—238.

[13] Nikanorov A.M., Tarasov M.G. Water ecosystems tolerance to acidulation: experimental modeling and buffer capacity calculation // In col.: *Ecological standardization and modeling of anthropogenic impacts on water ecosystems.* Edit. 2. — SPb., 1999. — P. 239—269.

[14] Nikanorov A.M., Lapin I.A. Evaluation of buffer capacity of freshwater ecosystem to heavy metals // *DAN USSR.* — 1990. — V. 314, No 6. — P. 1507—1510.

[15] Nikanorov A.M. *Hydrochemistry.* — L., 2002. — 447 p.

[16] Lapin I.A., Krasyukov V.N. Role of humic substances in processes of complex formation and migration of heavy metals in natural waters // *Water resources.* — 1986. — No 1. —P. 134—145.

[17] Braginsky L.P. Theoretical aspect of issue "standards and pathologies" in water ecotoxicology. — L., 1981. — P. 29—40.

[18] Ghilarov A.M. Correlation of organism and reductionism as basic methodological approaches to ecology // *General biology magazine.* — 1988. — V. 49, No 2. — P. 202—217.

[19] Barbault M. Partage des resources et organization des peuplements // *Bull. Ecol.* 1985. — V. 16, No 1. — P. 63—68.

[20] Biological monitoring of environmental pollution // *Proc. of the Fourth 1UBS Int. Symp. on Biomonitoring of the State of the Environment (Bioindicators),* 6—8 November, 1987, Tokyo, Japan / Eds. M. Yasuno, B. A. Whitton. — Tokai University Press, 1988. — 291 p.

[21] Breymeyer A.I. Ecosystem homeostasis — search for a definition // *Met. Zool.* — 1979. — V. 32. — P. 3—11.

[22] Braun J.N. On the relationship between abundance and distribution of species // *Amer. Nature.* — 1984. — V. 123, No 2. — P. 255—279.

[23] Disturbance and Ecosistems. Components of Response / Eds. H.A. Mooney, M. Gordon. — Berlin, Heidelberg, N. Y., Tokyo, 1983. — 292 p.

[24] Ecology of European Rivers / Ed. B. A. Whitton. — Oxford, London, Edinburg, Boston, Palo Alto, Melbourne, 1984. — 644 p.

[25] Edwards R.W., Brooker M.P. *The ecology of Wye.* —Hague, Boston, London, 1982. — 164 p.

[26] Ghilarov A.M., Timonin A.G. Relation between biomass and species diversity on marine and freshwater zooplankton communities // *Oikos.* — 1972. — V. 23, No 2. — P. 190—196.

[27] Kimmerer W.J. Diversity / stability: a criticism // *Ecology.* — 1984. — V. 65, No 6. — P. 1936—1938.

[28] Kolasa J., Biesiadka E. Diversity concept in ecology // *Acta Biother.* — 1984. — V. 33, No 3. — P. 145—162.

[29] Perspectives in Running Water Ecology / Eds. M.A. Lock, D.D. Williams. — N. Y., London: *Plenum Press.,* 1981. — 430 p.

[30] Ravera O. Considerations on some ecological principles // *Trends Ecol. Res.,* 1980, *Proc. NATO ARM and INTECOL Workshop Future and Use Ecol. Decade Environ. Onvain-la-Neuve,* 1983 — N. Y., London, 1984. — P. 145—162.

[31] Remmert H. Okologie. Dritte, neubearbeitete und erweiterte Au-flage // Berlin, Heidelberg. N. Y., Tokyo: *Springer-Verlag,* 1984. — 334 p.

[32] Rygg B. Effect of sediment copper on benthic fauna. — *Mar. Ecol. Progr. — Ser.* 1985. — V. 25, No 1. — P. 83—89.

[33] Seki H. Organic materials in aquatic ecosystems. *CRC Press Jnc. Boca Raton, Florida,* 1982. — 201 p.

[34] Stenseth N.C. The structure of food webs predicted from optimal food selection models: an alternative to Pimm's stability hypothesis // *Oikos.* — 1985. — V. 44, No 2. — P. 361 — 364.

[35] Trojan P. Units of homeostatic organization in terrestrial ecosystems // *Mem. Zool.* — 1979. — V. 32. — P. 13—23.

[36] Trojan P. Ecosystem homeostasis. — Nederlands, W. *Junk Publishers,* 1984. — 140 p.

[37] Polischuk L.V. Dynamical characteristics of zooplankton organisms populations. — M., 1986. — 128 p.

[38] Murdoch A. Distribution of major elements and metal in sediment cores from the western of Lake Ontario. — *J. Yreat Lakes Res.* — 1983. — V. 9, No 2. — P. 125—133.

[39] Reice S. Experimental disturbance and the maintenance of species diversity in a stream community // *Oecologia.* — 1985. — V. 67, No 1. — P. 90—97.

[40] Thorp J.H., Cothron M.I. Regulation of freshwater community structure at multiple intensibinds of dragonfly predation // *Ecology.* — 1984. — V. 65, No 5. — P. 1546—1555.

[41] Zhulidov A.V., Yemets V.M., Shevtsov A.S. Biomonitoring of heavy metals in natural reserves on basis of metal accumulation in bodies of big invertebrates investigation // *DAN USSR.* — 1980. — V. 252, No 4. — P. 118—120.

[42] Matsunagar K., Jgazashi K., Fukase S., Tsubota H. Behaviour of organically — bound irop in seawater of estuaries // *Estuarine, Coast and Shelf.* — 1984. — V. 18, No 6. — P. 615—622.

[43] Bronmark C. Freshwater snail diversity: effects of pond area, habitat heterogeneity and isolation // *Ecologia.* — 1985. — V. 67, No 1. — P. 127—131.

[44] Bukovsky I.V., Azovsky A.I. About relation and stability of species in community if marine ammophillous infusoria // *Zoological magazine.* — 1985. V. 64, No 6. — P. 813—819.

[45] Zhirmunsky A.V., Kuzmin V.I., Yablokov A.V. Critical levels of population systems development // *General biology magazine.* — 1981. — V. 42, No 1. —P. 19—37.
[46] Futuyma D. Community structure and stability in constant environments // *Amer. Natur.* — 1973. — V. 107, No 955. — P. 443—446.
[47] Kondratyev K.Ya., Gitelson A.A., Kamov I.Y. Method of interpretation of surface water remote sounding data in visible range of spectrum // *DAN USSR.* — 1988. — V. 303, No 2. — P. 333—337.
[48] Armand A.D. Stability (homeostasis) of geographic systems to different types of external effects // *Tolerance of ecosystems.* — M., 1983. — P. 14—32.
[49] Kindlmann P., Leps J. What is stability? //A mathematician's and ecologist's point of view // *Math. Res.* — 1985. — No 28. — P. 201—204.
[50] Nunney L. Short time delays in population models: a role in enhancing stability // *Ecology.* — 1985. — V. 66, No 6. — P. 1849—1858.
[51] Anderson J.M. Ecology and science about environment: biosphere, ecosystem, man. — L., 1985. — 165 p.
[52] Valter G. Vegetation on the globe // *Ecological-physiological characteristic.* V. 1. — M., 1968. — 552 p.
[53] Aleksandrova V.D. Studying of vegetative cover change // *Field geobotany.* — 1964. — V. 3. — P. 32—64.
[54] Sukachev V.N. Idea of development and phytocenology // *Soviet botany.* — 1942. — No 1—3. — P. 3—15.
[55] Sukachev B.H. About some basic phytocenology issues // *Issues of botany.* — 1950. — V. 1. — P. 12—154.
[56] Viktorov S.V., Vostokova E.A., Vyshivkin D.D. Modern issues of indicative geobotany // *Issues of geobotany.* — 1970. — Edit. 82. — P. 6—22.
[57] Jiller P. Structure of communities and ecological niche. — M., 1988. — 272 p.
[58] Abakumov V.A. Procaryotes and obligate-agamic protoxoas as indicators of the environment state and peculiarities populations thereof // *Problems of ecological monitoring and modeling of ecosystems.* V. 3. — L., 1980. — P. 24—50.
[59] Abakumov V.A. Monitoring of water qualibinds by hydrobiological indexes // *Development and implementation of biological monitoring methods at complex background stations.* V. 2. — Riga, 1983. — P. 11—32.
[60] Abakumov V.A. Laws of water biocoenoses change under the influence of anthropogenic factors // *Complex global monitoring of World ocean.* V. 2. — L., 1985, p. 273—283.
[61] Abakumov V.A. Chronobiological aspects of monitoring and significance of ideas of physical theory of space // *Labours of sov-fran symposium "Scientific bases of freshwater ecosystems biomonitoring"*— L., 1988. — P. 6—27.
[62] Abakumov V.A. About surveys and comparative evaluations of ecological systems state // *Problems of ecological monitoring and modeling of ecosystems.* V. 1. — L., 1978, p. 64—70.
[63] Abakumov V.A. Basic directions of water biocoenoses change under conditions of the environment contamination // *Problems of ecological monitoring and modeling of ecosystems.* V. 2. — L., 1979. — P. 37—48.
[64] Davies R.B. Hutton B. The effects of errors in the independent variables in linear regression // *Biometrica.* — 1975. — V. 62. — P. 383—391.

[65] Jones J.R., Smart M.M., Burroughs J.H. Factors related to algal biomass in Missouri Ozark Streams // *Verh. Jnt. Ver. Theor und Andew. Limnol.* — 1984. — V. 22, No 3. — P. 1867— 1875.
[66] Misztal M., Krupa D., Smal H. The chemical composition of bottom sediments and phytoplankton in the man-made Lake Zemborzyce near Lublin // *Acta Hydrobiol.* — 1984. — V. 25—26, No 2. — P. 123—133.
[67] Morel F. Iron uptake and phytoplankton growth // *Rev. Port. Quim.,* 1—2, 2-nd *Int. Conf. on Bioinorg. Chem. Abstr.* — Algarve, 1985. — P. 140—141.
[68] Falkner G., Stresser P., Craffins D. Phosphate uptake by blue-green algae during an algal bloom // *Verh. Jnt. Ver. Theor. und Andew. Limnol.* — 1984. — V. 22, No 1. — P. 195— 199.
[69] Hamilton-Taylor J., Willis M., Reymolds C.S. Depositional flurex of metals and phytoplankton in Windernere as measured by sediment traps // *Limnol. and Oceanogr.* — 1984. — V. 29, No 4. — P. 695— 710.
[70] Helmann H. Zum Berriff der Anreicherung in der Unweltschutz diskussion. Teil 11: Mechanismus und Bewertung der Anreicherung Von Spuren stoffen in Gewassem. — *Deusch. Gewassert Mitt.* — 1983. — BS 27, No 5—6. — P. 146—153.
[71] Jackson T.A. A biogeochemical Study of heavy metals in lakes and streams, and a proposed method for limiting heavy-metal pollution of natural waters // *Verh. Jnt. Ver. Theor und Andgew. Limnol.* — 1978. — V. 20, No 3. — P. 1945—1946.
[72] Menargues L., Lanza E. Speciation du cuire par redissolution anodique dans milien de culture de phytoplancton a differente stades de son developpement // *Vie Mar.* — 1983. — No 5. — P. 53—56.
[73] Les A., Walker R.W. Toxiciy and binding of copper, zinc, cadmium by the blue-green algae, chroococus paris // *Water, Air, and Soil Pollut.* — 1984. — V. 23, No 2. — P. 129—139.
[74] Olsson M. The environmental specimen bank at the Swedish museum of natural history // *US. Dep. Commer. Nat. Bur. Stand. Spec. Publ.* — 1985. — V. 106. — P. 26—33.
[75] Parker J.I., Stanlaw K.A., Marshall J.S., Kennedy C.W. Sorption and sedimentation of Zn und Cd by seston in southern lake Michigan // *J. Yreet Lakes. Pes.* — 1982. — V. 8, No 3. — P. 520—531.
[76] Vymazal J. Short-term uptake of heavy metals by periphyton algae // *Hydrobiologia.* — 1984. — V. 119, No 3. — P. 171—179.
[77] Goryunova S.V., Maximov V.N., Plekhanov S.E. Heavy metals absorption and excretion by macroalgae depending on physiological state thereof. Scient. Rep of the Highest school // *Biological science.* — 1984. — № 2. — P. 67—72.
[78] Briand F., Trucco R., Ramamoorthy S. Correlation between specific algae and heavy metal bindings in lakes // *J. Fish Res. Board Canada.* — 1978. — V. 35, No 11. — P. 1482—1485.
[79] Flatau G.H., Clement R.T., Ganthier M.J. Fixation dy cadmium par une. Pseudomonadacee marine vivante on tnee par lethanol ou le cyanure de potassium // *Chemosphere.* — 1984. — V. 13, No 12. — P. 1397— 1400.
[80] Skowronski T. Uptake of cadmium by Stichococcus bacillaris // *Cnemosphere.* — 1984. — V. 13, No 12. — P. 1385 —1389.

[81] Stary J., Havlik B., Prasilova J., Hunasova J. Cumulation of zinc, cadmium and mercury on the Alga Scenedes — mus obliquus // *Acta Hidrochim. et Hydrobiol.* — 1983. — BS. 11, No 4. — P. 401—408.

[82] Dre F. Ecology. — M.: *Atomizdat,* 1976. — 168 p.

[83] Pork M., Locc S. Bacteria and phytoplankton of the Yulimiste lake // *Hydrobiological researches* (Tartu). — 1979. — No 8. — P. 32—47.

[84] Boynton W.R., Hall C.A., Falkowski P.S. et al. Phytoplankton productivity in aquatic ecosystems. — *Physiol. Plant Ecol.* — 1983. — V. 5. — P. 305—327.

[85] Alimov A.F. Functional ecology of freshwater bivalve mollusks. — L.: *Science,* 1981. — 248 p.

[86] Smith R.E.H., Kallf. J. Phosphorus competition among phytoplankton Areply. — *Limnol. and Oceanogr.* — 1985. — V. 40, No 2. — P. 440—444.

[87] Sommer U., Kilham S.S. Phytoplankton natural community competition experiments: areinterpretation. — *Limnol. and Oceanogr.* — 1985. — V. 30, No 2. — P. 436—440.

[88] Les A., Walker R.W. Toxicity and binding of copper, zinc, and cadmium by the blue-green algae, Chroococcus paris // *Water, Air, and Sol Pollut.* — 1984. — V. 23, No 2. — P. 129—139.

[89] Anderson D. M., Morel F. M. Copper sensitivity of Gongaulax tamarensis // *Limnol. and Oceanorg.* — 1978. — V. 23, No 2. — P. 283—297.

[90] Barber R.T., Ryther J.H. Organic chelators: factors affecting primary production in the Cromwell current upwelling // *J. Exp. Mar. Biol Ecol.* — 1969. — No 3. — P. 191—199.

[91] Berhim F., Houba C, ReMPCle J. Cadmium toxicity and accumulation by Tetrahymena pyriformis in contaminated river waters // *Environ Pollut.* — 1984. — V. 35, No 4. — P. 315—329.

[92] Borgmann U. Determination of free metal ion concentration using bioassays // *Canadian J. Fish, and Aquatic Sci.* — 1981. — V. 38, No 8. — P. 999—1002.

[93] Borgmann U., Ralph K. M. Complexation and toxicity of copper and the free metal bioassay technique // *Water Res.* — 1983. — V. 17 No 11. — P. 1697—1703.

[94] Borgmann U., Ralph K. M. Copper complexation and toxicity to freshwater zooplankton//*Arch. Environ. Contam. and Toxicol.* — 1984. — V. 13, No 4. — P. 403—409.

[95] Canterford G.S., Canterford D.R. Toxicity of heavy metals to the marine diatom Ditylum brightwelli (West) Grunow: correlation between toxicity and metal speciation // *J Mar Biol. Assoc. U. K.* — 1980. — V. 60. — P. 227—292.

[96] Florene T.M. The speciation of trace elements in waters // *Ta lanta.* —1982. — V. 29, No 5. — P. 345—364.

[97] Giesy J.P., Newell A., Leversee G.J. Copper speciation in soft acid, humic waters: effects on copper bioaccumulation by and to Simocephalus serrulatus (Daphnidae) // *Sci. Total Environ.* — 1983. — V. 28. — P. 23—36.

[98] Jackson G. A., Morgan J.J. Trace metal-chelator interactions and phytoplankton growth in seawater media: Theoretical analysis and comparison with reported observations // *Limnol. and Oceanogr.* — 1978. — V. 23. — P. 268—282.

[99] Nelson A., Donkin P. Processes of bioaccumulation: the importance of chemical speciation // *Mar. Pollut. Bull.* — 1985. — V. 16, No 4. — P. 164—169.

[100] Ortner P. B., Kreader C, Harvey G. R. Interactive effects of metals and humus on marine phytoplankton carbon uptake // *Nature.* — 1983. —V. 301, No 5895. — P. 51—59.

[101] Pagenkopf G.K., Russo R.C, Thurston R.V. Effect of complexation on toxicity of copper to fishes // *J. Fish. Res. Board Can.* — 1974. —V. 31. — P. 462—465.

[102] Sposito G. Trace metals in contaminated waters // Environ. Sci. and Technol. — 1981. — V. 15, No 5. — P. 396—403.

[103] Sunda W., Gullard R.R.L. The relationship between cupric ion activity and the toxicity of copper to phytoplankton // *J. Mar. Res.* — 1976. —V. 34. — P. 511—529.

[104] Sunda W.G., Lewis J.A.M. Effect of complexation by natural organic Ugands on the toxicity of copper to a uncellular alga, Monchrysis lutheri // Limnol. and Oceanogr. — 1978. — V. 23. — P. 870—876.

[105] Winner R.W. The toxicity and bioaccumulation of cadmium and copper as affected by humic acid // *Aquat., Toxicol.* — 1984. — V. 5, No 3. — P. 267—274.

[106] Winner R.W. Bioaccumulation and toxicity of copper as affected by interactions between humic acid and water hardness//*Water Res.* — 1985. — V. 19, No 4. — P. 449—455.

[107] Wood A.M. Available copper ligands and the apparent bioavailability of copper to natural phytoplankton assemblages // *Sci. Total Environ* — 1983. —V. 28. — P. 51—64.

[108] Zitko P., Carson W.V., Carson W.G. Prediction of incipient lethal levels of copper to juvenile Atlantic salmon in the presence of humic acid by cupric electrode // Bull. Environ. Contam. and Toxicol. — 1973. — V. 10. — P. 265—271.

[109] Prokofyev A.K. Determination of physical-chemical and physical forms of trace elements in natural waters // *Successes of chemistry.* — 1983. — V. 52, edit. 3. — P. 483—498.

[110] Demina L.L. Form of heavy metals migration in the ocean. — M.: *Science,* 1982. — 120 p.

[111] Zhulidov A.V. Physical-chemical and physical state of metals in natural waters: toxicity for freshwater organisms // *Ecological standardization and modeling of anthropogenic impacts on water ecosystems.* — Edit. 1. — 1988, p. 78—82.

[112] Kulmatov R. A., Kist A. A., Karimov I. I. Neutron activation evaluation of elements distribution in the waters // *Mag. of anal. chem.* — 1980. — V. 35, edit. 2. — P. 254—259.

[113] Kulmatov R.A., Rachmetov U., Kist A.A. Forms of mercury, zinc and cobalt migration in natural waters. — 1982. — V. 35, edit. 3. — P. 393—398.

[114] Prokofyev A.K. Chemical forms of mercury, cadmium and zinc in natural environments // *Successes of chemistry.* — 1981. — V. 50, edit. 1. — P. 54—84.

[115] Lapin I.A., Krasyukov V.N. Role of humic substances in process of complex formation and migration of heavy metals and natural waters // *Water resourses.* — 1986. — No 1. — P. 134—145.

[116] Nikanorov A.M., Zhulidov A.V., Lapin I.A. and oth. Kinetic characteristics of the processes of metals accumulation by biotic and abiotic components of freshwater ecosystems // *DAN USSR.* — 1989. — V. 305, № 5. — P. 1274—1276.

[117] Borgmann U., Charlton C.C. Copper complexation and toxicity to Daphnia in natural waters // *J. Great Lakes Res.* — 1984. — V. 10, No 4. — P. 393—398.

[118] Butler E. I., Tibbits S. Chemical survey of the Tamar Estuary. J. Properties of the waters // *J. Mar. Biol. Assoc. U. K.* — 1972. — V. 52. — P. 681—699.

[119] Guminski S. Outline of the history of studies on the effect of humic compounds on algae // *Oceanologia (PRL).* — 1983. — V. 17. — P. 9—18.

[120] Pagenkopf G.K., Russo R.C, Thursion R.V. Effect of complexation on toxicity of copper to fishes. — *J. Fish. Res. Board Can.* — 1974. — V. 31. — P. 462—465.

[121] Mur J.V., Ramamurti S. Heavy metals in natural waters. Monitoring and evaluation of the influence. — M.: *Mir,* 1987. — 195 p.

[122] Lapin I.A., Krasyukov V.N., Zhulidov A.V. Method of natural waters organic matter complex formation capacity determination // *Thes. rep. at All-union school "Influence of industrial enterprises on the environment"— Zvenigorod,* 1984. — P. 113—114.

[123] Linnik P.N., Nabivanets B.I. Forms of metals migration in fresh surface waters. — L.: *Gidrometeoizdat,* 1986. — 272 p.

[124] Iwase K., Kobayashi K., Haraguchi H., Fuwa K. A new method for measurement of complexing capacity of natural water using gel chromatography // *J. Chem. Soc. Jap. and. Ind. Chem.* — 1983. — No 1. — P. 169—171.

[125] Mantoura R. F. C, Riley J. P. The use of gel filtration in the study of metal binding by humic acids and related compounds // *Anal. Chim. Acta.* —1975. — V. 78. — P. 193—200.

[126] Lapin I. A., Krasyukov V. N. Methods of humic substances determination in natural //*Thes. rep. Of the 3rd Meeting of oceanologisis counc. Section "Ocean physics and chemistry".* — L., 1987. — P. 131 —132.

[127] Manual on surface waters chemical analysis. — L.: *Gidrometeoizdat,* 1977. — 542 p.

[128] Edigarova I.A., Krasyukov V.N., Lapin I.A., Nikanorov A.M. Complex forming capacity of dissolved organic matter of natural waters // *Water resources.* — 1989. — No 4. — P. 122—129.

[129] Nikanorov A.M., Zhulidov A.V., Yemets V.M. and oth. Peculiarities of conjugated mercury accumulation in the body of water invertebrates and bottom sediments at initial stage of metal accumulation in river ecosystems // *DAN USSR.* — 1982. — V. 264, No 4. — P. 1022—1024.

[130] Campbell P.G.C, Tessier A. Geochemistry and bioavailability of trace metals in sediments // *Aquatic Ecotoxicology: Fundamental concepts and methodologies. V. 1 — Boca Raton, Florida CRC Press Inc.*, 1989. — P. 125—148.

[131] Kryshev I.I., Sazykina T.G. Mathematical modeling of radionuclides migration in water ecosystems. — M.: *Energoatomoizdat,* 1986. —152 p.

[132] Lapin I. A., Gekov V.F., Krasyukov V.N. Studying of phase-dispersion content of humic acids as a result of the solution mineralization. Sedimentation mechanism // *Thes. of rep. of All-Union meeting "Cycle of matter and energy in water basins" Listvennichnoe,* 4—6 Sept. 1985. — Irkutsk, 1985. — P. 115—116.

[133] Reuter J.H., Perdue E.M. Importance of heavy metal-organic matter interactions in natural waters. — *Geochim. Cosmochim. Acta,* 1977. — V. 41. — P. 325—334.

[134] Trunov N.M., Teplyakov Y.V., Tavlinova T. A. and oth. Complex of technical means for full-scale ecological modeling at water bodies // *Ecological standardization and modeling of anthropogenic impacts on water ecosystems.* — 1988. — Edit. 1. — P. 17—23.

[135] Guide on methods of surface water and bottom sediments hydrobiological analysis по / Under edit. of V. A.Abakumov. — L.: *Gidrometeoizdat,* 1983. — P. 87—91.

[136] Kimstach V.A., Lapin I.A., Tchudinov E.G., Varvan-ina G.V. Application of a complex of physico-chemical methods to study the forms of heavy metal occurence in natural waters. — Abstracts *Intern. Symp. «New Sensors and Methods for Environ». Characterization (SMEC).* — Kyoto, Jap., 1986, 53 p.

[137] Izrael Y.A., Nazarov I.M., Filippova L.M. and oth. Acid rains. — L.: *Gidrometeoizdat,* 1983. — 206 p.

[138] Lefohn A.S., Brocksen R. E. *Proc. Mont. Acad. Sci.* — 1984. — V. 43. — P. 74—98.

[139] Lucas A.E // *J. Exper. Bot.* — 1977. — V. 28, No 107. — P. 1321—1336.

[140] Charles D.F. *Vern. Unt. Var. Theor. und Angew. Limnol.* — 1984. — III. Lakes I. North America. — P. 559—566.

[141] Tam S.C, Williams R.S.P. // *J. Inorg. Biochem.* — 1986. — V. 26. — P. 35—44.

[142] Oliver B. G., Kelso J. R. M. // *Water. Air. Soil. Pollut.* — 1983. — V. 20. — P. 379—389.

[143] Faust S.D. // *J. Environ. Sci. Health.* — 1983. — V. A18, No 5. — P. 701—711.

[144] Turk J.T. Effects of emissions from oil-shale reports to the atmosphere // *U. S. Geol. Surv. Profess. PAP.* — 1987. — No 1310. — P. 117—122.

[145] Batoyan V.V., Zaytsev N. K., Kucheryaeva V. V. and oth. Quantitative evaluation of aquasystems tolerance to acidic and alkaline contamination // *Vestnik MSU, ser. geogr.* — 1989. — No 5. — P.60—67.

Chapter 11

Distribution of Heavy Metals and Evaluation of Maximum Permissible Ecological Loads in Model Freshwater Ecosystems

11.1. Kinetics of Heavy Metals Distribution

It is well known that problems with water bodies' contamination appear when inflow of contaminants exceeds the capacity of water ecosystems to disseminate and dissipate it. From the point of view of the influence on the ecosystems, two basic types of contaminants are singled out [1, 2]:

1) Stable non-decomposable or hardly decomposable compounds, such as heavy metals to salts thereof, phenol compounds with long chain, chloroorganics, radionuclides, etc.
2) Unstable compounds decomposable during biochemical processes

Substances of the first group either do not break up at all or disintegrate very slowly and are redistributed in the ecosystems as more are coming in, becoming the sources of secondary contamination. For organic matters, factors of water self-purification rate in laboratory microcosms [3] were obtained and are still used for the purposes of forecasting, although conditions of laboratory experiments are far away from real conditions of natural ecosystems existence.

In 1990, the Hydrochemical Institute studied the influence of copper and other heavy metals, specified below, on lake ecosystems through mesocosms. These experiments continued the works in the sphere of full-scale modeling, making it possible to obtain and accumulate information *in situ* at water bodies [4—6]. The following tasks were established:

- Qualitative evaluation of water self-purification rate (rate of metal content decrease in the water) at natural conditions
- Calculation of kinetic sorption parameters, copper sedimentation and accumulation of the components in the water body ecosystem
- Study of contribution of each specified process determining the rate of heavy metals content in the water by means of model ecosystems of different dimensions

- Evaluation of response of model biocoenoses state to heavy metals impact in terms of range of physical, chemical and biological parameters

During experimental research, especially in full-scale ones, quite a great initial heavy metals content in the water is used, because the possibility of successful water bodies' conditions reproduction in the mesocosms is time limited. Thus, the influence of copper on model ecosystems was studied at initial concentrations of 0.15, 0.50, 3.0 and 5.0 mg/dm^3 [7—10].

Experiments in mesocosms were conducted at a small lake in the bottomland of the Don River in the Rostov region. Ecosystems of five mesocosms that "have cut out" a volume of water in the form of a cylinder with a diameter of 2-3 m, jointly with bottom sediments having an open surface, were the object of the research. Polyethylene film fixed on special foundations of stainless steel deepened into bottom sediments was used as the isolating material. The volume of each mesocosm, the initial content of copper and the conditions of the experiment are shown in Table 11.1. The depth of the lake at the point of mesocosms installation was between 1.8-2.0 m.

Table 11.1. Characteristics of mesocosms

Mesocosm	Volume of water, m^3	Initial concentration of copper after its one-time addition and water mixing, mcg/dm^3
I	4.0	145
II	4.8	140
III	4.8	29[1]
IV	4.7	Reference
V	12.7	144

[1] Increase of copper concentration as a result of daily addition.

In mesocosms I, II and V, copper ($CuSO_4$) was applied at the beginning of the experiment for the creation of an initial concentration of about 150 mkg/dm^3 on conversion to copper (II) ions. Mesocosm I was closed from below and did not contain bottom sediments and superior aquatic algae, while mesocosm IV was the reference one. Copper was added into mesocosm III every day for 19 days, and its concentration in the water changed by 29 mkg/dm^3 after mixing. It was made for biocoenosis tolerance evaluation and rate of water self-purification in regime of the constant load. Curve *1* on Figure 11.1 shows the suppositional increase of the copper content if there are no processes leading to its decrease. Experiments with the mesocosms lasted for 35 days, beginning from September 26th to October 30th in 1990.

Water samples taken integrally in terms of depth were divided into two parts: one part was filtered through a membrane filter with a pore diameter of 0.45 mkm. Afterwards, the content of gross forms of copper in filtered and unfiltered water samples was determined by atomic absorption method. The samples were taken for determination of copper content in bottom sediments and macrophites (hornweed) by means of the same method. Immediately on site, such parameters as content of dissolved oxygen, pH, water temperature and intensity of light flux by the water surface, the latter of which determines the influx of energy into the ecosystem, were measured. The following water indices were measured as well: BOD_5;

dissolved organic carbon; organic, mineral and total nitrogen; NO_2^-, NO_3^-, NH_4^+; phosphorus organic and its total; PO_4^{3-}; product, biomass and species composition of phytoplankton; amount and mass of bacterial plankton and the biomass and species diversity of zooplankton. Copper variation in components of the model ecosystem is specified on Figure 11.1 and 11.2. Curves *1*, *5* (see Figure 11.1) and *8* (see Figure 11.2) are approximations by linear functions and parabolic dependencies are shown in curves *2-4* (see Figure 11.1) and *6*, *7* (see Figure 11.2).

Curves *1-5* (see Figure 11.2) are related to the exponential dependencies shifted towards the positive direction on the y-axis. On basis of obtained data, kinetic factors of water self-purification rate (K_w), rates of copper accumulation in bottom sediments (K_b) and in macrophites (K_m) are calculated (Table 11.2).

During calculation of the specified factors, the equation of reaction of the first order [11] is:

$$K = (1/t)\ln(C_0/C_t) \quad (1)$$

where C_0 is the initial concentration of copper in the examined environment (mkg/dm^3, mg/kg, mkg/g) C_t is the copper concentration in time, t.

Comparing the results obtained in mesocosms I and II revealed considerable differences in the rate of copper content in the water diminution. Excessive K_w in mesocosms II and V by 7-9 times in comparison to mesocosm I indicates the significance of bottom sediments and superior aquatic algae—important factors determining metal ions removal from the water. Slower copper concentration increase in mesocosms II and III was noted (curves *8* and *5* on Figures 11.2 and 11.1, respectively) in comparison with mesocosm V (curve *7* on Figure 11.2). This may be explained by the fact that samples in mesocosm V were taken from the area of bottom sediments free from macrophites. In mesocosms II-IV, such areas were almost absent and after the sharp drop of the temperature on the sixth day of experiment, the bottom was quasi screened by fossil plants. Factors $K_в$ for copper in mesocosms II and V are comparable to the decay factor (rate of substance content in the water diminution) of the intermediate group of organic matter ($0.05 \leq K \leq 0.30$) at the boundary with the group of biologically rigid substances ($K \leq 0.05$) [11]. In the paper [12], where a similar approach for the evaluation of copper (II) redistribution in the components of the ecosystem (the Usman River, the Voronezhsky biospheric reserve) was used, the following values for factors were obtained: K_w = 1.03 day^{-1}, K_m = 0.3 and K_b = 0.47 day^{-1}. These values exceeded the average value, K = 0.07 day^{-1}, by almost 15 times; this value was obtained for mesocosms II and V.

Table 11.2. Factors of competing processes rates in the mesocosms (day^{-1})

Mesocosm	K_w	K_b	K_m	Note
I	0.008	—	—	Mesocosm without bottom sediments
II	0.057	0.09	0.11	Mesocosm with volume of 4.8 m^3
V	0.075	0.08	—	Mesocosm with volume of 12.7 m^3

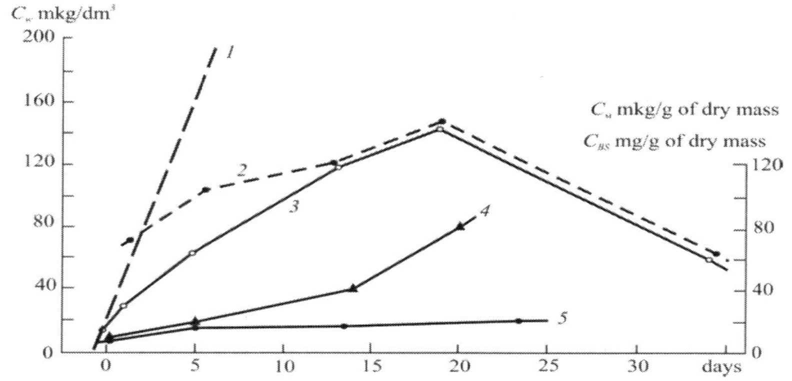

1—suppositional progress of copper concentration in the water (C_w) at daily addings, *2* and *3*—in unfiltered and filtered water samples, **accordingly** (C_w), *4*—in macrophites (hornweeds) ($C_м$), *5*—in bottom sediments (C_{BS})

Figure 11.1. Dynamics of copper concentration in mesocosm III.

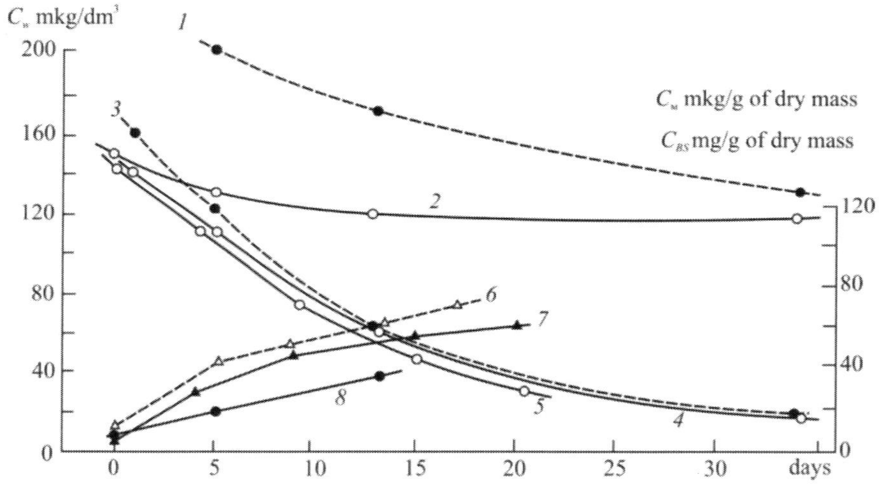

Figure 11.2. Dynamics of copper concentration in components of model ecosystems 1 and 2—accordingly, in unfiltered and filtered water samples (C_w) in mesocosms I, 3 and 4—the same in mesocosm II, 5—**in the filtered water samples** (C_w) in mesocosm V, 6—in macrophites (hornweed) ($C_м$) in mesocosm II, 7 and 8—in bottom sediments (C_{BS}) in mesocosms V and II, accordingly.

Obtained during the experiments in mesocosms I, II and V, factors of copper decrease rate may be used in forecasting water bodies' conditions at known contaminants influx thereto. Such an approach makes it possible to evaluate the different roles of separate factors in aggregate way, characterizing the self-purification capacity of the water body. The results of the research may be used for determination of load on a water body in order to obtain such contaminants content, which, taking into account desaturation, would not exceed certain critical value, $C_{кр}$, established by standards and regulations of certain water consumers (MPC).

As far as substance content in the water (not taking into account the desaturation) depends on the rate of its incoming into the water body and ability of the body to self-purify, there exist certain maximal permissible levels of load on the water basin, W_{CR} [13]. For hardly decomposable substances:

$$W_{cr} = K_w C_{cr} \quad (2)$$

where W_{cr} is the critical level of contaminants influx rate to the water body (mg/dm^3· day)) and C_{cr} is the MPC of the contaminant in the water body (mg/dm^3).

Having calculated $K_в$ on the basis of the results of full-scale experiments and knowing MPC for copper (1 mkg/dm^3), W_{cr} for water basin of this type may be calculated. In the present example, $W_{cr} = 0.07$ mkg/(dm^3· day), therefore, the daily growth of copper content in the specific point of the water basin shall not surpass this value. Having in possession the reliable ways of contaminants desaturation in the water basin, the calculation of MPC may be made by taking into account its desaturation and self-purification (on which existing schemes of MPC standards are actually based).

The purpose of the experiment in mesocosm III was to study the processes of copper migration in a regime of constant load. Copper possesses considerable biological activity and in case of its inflow in small amounts, the effect of vital activities processes stimulation may be observed. In cases where the inflow exceeds a certain critical level, disturbances of ecological structure are possible. Irreversible disturbances may occur in cases of certain values exceeding C_{cr} exceeding.

Analysis of measured parameters dynamics evidences the absence of any notable changes in mesocosm III in comparison with the lake and reference mesocosm, especially in terms of its chemical values. A considerably greater mass of phytoplankton was observed in the first 20 days in comparison with the lake, reference mesocosm and other experimental mesocosms. On the 20th day, the biomass of zooplankton was two times less than on the lake and reference mesocosm (60.3 and 129.5 mg/m^3, accordingly) and was comparable with the mass in mesocosm I (58.4 mg/m^3), where the elevated copper content in the water persisted (curve *2* on Figure 11.2). The share of dead specimens in mesocosm III was two times greater than in the reference mesocosm and mesocosm II (during the period of the load increase). In the analysis of the bacterial plankton state, a sevenfold increase of its biomass in comparison with the lake and reference mesocosm may be noted (7.43, 0.92 and 0.77 mg/dm^3, accordingly). In 1991, the experiments were continued in accordance with this scheme with an increase of load and time of action. Herewith, the list of controlled parameters was extended by means of functional indices of plankton communities' state as well as by means of bioindication methods able to reveal changes in the metabolism of communities at the cellular level [14].

For all mesocosms, except for mesocosm I, the average growth of zooplankton biomass up to 700 mg/m^3 at the end of the experiment (in the lake up to 27.6 mg/m^3) at the initial level of 1.5 mg/m^3 was typical. In mesocosm I, the growth was insignificant and the percentage of dead specimens in this mesocosm was to a considerable extent greater (by 3-4 times in comparison with the reference mesocosm and lake). Only some species of algae disappeared from the phytoplankton community of mesocosm I.

Comparison of oxygen content, BOD_5, and other chemical parameters shows that there were no significant changes in the mesocosms. Production of phytoplankton measured in daily exposition by the oxygen method in all experiments was close to zero.

Using traditional methods of chemical and biological parameters measurement and monitoring, it is possible to study such processes as, for example, the influence of specific toxicant content on plankton communities. New approaches are required, however, for the evaluation of an ecosystem as a unitary whole at least at the level of its basic functional elements.

The conducted study of copper migration and distribution in freshwater ecosystem at the conditions of full-scale modeling revealed [12] that the content of metal in the water is conditioned not only by potential capabilities of accumulation capacity of biota, but also by the deposing capacity of bottom sediments. Kinetics of these processes plays a great role in copper distribution between biogenous and abiogenous components of the ecosystem. Herewith, it shall be noted that the results of the full-scale experiment [12] and laboratory research [15] demonstrate the proximity of constant values of the rates of processes of copper accumulation by biota and its binding by a high-molecular component of dissolved organic matter (DOM) mainly by humic acids (HA) and fulvic acids (FA) into stable complex compounds. Such proximity is evidence of the comparability of these processes in the water medium. Moreover, specially planned laboratory experiments [15] showed that the presence of natural organic matter, for example, HA, in the water leads to a threefold reduction of accumulation degree of *Lymnaea Stagnalis*, which perfectly complies with the results of numerous works [16—21]. This reasonably proved for different species of hydrobionts the considerable decrease of heavy metals toxicity and bioaccumulation in water medium in the presence of HA, FA and other high-molecular complex forming substances.

Electronic structure, ionic radius and physical-chemical parameters are singular to each metal [22] and the condition type of its biochemical behavior in the environment [23, 24]. There are groups of metals, however, possessing close chemical and biological activity, migration properties and toxicity with regard to biota [23, 24]. For example, a group of alkali and alkali-earth metals (sodium, calcium, magnesium) are macrocomponents of natural waters chemical composition, group of biometals or microelements (iron, manganese, copper, zinc, cobalt, molybdenum, etc.) and groups of metals coming into the natural waters mainly with wastewaters of industrial enterprises (lead, chromium, nickel, cadmium, mercury, etc.).

The study of biochemical processes determining behavior of metal in a freshwater ecosystem [25] was continued by similar experimental research for *lead, cadmium and mercury* [26]. The selection of these metals is conditioned by typical belonging thereof to a group of technogenic metals contaminating freshwater ecosystems through discharges of wastewaters, atmospheric precipitations, etc.

As in the case of copper behavior [12], the active experiment in model ecosystems—mesocosms—was used as a methodological approach. Herewith, it shall be noted that at the modern stage of its development, mesocosms cannot be considered absolutely reliable reflections of natural systems. Therefore, an important issue is the interpretation of data obtained in the mesocosms. Even before the issue about the meaning of collected data during the experiment arises, it shall be decided which kind of data shall be obtained. In our specific case, the following tasks were set during the experiments with mesocosms conducting:

1) To study basic directions and intensity of the processes regulating heavy metals distribution and migration in freshwater ecosystem and determining inflow thereof into biotic and abiotic components
2) To obtain information reflecting the kinetics of the studied processes and necessary for building mathematical models of heavy metals behavior in freshwater ecosystem at a one-time discharge, creating the basis for deeper interpretation of experimental results and technique of freshwater ecosystems buffer capacities calculation in terms of a specific metal

Experimental studies of lead and cadmium distribution and migration were conducted at Lake Krivoe (Rostov-on-Don). A similar experiment in terms of mercury was conducted at the Isakovskoe water storage basin (Komunarsk, Luganskaya region). A soft mesocosm structure with a square section, 2 x 2 m, of polyethylene film fixed at the top on foam plastic floats and at the bottom part on a metal insulated foundation was installed in the layer of bottom sediments. Installed in the water bodies at a depth of 1.7-2.0 m, the mesocosms were isolated by means of soft walls of polyethylene film with a water volume of 7-8 m^3 jointly with the layer of bottom sediments. In case of necessity, bottom sediments were isolated by a polyethylene screen.

Processes of biochemical behavior of lead and cadmium in freshwater ecosystems were studied in five mesocosms:

I Reference
II With addition of 200 mkg/dm^3 of lead
III With addition of 200 mkg/dm^3 of lead (with isolated bottom sediments)
IV With addition of 120 mkg/dm^3 of cadmium
V With addition of 120 mkg/dm^3 of cadmium (with isolated bottom sediments)

During the process of the experiment, the following parameters were measured in the integral water samples: alkalinity, pH, Eh, content of dissolved oxygen, calcium, magnesium, total and organic nitrogen, nitrates, HA and FA. Techniques of the indicated parameters determination are specified in papers [27, 28]. Water samples for assay were taken every 14-24 hours over 6-7 days. The samples were filtered through a membrane filter with pores of 0.45 mcm. The total content of examined metal as well as its content in the ionic and bound DOM forms was determined in the filtrate. Ionic and bound forms of lead were extracted and disintegrated according to the analysis diagram of heavy metals forms [29, 30] with the usage of DEAE- and KM-cervacels.

After appropriate preparation, the content of model metals in samples of suspended material, biota and bottom sediments was determined through atomic-absorption spectrophotometry with the AAS-3 [28] device.

Composition of Lake Krivoe water during the period of experiment in August is specified in Table 11.3.

Table 11.3. Water parameters of the Krivoe Lake

T°C	pH	Concentration, mg/dm^3						
		O_2	BOD_5	COD	NO_2^-	NO_3^-	NH_4^+	N_{org}
23.2	7.6	5.8	2.75	107.0	0.10	–	1.82	1.84
P				CO_2		HCO_3^-	HA	FA
total	mineral	organic						
0.05	0.01	0.04		–		117.5	0.05	0.40

The natural background of cadmium and lead content in Lake Krivoe amounted to 5.4 and 0.3 mkg/dm^3 accordingly.

Observed changes of ionic and bound lead forms content in mesocosms II and III during the entirety of the experiment make it possible to judge tendencies of metal distribution in a model freshwater ecosystem after one-time discharge (Figure 11.3). At the conditions of experiment, the content of lead ionic forms drops sharply. Even in the first 17 hr. up to 86.5% of ionic lead passes to the bound state or is accumulated by bottom sediments and biota. The decrease of the lead ionic form concentration to a constant level not exceeding 0.5-0.8 mkg/dm^3 is observed during the second and third days of the experiment. The process of lead binding with DOM may be provisionally divided into three stages. At the first stage, during 10-17 hr., the content of bound forms sharply rises (by 70.5%). At the second stage, which takes about three days, the content of lead forms, bound with DOM (mainly with HA and FA), remains practically the same. At the third stage, the decrease of content thereof takes place.

Decrease of bound lead forms content is mainly conditioned by the decrease of DOM itself (reduction of HA and FA by 0.01 and 0.10 mg/dm^3 of carbon, accordingly). By the end of the experiment, the relative HA and FA content in the bound forms of lead increases from 70 to 92%.

A more complicated relation is typical for lead accumulation by biota (only metal coming into phytoplankton is examined). During the experiment, lead coming into phytoplankton was steadily rising (by 30.5% of initial concentration) due to its sorption or accumulating capacity.

As follows from experimental data, lead accumulation in bottom sediments is the most intensive during the first 60 hr. Afterwards, the process has been stabilized. Lead coming into bottom sediments insignificantly increased at the final stage of experiment, most likely, due to sedimentation of the suspended material (mineral suspended materials and detritus) (see Figure 11.3).

Heavy metals are accumulated in the upper layer of bottom sediments mainly owing to the following mechanisms:

1) Sedimentation of suspended material
2) Exchange processes taking place in the benthic layer of the water at the boundary water—bottom sediments

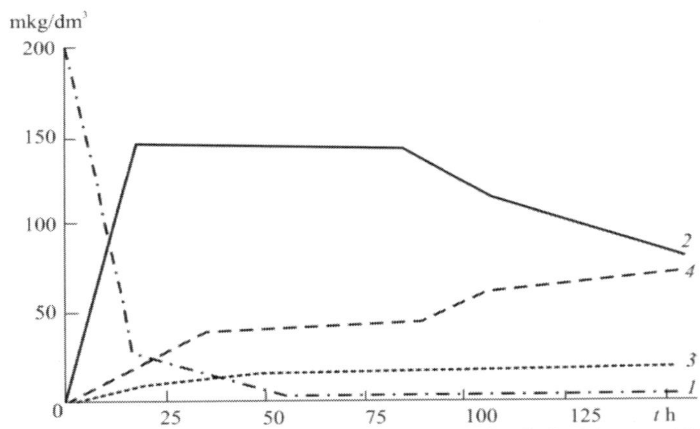

2—forms bound with DOM; *3* and *4*—forms accumulated accordingly by biota and bottom sediments

Figure 11.3. Distribution of concentration (mkg/dm^3) of ionic (*1*) and bound (*2-4*) forms of lead in mesocosms II and III after one-time discharge.

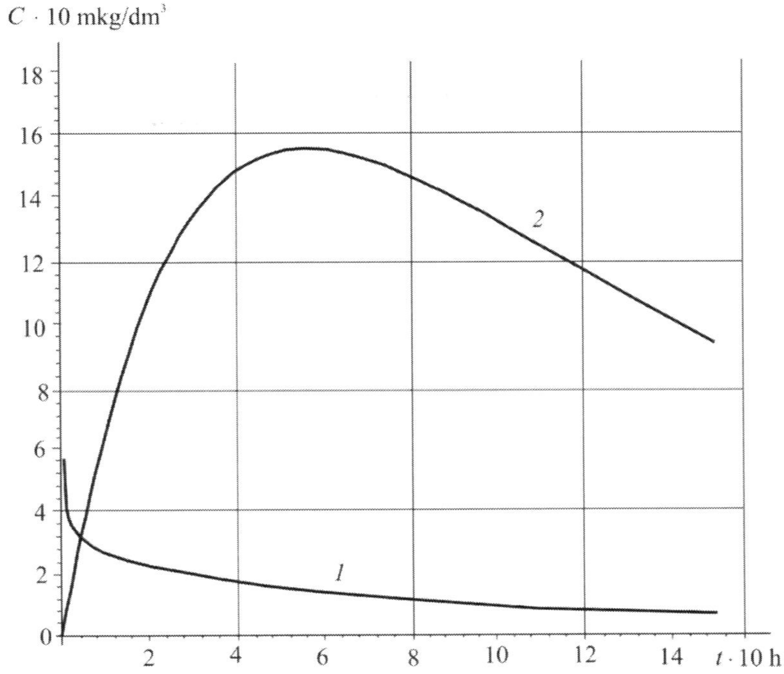

Figure 11.4. Graphic representation of mathematical data from Figure 11.3.

The mathematical processing of data from Figure 11.3 makes it possible to represent the dependencies (curves *1* and *2*) in graphic form (Figure 11.4) (hereinafter, the scale of coordinate axes is reduced by 10 times) and describe them with the following equations:

curve *1*

$$C_1(t) = (2.5 + 0.00023K_{0.8}(0.0001(t - 0.001)))\exp(-0.1t) \quad (3)$$

curve *2*

$$C_2(t) = 8t\exp(-0.2t + 0.002t^2) \quad (4)$$

Processes of interaction taking place at the boundary of two phases and leading to the transition of dissolved lead forms from the water to the top layer of bottom sediments are evaluated during comparison of the experiment results obtained in mesocosms II and III (Figure 11.5).

On Figure 11.5, curves *2* and *4* are exponentials, and the equation of curves *1* and *3* are similar to equation of the curve *1* on Figure 11.3. The general structure of these dependencies is described by the function[22]:

$$C(t) = (A + BK_\nu(m(t + n)))\exp(-at) \quad (5)$$

[22] In the present section of the monograph, the functions of Macdonald of the second series are constantly mentioned. One can find information about such functions in a reference book by Beytmen G., Erdeyn A. Special functions. – M., 1978.

where A is the shift of curve $C(t)$ along the y-axis, B are the initial ordinates of the function $C(t)$, m represents the decay of ordinates of the function $C(t)$, n is the shift of curve $C(t)$ along the x-axis, a is the additional decay of the function $C(t)$ and v is the distance from function $C(t)$ to the y-axis.

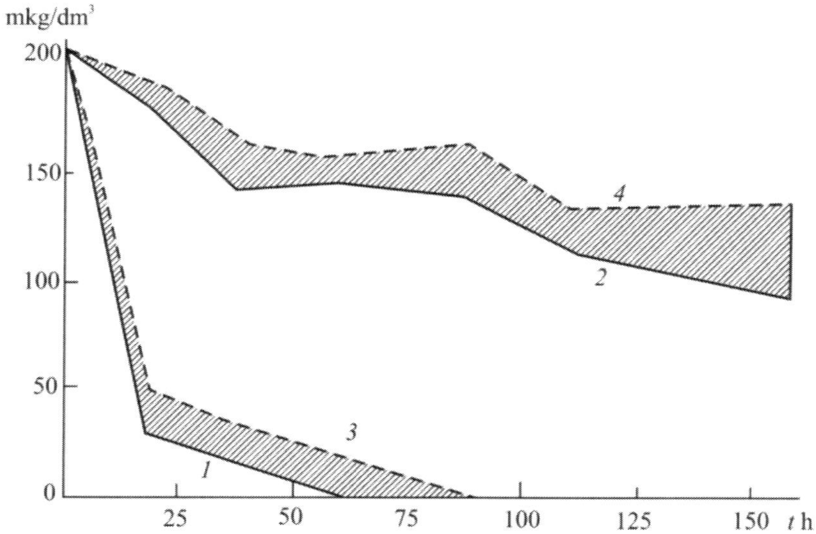

1 and *2*—ionic forms; *3* and *4*—forms bounded with DOM

Figure 11.5. Dependence of lead forms absorption from the solution by bottom sediments in mesocosm II (*1, 3*) and mesocosm III (*2, 4*).

Lead absorption from the solution is an intensive process mainly contributing to the accumulation of this metal in the upper layer of bottom sediments (up to 85% of accumulated lead). In the first part of the experiment, the major part of the lead coming into the bottom sediments is represented in ionic form, which is actively sorbed on particles of clayey minerals and active silt. In the second part of the experiment, when ionic forms are almost fully absent in the solution, the transition of the lead forms is continued by means of forms bound with DOM. This is determined by processes of organic complexes sorption on particles of bottom sediments and the influx thereof to silty waters due to change of oxidation-reduction potential at the boundary of a phase's division.

Obtained during the experiment, data was used for the evaluation of kinetics of the processes of *lead* distribution in a freshwater ecosystem. The following values were calculated through mathematical statistics:

$$K_B = 0.11 \cdot 10^{-3},\ K_{DOM} = 0.18 \cdot 10^{-3};\ K_{BS} = 0.20 \cdot 10^{-3} ч^{-1}$$

The results of the experiment, describing *cadmium* distribution and its migration in the model freshwater ecosystem (Figure 11.6) made it possible to evaluate the intensity of the set of inter-water basins processes having an influence on the behavior of the present metal. At the conditions of the model experiment, a reduction of cadmium ionic forms content in time corresponds to a relation close to exponential. Even during the first hours from the beginning, more than 60% of cadmium ions are distributed between DOM, bottom sediments and biota. It shall be noted that, as in the case with lead, DOM plays the most important role in metal ions content decrease. A practically full absence of the most toxic ionic form of cadmium in the water of model ecosystem was observed 5 days after the beginning of the experiment. Comparatively fast, cadmium binding by HA and FA contained in the water conditioned the presence of approximately 46% of bound metal forms after 11 hr. Furthermore, the content of bound forms of cadmium remained practically at the same level until the end of experiment. Cadmium influx into biota was steadily increasing during the experiment as a result of which up to 27-38% of cadmium was accumulated by biotic components of the model ecosystem.

As follows from the data of model experiment (see Figure 11.6), most active cadmium was accumulated by bottom sediments during the first 2 days. Furthermore, the intensity of this process was decreasing and metal was coming into bottom sediments mainly by means of suspended materials sedimentation.

From the position of a mathematic description of the processes, represented on Figure 11.6, the following shall be noted. The structure of curve *1* on Figure 11.6 is similar to function (5), specified on Figure 11.5. Curves *3* and *4* on Figure 11.6 represent parabolas with a horizontal axis. Function *2* is approximated by the relation of the algebraic function (Figure 11.7):

$$C(t) = \frac{12x^{1.7}}{x^2+x+1} \tag{6}$$

The set of algebraic functions relations is immense and this set may be described as the most general dependence:

$$C(z) = \frac{F(z)}{G(z)} \tag{7}$$

where $F(z)$ and $G(z)$ are algebraic functions.

The analytical structure of the diagrams represented on Figure 11.8 is the same as those on Figure 11.5.

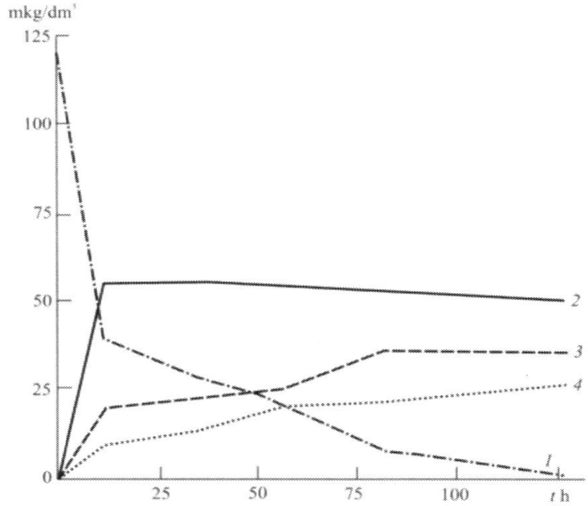

2—forms bound by DOM; *3* and *4*—forms accumulated by biota and bottom sediments, accordingly

Figure11.6. Distribution of concentration (mcg/dm^3) of ionic (*1*) and bound (*2*-*4*) forms of cadmium in mesocosms IV and V after a one-time discharge.

The processes of cadmium dissolved forms depositing are examined during comparison of data obtained by experiments in mesocosms IV and V (see Figure 11.8). Absorption of ionic forms of cadmium began from the first measurement and was gradually increasing. The unevenness of cadmium sorption by particles of the upper layer of bottom sediments is probably related with ionic form of metal accumulation in the boundary benthic layer. Cadmium forms, bound with DOM, were absorbed more actively. At certain stages of the experiment, the concentration of organic forms, accumulated by bottom sediments, amounted to 20 mkg/dm^3.

Modeling data obtained during the full-scale experiment made it possible to calculate constant rates of processes of cadmium coming into DOM, biota and bottom sediments:

$$K_{DOM} = 0.11 \cdot 10^{-3},\ K_B = 0.03 \cdot 10^{-3},\ K_{BS} = 0.02 \cdot 10^{-3}\ h^{-1}$$

During the study of biochemical behavior of *mercury* in a model freshwater system, the scheme of the experiment followed those with lead and cadmium.

It is conditioned first of all by the fact that mercury has a specific position in the range of standardized priority heavy metals due to its elevated toxicity with regard to hydrobionts; large variety of forms of being in water medium, including the forms of organic and inorganic complexes; mercury-organic compounds (metal-, dimetylmercury and other derivatives) and its ability to be accumulated in biotic and abiotic components of water ecosystems [31—33]. Therefore, the processes of conduction in the mesocosms changed in comparison with the similar mercury distribution and migration in a model freshwater ecosystem and were examined in four mesocosms with different initial mercury contents:

I Reference (without addition of metal)
II With addition of 5 mkg/dm^3 of mercury

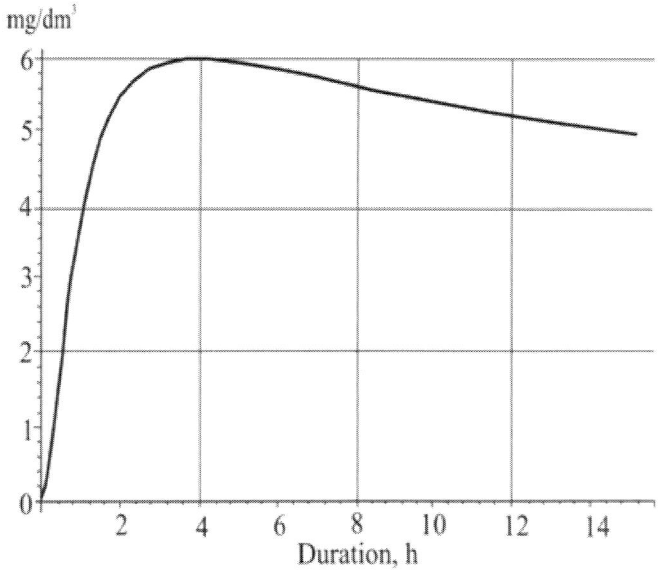

Figure 11.7. Variation of concentration (mg/dm^3) of bound cadmium forms in mesocosms IV and V, calculated by formula (6).

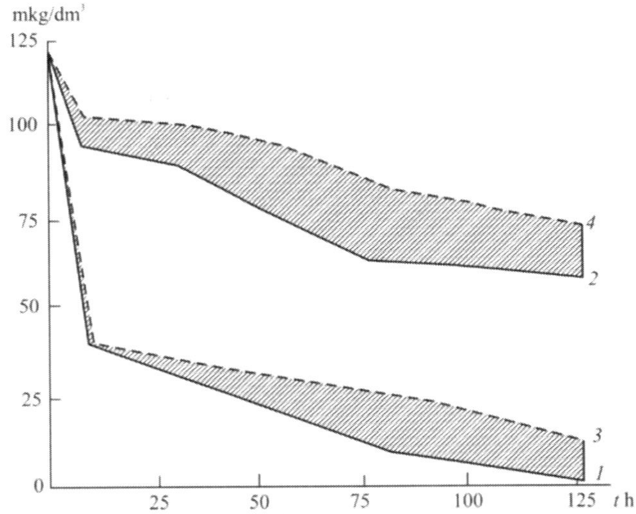

1 and *2*—ionic forms; *3* and *4*—forms, bound with DOM

Figure 11.8. Dependence of cadmium forms absorption from the solution by bottom sediments in mesocosm IV (*1, 3*) and mesocosm V (*2, 4*).

III With addition of 25 mkg/dm^3
IV With addition of 50 mkg/dm^3

During full-scale modeling, hydrochemical and hydrobiological parameters are measured in integral samples: pH and Eh, content of dissolved oxygen, calcium, magnesium, total nitrogen, organic nitrogen, nitrates, total phosphor, HA, FA, alkalinity and biomass of phytoplankton [27, 28]. The content of organic and inorganic mercury forms in the water, biota and bottom sediments is determined by method of cold vapor on the device, "Julia-2" [34]. It shall be noted that mercury compounds, reduced to free mercury by chloride of tin in acid the medium, and were related to inorganic forms. The samples were treated with a strong oxidant (HNO_3, + $KMnO_4$) before the procedure of reduction during the determination of total dissolved mercury (for disintegration of organic forms). During the period of the experiment, the water from the Isakovskoe water storage basin was described by hydrochemical and hydrobiological indices specified in Table 11.4 and bottom sediments were mainly represented by the coarse-detrital silt.

Variation in the content of general dissolved organic and inorganic mercury forms in mesocosm II (Figure 11.9*a*) is evidence of the tendency to decrease the content of dissolved metal forms. It is significant that the inorganic mercury form almost fully disappears in the initial hours.

Table 11.4. Parameters of water from the Isakovskoe water storage basin

T°C	pH	Concentration, mg/dm^3								
		O_2	COD	NO_2^-	NO_3^-	NH_4^+	P_{min}	Fe_{hen}	HA	FA
15.1	7.2	7.7	167.0	0.05	0.23	0.37	0.10	0.05	0.12	1.56

Herewith, it shall be noted that the process of mercury complexes formation with organic ligands presented in natural forms (up to 87% of dissolved mercury passes to organic form) runs most intensively. The subsequent time period is characterized by a decrease of dissolved mercury forms content (stable concentration of approximately 0.3 mkg/dm^3 is attained) and accordingly by its increase in suspended particles, bottom sediments and biota (see Figure 11.9). At the end of experiment ($t > 4$ days), the content of mercury in the solution slightly grows which is most likely related to its exchange at the boundary water—suspended materials and water—bottom sediments, leading to mercury excretion to a water medium. The possibility of organic mercury forms emergence as a result of emission thereof by planktonic organisms shall not be discarded.

In Figure 11.9*b*, data for mesocosm III is specified (initial metal concentration is 25 mkg/dm^3). In general, the nature of dependence of dissolved mercury forms variation is similar to the one observed in mesocosm II. It shall be noted, however, that in this mesocosm, mercury is actively emitted to the solution of bottom sediments, suspended materials and plankton. At the end of the experiment, the relative content of dissolved forms of mercury reaches 75% of initial amount.

Repeated twice, the increase of the initial mercury load on model freshwater ecosystem (50 mkg/dm^3) had no significant influence on the nature of content of dissolved metal forms variation (see Figure 11.9*c*). According to experimental data of mesocosm IV, however, by far a considerable amount of metal passed to suspended materials, bottom sediments and biota (approximately 60% of introduced mercury). In our opinion, this fact is conditioned by more active sorption of dissolved mercury forms from the water by means of the high initial content. In particular, mercury content in suspended materials (comparing the data of experiments in mesocosms II and IV) increases by 3.0-3.5 times and the degree of accumulation by benthal deposits by 3-4 times. In Figure 11.10, the mathematical processing of data from Figure 11.9*a* is presented (curve *1*—Figure 11.10*a*, curve *3*—Figure 11.10*b*) as well as its function (receding from the coordinate axis):

$$C(t) = 0.05 + K_0(3t) \tag{8}$$

where $K_0(3t)$ is the Macdonald function of the second genre of zero order

$$C(t)=\begin{cases} 0.22(10t-1.4t^2+5\sin^2(0.8t)),\ 0 \le t \le 7 \\ 0.4-0.2(t-7)\exp(-0.9(t-7)),\ t > 7 \end{cases} \tag{9}$$

The toxic impact at a variation of initial mercury content in the water of the model ecosystem was evaluated on the basis of dynamics of phytoplankton biomass variation (Figure 11.11).

According to the data provided by the experiment, the dynamics of phytoplankton biomass in the water of the water storage basin and reference mesocosm, regardless of its similar nature, has some differences conditioned most likely by the "mesocosm effect." The phytoplankton biomass varied by a more considerable degree after the mercury was introduced and impacted the minimal load by the contaminant (5 mkg/dm^3) on the model ecosystem was characterized first by a sharp reduction of plankton biomass (by 24%) and then by its increase up to full restoration with regard to the reference mesocosm. A maximal load provided by mercury (50.0 mkg/dm^3) caused irreversible (at the conditions of experiment) reduction of phytoplankton biomass as a result of the mass dying off of planktonic organisms. The latter perfectly complied with the variation of the dissolved oxygen content. On this basis, we may suppose that for the examined model ecosystem, the maximal mercury load shall be less than the minimal concentration used in the experiment (5 mkg/dm^3).

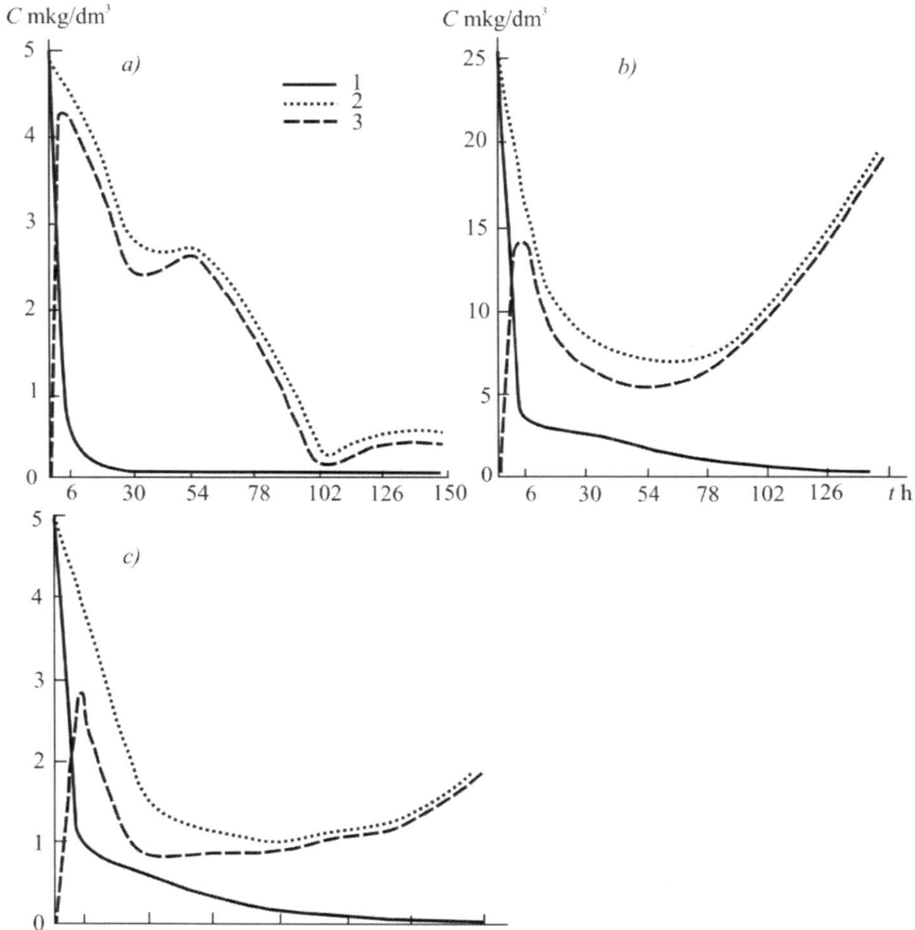

1—ionic form; 2—forms bound with DOM; 3—total mercury

Figure 11.9. Progress of concentration of total dissolved organic and inorganic mercury forms in mesocosms II (*a*), III (*b*) and IV (*c*).

Mathematical processing of data specified on Figure 11.11 (curve *2*) is presented in Figure 11.12*a*.

Curve *2* on Figure 11.11 represents the decaying aperiodic oscillating process. One cycle of the oscillating process was revealed during such an experiment. The other cycles are not visible due to the quick decay of oscillations and concentration passing to a stationary regime. The general view of decaying oscillation of this type is described by special function called the "integral sine":

$$Si(t) = \int_0^t \frac{\sin x}{x} dx \qquad (10)$$

On Figure 11.12*a*, the dependence $C(t) = \mathrm{Si}(0{,}8t)$ at $2 < t < 25$ is shown as an example.

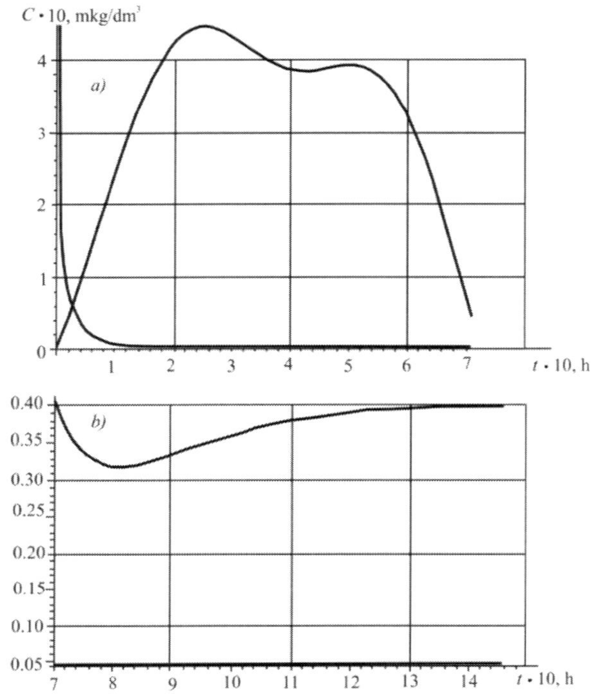

Figure 11.10.

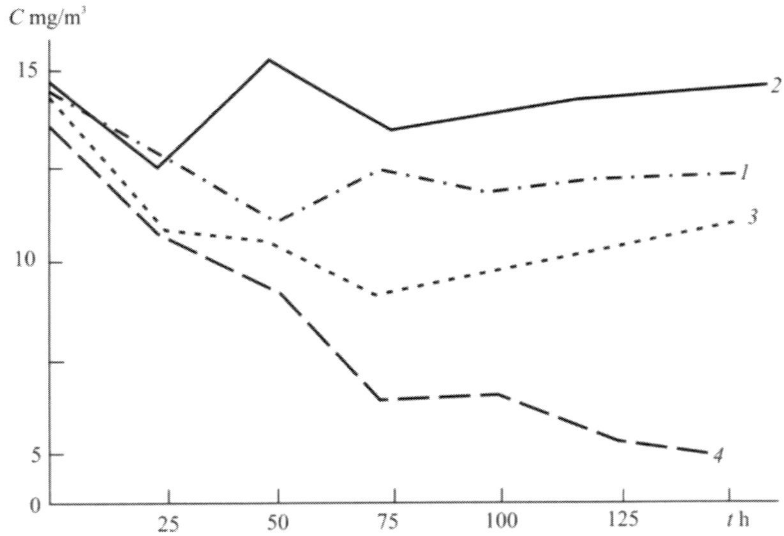

Figure 11.11. Dynamics of phytoplankton biomass in water storage basin (*1*), in the reference mesocosm (*2*) and in mesocosms II (*3*) and IV (*4*).

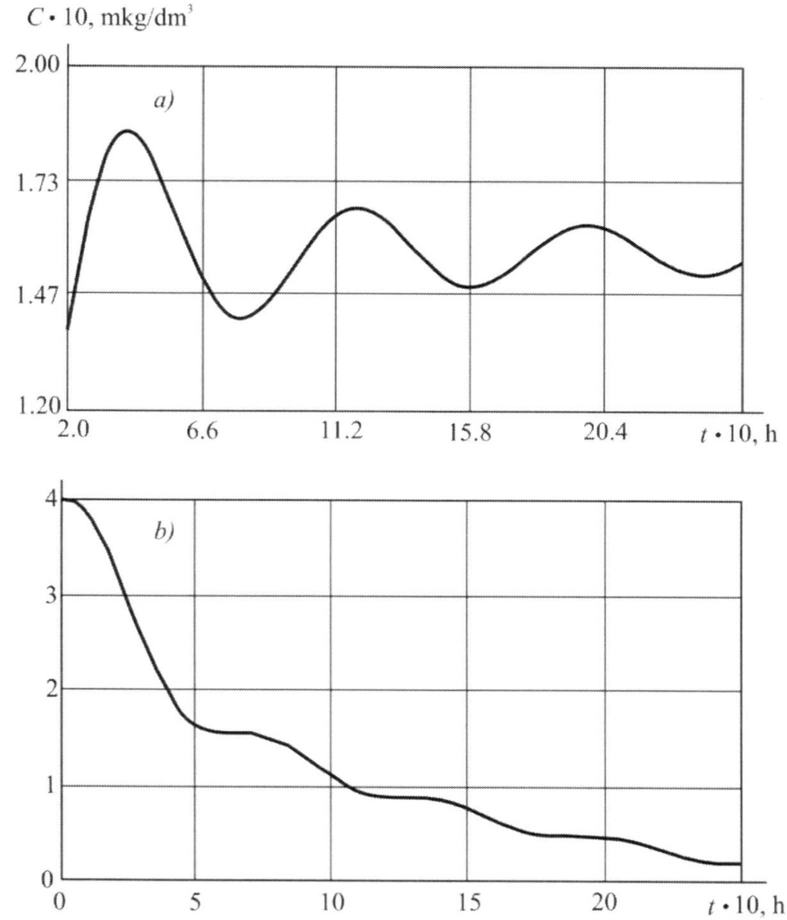

Figure 11.12.

Such a function decays slower than the experimental curve *4* (see Figure 11.11). Therewith, it has several cycles of oscillating processes.

The diagram of function *1* on Figure 11.11 is approximated by the integral sine as well.

Curves *3* and *4* on Figure 11.11 describe aperiodic oscillating processes with parabolic modulation. The approximation of such dependencies is carried out by means of higher transcendental function—integral cosine, taken with reverse sign and shifted upward along the y-axis:

$$4 - Ci(t) = 4 - \int_0^t \frac{1-\cos x}{x} dx \tag{11}$$

The diagram of this function is specified on Figure 11.12*b*.

As in case with other toxic metals, mercury content in the water of an examined freshwater ecosystem exceeding MPC, at least by one order, did not cause serious threat to the system's functioning, which is conditioned by buffer properties of the ecosystem. The main contribution to buffer properties of freshwater ecosystem is made by DOM and bottom sediments (see Figures 11.9 and 11.13).

Up to 87% of ionic mercury form was bound by DOM (HA and FA) to form strong, high-molecular complex compounds, which are not toxic for hydrobionts. The upper layer of bottom sediments accumulated 52% of the introduced mercury and became a significant factor of excretion of ionic forms and bound with DOM mercury forms. The content of mercury, coming from the water to biota became maximal 2.5 days after the beginning of experiment. It shall be noted, however, that only 30% of mercury was accumulated by phytoplankton. Dependencies, specified in Figure 11.13, represent the relation of algebraic functions (curves *1* and *3*) and the Macdonald function shifted along the x-axis (curve *2*); the dependence curve *4* is similar to the one specified in Figure 11.9*c*.

Kinetics of *mercury* coming from the water to biotic and abiotic components of the model freshwater ecosystem are quantitatively expressed by the rate indices of these processes and calculated on the basis of the experiment results:

$$K_B = 0.1\cdot 10^{-3},\ K_{DOM} = 0.20\cdot 10^{-3},\ K_{BS} = 0.45\cdot 10^{-3}\ h^{-1}$$

The obtained results make it possible to consider that kinetics of the reaction of HM ions with DOM complexes formation in natural waters is a complex process, which may be related to the reaction of the second order. Kinetic parameters of complexes formation are determined by a range of factors. For example, the factor of complexes formation rate, *K*, for each of the studied metals is mainly conditioned by the chemical nature of the metal itself as well as its composition and properties of HA and FA. The tendency for *K* to decrease is characteristic for the range of studied HM (Cu > Pb > Cd), observed in laboratory experiments as remaining at full-scale conditions of water ecosystems as well. Studies of external factors of influence (pH, ionic power of the solution) conducted on the example of Cu showed in model solutions and samples of natural water that the value *K* varies within quite a narrow range and its variations do not exceed one order. This fact made it possible to use *K* and the rates of reaction of HM complexes formation with HA and FA during the building of models of metals behavior in freshwater ecosystems. The nature of the experimental kinetic curves (Figure 11.14) obtained for samples of natural water at the conditions of full-scale modeling (in terms of the samples of model solutions) are evidence of the fact that the binding of HM ions by HA and FA is most intensive in the initial hours of the experiment, and the state of equilibrium, as a rule, is attained 48 hr. after the beginning of the reaction.

Thus, experimental research concerning the study of the biochemical behavior of lead, cadmium and mercury in model freshwater ecosystems makes it possible to evaluate the direction and intensity of metals distribution processes between biotic and abiotic components of an ecosystem.

As follows from the results of all conducted experiments, the *major part of introduced toxic metals is bound by DOM.* In spite of the relatively low content of complex forming DOM components in the water of Lake Krivoe during the whole experiment, *up to 50-60% of lead and more than 45% of cadmium remained in the solution in the form of complex compounds with DOM (HA, FA, polypeptides, steroids, etc.) A more complicated process of metal interaction with DOM was observed for mercury; first, more than 85% of mercury passed to organic forms but then its relative content decreased considerably.* This is conditioned first of all by the large variety of organic forms of mercury, its high reaction activity and tendency for adsorption. On the example of the experiments conducted at the

Isakovskoe water basin, such great potential possibilities of DOM were established, that even a 10-fold increase of initial mercury content had insignificant influence on the relative content of organic forms in the solution.

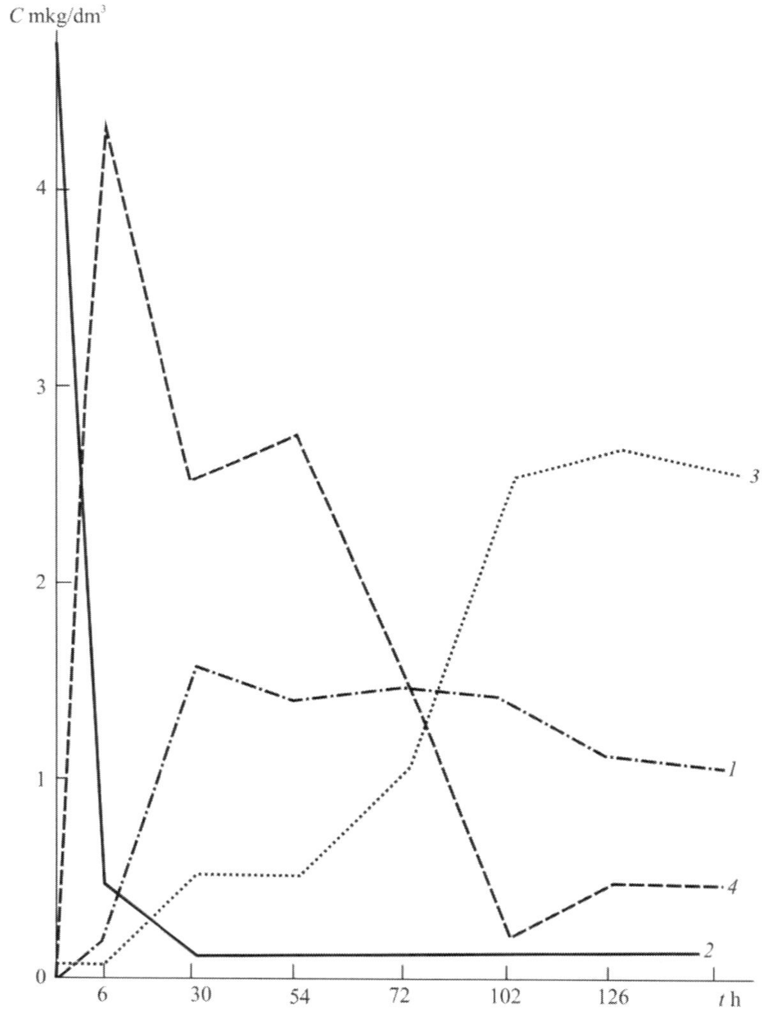

2 and *3*—forms, bound with DOM and bottom sediments accordingly; *4*—forms, accumulated by biota

Figure 11.13. Distribution of ionic (*1*) and bound (*2-4*) mercury forms in mesocosm I.

Analysis of the results presented on Figure 11.5 and 11.8 showed that organic forms of lead and cadmium possessed high sorption activity and easily passed to the upper layer of bottom sediments (up to 15-30% of organic forms came into bottom sediments). Besides, the formation of high molecular and strong complex compounds with DOM impedes metals bioaccumulation and reduces its toxicity. Therefore, the presence of DOM in the water medium regulates the content of biologically active forms of metal in the water as well as favors the metals excretion from the water medium and burying thereof in bottom sediments. On this basis, it may be supposed that for *lead, cadmium, mercury and copper [12], DOM is the most important factor in determining buffer capacities of a water ecosystem.*

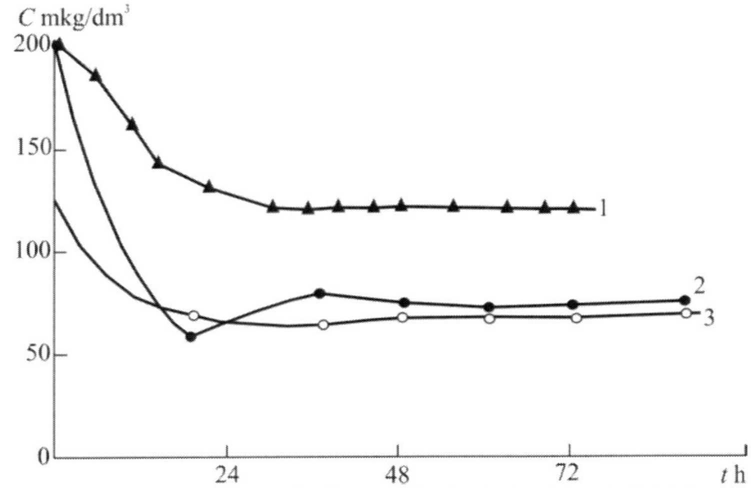

1—Cu (the Usman River); *2* and *3*—Pb and Cd, accordingly (Lake Krivoe). Initial concentration of Cu, Pb and Cd is 100, 200 and 120 mkg/dm^3, accordingly.

Figure 11.14. Variation of HM ionic forms concentration during complexes formation with DOM at the conditions of full-scale modeling in mesocosms.

In all conducted model experiments, the incoming of studied metals into bottom sediments was increasing close to a logarithmic nature. Up to 10% of lead, 20% of cadmium and 30-50% of mercury passed from water to bottom sediments. Herewith, *at the first stage of the experiment, the most intensive increase of metal content in the upper layer of bottom sediments was determined by sorption processes at the boundary of division water— bottom sediments. At the second stage, sedimentation of minerals suspended materials and detritus.*

It shall be noted that depositing capacity is conditioned not only by mineralogical, chemical and granulometric composition of bottom sediments but also by metal content in the water. Thus, for example, in the case of initial mercury content in water variation of the upper layer of bottom sediments, it becomes supersaturated with mercury owing to which the reverse transition of metal from bottom sediments to the water is observed (see Figure 11.13). Thus, the conducted research is evidence of the fact that *the second important abiotic factor, conditioning buffer capacity of freshwater ecosystems for lead, cadmium and mercury, is bottom sediments.*

The biological factor (or factors) is the most important and, most likely, dominates among the factors determining buffer capacities of freshwater ecosystems. Nowadays, however, there are no practical approaches making it possible to take into account the potential possibilities of all ranges of interrelated biotic components of freshwater ecosystem to reduce the impact of heavy metals. This fact considerably complicates the evaluation of buffer properties of biotic components in the conditions of full-scale research and model experiments. Within the framework of conducted research of biochemical behavior of lead, cadmium and mercury in the model freshwater ecosystem, we only studied kinetics of these metals incoming into phytoplankton. As follows from the data provided by the experiment, the accumulation of lead and cadmium by phytoplankton has graduated, increasing its nature. In the process of the experiment, up to 30% of lead and 27% of cadmium passed to biota.

The dynamics of mercury accumulation by phytoplankton is more complicated. In the beginning of the experiment, the mercury content in biota rose sharply, afterwards it was decreasing. Noted variations, in our opinion, are conditioned by dynamics of plankton biomass (see Figure 11.11). The decrease of biomass of planktonic organisms at the first stages of the experiment was determined by high values of mercury content in biota; the increase of phytoplankton biomass at the second stage led to its decrease thereof.

As is obvious, values of rates of lead, cadmium and mercury incoming to biotic and abiotic components of a freshwater ecosystem give an idea about kinetic regularities of the examined processes and intensity thereof.

11.2. Evaluation of Maximal Permissible Ecological Loads

Results of conducted model experiments with the usage of mesocosms make it possible, along with the evaluation of kinetic characteristics of contaminants distribution in the water, to develop methods of maximal permissible ecological loads (MPEL) determination while taking into account the individual peculiarities of specific water bodies. Development of these methods is a very challenging task that shall be solved by joint endeavors of numerous research teams. In the future, it will be solved for sure. The fragment of the specified great and challenging task is a trial to use the results of modeling for ecological standardization purposes.

Results of the experiments conducted by the Hydrochemical Institute from 1990-1992 may serve as illustration of the suggested approach [35]. In Table 11.5, the data from modeling in the mesocosms installed at the Kirzhach River in 1991 is specified. The experiment lasted for 14 days. Mesocosms with unified structures, making it possible to isolate volume of water from the surface to the bottom, served as model ecosystems. Polyethylene film fixed in demountable stainless steel foundations deepened into bottom sediments served as the isolating material. Characteristics of mesocosms and initial conditions of the experiments are specified in Table 11.5. Copper ($CuSO_4$) in the regime of constant load, *W* (daily application), was applied into the mesocosm as a model contaminant. The experiment results make it possible to *evaluate the load on the ecosystem in terms of two criteria: the system's ability to self-purify and copper MPC at the end of experiment.* Figure 11.15 shows the results of the experiment. The factor of water self-purification rate, K_w, is calculated on the basis of the equation of the first order [36]. Three zones are singled out on the figure:

I. In this diapason of loads, copper inflow increase is accompanied with the increase of the system's ability to accumulate it into other components.
II. Increase of the load is accompanied with a quick decrease of the system's ability to self-purify water.
III. Abilities of the system to self-purify are fully exhausted and the increase of load has a negligible influence on the self-purification rate.

Results of mathematical processing of the data provided in Figure 11.15 are specified in Figure 11.16.

The functional approximation (by shooting method) is described by equation:

$$K(x) = \frac{0.030+0.027x}{1+0.03x^4} \quad (12)$$

with an approximation error $\Delta = 2.5\%$.

Table 11.5. Results of the experiments with copper in mesocosms

Mesocosm number	Volume, m^3	Load, mkg/ ($dm^3 \cdot$ day)	Copper concentration in the water at the end of experiment, mkg/dm	Factor of water self-purification rate K, day^{-1}
1	1.00	2	6	0.110
2	1.05	20	43	0.130
3	1.10	30	98	0.140
4	1.10	40	134	0.100
5	1.10	50	192	0.090
6	1.20	80	355	0.080
7	1.20	100	4,640	0.078

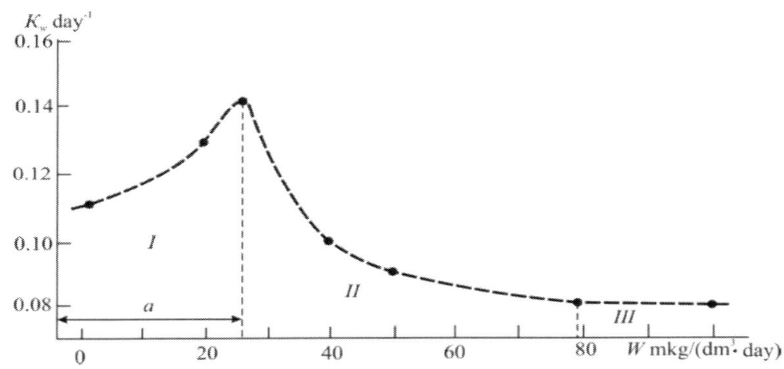

a—diapason of permissible loads on the basis of data provided by a two-week experiment.

Figure 11.15. Influence of outside load on the capacity of the system to self-purify water.

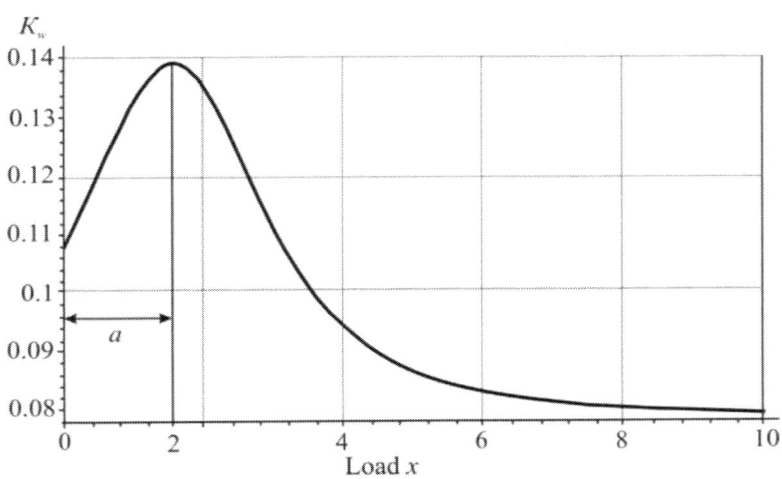

K_B and x are dimensionless

Figure 11.16.

In general, the dependencies of this type represent the function:

$$K(x) = \frac{\alpha+\beta x}{1+\gamma x^{\mu}} \quad (13)$$

where parameters $\alpha > 0, \beta \geq 0, \gamma > 0, \mu > 0$ are determined by shooting method.

From Figure 11.17, it is clear that within the whole diapason of the model loads, an increase of copper content in the water was observed. Even at a minimal load (mesocosm 1), the copper content in the water exceeds MPC. Results of mathematical processing of the diagram are specified in Figure 11.18. The concentration of copper by the end of the experiment is approximated by the linear function:

$$C(x) = 2.5x \quad (14)$$

with an approximation error $\Delta = 0\%$.

The loads diapason is marked out (zone *a* in Figure 11.15), the exceeding of which for a specific water body will lead to the copper content in the water exceeding MPC.

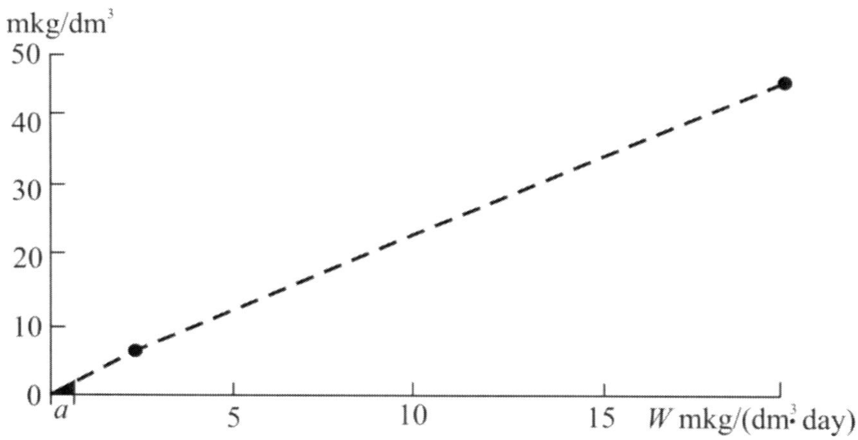

Figure 11.17. Influence of constant load *W* on copper concentration in the water by the end of experiment.

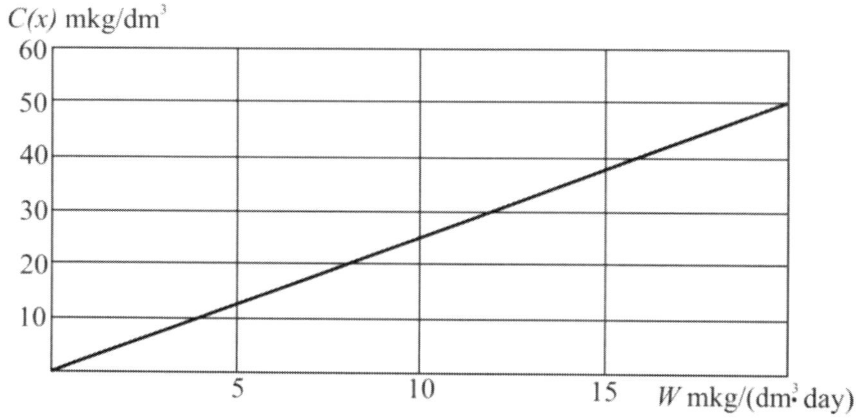

Figure 11.18.

Taking into account the evaluations obtained during the experiment, it is possible to determine certain maximal permissible (critical) levels of load, *W*, with the formula:

$$W_{cr} = K_{в}C_{cr} \tag{15}$$

where C_{cr} is the MPC of contaminant in the water.

In taking the maximal value of $K_{в} = 0.14$ at MPC for copper in the water 1 mkg/dm^3, we will obtain: $W = 0.14$ mkg/(dm^3 · day). Therefore, the copper incoming to the unit of the system volume at a specific regime of load shall not exceed this value.

Segment *a* in Figure 11.15 may be considered a *useful resource* of self-purifying ability, corresponding to zone I. It shall be noted that values of kinetic self-purification parameters were obtained at the conditions of the two-week experiment. In long-terms of the experiment, these parameters may slightly differ. The issue about how the intensity of the self-purification processes varies depending on the duration of impact at the constant load is not studied sufficiently.

On Figure 11.19*a*, the results of influence on the river ecosystem provided by biogenic substances (nitrogen and phosphor) in the regime of constant load are specified. The diagrams were built on the results of the experiments in five mesocosms, one of which was reference. Different amounts of complex mineral fertilizer were added on a daily basis in four mesocosms (volume of water was about 4.5 m^3).

The experiment lasted for 15 days. Figure 11.19*a* provides boundaries marking olygothrophic, mesotrophic and eutotrophic levels on the basis of data from paper [12]. At load *W*, equal to 2 mg/(dm^3· day) of phosphorus and 2.2 mg/(dm^3· day) of nitrogen, the system is in an olygotrophic zone by the end of experiment (see Figure 11.19*a*). Furthermore, with an increase of the load by 2, 3 and 4 times, the system gradually passes to the mesotrophic zone and in terms of phytoplankton biomass—to the eutotrophic.

Processing the diagram specified on Figure 11.19*a*, on basis of data of different plankton type biomass measurements and Table. 11.6, with the usage of the least-square method, made it possible to obtain approximate analytical expressions of the functions (Figure 11.20): 1 and 2—$u(W)$ and $v(W)$ (linear functions) and 3—$z(x)$ (parabolic relation), which have the following form:

$$u(W) = 0.20 + 1.11W \tag{16}$$

$$v(W) = 0.26 + 0.64W \tag{17}$$

$$z(W) = 2.49 - 0.97W + 0.13W^2 \tag{18}$$

The results of the mathematical processing of the diagram provided on Figure 11.19*a* are specified on Figure 11.20*a*.

In terms of bioproductional characteristics (see Figure 11.19*b*) at small loads, the increase of gross production of photosynthesis is observed (P_{gr}), which may be considered a subsidium effect [1]. Furthermore, in case of *W* increase, the value of P_{gr} decreases. Dispersion is evaluated by three of the latest parameters' measurements at the end of the experiment.

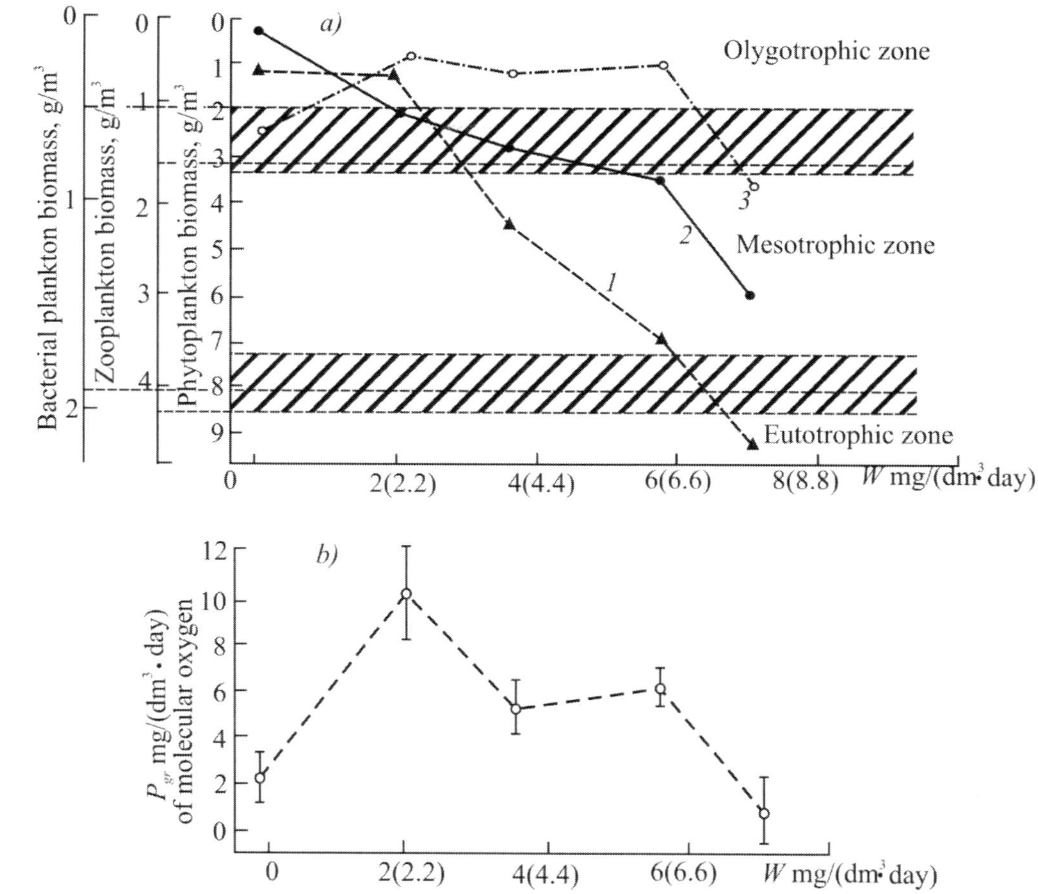

1 — phytoplankton; 2 — zooplankton; 3 — bacterial plankton.

Figure 11.19. Influence of constant load on conditions of an ecosystem evaluated by hydrobiological indices.

Table 11.6.

No.	Biomass, g/m³			W mg/(dm³· day)
	bacterial plankton	zooplankton	phytoplankton	
1	1.2	0.2	2.5	0
2	1.3	2.0	0.9	2
3	4.3	2.7	1.4	4
4	6.9	3.3	1.0	6
5	9.5	6.0	3.6	8

After mathematical processing of the diagram on Figure 11.19*b*, it takes the form of Figure 11.20*b*, and the functional approximation (by shooting method), on basis of its processing, is expressed by the equation:

$$P(x) = 2 + 17.117x - 8.792x^2 + 1.54x^3 - 0.090x^4 \quad (19)$$

Approximation error, $\Delta = 13\%$, is permissible since the confidence interval, amounting to 30% of the function $P(x)$, is included in the confidence interval of the measurements.

The general view of relation (19) has forms the multinomial:

$$P(x) = \sum_{k=0}^{n} a_k x^k \tag{20}$$

in which the degree of multinomial n and its factor a_k are determined in accordance with the experiment data obtained by different methods (least square method, Lagrange method, shooting method, etc.).

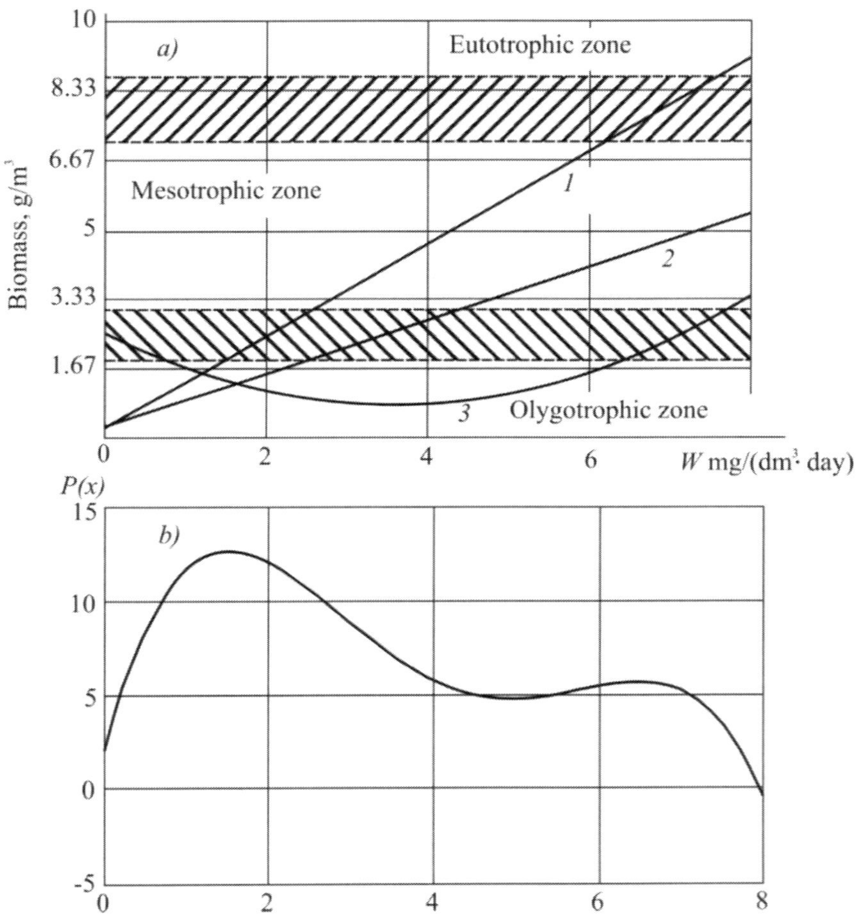

Figure 11.20.

Results of nitrogen and phosphor impact in the experiment may be evaluated by the dependence of its content at the end of the experiment on the load or by the capability of water to self-purify as well.

In general, the scheme of MPEL determination during the experiments on model ecosystems on the basis of the ecological-hygienic approach is shown on Figure 11.21.

After mathematical processing, this diagram takes the form represented on Figure 11.22.

The equation of this function is as follows:

$$f(W) = 50arctg\frac{0.01W^2}{1+0.01W^3} \quad (21)$$

or in general view:

$$f(W) = Barctg\frac{aW^{\mu}}{1+bW^{v}} \quad (22)$$

where B is the height of the curve, a is the slope of the curve horizontal, μ is the initial curvature of the curve and v reflects the curvature of the curve after passing its maximum value.

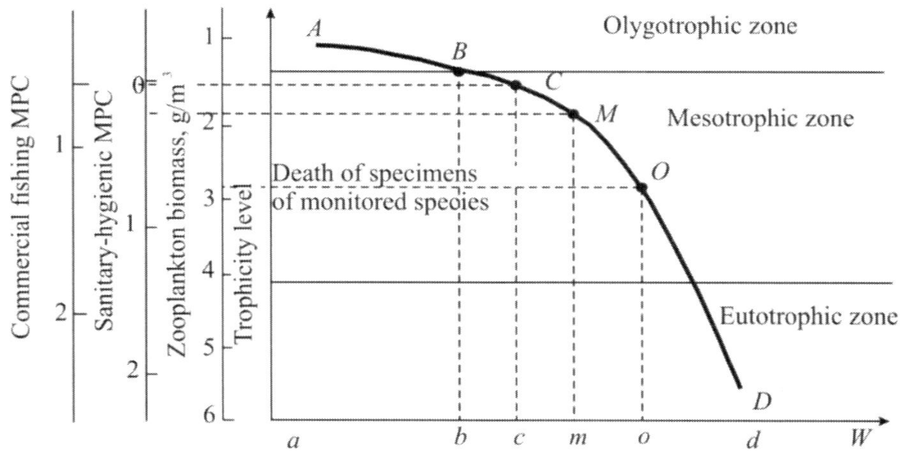

Figure 11.21. Determination of MPEL on the basis of experiments on model ecosystems.

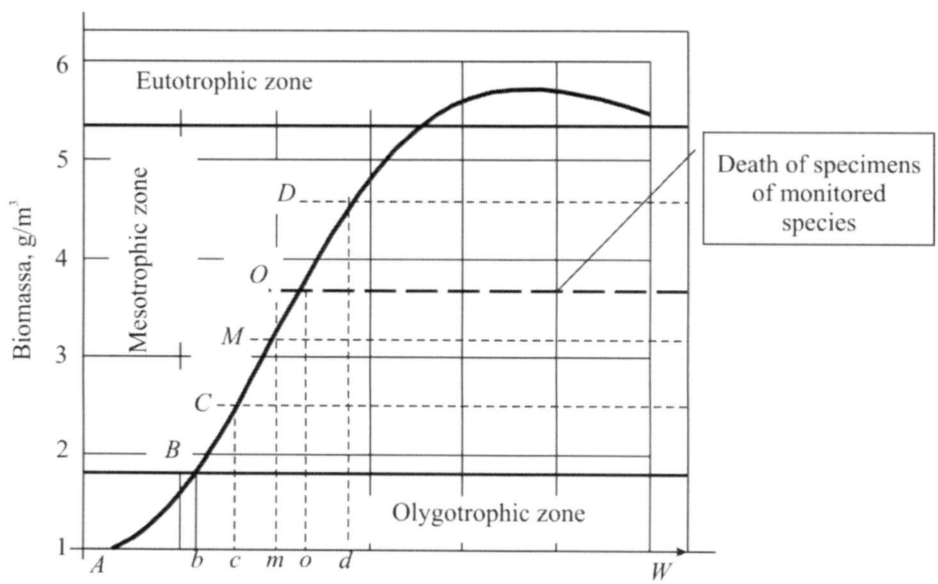

Figure 11.22.

Let us suppose that the contaminant from nearby farmlands flows into a certain water body, for example, into a small lake at the flood plain of the Don River, in which the experiments in mesocosms were conducted. Ecological standards are always stricter than commercial fishing or sanitary-hygienic ones [37, 38]. Curve *AD* on Figure 11.21 shows the dependence of ecosystem conditions on load *W* in any of the selected systems of evaluation. Depending on the task, different values may be protracted along the y-axis, for example, MPC, trophicity level or any index of water quality (i.e. the y-axis may reflect conditions of the ecosystem in selected systems of evaluations).

If a requirement is specified to the water basin, conditioned by its commercial fishing significance, then the commercial fishing MPC will become criteria of quality. On the basis of the results of the experiment with a certain contaminant at a specified load regime, it may turn out that segment *AC* on the curve *AD* corresponds to the condition of commercial fishing MPC non-exceedance. In this case, the load shall not exceed the values marked by segment *ac* on the x-axis.

In terms of hydrobiological indices, for example, on the trophicity scale, the load diapasons at which the water basins remains in olygotrophic state, are singled out. Therefore, in case its transition to the mesotrophic state is impossible, the maximal load shall be less than that of commercial fishing MPC (segment *ab*).

Definite requirements to water bodies may also be specified from the point of view of its biological integrity preservation, for example, the nonadmittance of the death of specimens of certain kinds of a community. Herewith, the load on the water basin shall not exceed the value on segment *AO* on the x-axis. The range of other requirements may be specified to the water body, for example, to establish maximal levels of contaminants content in plants and mollusks.

Thus, *controlling during the experiment the range of selected chemical and biological parameters notion of "conditions of ecosystem" may be filled with a specific sense in different evaluation systems.*

The selection of the load regime depends on the conditions of contaminants incoming into water body, i.e., on conditions that are modeled in the experiment. The regime of the contaminants incoming reflects two antagonistic mechanisms of stability maintenance in the ecosystem [1]. *The scheme of the experiment with constant load W has priority for the solution of standardization tasks.* During implementation of such a scheme, load regimes not leading to the contaminant MPC exceedance in water and worsening of the ecosystem conditions in terms of selected criteria, may be selected. Herewith, the reliability of these evaluations at set terms of the experiment (for example, with the term of 1 month) shall be taken into account. The degree of accuracy of these evaluations, obtained by means of model ecosystems, is related to the issue of evaluation of the model and its object similarity, which is examined in Chapter 9.

Realization of the experiment by the scheme of the *one-time load W* is to a greater extent related to the *forecasting of the contaminants' impact consequences,* since after the calculation of dissolution, the content of these substances shall not exceed MPC. For example, in the case of wastewaters containing phenol discharge to the Seversky Donets River in the area of Rubezhnoe, its concentration in the water after quite thorough mixing amounted to about 30 MPC. After the calculation of dissolution, it is clear whether the discharge of any amount of the substance into the water body is permissible or not. Here, the model experiments may serve the purposes of forecasting possible consequences. On the results of these experiments, it is possible to forecast changes in the water body state after 10 days, 20 days, a month, etc. It is also possible to determine the contaminants content in the water, bottom sediments and hydrobionts after specified periods of time.

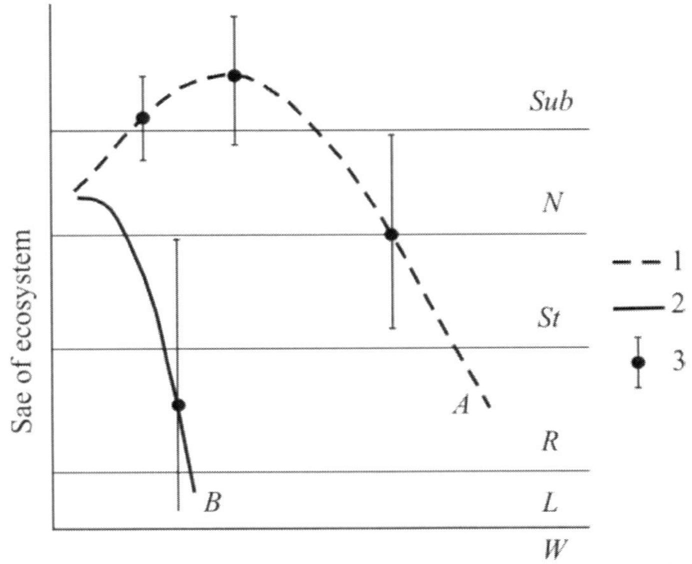

Sub—increase of productivity, *N*—diapason of normal functioning, *St*—suppression of the functions (stress effect), R—structural reconstructions (replacement), *L*—death, *1*—useful incomings, *2*—toxic incomings, *3*—relative dispersion

Figure 11.23. Curves of subsidium and stress.

In terms of self-purification processes, the intensity and consequences for the ecosystem is directly influenced by the type of the contaminants. Different types thereof cause different ecological effects. There are different classifications of contaminants from the point of view of hydrobionts' influence on the process of decomposition thereof. There are two main groups of these contaminants [37]:

1) Easily or hardly decomposable organic compounds
2) Organic compounds adsorbed by the organisms without decomposition and non-decomposable mineral compounds of different toxicity

Substances of the first group may be decomposed in the process of the water organisms' living activity and may be used by them. Incoming thereof may serve as subsidium for the system. The diametrically opposed effect on hydrobiocoenoses is caused by the substances of the second group, which may not be included into biochemical cycles and are toxic for the ecosystem.

Figure 11.23 specifies the concept of the load gradient from subsidium to stress for the substances of the first and the second group proposed by Y. Odum [1]. If the disturbance on the input is harmful, the reaction of the system shall be negative. In the case of toxic substances influx (curve B), conditions of the ecosystem[23] are gradually worsened and the

[23] This term is not definitely determined in the paper of Y. Odum.

probability of the community replacement by another one is quite high or the system may die off entirely. If useful substances or energy come to the input of the system (curve *A*), then the productivity may increase. Y. Odum called this system "subsidized." It is clear from curve *A* that if the system uses these incomings, its conditions (basic functions, for example, productivity) may improve—the subsidium effect on Figure 11.21 in the experiment with nitrogen and phosphorus. Upon further increase of the incomings, these functions may come to be suppressed.

Determination of MPEL on water bodies requires the development of the criteria of permissible biological structure changes. Maximal incoming is determined by the amount of contaminant that comes into the water body per unit of time without damage to its residents [37]. In some papers, the term "biological integrity" is used in this context, under which the "preservation of necessary aboriginal biota vary at all levels of ecological structure hierarchy," The most important task in the solution of the ecological standardization problem is the search for a one-dimensional description of the ecosystem state. Herewith, the issue of determination of this one-dimension scale or integral parameter of the state remains open. Experiments on model ecosystems make it possible to pass to quantitative research of the processes in the ecosystems and provide, in prospective, the material for solution to the problem.

REFERENCES

[1] Odum Y. Ecology basis. — M.: *Mir,* 1986. — 556 p.

[2] Tushinsky S.P., Shinkaz G.P. Environmental waters pollution and protection // *Environment conservancy and water resources reproduction.* — 1982. —No 12. — 199 p.

[3] Kaplin V.G. Evolution of organic matter in environmental waters. — Irkutsk: Irkutsk uni. pub., 1973. — 46 p.

[4] Nikanorov A.M., Tarasov M.G., Trunov N.M., Klimenko O.A. and others. The problem of surface waters quality and full-scale ecological modeling // *Ecological standardization and modeling of the anthropogenic impacts on aquatic ecosystems.* — 1988. — No. 1. — P. 5—9.

[5] Nikanorov A.M., Trunov N.M. Full-scale modeling as the most important element of the ecological modeling of the anthropogenic impacts on aquatic ecosystems // Integrated global monitoring of the environment biosphere state. — M.: *Hydrometeoizdat,* 1985. – P. 38—39.

[6] Nikanorov A.M., Trunov N.M. and others. Full-scale modeling of the main process of the polluting substance and aquatic ecosystems interaction // *Ecological standardization and modeling of the anthropogenic impacts on the aquatic ecosystems.* — 1988. — No. 1. — P. 9—17.

[7] Ivanova A.A., Kaplin V.G., Goncharova T.O. Transformation of the metal compounding in the environmental waters // *Transactions of the IV All-Union hydrological congress.* V. 9. — L., 1976. – P. 44—53.

[8] Nikanorov A.M., Pantykhin Y.V., Zavolzhensky M.V. Some methodological issues of the ecological regulation // State and prospects of the methodological basis

development of the surface water chemical and biological monitoring // *Thesis of the XXIX All-Union hydrochemical cong.,* Rostov-na-Donu. 28—29 oct. 1987, v. 2. – P. 127—129.

[9] Oparina O.P. Sanitary protection of the waters reservoirs from industrial wastewaters pollution. — M., 1949. — 201 p.

[10] Steets W.D. Toxicity studies of metal-fimshed // *Sew. and Ind. Wastes.* — 1957. — P. 21—23.

[11] Zenin A.A., Sergeeva O.V., Zemchenko G.N. The coefficients of the transformation (disintegration) of polluting substance in water // *VNIIGMI overview* — IDC. — 1977. — No. 1. — 43 p.

[12] Nikanorov A.M., Lapina I.A., Gekov V.F. and others. Calculation of the buffering capacity of freshwater ecosystems to heavy metals // *Ecological standardization and modeling of the anthropogenic impacts on the aquatic ecosystems.* — 1988. — No. 1. — P. 70—78.

[13] Svirzhev Y.M., Logofet D.O. Sustainability of the biocommunity. — M.: *Sience,* 1978. — 352 p.

[14] Bryzgalo V.A., Fedorova L.C, Horuzhaya T.A. and others. The usage of the biota functional state data in the surface waters of bioindication // *Problems of aquatic toxicology, biological testing and water quality management.* — L., 1986. – P. 107—118.

[15] Nikanorov A.M., Zhulidov A.V., Lapin I.A. and others. Kinetic characteristics of the processes of metals accumulateon by biotic and abiotic components of freshwater ecosystems // *USSR DAN.* — 1989. — V. 305, No 5. — P. 1274—1276.

[16] Edigarova I.A., Krasyukov V.N., Lapin I.A. and others. Complexing ability of the dissolved organic matter of the environmental waters // *Water resources.* — 1989. — No 4. — P. 122—129.

[17] Linnik P.N., Nabivanets B.I. Forms of metal migration in the fresh surface waters. — L.: *Hydrometeoizdat,* 1986. — 272 p.

[18] Gjessing E.T. The effect of aquatic humus on the biological availability of cadmium // *Arch. Hydrobiol.* — 1981. — V. 91, N 2. — P. 144—149.

[19] Hodson P.V., Borgmann U., Shear H. Toxicity of copper to aquatic biota. *Copper in the environment. Part 2. Health effects.* — N. Y., 1979. – P. 307—372.

[20] Leckie J.O., Davis J.A. Aqueous environmental chemistry of copper // Copper in environment. Part 1. Ecological cycling. — N. Y., 1979. – P. 89— 211.

[21] Pagenkopf G.K. Zink speciation and toxicity to fish // *Zinc in the environment. Part 2. Health effects.* — N. Y., 1980. – P. 271—277.

[22] Umland F., Yansen A., Tirich D. and others. Complex compounds in analytical chemistry. — M.: *Mir,* 1975. — 536 p.

[23] Vernadsky V.I. Chemical composition of the Earth biosphere and its environment. — M.: *Sience,* 1987. — 339 p.

[24] Vinogradov A.K. Ecotesting and ecological forecast of the water pollution consequences // *Experimental aquatic toxicology.* — 1984. — No. 9. — P. 68—77.

[25] Krasyukov V.N., Lapin I.A. The usage of the exclusion chromatography method for the study of humic substances in the solution // *Soil sience.* — 1990. — No 7. — P. 48—57.

[26] Nikanorov A.M., Lapin I.A., Krasyukov B.N., Sedush B. P. The study of the plumbum, cadmium and mercury allocation and migration in the freshwater ecosystems // *Ecological standardization and modeling of the anthropogenic impacts on the aquatic ecosystems.* — 1999. — No. 2. — P. 222—239.

[27] Manual for the hydrobiological methods of surface water and sediment analysis / Ed. B.A. Abakumov. — L.: *Hydrometeoizdat,* 1983. — 238 p.

[28] Manual for chemical analysis of the ground surface waters. — L.: *Hydrometeoizdat,* 1977. — 365 p.

[29] Kimstach V.A., Lapin I. A., Tchudinov E.G. et al. Application of a complex of physico-chemical methods to study the forms of heavy metal occurrence in natural waters // *Inter Symp. «New Sensors and Methods for Environ. Character (SMEC)»,* Nov. 10—12, 1986, Kyoto, Jap. – P. 3—10.

[30] Kocharian A., Malutin A., Lapin I.A. et al. Study of humic organic substances and heavy metals in the Ivankovo reservoir waters // *Toxicolog. Environ. Chem.* — 1989. — V. 24. — P. 83—93.

[31] Varshal G.M., Kosheeva I.Y., Sirotkina I.C. and others. The study of organic matter of surface waters and their interaction with metal ions due to migration of polluting substances in the environmental objects // *Organic and prospecting geochemistry.* — M.: *Sience,* 1982. – P. 202—212.

[32] Komarovsky F.Y., Polischuk L.R. Mercury and heavy metals in the aquatic environment: migration, accumulation and toxicity to the aquatic organisms // *Hydrobiological journal.* — 1981. — V. 17. — P. 71—83.

[33] Prokofyev A.K. Chemical formulas of mercury, cadmium and zinc in the natural aquatic environments // *Chemistry achievements.* — 1981. — V. 50, no. 1. — P. 54—84.

[34] Zelukova Y.V., Bezlutskaya I.V., Novoselova M.M. The appliance of the "cold steam" method for determination of mercury in the environmental studies // XIV Mendeleev Congress on General and Applied Chemistry. Abstracts, overviews and reports. V. 2. — M., 1989. – P. 463—466.

[35] Nikanorov A.M., Teplyakov Y.V. Evaluation of the maximum load on freshwater ecosystems by the full-scale modeling methods // *Ecological standartization and modeling of the anthropogenic impacts on aquatic ecosystems.* — 1999. — No. 2. — P. 212—221.

[36] Hydrochemistry reference book / Ed. A.M. Nikanorov. — L.: *Hydrometeoizdat,* 1989. — 392 p.

[37] Kamshilov M.M. Ecological aspects of water objects pollution and essential ways to eliminate it // *Hydrobiological journal.* — 1979. — No 1. — P. 56—59.

[38] Krasovsky G.N., Elpiner L.I., Beim A.M. and others. Principles of the ecological and hygienic regulation of the water quality on the water bodies // *Water resources.* — 1982. — No 2. — P. 3—19.

Chapter 12

COMPLEX ASSESSMENTS OF INLAND SURFACE WATER QUALITY MONITORING

12.1. COMPLEX WATER QUALITY INDICES IN HYDROCHEMISTRY

One of the most important problems, which is investigated by lots of scientists, is the development of surface water quality monitoring methods using conditional parameters and taking into account different features. The author of this scientific paper and V.P. Emelyanova show the results of long-term investigations into this problem in this chapter. This complex assessment is difficult because of the necessity to take into account at the same time different features of water bodies, their formation conditions, consumption types, variety, etc. The difficulty of this work nowadays, as well as several decades ago, is that no lawful document, accepted by the scientific community or ratified by law and describing the particular surface water quality assessment method, recommended for hydrochemical monitoring systems, has appeared during the last fifty years.

All the elements, typical for any cognitive activity, e.g., problem definition, objects determination, definition of methods and procedures for solving the problem, are also specific for the process of surface water quality and its pollution level assessment as well as for the process of assessment generally. The most widespread problem in the sphere of water pollution level and quality assessment is the detection of any types of water use grade. The *objects* here are the ecosystem of the water body, the chemical constitution of the water, and man, his health, human environment, productive activity, etc.; the *methods* and *procedures* are the means and techniques of assessment, forming the assessment methods. The *resultant* is the pollution level resulting from the assessments, summary and recommendations.

Such terms as “water quality assessment” and “pollution level assessment” result from the definition of terms such as “assessment,” “water quality” and “water pollution,” which are explained in the references. According to this fact, we will define the phrase “surface water quality assessment” as “the surface water quality meeting water users’ requirements, expressed by any possible metrics.” The phrase “surface water pollution level assessment” will be defined as the identification by any possible metrics of surface water composition and properties and their difference from criterion value, showing their suitability to be used [1].

The surface water quality assessment research basis is formed by two groups of hydrochemical indices: natural and specific estimate indicators, which are interrelated and conjugated (Figure 12.1).

Natural indices are the ingredients or surface water chemical composition parameters, which are defined analytically. They can be of three types: differentiated, compound and integrated. *Differentiated* indices are water quality components, characterizing only one feature of water (e.g., sulfate of other ions concentration). Surface water quality assessment based on the separate ingredients' analysis is the most widespread and generally accepted assessment type nowadays. The aim of this very assessment is to identify quality changes by comparing absolute values of separate chemical ingredient concentrations with the standard value.

As a rule, *compound natural* indices unite information about lots of (groups of) substances (e.g., phenols, oil products, chemical oxygen demand, etc.). The *integrated natural* index is a water quality parameter characterizing a group of features of the same kind and specified by a complex of physical-chemical, and sometimes even biological, processes (e.g., pH, stream-dissolved oxygen, etc.). Human origin pollutants that cannot be found in nature itself (e.g., compound organic matters) and identified by analytical methods are also included in this group.

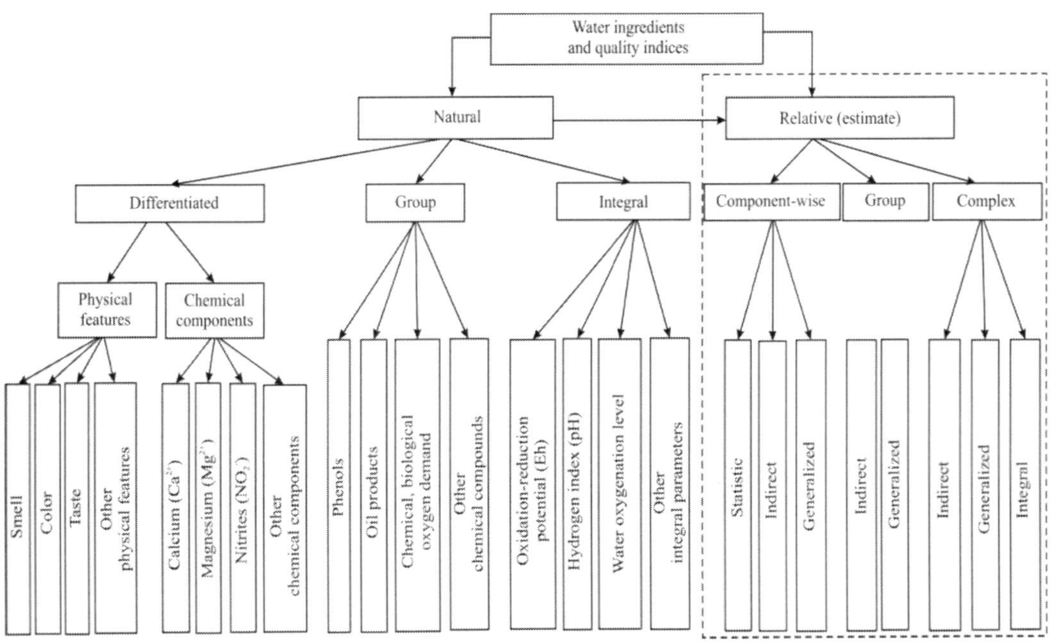

Figure 12.1. Water quality parameters hierarchy in hydrochemical research.

The group of natural indices is used in hydrochemistry for differentiated water quality assessment by comparing these indices with the standard values. Besides their fundamental function, which is surface water quality and pollution level element-by-element assessment, they are the basis for the estimated parameters definition.

Specific estimate indicators of surface water quality or pollution level are calculated through the analytical method by means of analytical data transformation, i.e., these indicators are formalized. As a rule, they show information different from the original; they complete and upgrade it. A big range of different estimate indicators can be worked out, because this quantity depends only on the physical factors, formalization degree, form of expression, grade, etc., of these indicators. In this complex hierarchy of chemical composition and surface water features, the relative measure of water pollution or quality comes laden with elements of assessment or fully clustered with a rating class. The bulk of relative measures are somewhat estimated. This clearly distinguishes the estimated hydrochemical indices from the measured ones. Natural characteristics of water structure and its features, even the most detailed and correct, but not correlated with a person's heath, his needs and activity, can be neither useful nor harmful. One and the same water condition, characterized by the same natural indices features, can be appraised differently depending on whom it is assessed for and with what purpose it was assessed. During all evaluative research, the assessment of something is a ratio set for it. This ratio set always expresses a subjective attitude. This is one of the most important theses of the estimation theory [2]. It is impossible to assess water body pollution level or quality without reference to someone or something. One and the same water object, its features or conditions of assessment can be multi-valued, but its estimation is always single-valued. Therefore, we can say that assessments are always relative and results of estimations are absolute, as well as their indices.

The group of nominal (relative) indices is commonly used in hydrochemistry, but their introduction is limited by special spheres of use. Being a special type of hydrochemical information, obtained by analytical data processing, relative water quality indices fundamentally change and upgrade basic data by adding additional information.

Taking into account the character of hydrochemical information, shown by the relative indices, we can subdivide them into three types: component-wise, group and complex ones. Taking into account the level of generalization and formalization, we can subdivide the relative indices into four sub-types: statistical, indirect, generalized and integral. According to the forms of expression, they are subdivided into coefficients, indices, codes and classifications (as well as types, levels, categories, etc.).

Component-wise relative indices show water pollution by analyzing its separate components. It makes it possible to assess water quality in the form of a set of its characteristics. Differentiated indices do not give single-value estimation and can refer to one and the same water body pollution level as different classes. This type of relative indices is traditional and has a low level of generalization (first level, less frequently, a second one). Such indices are subdivided into three sub-types in their turn: *statistic, indirect,* and *generalized.* Such statistic characteristics as arithmetic averages, mean weighted values, error of mean, dispersion, coefficient of skewness, median, etc., are the most widely used ones. Later, however, such indirect measures as, e.g., maximum permissible concentration exceedance, frequency of water pollution cases, etc., appeared and now they are also widespread. These indices are easily accessible and have specific content, so they augment and enlarge the informational capacity of the initial analytical data.

Generalized relative indices came into general use in hydrochemical research much later. They express surface water pollution level through conditional, numerical indices, calculated through the estimation method and relatively show any given water body's condition. As a rule, these are the indices of the first and second formalization levels. The subtype of the generalized relative indices is typical for each basic type and can be found in component-wise, group and complex assessments as well. For example, relative and maximum permissible flow filling with pollutants and hydrological-hydrodynamical indices of river flows and water bodies pollution level are used for generalized component-wise surface water pollution level assessment with low relevancy level by mixture of pollutants.

Relative group characteristics give an opportunity to assess separate groups of homogenous chemicals. They can be developed and used for chemicals combined by the following characteristics, e.g., minerals, playing a full role in aquatic organisms activity (biogenous—mineral forms of nitrogen, phosphorus, silicon and ferrum); dissolved gases (oxygen, nitrogen, carbon dioxide, hydrogen sulfide, etc.); microelements (chemicals and their compounds, which appear in natural waters only in low concentration of several micrograms and not more than 1 dm^3); cations (Li^+, Rb^+, Cs^+, etc.); anions (Br^-, I^-, F^-, B^{3-}); heavy metals ions (Cu^{2+}, Ag^+, Pb^+, Co^{2+}, Fe^{2+}, Ni^+, etc.); amphoteric complexing substances (Cr, Mo, V, Mn); principle ions (Cl^-, SO^{2-}, HCO_3^-, CO_3^{2-}, Ca^{2+}, Mg^{2+}, Na^+, K^+); radioactive elements, etc. Such indices give sector-specific water quality assessment. Relative group indices can be of two sub-types: indirect and generalized. For example, the modulus coefficient of biogenic elements washout can be an indirect characteristic of nitrogen mineral forms content in water, and indices of ion sink microelements can play the role of generalized assessments.

Depending on the specific target, water quality can be assessed according to its several components not only according to the groups of homogenous substances, but also according to their different complexes. If the relative index includes a great number of indices, including heterogeneous, and helps to assess water quality, it is called a "*complex index*" of water pollution level. Complex indices are the subtype of relative indices, allowing them to obtain single-valued qualitative and quantitative (numerative and verbal) characteristics and surface water pollution level assessment on the basis of taking into account the list of different components and their chemical composition at the same time. All types of surface water quality or pollution level complex indices are evaluative ones.

Complex assessment of surface waters pollution level (quality) translates into sequential actions that are focused on getting the idea of water pollution level or water quality, on the basis of chosen water composition or features combination, in the interests of special types of water need or water use. The new relative assessment type of complex indices occurring in hydrochemistry was determined by the necessity to plainly define water pollution character and its levels because of the growing quantity of anthropogenic water bodies, pollutants flowing into water or appearing in it and by necessity, to range water in accordance to this feature.

As a rule, all complex indices are highly informative. Natural hydrochemical characteristics or relative ones, but with much lower formalization levels, are the basis for obtaining these indices. Much attention is devoted to the problems of obtaining and using such integral assessments in the sphere of ecological research in Russia and abroad. Integral indices are the surface water quality and pollution level assessment types, amending original information of a wide range of informal hydrochemical indices, including heterogeneous ones, with the aim of obtaining single-valued assessment. As a rule, these indices are of a high formalization level. They aggregate and integrate information in accordance with the bulk of hydrochemical characteristics—natural, and less frequently, relative (of the first or second formalization level), and in some cases, of both of these types.

Compared to other types of assessments, the integral ones are much more informative and have a higher level of generalization. Integral assessments are more logical and practical for some types of hydrochemical information use. One of the basic ideas of the integral indices is bringing up the analytical data of different water chemicals and their physical aspects together to formulate them by one scalar value. The basic type of computation algorithm, used for such assessment research, is a hierarchic schematic diagram. This means that some relative indices of the first (lowest) level are united with the indices of the second level by some logistic antecedent or mathematical expression; the second level indices are united with the indices of the third level, etc. All the indices of the third and following levels are called "integral."

According to the form of expression. the complex assessments are subdivided into coefficients, indices and classifications. Water bodies' pollution level coefficients are complex assessments of the first or, rarely, the second generalization level.

The most informative ones are *pollution level* (or quality) *indices* of the surface water. The index is a relative numerical value, characterizing heterogeneous populations of surface water components and chemical compounds quantitatively and clearly. The index method of the surface water and their quality estimation is reduced to the reception of a number (index) on a complex of chosen parameters and operating by it for the decision of various applied problems. Various indices can be developed and used in hydrochemistry with the purpose of studying the quality of surface water. Thus, however, it is important to observe conformity of a method and the purpose of choosing an index for problems, which the user is faced with during his work. As a rule, indices are the formalized water pollution parameters, generalizing wider groups of natural indices with more variety and greater objectivity, than the coefficients considering the various sides of the estimated object. That is why they have a much more complex structure.

This type or complex assessments provides versatile and adequate water quality valuation. Complex assessments include A.S. Shine's water quality index, the sanitary water quality index, the combinatorial water pollution index, etc. The oldest and at the same time, most widely used in practice, are the water pollution level or quality classifications. Classification of surface water pollution (or quality) is a ranking of pollution or quality indices according to certain common features and their system formation. Sometimes the results of a water body's condition in natural research are used as basic signs for waters. In other cases, the concentration of components correlated with standards, or various formalized estimations, obtained by an analytical method, such as points, indices, etc., are used as these basic signs. Besides, classifications based on separate chemical components, on separate relative indices and on their sets also exist nowadays.

The classification procedure usually consists of comparing calculated values of the basic features (concentration, relative indices, etc.) with evaluative scale steps of quantitative characteristics corresponding to certain water quality or pollution level classes. The choice of classification scale divisions number is very important, however, this choice is very often made voluntarily and is not based on some subjective or objective reason. In the majority of known classifications, water quality or its pollution level are subdivided into five to six classes, which allow ranges from pure or very pure to dirty or very dirty, etc. Interval scales precede all used scales. Limiting values of measurements results, and also values of relative characteristics, very often vary considerably, despite the duration and depth of a problem on the substantiation of an estimated gradation choice for investigation.

12.2. METHODOLOGICAL SCHEME OF COMPLEX ASSESSMENT DESIGN

It is well known that water bodies' water composition is a condition that varies greatly. Assessment and criteria applied for this assessment also have multiple characteristics. A variety of assessment process elements (which include problem, means and procedures, assessment result and their various combinations) defines difficult assessment research methods and techniques interweaving while it explains and justifies the existence of some various accuracy assessment methods. The total number of existing complex assessments and their estimation methods is great enough. Their methodological bases, however, should not differ fundamentally. They should be limited, exact and clearly defined.

Works carried out at the Hydrochemical Institute on analytical information generalization forms improvement allowed us to substantiate basic elements and stages of water quality assessment, to ascertain their hierarchy and to define the methodological scheme of surface water quality and pollution level complex assessments (Figure 12.2). We will briefly characterize each of the specified scheme's structural parts.

Definition of the complex assessment problem. The clearly defined problem of an assessment and basic assumptions are necessary in any estimative research. In this regard, the initial stage of works on method investigation should be accompanied by the detailed problem defining its substantiation.

Specifying assessment problems, the researcher defines the purpose of an assessment, assessment consumer, prominent and unique features of the assessed body, assumptions accepted in calculations, conditions of the assessments use, etc. In the case of the general problem defining (to assess water bodies' water quality), various methods of an assessment should improve and specify it by separate, private statements.

There is lots of surface water and their pollution level assessment aims. They can include the all-round analysis of water body pollution level to define its ecological condition; a rough estimation of its quality would define the suitability of water for supplying it to population, for different kinds of industry branches, for irrigation and recreation; reveal high-priority water bodies, administrative areas and regions demanding additional financing for water security works carrying out and would look at surface waters sources of pollution identification, etc. The types of specific water components (chemical, hydrobiological, bacteriological, etc.),that will be taken as the basis of an assessment, and also the number and the nomenclature of signs by which each kind of property, their criteria, meanings, etc., will be assessed, depends on the aim of the assessment.

Taking into account the consumer's requirements plays a great role in the assessment of the problem's definition. There are a great number of surface water quality assessment consumers. Both physical and legal bodies of various ranks (private persons, companies, enterprises, administrative bodies of different levels, research establishments and governmental bodies) can be these consumers. Each of them needs assessments of different accuracy, degree of formalization or specification, volume of financial expenses for assessment, etc.

Assessment problems specification also includes taking into account the specific features of the assessed water body, e.g., its hydrological characteristics, hydrometeorological and other associating environmental conditions and water formation factors, data on volumes and character of anthropogenous influence, etc.

Information on the complexity of the structure and properties of surface waters to be assessed is not full data about the processes occurring in water layers and mechanisms of the pollutants' combined influence on human bodies and biota, because chronic and sharp action of pollutants are not enough for assessment, so these and many other factors force researchers to accept certain assumptions during the assessment process, which should be clearly defined at a stage of a complex assessment problem definition.

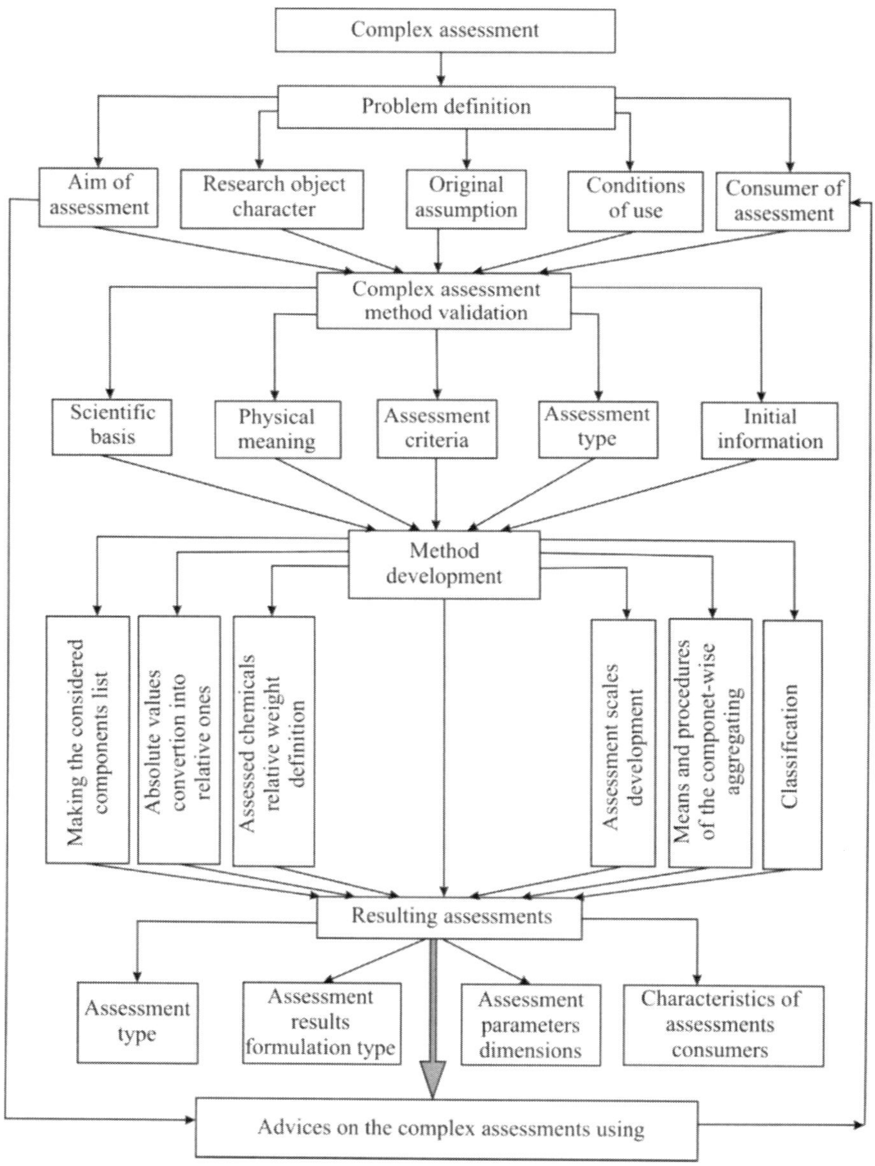

Figure 12.2. Methodology of inland surface water pollution level (quality) complex assessments development.

Except for the aforesaid, it is necessary to define conditions of a specific kind of an assessment for at this stage. It is also necessary to consider the area of the prospective method and its practical use and qualification by experts who will use it to define the information required for calculations; to find out whether the volume and quality of the initial data are sufficient, whether some special means are necessary for calculations or can be made manually; to define the computation algorithm degree of complexity, etc. Absence of accuracy in the definition of a sphere and conditions of the developed complex parameter's prospective application or assessment system can lead to errors at subsequent stages of the methods development and during its usage.

Validation of a complex assessment *method* is the next obligatory and rather important stage. The essential basis of an assessment are those key ideas that make the scientific base of a method under development predetermine its physical sense and type of the assessment result indices. Choosing the assessment's basic criteria and defining requirements for the initial data, according to which calculations will be carried out later, can also be referred to as the basic method provisions. In various assessment methods principles, their initial data can differ greatly, and despite the great variety of water bodies, they can be polycomponent, while having a considerable quantity of possible external influences and other factors, mentioned above. In our opinion, the following aspects are really important: the degree of surface waters quality abnormalities in comparison to a basic condition (background, standard, reference, previous information, etc.), complex character of surface waters pollution level, kind of pollutant hazard (toxicological, sanitary-toxicological, epidemiological, organoleptical, etc.), prevalence of this or that pollutant, water flow load by representative substances, etc. Original ideas and decisions can be offered and used on each of these and other aspects.

At any stage of consideration, water quality assessment depends on the accepted *basic parameters of the system*. The choice of assessment criteria is defined by the correlation of the object and the subject of an assessment in close connection with the purposes facing the developer and its principal method. The following criteria can be used for the assessment basis: criteria chosen on the basis of the natural data (background, extreme, characteristic values of pollutants concentration) and any standard values, including maximum permissible concentration, and also concentration values, averaged in the spatial relation (for example, regional), in the time relation (for example, long-term averages for any interval of time), relation according to the body (for example, for the water body as a whole, river basin, etc.). Assessment criteria can also differ in accordance with their being obligatory to execution, i.e., their being standard or not, their being connected with actual water usage types. It is possible to use the state and departmental specifications, literary data, scientific works and personal experience of authors and experts in these assessments.

Final assessment types also vary. They can be expressed as factors, sets of factors, codes, classifications and coefficients systems. Surface water pollution level assessment can also be expressed in the qualitative (verbal) form.

By working out assessment complex methods and their use, it is necessary to define the requirements for the initial data clearly. This basic initial data is as follows:

- Surface waters chemical composition testing (time frames, area and frequency of testing, water samples analysis and storage methods used in each case)
- Analytical data used can be obtained as a result of any official or special supervision made by scientific, industrial and other organizations, but an essential condition of

carrying out water sampling and its chemical analysis is a single methodological foundation, allowing for the exclusion of distinct influences of accuracy, sensitivity and other characteristics of the used data
- Nature of pollutants, e.g., fugitive phenols, ferrum, ferriferrum, etc., should be clearly defined
- List of representative pollutants and their quantity data should depend on the source of surface water pollution, aim of assessment and its meaning
- It is necessary to specify the limit of initial data needed as a concrete quantity.

If the specified requirements are not followed, the result of the assessment cannot be correct.

One of the most difficult stages of complex assessment design is a *method development*. The basis of any complex assessments methods development should be samples of data formalization ways, their assessment techniques from the point of view of production positions, the aggregation of the transformed data, means and procedures applied at all abovementioned stages and the kind of assessment obtained, its form of expression and grade. The use of these conditions, their choosing and implementation vary a lot.

There are, however, a number of basic, fundamental procedures that are part of an integral, constructive component of various surface water quality or pollution level assessment methods. First of all, following the basic, fundamental procedures can be mentioned:

- Obtaining a single-valued, integral estimate on the basis of the entire scope of components and water quality parameters assumes the process of great hydrochemical information quantity concentration and compression until the scalar value. The chemical composition components, on which water pollution level is complexly assessed, however, cannot be compared with each other for a number of reasons. It is necessary to convert absolute values of components into comparable relative values and express all the characteristics through the general relative values to solve this problem. The approaches used by various researchers for this purpose differ a lot. The most widespread method is one where converting separate samples data (concentration, absolute values) into dimensionless units or points is used. The basic idea is to substantiate the equality or conformity of points with separate components concentration and water quality changes. Converting concentration into excess multiplicity of their maximum permissible values, etc., is used in lots of methods.
- There can be some problems during the complex assessments development, because some components and features are contained in water in different quantities. This factor should definitely be considered. As a rule, only main or major factors that have the greatest impact on any water use type are selected for water quality assessment. Other factors, having less significant influence in the case under consideration, are not considered during the assessment. Thus, sampling is a primary selection of the parameters significance.

This components selection, however, is insufficient because the influence of the selected major factors of an assessment is unequal too. There is a need to carry out the procedure of repeated parameters selection directly in the course of the assessment, to find out the meaning, importance of various water components and parameters. It is obvious that there is a need to use relative values. The ways of substantiation and definition can be different.

There are some frequently used ways of relative values definition: expert, statistically disturbed, mixed and cost.

The following variants are usually used in hydrochemical practical works: the relative weight of each component considered during water components or chemical composition index assessment is accepted to be a unit; the relative weight of the most the important components increases, while the relative weight of minor parameters decreases by the determined quantity of times and the relative value definition is made by means of expert assessments method and correlation indices on the basis of authors of long-term experience in accordance with the danger degree depending on the fishery maximum permissible concentration on the basis of pollution danger curve tracing. Unfortunately, the accurate theoretical substantiation of some pollutants' relative values definition procedure is not developed yet.

It is possible, however, to agree that the closer relative values and their recognition should connect with the necessary assessment accuracy, and the more substantiated the manner of their definition, the more exact and responsive the method of assessment.

The *evaluating component* is a distinctive feature of water quality or pollution level complex assessments. There should always be measurement results or their relative quantitative characteristics converted into assessment categories during methods development. This is one of the most important moments of assessment research. Results of natural observations, personal experience, expert assessment method, etc., can be used for such conversion regularities setting.

Evaluative scales or, more often, interval scales creation is a widespread way. Linear or curvilinear dependence between measurements and assessment can be used during their creation.

As was mentioned above, to assess water quality or pollution level means to define compliance of water quality with water use requirements or surface water chemical parameters and features differences from their standard values. The degree of compliance or difference can be different: either small, or considerable, or intermediate, etc.

All ranges of absolute or relative values of considered priority parameters should be separated into groups or intervals of values corresponding to any waters quality or its pollution level. These intervals are later defined by certain numbers and notional (verbal) description. Numbers and their qualitative characteristics are showing differences between intervals in an interval evaluative scale. Rating scales thus can be equal and unequal. Scales with unequal intervals are quite often used because assessment changes are numerous even if some values change proportionally.

One of problems of water quality complex criteria definition, based on many partial components, is to choose the way of assessments aggregation. The aim of assessments aggregation is to solve the problem of initial surface waters quality hydrochemical information reduction, because it is very difficult to formalize this information. As a rule, mathematical apparatus techniques, substantiated by some concepts, are used for such aggregation.

Various authors suggest both difficult and simple mathematical techniques. Different types of assessments are used for information reduction and obtaining complex assessments.

Arithmetic, geometrical, harmonic, weighed or unweighed averages are much more often used in these cases. All these assessments, however, are rather rough.

Besides the averaging methods, linear or curvilinear functions are used for standard parameters reduction and for obtaining a single, generalized parameter as a result. On this stage weighing, penalizing, preference functions, etc., can help to solve some problems of surface water pollution level and should be used in addition to the abovementioned methods. Some of these problems are as follows: differences in values of some component concentrations, influence of some components' very low values, components with dangerously high levels of concentration, etc.

The results of complex assessment are certain overall estimates. Overall estimates *forms of expression* are usually as follows: qualitative (verbal), quantitative (numerical), expressed by a system of several numerical parameters logically united or expressed by a system of interconnected numerical and verbal estimates formed as a certain notional sequence. The verbal form of expression, for example, is as follows: good, bad, suitable, unsuitable, etc.; the numerical one is expressed in points, ranks, etc.

As a rule, numerical estimates are just a different way of expressing verbal estimates, but, at the same time, numerical estimates have some advantages. They allow the brief expression of the results, which help to quickly carry out comparisons. Zero dimensions of points give the chance to compare estimates of incommensurable parameters with different qualities.

According to their *types, assessments* can be subdivided into: coefficients; sets of coefficients; generalized parameters; sets of generalized parameters; codes (simple quantities or number sets or formulae); code groups; code systems; coefficient, parameter and code systems; classifications (classes, types, categories, digits, ranks, groups, subgroups); etc.

All the parameters of a complex assessment can have *dimension*, be defined by points, percents, natural units, etc.

Any complex assessment indices should have its own consumer properties. Informativeness, accuracy, assessments delicacy, their cost and economic gain (i.e. their profitability) are among these properties. Authors of surface water quality or pollution level complex assessments methods should support these assessments with the abovementioned data. There are no methods of complex assessments efficiency indication nowadays. It is necessary to work them out. Even nowadays, however, we can comfortably affirm that all complex assessments are highly informative regardless of the fact that information capacity of separate complex assessments can differ a lot. The informativeness of complex assessment indices depends on many factors, the main ones being: method development, character and volume of the information used for it and also principles taken as a basis for the assessment. Complex assessments informativeness directly depends on the list of pollutants taken into account during the assessment, on their informativeness, total quantity, number of characterized features (chemical, physical, bacteriological, etc.) and quantity of monitoring activities. Various methods of surface waters pollution level assessment yield the results of various accuracy. The methodical base of initial data gathering affects the *accuracy* of complex assessments. Assessments can be rough, meant for approximate results and specified, but the character of their specification can vary a lot. It is very important to check assessment aims to see if they correspond with the accuracy demanded. For example, while comparing any kind of water body with different pollution levels, complex assessments may have equivalent analytical data and it is necessary to understand the complex indices that they come from, allowing for much more accurate results (assessments) to be used for the nonpolluted areas. The *responsiveness* of surface water pollution level complex assessment parameters depends on their estimation method development, aggregation type of initial data, evaluative scale range, etc. For example, values from 1 to 5 or, more rarely, from 1 to 10, are used in methods developed in Russia and values from 0 to 100 are usually used in methods used in the USA, Great Britain, Spain, Canada and some other countries.

A considerable range on the evaluative scale gives an opportunity to fix more differences in a polluted water body condition, i.e., to increase an assessment's responsiveness. Complex assessment responsiveness is to be calculated in accordance with the following data: final complex assessment reaction for initial analytical data diversity, indices' susceptibility to sub-indices values changes, ability of an index to show even insignificant changes in surface waters quality, index responsiveness in case of one quality variable changes, etc.

From the point of view of economics, it is very important to count the price of the supposed assessments and the expected economic effect, i.e., cost of the effectiveness of the complex assessments. Cost of the complex assessments made on the basis of different methods can vary a lot, because it depends on the complexity and amount of necessary calculations, technical equipment used, time input, etc. Spheres of application, degree of flexibility, susceptibility characteristics, accuracy and reliability of complex indices can also vary a lot and this fact defines the variety within economic effects.

The final stage of the surface water quality and pollution level complex assessment is its adequate and profitable *use in practice*. Accuracy in the method selection, its correspondence with the challenge to consumer, ability to realize expediency of the method used on the basis of real possibilities (basic data and equipment availability, staff qualification, etc.) plays a great role during practical work with complex assessments. It is necessary to observe the conditions of the practical method in use, formulated on the stage of its development, in order to use complex assessments correctly. The user of the method should observe requirements of the initial data, provide correct calculations, correct understanding of the assessment result's aim and meaning and their correct explanation. At the end of the work, the results of the assessment should be compared with the assessment purpose in view, i.e., return coupling in Figure 12.1. Basic results of the surface water quality or pollution level complex assessment are water bodies condition analysis using calculated complex assessment indices with regard to the researcher's tasks, and also conclusions made on the basis of this analysis. It is advisable to combine a new comprehensive approach with the traditionally used ways of grade assessment based on pollutant priority during this analysis.

Complex indices can be used for a water body's condition assessment and some comparing of problems solution, tendencies, surface water (transits, area, whole water body, river or hydrographic basins, etc.) quality or pollution level studying. It is very comfortable to use complex indices to solve different problems of analysis, assessment, classification, mapping and surface waters pollution level prediction.

12.3. New Method of Water Pollution Level Complex Assessment

Taking into account principles stated in the sub-items 12.1 and 12.2, A.M. Nikanorov and V.P. Emelyanova, working at the Hydrochemical Institute, developed a new method of water pollution level complex assessment that is supposed to help in processing surveillance information from the Russian Federal Service for Hydrometeorology and Environmental Monitoring. In our opinion, it can also be used by other ministries and departments, solving problems of surface water quality control. The basic framework of the abovementioned method is the combination of the differentiated and, at the same time, complex approaches to a quality assessment based on the use of relative indices, allowing them to solve the set problem by taking into account water bodies and their hydrochemical mode features. This method includes a system of 18 assessment formalized indices allowing analysis and characterization of a water body's condition. It allows us to assess water pollution level by scalar value, taking into account a wide range of components and water quality parameters in order to classify water according to its pollution level and to prepare easily understandable analytical information to provide to state authorities and interested organizations. The method of complex parameters calculation helps to formalize the process of analysis, generalization and assessment of information on a water chemical composition and to convert it into relative indices that characterize a water body's quality and pollution level in an integrated manner. The functional feature of the method is the connection of a water chemical composition component-wise analysis with the following assessment components integration on the second stage along with water quality parameters recognition. The functioning method includes six main stages (blocks), presenting a sequential processing of the analytical material (Figure 12.3).

Block 1. Initial Data Accumulation

First of all, it is necessary to choose the interval of time for water assessment. The results of a water sample's chemical analysis are to be used as initial data and the essential condition of the analysis being effective is an integrated methods-methodology system of water sampling and its chemical analysis. The list of components and indices of water quality, according to which complex indices will be calculated, is to be made before the calculations begin. It is necessary to match all initial data in accordance with the information quantity for each component, quantity of pollutants under consideration, their list, initial data accuracy and expected results accuracy. Various components and quality factors can be used for the complex assessment indices of surface water pollution level definition; this variety depends on the aim of assessment, presence/absence of water chemical analysis data and sampling methods. For example, it is necessary to use components and quality factors typical for the majority of surface waters of the Russian Federation to finish the database for administrative bodies. Lists of 15 chemicals are usually used in surface waters quality Year-Books (official data provided by the Public Service for Environmental Monitoring). These chemicals lists include: dissolved oxygen, biochemical oxygen demand—5 day test, chemical oxygen demand, phenols, oil-based products, nitrite nitrogen, nitrate, ammonium nitrogen, total iron, copper, zinc, nickel, manganese, chlorides and sulfates. In some cases, some specific pollutants can be added to this list (synthetic surface active substance, pesticides, lignin, fluorides, etc.).

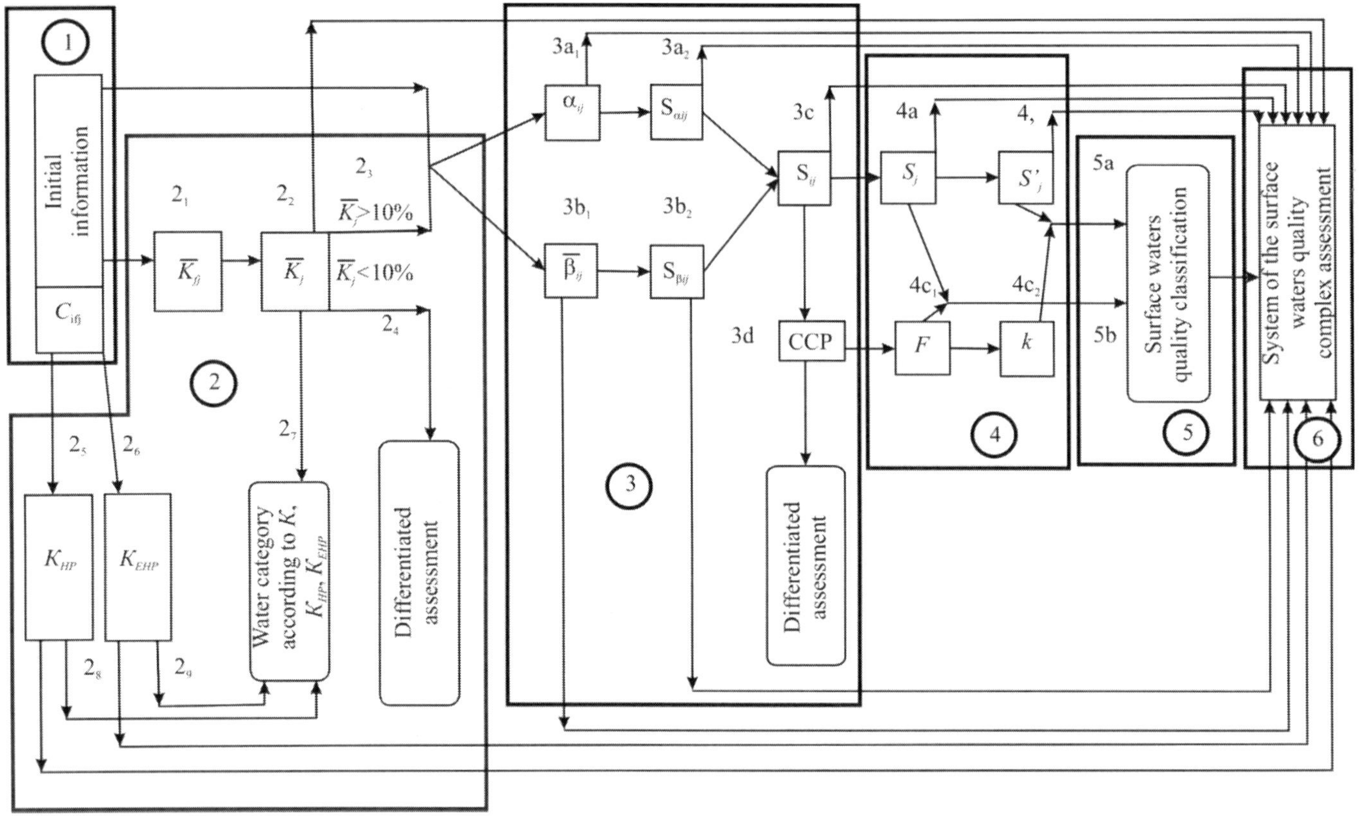

Figure 12.3. Complex assessment of the surface water's pollution level.

Block 2. Water Pollution Complexity Assessment

Preliminary assessment of surface water pollution level is made on the basis of the chosen initial data with the help of the water pollution level complexity coefficient.

The water pollution level complexity coefficient (*K*) is the ratio between pollutants quantity, which exceeds the maximum permissible concentration (MPC) and the total number of pollutants taken into account during assessment. Complexity coefficient, *K*, is measured in percents with a changing range between 1-100%.

In case of water deterioration, this coefficient shows all the spectrum of anthropogenic impact on a water body. The *K* value increases as new pollutants increase in water. This block includes the following actions:

2_1—definition of water pollution level complexity coefficient in K_{fj} sample (action 2_1 on the diagram)

2_2—definition of complexity coefficient means value on the interval of time at the point K_j

2_3—if $K_j > 10$ %, the complex assessment of water quality is used and all the complex values are defined

2_4—if $K_j < 10$ %, water pollution depends on separate components and it is necessary to make differentiated sampling and research

2_5, 2_6—it is necessary to make similar water pollution level complexity coefficient calculations on the basis of values of concentrations corresponding to the high and extremely high pollution levels to define high and extremely high pollution levels frequency. K_{HP} (high pollution level coefficient) and K_{EHP} (extremely high pollution level coefficient) values growth corresponds to increasing of high and extremely high pollution concentration levels

2_7, 2_8, 2_9—water pollution categories are defined in accordance with each *K*, K_{HP} and K_{EHP} value. Therefore, three categories can be defined: category I is defined according to pollution level complexity, category II is defined according to the high pollution level complexity and category III is defined according to the extremely high pollution level complexity. Water with a pollution level higher than MPC or high or extremely high pollution levels, defined according to the separate method, are included into category I.

Waters with a pollution level higher than MPC or high or extremely high pollution level if assessed by several pollutants are included into category II. Waters with pollution level higher than MPC or high or extremely high pollution level if assessed by a set of pollutants are included into category III.

Water pollution level categories, defined according to *K*, K_{HP} and K_{EHP}, take into account the various consequences of aquatic ecosystem pollution. That is why it is necessary to use them jointly.

These characteristics complement each other. In case of categories mismatch, the situation should be understood in two ways: chronic pollution during the better part of a year according to *K*, and additionally accidental pollution according to K_{HP} or the emergency situation according to K_{EHP}.

Block 3. Water Quality Relative Assessment on the Basis of Each Considered Component

Assessment is based on two parameters very important in hydrochemistry—standard violations recurrence and excess of standards surpassed. This block includes the following actions:

3a1—water pollution occurrence frequency, α_{ij}, definition

3a2—particular assessment point definition according to the α_{ij} value and water classification (single, unstable, stable, typical) (Table 12.1). The value of this point can differ from 1 to 4. This definition is the first stage of water pollution level classification.

3b1—means the value of MPC excess definition according to the sample analysis results of the samples, where this excess was discovered ($\overline{\beta_{ij}}$).

3b2 — particular assessment point (β_{ij}) is defined in accordance with the mean value of MPC excess ($S_{\beta_{ij}}$) and water classification in connection with its pollution level (low, medium, high, extremely high) (Tables 12.2 and 12.3) [2]. The value of this point can differ from 1 to 4. This action is the second stage of the pollution level classification.

3c—generalized assessment point (S_{ij}) is defined as a result of particular assessment points of frequency and multiplicity; it can be ranged from 1 to 16. The higher degree of water pollution by a single pollutant corresponds to a larger value. This action is the third stage of classification, and is based on that fact that participation of single chemicals in general water bodies pollution can be defined by various combinations: high concentration during a short period of time, low concentration during a long period of time or by other possible combinations of concentrations and their recurrence (Table 12.4). Their calculation is kept not in parallel way, but through a uniform parameter that is called the "generalized assessment point." Assessment of complex gradations from the point of view of pollution level: "single low level" (can also be defined as "slightly polluted"), "single high level" and "very high level," and also "stable" and "typical low level" (can be defined as "dirty"), etc., is made on the basis of complex gradations analysis and qualitative (notional) characteristics. Five assessment gradations can be defined: conditionally pure, slightly polluted, polluted, dirty, extremely polluted.

3d—critical parameters of water pollution (CPP) are chosen from the total quantity of the ingredients according to the following condition: generalized assessment point value is 9 or more ($S_{ij} \geq 9$), i.e., pollution by this very ingredient is of high or very high level. The set or list of critical pollution parameters is used for water pollution analysis.

Table 12.1. Water bodies classification according to pollution recurrence

Recurrence, %	Water bodies pollution characteristics according to pollution recurrence	Particular assessment point of recurrence ($S_{\alpha_{ij}}$)	Particular assessment point value for 1% of recurrence
[11; 10)	Single	[1; 2)	0.11
[10; 30)	Unstable	[2; 3)	0.05
[30; 50)	Stable	[3; 4)	0.05
[50; 100)	Typical	4	—

1 At recurrence rate less than 1, we have the equation $S_{\alpha_{ij}} = 0$.

Table 12.2. Classification of water bodies in accordance with their pollution level

Excess of standards surpassing 1	Level of pollution	Particular assessment point according to the excess of standards surpassing ($S_{\beta_{ij}}$)	Particular assessment point value for unit of excess of standards surpassed
[1; 2)	Low	[1; 2)	1.00
[2; 102)	Medium	[2; 3)	0.125
[102; 503)	High	[3; 4)	0.025
[503; + ∞)	Very high	4	0.025

1 The following gradations of pollution level multiplicity are used for the dissolved oxygen: (1; 1, 5]—low; (1, 5; 2]—middle; (2; 3]—high; (3; + ∞]—very high. If the concentration of dissolved oxygen in the sample equals 0, it is necessary to take it up 0.01 mg/dm3.

2 The abovementioned multiplicity levels are characterized as "high pollution level" (HP) of the water body and used for the bulk of pollutants of the 3rd and 4th hazardous classes. For those pollutants, whose high pollution criterion differs from 10, this number 10 should be changed into high pollution level criteria defined according to Table 12.3.

3 The abovementioned multiplicity levels are characterized as "extremely high pollution level" (EHP) of the water body and used for the bulk of pollutants of the 3rd and 4th hazardous classes. For those pollutants, whose high pollution criterion differs from 50, this number 50 should be changed into high pollution level criteria defined according to Table 12.3.

Table 12.3. Criteria of water bodies with high and extremely high pollution level definition according to hydrochemical parameters

Ingredients and water quality parameters 1	MPC excess multiplicity	
	HP	EHP
1—2 hazardous classes	from 3 to 5	≥ 5
3—4 hazardous classes, except oil-based products, phenols, copper, total iron	from 10 to 50	≥ 50
4 hazardous classes—oil-based products, phenols, copper, total iron	from 30 to 50	≥ 50

[1] High pollution level for dissolved oxygen is 3 mg/dm^3 or less, extremely high pollution level is 2 mg/dm^3 and less, water of high pollution level for biochemical oxygen demand with a 5 day test is 10 mg/dm^3 and more, water of extremely high pollution level for biochemical oxygen demand with 5 day test is 40 mg/dm^3 and more.

Table 12.4. Possible variations of watercourses quality condition according to the single ingredients and pollution parameters

No.	Watercourse pollution data			Pollution level assessment
		quantitative, points		
	Qualitative	Single (products of numbers)	Total points	
1	Single low level of pollution	1 · 1	1	Slightly polluted
2	Single medium level of pollution	1 · 2	2	Polluted
3	Single high level of pollution	1 · 3	3	Dirty
4	Single very high level of pollution	1 · 4	4	Dirty

Table 12.4. (Continued)

No.	Watercourse pollution data			Pollution level assessment
		quantitative, points		
	Qualitative	Single (products of numbers)	Total points	
5	Unstable low level of pollution	2 · 1	2	Polluted
6	Unstable medium level of pollution	2 · 2	4	Dirty
7	Unstable high level of pollution	2 · 3	6	Very dirty
8	Unstable very high level of pollution	2 · 4	8	Very dirty
9	Stable low level of pollution	3 · 1	3	Dirty
10	Stable medium level of pollution	3 · 2	6	Very dirty
11	Stable high level of pollution	3 · 3	9	Highly polluted
12	Stable very high level of pollution	3 · 4	12	Highly polluted
13	Typical low level of pollution	4 · 1	4	Dirty
14	Typical medium level of pollution	4 · 2	8	Very dirty
15	Typical high level of pollution	4 · 3	12	Highly polluted
16	Typical very high level of pollution	4 · 4	16	Highly polluted

Block 4. Calculation of Water Pollution Level Combinatoric Index

It is necessary to take into account the conditional assumption of pollutants' mutual effect additivity if these pollutants appear simultaneously. This block includes the following actions:

4a—combinatoric index of the water pollution level (WPLCI, S_j) is calculated as a sum of generalized assessment points

4b—specific combinatoric index of the water pollution level (SWPLCI, S'_j) is identified, and if waters have different pollution levels, this index can vary from 1 to 16; the higher the index value is, the lower water quality

4c, $4c_1$—reserve coefficient, k, is identified according to the number of CPP (F)

Block 5. Water Quality and Pollution Level Classification

General classification can be made only after a differentiated analysis of water quality ingredients parameters has been done in each block. This general classification includes 5 quality classes:

st class—conditionally pure
nd class—slightly polluted
rd class—polluted
th class—dirty
th class—extremely polluted

If the pollution (by a set of pollutants) level of water increases, the class number also increases.

Long-term research of the Russian Federation surface waters pollution level made by the Hydrochemical Institute showed that the majority of water bodies in some regions of Russia can be included in the 3rd or 4th classes. In these cases, their individual distinctions (changes of water quality on a current moment or during a long period of time) can be flattened out. The 3rd and the 4th classes of the general classification are subdivided into two and four categories to provide more detailed water quality assessment. In this case, classification has the following structure:

1 class—$(1\cdot N_j\cdot k]$, conditionally pure
2 class—$(1\cdot N_j\cdot k;\ 2\cdot N_j\cdot k]$, slightly polluted
3 class—$(2\cdot N_j\cdot k;\ 4\cdot N_j\cdot k]$, polluted
  category a)—$(2\cdot N_j\cdot k;\ 3\cdot N_j\cdot k]$, significantly polluted
  category b)—$(3\cdot N_j\cdot k;\ 4\cdot N_j\cdot k]$, very polluted
4 class—$(4\cdot N_j\cdot k;\ 11\cdot N_j\cdot k]$, dirty
  category a)—$(4\cdot N_j\cdot k;\ 6\cdot N_j\cdot k]$, dirty
  category b)—$(6\cdot N_j\cdot k;\ 8\cdot N_j\cdot k]$, dirty
  category c)—$(8\cdot N_j\cdot k;\ 10\cdot N_j\cdot k]$, very dirty
  category d)—$(10\cdot N_j\cdot k;\ 11\cdot N_j\cdot k]$, very dirty
5 class—$(11\cdot N_j\cdot k;\ +\infty]$, extremely polluted

In practice, the classes' subdivision can be based on WPLCI and SWPLCI values. Block 5 includes the following actions:

5a—if WPLCI values are used, water pollution level is classified by taking into account the number of pollutants used for complex assessment and number of complex water pollution parameters (CPP) (Table 12.5)

5b—if SWPLCI values are used, water quality and pollution level is classified by taking into account number of CPP (Table 12.6)

Formalized parameters, intermediate and basic (Table 12.7), are defined according to the used method. The bulk of the intermediate parameters are widely used for further calculations of the basic parameters.

Water quality assessment can be carried out using all the sets of parameters and also using separate groups or single characteristics. Qualitative notional characteristics corresponding to the numerical values can also be used for an assessment.

At the end of the surface waters pollution level complex assessment description, it is necessary to make a brief description of this system's advantages.

The following features can be characterized using complex assessments:

- Section line, water quality observing station, area, water body, basin, region, etc.
- Certain period of time: twenty-four hours, month, quarter, hydrological season, year, several years, etc.
- Water bodies' current condition, tendencies of its condition changing during the specified period of time

Block 6. Complex Assessment of Surface Waters Pollution Level

Table 12.5. Watercourse quality classification according to the combinatoric index value

Water quality class	Water quality category	Water pollution level	Specific combinatoric index of the water pollution level					
			CPP is not taken into account	CPP is taken into account				
				1 CPP (k = 0.9)	2 CPP (k = 0.8)	3 CPP (k = 0.7)	4 CPP (k = 0.6)	5 CPP (k = 0.5)
1	–	Conditionally pure	$1 \cdot N_j$	$0.9 \cdot N_j$	$0.8 \cdot N_j$	$0.7 \cdot N_j$	$0.6 \cdot N_j$	$0.5 \cdot N_j$
2	–	Slightly polluted	$(1 \cdot N_j; 2 \cdot N_j]$	$(0.9 \cdot N_j; 1.8 \cdot N_j]$	$(0.8 \cdot N_j; 1.6 \cdot N_j]$	$(0.7 \cdot N_j; 1.4 \cdot N_j]$	$(0.6 \cdot N_j; 1.2 \cdot N_j]$	$(0.5 \cdot N_j; 1.0 \cdot N_j]$
3	–	Polluted	$(2 \cdot N_j; 4 \cdot N_j]$	$(1.8 \cdot N_j; 3.6 \cdot N_j]$	$(1.6 \cdot N_j; 3.2 \cdot N_j]$	$(1.4 \cdot N_j; 2.8 \cdot N_j]$	$(1.2 \cdot N_j; 2.4 \cdot N_j]$	$(1.0 \cdot N_j; 2.0 \cdot N_j]$
	Category a)	Significantly polluted	$(2 \cdot N_j; 3 \cdot N_j]$	$(1.8 \cdot N_j; 2.7 \cdot N_j]$	$(1.6 \cdot N_j; 2.4 \cdot N_j]$	$(1.4 \cdot N_j; 2.1 \cdot N_j]$	$(1.2 \cdot N_j; 1.8 \cdot N_j]$	$(1.0 \cdot N_j; 1.5 \cdot N_j]$
	Category b)	Very polluted	$(3 \cdot N_j; 4 \cdot N_j]$	$(2.7 \cdot N_j; 3.6 \cdot N_j]$	$(2.4 \cdot N_j; 3.2 \cdot N_j]$	$(2.1 \cdot N_j; 2.8 \cdot N_j]$	$(1.8 \cdot N_j; 2.4 \cdot N_j]$	$(1.5 \cdot N_j; 2.0 \cdot N_j]$
4	–	Dirty	$(4 \cdot N_j; 11 \cdot N_j]$	$(3.6 \cdot N_j; 9.9 \cdot N_j]$	$(3.2 \cdot N_j; 8.8 \cdot N_j]$	$(2.8 \cdot N_j; 7.7 \cdot N_j]$	$(2.4 \cdot N_j; 6.6 \cdot N_j]$	$(2.0 \cdot N_j; 5.5 \cdot N_j]$
	Category a)	Dirty	$(4 \cdot N_j; 6 \cdot N_j]$	$(3.6 \cdot N_j; 5.4 \cdot N_j]$	$(3.2 \cdot N_j; 4.8 \cdot N_j]$	$(2.8 \cdot N_j; 4.2 \cdot N_j]$	$(2.4 \cdot N_j; 3.6 \cdot N_j]$	$(2.0 \cdot N_j; 3.0 \cdot N_j]$
	Category b)	Dirty	$(6 \cdot N_j; 8 \cdot N_j]$	$(5.4 \cdot N_j; 7.2 \cdot N_j]$	$(4.8 \cdot N_j; 6.4 \cdot N_j]$	$(4.2 \cdot N_j; 5.6 \cdot N_j]$	$(3.6 \cdot N_j; 4.8 \cdot N_j]$	$(3.0 \cdot N_j; 4.0 \cdot N_j]$
	Category c)	Very dirty	$(8 \cdot N_j; 10 \cdot N_j]$	$(7.2 \cdot N_j; 9.0 \cdot N_j]$	$(6.4 \cdot N_j; 8.0 \cdot N_j]$	$(5.6 \cdot N_j; 7.0 \cdot N_j]$	$(4.8 \cdot N_j; 6.0 \cdot N_j]$	$(4.0 \cdot N_j; 5.0 \cdot N_j]$
	Category d)	Very dirty	$(10 \cdot N_j; 11 \cdot N_j]$	$(9.0 \cdot N_j; 9.9 \cdot N_j]$	$(8.0 \cdot N_j; 8.8 \cdot N_j]$	$(7.0 \cdot N_j; 7.7 \cdot N_j]$	$(6.0 \cdot N_j; 6.6 \cdot N_j]$	$(5.0 \cdot N_j; 5.5 \cdot N_j]$
5	–	Extremely polluted	$(11 \cdot N_j; +\infty]$	$(9.9 \cdot N_j; +\infty]$	$(8.8 \cdot N_j; +\infty]$	$(7.7 \cdot N_j; +\infty]$	$(6.6 \cdot N_j; +\infty]$	$(5.5 \cdot N_j; +\infty]$

Table 12.6. Watercourse quality classification according to the specific combinatoric index value

Water quality class	Water quality category	Water pollution level	Combinatoric index of the water pollution level					
			CCP is not taken into account	CCP is taken into account				
				1 CCP (k = 0.9)	2 CCP (k = 0.8)	3 CCP (k = 0.7)	4 CCP (k = 0.6)	5 CCP (k = 0.5)
1	–	Conditionally pure	1	0.9	0.8	0.7	0.6	0.5
2	–	Slightly polluted	(1; 2]	(0.9; 1.8]	(0.8; 1.6]	(0.7; 1.4]	(0.6; 1.2]	(0.5; 1.0]
3	–	Polluted	(2; 4]	(1.8; 3.6]	(1.6; 3.2]	(1.4; 2.8]	(1.2; 2.4]	(1.0; 2.0]
	Category a)	Significantly polluted	(2; 3]	(1.8; 2.7]	(1.6; 2.4]	(1.4; 2.1]	(1.2; 1.8]	(1.0; 1.5]
	Category b)	Very polluted	(3; 4]	(2.7; 3.6]	(2.4; 3.2]	(2.1; 2.8]	(1.8; 2.4]	(1.5; 2.0]
4	–	Dirty	(4; 11]	(3.6; 9.9]	(3.2; 8.8]	(2.8; 7.7]	(2.4; 6.6]	(2.0; 5.5]
	Category a)	Dirty	(4; 6]	(3.6; 5.4]	(3.2; 4.8]	(2.8; 4.2]	(2.4; 3.6]	(2.0; 3.0]
	Category b)	Dirty	(6; 8]	(5.4; 7.2]	(4.8; 6.4]	(4.2; 5.6]	(3.6; 4.8]	(3.0; 4.0]
	Category c)	Very dirty	(8; 10]	(7.2; 9.0]	(6.4; 8.0]	(5.6; 7.0]	(4.8; 6.0]	(4.0; 5.0]
	Category d)	Very dirty	(10; 11]	(9.0; 9.9]	(8.0; 8.8]	(7.0; 7.7]	(6.0; 6.6]	(5.0; 5.5]
5	–	Extremely polluted	(11; + ∞]	(9.9; + ∞]	(8.8; + ∞]	(7.7; + ∞]	(6.6; + ∞]	(5.5; + ∞]

Table 12.7. The list of surface water pollution level assessment parameters

Character of information, provided by the parameter	Calculation assessment parameters	
	Intermediate	Basic complex
Each ingredient for one result of analysis	Ingredients concentration exceeds 1 MPC, Ingredients concentration exceeds concentration of pollutants corresponding to HP level, Ingredients concentration exceeds concentration of pollutants corresponding to EHP level	
Each ingredient for all the results of analysis during a certain period of time	Average ingredients concentration exceeds 1 MPC in samples, where these ingredients were found Cases of MPC exceeding recurrence Average annual concentration	Particular assessment points according to MPC exceeding Particular assessment points according to MPC exceeding recurrence Generalized assessment points
Complex of water chemical composition ingredients during a certain period of time	Total quantity of ingredients used for water quality assessment Number of ingredients violating standard requirements Assurance coefficient	Complexity coefficient of water pollution Complexity coefficient of high and extremely high pollution level Critical parameters of water pollution level (number and list of them) Water pollution level combinatoric index Specific combinatoric index of water pollution level Water quality class

Moreover, they help to solve different problems of qualitative and quantitative characteristics, helping to make necessary calculations.

The system of coefficients sufficiently provides the possibility of processing, preparing and supplying a regime hydrochemical information and can be used to solve operational problems. It also helps prepare various forms of reports, which are highly informative and at the same time available for wide usage.

12.4. Complex Assessment System Based on Hydrological Parameters Improvement

The complex assessment system improvement means taking into account hydrologic and hydrodynamic parameters (including parameters of water discharge variability) and also hydrochemical assessments compatibility with hydrobiological, hygienic assessments, etc.

Several ways are offered for hydrological parameters analysis. They are as follows:

1. Using concentrations conversion into 95% frequency values during complex assessment to make accurate water quality assessment.

2. Making complex assessment based on a water regime phase.

3. Using weighted average concentration values of water discharge for indices calculation. In case of constant sources of water body pollution, it is necessary to construct a hydrograph and plot on special signs corresponding to the dates of the chemical analysis sampling. Then, the dependence diagram of concentrations from water discharge is to be constructed. Values of expense provisions used for sampling are not used for expense provisions values curve. Later, concentration provisions are calculated according to the expense provisions. The next stage is the creation of the concentration provisions curve. On its basis, it is easy to get all characteristics necessary for complex assessments. In this case, the average annual concentrations are to be replaced by weighted average values. If there are new sources of pollution, their expenses and polluted waters composition are to be registered and later, new curves, including information about these pollutants and provisions, are to be created. In cases when the number of measurements of components concentration or quality parameters is insignificant (three to four samples), it is desirable to have additional signs to create the curve of provisions. If concentrations and water discharges are connected with each other, concentration values corresponding to critical points of the hydrograph are not taken into account on the curve. Thus, the prolongation of the number of concentrations can be carried out.

4. Combining water pollution level complex assessment and integral hydrologic and hydrodynamic parameters, defined by the State Hydrological Institute.

In this case, water pollution level is assessed according to the computation scheme of hydrologic and hydrodynamic parameters, but combinatoric and specific combinatoric indices of water pollution level, calculated for a definite period of time (month, year, hydrological season, etc.), are used for this calculation instead of certain substances' concentration data. A chronological diagram based on WPLCI data and characterizing the water pollution index changes during periods of the year is to be marked on the hydrograph, taking into account everyday water discharge. A straight, borderline parallel to axis of abscissas is drawn on the basis of criteria values of the water pollution level index, corresponding to water quality changing from one class to another one and defined by the abovementioned method. This straight line makes it possible to break a curve of SWPLCI changes into separate sections corresponding to the periods of conditionally pure, slightly polluted, polluted, dirty and extremely polluted drains. Having removed the duration (in days) and relative volume (in m^3) data typical for slightly polluted drain and other gradations of water pollution degree from the diagram, it is possible to assess water bodies' qualitative condition taking into account the complex of specific, priority pollutants that are important for the researcher in conditions of the specific hydrologic and hydrodynamic situation of water body.

The choice of the water quality method is based on several factors, including the aim of assessment, presence/absence of initial data, specific characteristics and water pollution level, etc.

12.5. PROSPECTIVE LINES OF THE WATER QUALITY COMPLEX ASSESSMENTS DEVELOPMENT

After analyzing and summarizing modern papers on this subject, it is possible to confirm that the problem of complex assessments is serious enough and demands us to take into account most various features of a water body.

Nowadays on this stage of scientific development, it is important to consider the following facts:

- Theoretical and practical urgency of the surface water quality complex assessment problem and possibility of its further improvement are not impeachable
- Complex assessment systems should include the whole range of different surface water quality characteristics

As we already know, the system of water quality complex assessment consists of several blocks among which the basic one is the block of chemical parameters, including methods of assessment based on hydrochemical parameters and supplemented by certain hydrodynamic parameters.

The second block of the biological parameters is based on using methods of water quality assessment according to their hydrobiological and toxicological parameters. Toxicological analysis is based on methods of water toxicity determination. Unlike the classical hydrobiological approaches based on methods of bioindication, methods of biotesting are used for an integrated assessment of water toxicological conditions. These methods are based on comparing test reactions of hydrobionts (toxicity parameters on survival rate, fertility, cells increase, etc.) in the polluted zone of a water body and a pure zone. Biotesting allows us to characterize the toxic influence of all the sets of pollutants that can be found in an aquatic ecosystem.

In the future, the block of radioactive pollution assessment of water bodies, characterized by a higher level of radioactivity, should be included in the complex assessment of surface water pollution level.

Tests of nuclear weapons, nuclear power stations accidents (particularly, the Chernobyl NPS) and the accumulation (transportation, processing) of radioactive waste have caused global problems for the environment in terms of radioactive pollution. The Earth is very often polluted by enormously energetic land and air nuclear explosions accompanied by a large number of radioactive products forming precipitation in the troposphere (stratosphere). These precipitations consist of isotopes of strontium (^{90}Sr), caesium (^{137}Cs), zirconium, niobium, etc.

Surface waters pollution by radioactive substances is too dangerous.

The most dangerous sources of surface waters pollution are the nuclear power stations using water bodies as coolers. Radionuclides emitted into water bodies with waste waters are usually absorbed by hydrobionts and bottom sediments, so new secondary sources of pollution appear in aquatic ecosystem. Thus, the isotopes accumulation in biota and bottom sediments coefficient depends on the set of natural and anthropogenous factors defining variability of its values in aquatic ecosystems in different regions.

The system of radiation monitoring developed in Russia will allow us to provide further supervision and water bodies' radioactive pollution assessment in some regions by using this data in the system of water quality complex assessment.

To summarize the abovementioned information, we will make a special scheme of surface waters quality complex assessment, shown in Figure 12.4.

Taking into account all the parameters listed above, it will be easier to define the degree of real ecological danger of polluted waters for different types of water use.

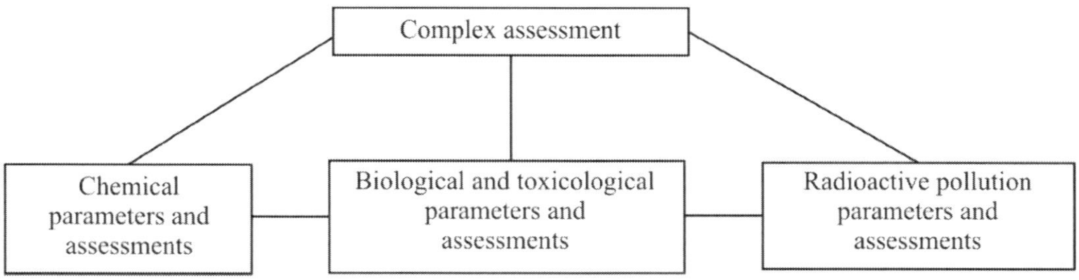

Figure 12.4. Complex assessment of inland surface waters quality.

REFERENCES

[1] Hydrochemistry information guide/ Edited by A.M. Nikanorov. — Leningrad: *Hydrometeoizdat,* 1989. — 250 p.

[2] Nikanorov A.M. Hydrochemistry. — St-Petersburg: *Hydrometeoizdat,* 2001. — 447 p.

[3] Nikanorov A.M., Khoruzhaya T.A. Global Ecology. — Moscow: *Publishing house «Prior»,* 2001. — 285 p.

SECTION IV. MATHEMATICAL METHODS OF HUMAN IMPACT ON FRESHWATER ECOSYSTEMS ASSESSMENT

Several years ago (in the early nineties of the twentieth century) the problems of normalizing human and other[24] impacts on aquatic ecosystems were intensively studied theoretically and experimentally in Russia. The present section of this monograph is devoted to the theoretical problems and some mathematical considerations connected with the definition of water bodies impact quantitative norms.

Definition of maximum allowable and other ecological norms of the impact on an aquatic system or water bodies is a very difficult problem because of the following reasons:

- Development of existing norms of impact on water bodies is based on different theoretical-experimental conditions.[252] Thus corresponding standards sometimes differ too much.
- There are no quantitative definitions of some basic characteristics of ecological normalizing. Maximum allowable impact on an ecosystem, the ecosystem's quantitative adaptation, influence of the ecosystem and some other concepts appear in scientific and practical works, but there are no accurate definitions for them, and each author define them in his own way.
- The term "normalizing the impact" on aquatic ecosystems and terms like "normal functioning" of ecosystems or "ecological prosperity," etc., need an accurate quantitative specification.

The present section, worked out together with Physics and Mathematics Ph.D., M.V. Zavolzhenskiy, is based on the general ideas of aquatic ecological systems functioning and on mathematical methods for the quantitative definition of the abovementioned and some other ideas. An axiomatic approach for ecological norms' definition is developed. Accurate parameters of frequency impact on oscillatory processes are defined. One of these parameters, connected with single-frequency oscillation, is coordinated with the results of experiments run in various areas of natural sciences, such as chemistry, biology, physics, medicine, etc.

---

[24] E.g., disasters: droughts, floods, earthquakes, volcanic eruption.

[25] See Chapter 3 of the present monograph.

Chapter 13

FUNDAMENTAL PROPOSITIONS OF MATHEMATICAL DYNAMIC HYDROLOGY

13.1. MATHEMATICAL AXIOMATICS

Supported mathematical and biological axiomatics are used in the present work to define quantitative characteristics of terms connected with ecological normalization. Two axioms are used in this paper.

The mathematical axiom is based on the wide range of scientific observations, which proves that the majority of ecological systems have elements that function on the basis of the so-called principle of impact-reaction. Ecosystem dynamics is also based on this principle (e.g., predator-victim, resources-consumers, parasite-owner, etc.). Each environment, including aquatic[26] and ecological ones, exists thanks to the biological organisms' competition and is a system of the impact-reaction. In these conditions at least, the term "impact" is the mathematical one but the basis of the mathematical axiom is the idea that the term "reaction" is also a mathematical one. In essence, the first axiom can be defined as follows: *water body dynamics can be described by mathematical terms of general dynamical systems.*

This means that if $X = \{x_1, x_2, ..., x_n\}$ is a vector, describing the condition of the ecosystem and $F = \{f_1, f_2, ..., f_m\}$ is a vector characterizing the impact on this system during some period of time, t, we can get the following mathematical relation:

$$\mathbf{X} = \mathrm{X}(t, \mathbf{F}) \tag{1}$$

X is a nonlinear operator, which is defined when the dynamical system,

$$\frac{d\mathbf{X}}{dt} = A(\mathbf{X}, \mathbf{F}, t) \Leftrightarrow \mathbf{X} = \mathbf{X}(t, \mathbf{F}) \tag{2}$$

---

[26] There is no doubt that water is a "living" organism not only from the point of view of biology, but also dynamically, because water body characteristics are factors of change and chemical and biological reactions change depending on time period.

characterizing this system as the impact-reaction as defined. Operator *A(...)* is defined by the water body dynamics' character. Relation (2) should take into account the initial and intermediate conditions providing the water body's existence. As a rule, operator *A* is nonlinear and includes lots of components. The powerful mathematical system for the operator equation's properties analysis was developed. This system includes simple balance formulae and also includes equations in fractional derivatives on a system of coordinates, time and the nonlinear integrated equations with parameters.

The possibility of complete water bodies' mathematical description is often called into doubt. The following short description of mathematical ecology history will probably remove these doubts.

First, the successful attempt of modeling the system predator-victim was made by Vito Volterra (1860-1940).[27] When Volterra recorded the progress of such a system as two-componential differential equations of type (2), and analyzed the solution of the system, he found the conditions for making the system predator-victim counterbalanced. Lots of scientists, Volterra's followers, made attempts to model various ecosystems. There was even an attempt to make a model of the whole terrestrial biosphere dynamics (global modeling) under various initial conditions. Global models were made by the Roman club participants. There are also Latin American and Asian global models.

Unfortunately, the result of the global modeling disappoints us. All model developers come to one conclusion; if people continue to develop and use natural resources as they did in the middle of the XX century, the global disaster will not be prevented. This result inefficiency is obvious; a biosphere will not be ruined no matter how actively resources will be used. It is connected with that fact that with the course of time, in cases of a global disaster's absence, the population growth rate will decrease in the process of the planet natural resources' expenditure and will be defined by the quantity of natural resources (Figure 13.1). Resource limitation will slow down the population growth rate.

Apocalyptic results of global modeling were refuted by Russian scientists [1], who proved that if management (water quality improvement, taking into account technical progress consequences, normalizing impact on ecosystems, etc.) will be included into differential dynamics equations (2) (all known global models were based on such dynamics), global disaster can be prevented.

Further mathematical modeling of the ecosystems dynamics showed that researching ecosystems this way can be considerably advanced. Well-known, Belousov-Zhobotinnskiy's chemical reaction [2] is the result of multicomponent chemical processes modeling. Bryusselator, "clock reaction," [3], was developed as a result of nonlinear chemical reactions modeling in liquid solutions. First, mathematicians offered the equations system, describing chemical reaction with the exact periodic change of solution color and then chemists reproduced this reaction.

When working at the Hydrodynamical School, the "nonlinear tasks of the hydrodynamic stability theory," named after Academician G. I. Petrov, Academician A.S. Monin made a report "On hydrodynamic environmental problems" in 1992. Mr. Monin provided the data collected in all branches of science (history, archeology, geology, mathematics, etc.), for the whole period of these branches' development. That data was about Earth's environmental development. He also made a diagram proving that the development of our biosphere during

[27] V.Volterra monograph "Mathematical Theory of the Struggle for Existence," published in the USSR in 1977.

all geological periods varied from its average situation. The amplitude of such variances has a rather serious disorder. That is why our biosphere already underwent disasters that were much more dangerous than disasters nowadays. Its average condition, however, remains unchanged. Mr. Monin's diagram also shows that the present-day biosphere develops slowly but will develop faster in future. In the conclusion to his report, Mr. Monin provided a multidimensional nonlinear system of possible global biosphere modeling, consisting of six blocks (demography, ecology, energy, industry, agriculture and transport). According to Mr. Monin, the problem of chaos in mankind's development is based on the absence/presence of a special factor, the so-called "attractor," to solve the problem of chaos.[28]

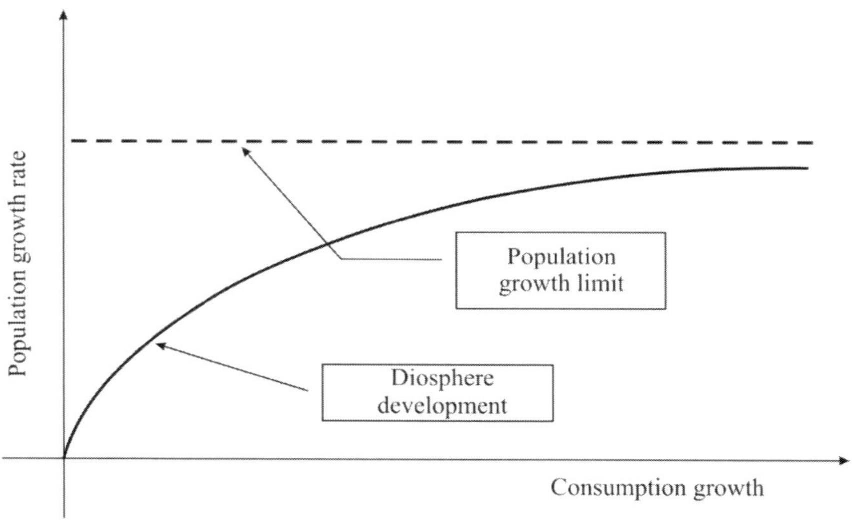

Figure 13.1.

Probably, this peculiar attractor exists in the Academician Monin's system. As was already mentioned before, however, such systems should include damping. Even with damping, this peculiar attractor will collapse like fluctuations in a transverse current. This information is shown in Mr. Monin's global diagram.

Nowadays, mathematical modeling of biological and chemical processes is developed enough. Such subdivisions of science, as mathematical chemistry, mathematical biology, mathematical ecology and mathematical genetics already exist. Mathematical models are also widely used in medicine, psycho-physics and economics. Synergetics [4] is one of the best sections of mathematics uniting many divisions of science. Cybernetics and computer engineering allow us to carry out both mathematical modeling of water bodies and normalize the impact on aquatic ecosystems.

A very important conclusion can be made on the basis of the abovementioned data: *in order to describe the ecosystem by means of a mathematical model, it is necessary to make the reaction of the ecosystem on the impact the function of this impact.*

---

[28] The peculiar attractor is a multidimensional surface in a phase space of nonlinear differential equations solving systems, which includes all the system solutions at any initial condition and does not get the surface. That is why there is chaos for this attractor equation system's solution.

Numerous research studies and experiments with biological objects with sensor-based subdivisions of "living" systems, with water objects, etc., really do confirm that the reaction of an ecosystem on some impact is an impact function. This is an experimentally proven fact, which was stated in equation (1)

13.2. Ecological Axiom

Now we will define an ecological axiom. The term "ecological axiom" in this case is conditional because we will take into account all the processes taking place in water bodies. The term "nonmathematical axiom" will be more exact in this case (as equation (1) shows a mathematical axiom).

The common feature necessary for the ecological system's proper functioning is the stability of processes occurring in this system during a long period of time. Stable systems are dynamically balanced. On the point of view of differential dynamics (2), stability means the balance (2) of processes described by the equations. It is not, however, enough to have the balance of processes occurring in water to make a model of aquatic ecological systems. On the one hand, the sign of dynamic balance stability criteria imperfection is ambiguity of the stability definition; there are some definitions of this term (for example, Mr. Lyapunov's definition of stability and Poincare's definition of this term). On the other hand, such terms as ecosystem capacity, assimilation capacity, buffer capacity, maximum allowable load, adaptation to impact, etc., are used nowadays to describe aquatic ecosystems. The processes appearing during the description of nondisastrous phenomena in water bodies are stable *a priori*. That is why it is necessary to take into account only stable solutions of differential system (2). Thus, it means that the adaptation is an internal property of the ecological system described by the equation (2). If this condition is not fulfilled, the system (2) is defined incorrectly, because systems incapable of adapting to impacts will collapse, and the equation's nonadaptive solution should be excluded.

Nowadays four approaches to the system's adaptive reaction on the impact [5] definition are known: thermodynamical adaptive reaction, cybernetic, biological and physiological ones. Let's say that the equations $X = X(t, F)$ (see (1)) takes into account not only the stability of the aquatic system's functioning adaptive dynamics, but also the four aspects of adaptation, including their direct and indirect relations. Thus the element:

$$dX = X(t, F)dF$$

contains information about the water environment's local adaptive reactions on the impact. Then, the functional relation of F and time, t, yields:

$$N(t, F) = \int_0^F X(t, \Phi)d\Phi$$

which characterizes the aquatic system's stable adaptive reaction on the impact concerning all multiplicities of F. It is supposed that all components of vector $F(t)$ are single-valued in the abovementioned formula. If the impact on water is parametrizable as F = F(t,b), the abovementioned formula is integrated by the parameter b.

Let us use the average value of all possible values of N(t,F) in the adaptive reaction, t:

$$S(F) = L_t N(t, F) \tag{3}$$

L_t is a linear operator, and its use is defined by the water body functioning. For example, if the water body becomes stable in some period of time ($t \rightarrow \infty$), we can get the following equation:

$$L_t \psi(t) = \lim_{t \to \infty} \psi(t).$$

If this body functioning is synchronized with the basic period, T:

$$L_t \psi(t) = \frac{1}{T} \int_0^T \psi(t) dt$$

There are some other ways of operator L_t functioning (average values, mathematical expectations, etc.).

Later, the value S (F) defined by formula (3) will be called the "water body adaptive reaction" on the impact, F, as it increases.

The ecological axiom states that the *maximum allowable impact on an aquatic ecosystem corresponds to that impact value F*, which helps to get the highest value of S (N) adaptive reaction on F.*

The ecological axiom means that $\| F \| > \| F^* \|$ impact increase will lower the adaptive stability of an aquatic ecosystem.

Normalization of the impact on the aquatic ecosystem is to be taken as definition of impacts:

$$\| F \| \leq \| F^* \|$$

which do not break the maximum adaption conditions of an aquatic ecosystems functioning. These aquatic ecological systems will be called *systems with normal functioning conditions*, i.e., conditions not breaking adaptive stability of the water ecological system.

If F_0 is the first (smallest) root of adaptive reaction that differs from zero (i.e., first root of $S(F_0) = 0$ equation), the ecological capacity of the aquatic ecosystem will be defined as the multitude of impacts:

$$0 \leq \| F \| \leq \| F_0 \|$$

In other words, from the point of view of adaptation, differential dynamics ecosystem capacity is the multitude of impacts not breaking the adaptive stability of the ecological water system.

Notice that the differential operator A can be continuous and also differential. Moreover, operator A can be constructed on the basis of probability calculus and mathematical statistics methods.

Oscillatory processes are the simplest ones in the ecosystems.

Chapter 14

NORMALIZING THE IMPACT ON OSCILLATORY PROCESSES IN AQUATIC ECOLOGICAL SYSTEMS

We will use the axiomatic approach to normalize the periodical impact on the aquatic system described in Chapter 13. There is a simple oscillatory process going on in one system; this process is based on the periodic, external impact (e.g., periodic cleaning of the water body if pollutants enter it with the frequency ratio of Ω: $f(t) = \cos(\Omega t)$). First, water quality was equal to a quality units, and the number of pollutants was equal to b units. The water quality of such a water body will be described by Cauchy initial-value problem for the second-order differential equation:

$$\frac{d^2x}{dt^2} + \omega^2 x = \cos(\Omega t), x(0) = a, \left.\frac{dx}{dt}\right|_{t=0} = b \tag{1}$$

The solution for this equation (1) is expressed by the function:

$$x(t, \Omega) = a\cos(\omega t) + \frac{b}{\omega}\sin(\omega t) + \frac{\cos(\Omega t)}{\omega^2 - \Omega^2} \tag{2}$$

Now we will define the water body's adaptive reaction on the periodic impact of water quality improvement actions with Ω frequency. Here, we will use formula (3), mentioned in section 13.2 of the present paper. In this case, the process of the water body quality changing is periodic and defined as $T = 2\pi/\omega$. That is why operator L_t is an average for the T period. We can get an adaptive reaction, S(Ω), of the water body when integrating function (2) with the limits of time 0-T and the frequency rate 0-Ω:

$$S(\Omega) = \frac{\omega}{2\pi}\int_0^{\Omega} \frac{\sin\left(\frac{2\pi\tau}{\omega}\right)d\tau}{\tau(\omega^2 - \tau^2)} \tag{3}$$

In this case, the adaptive reaction (3) is expressed by the higher transcendent function—integral sine Si(Ω). It should be mentioned that formula (3) can be reduced if the time is measured by pollution impact frequency units, ω, and the water quality is this frequency squared. This means that equation (1) will now include such variables of time ($t' = t/\omega$) and water quality ($x' = t/(\omega^2)$). Then, formula (3) will be as follows:

$$S(\Omega) = \frac{1}{2\pi}\int_0^{\Omega} \frac{\sin(2\pi\tau)d\tau}{\tau(1-\tau^2)} = \frac{2Si(2\pi\Omega) - Si(2\pi(\Omega-1)) - Si(2\pi(\Omega+1))}{4\pi} \quad (4)$$

Function $S(\Omega)$ is shown in Figure 14.1. In the case of such variables as x and t, $S(\Omega)$ function will get its maximum at the frequency rate $\Omega = 0.5\omega$. The maximum is $S(0.5\omega) = 0.309\omega$. The first time this function will get a zero value is at the frequency rate $\Omega = 1.323\omega$.

This function, shown in Figure 14.1, looks like a typical adaptive reaction before its first zero value. That is why the function $S(\Omega)$ shown above really characterizes the adaptive features of an aquatic system's oscillatory processes.

Thus, *aquatic environment adaptation for periodic impact is maximal at the frequency rate equal to half of the resonance frequency rate.* This is a principal conclusion, because resonance impact ($\Omega = \omega$) increases the water quality oscillatory amplitude and has no relations with its adaptation features. Resonance frequencies often break oscillatory processes and sometimes even the system as a whole. Breaking the aquatic system adaptation will happen when the external impacts rate is equal to 1.323 of the resonance frequency system.

If the water self-purification frequency (or water protection measures frequency) is equal to ω, and the water body pollution frequency is equal to Ω, the most adaptive self-purification frequency (or water protection measures frequency) is the frequency (approximately) equal to aquatic ecosystem pollution frequency multiplied by two ($\omega \sim 2\Omega$).

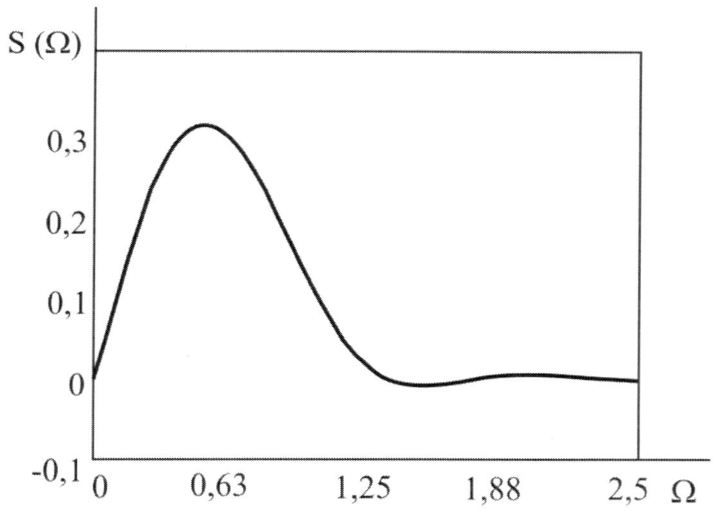

Figure 14.1.

This example shows that *to stabilize an aquatic system's adaptation to periodic impacts, it is necessary, first of all, to observe and normalize not the quantity (amplitude) but the frequency of external impact on the aquatic ecosystem.*

The last conclusion is correct for ecological systems in case of their normal functioning, while the exclusions here are disasters and nonperiodic processes.

It is also necessary to point out that before multiplying the periodic spectrum (usually $S(\Omega) = 0$ if $\Omega = 0.5\omega n$ and $n = \pm1, \pm2,...$), synchronous oscillatory processes were previously identified as a result of experiments in several scientific fields such as biology, medicine, chemistry, physics (e.g., radio elements half-decay period), mechanics, astronomy, etc.

Multiplying the periodic process, however, was a serious scientific achievement or even a discovery [6] in each of these sciences.

The method of normalizing the frequency of periodic impacts on an aquatic ecosystem is a very good demonstration of multiplying periodic spectra's existence in any oscillatory system that is close to a mathematical pendulum according to its dynamics.

The $S(\Omega)$ curve of higher frequencies of the external impact on an aquatic system is shown in Figure 14.2. Both figures (see Figures 14.1 and 14.2) characterize the multiplying periodic spectrum defined in this article.

It is also very important to notice that deviations occur in all (hydrobiological and hydrochemical) processes taking place in the aquatic ecological systems. In most cases, the oscillatory processes in the aquatic ecosystems are of a higher priority. That is why normalizing frequencies of impact on oscillatory processes is one of strategic problems of ecological normalizing.

The ecological capacity of the first oscillatory process phase is the value $\Omega = 1.323\omega$. The diagram interval [0; 1.323] in Figure 14.1 shows a unicomponent ecosystem's adaptive reaction on the oscillatory impact. It has a specific form of the ecosystem reaction curves, which has its cutoffs. Later, we will show that the experimentally defined forms of many hydrochemical and hydrobiological systems' reaction on impact are identical. It is important to note that such a form for the reaction also remains in cases when the impact on hydroecosystems is not oscillatory. We will first give, however, examples of the model's adaptive reactions similar to those represented in Figure 14.1 on the interval 0-1.323 of the system's resonance frequency.

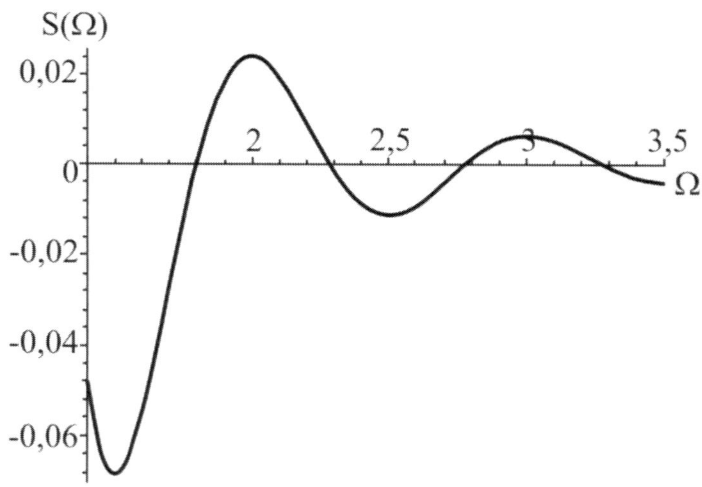

Figure 14.2.

Chapter 15

MODEL ADAPTIVE REACTIONS

Dependencies with one maximum do not only have local character, connected with one or several processes in ecological systems, but also have global character, typical of the whole system. Moreover, global dependencies with one maximum can be found in:

ecology (Ollie principle [7]),
biology (tolerance principle),
psycho-physics (Zabrodin Law [8]),
physics (e.g., Maxwell distribution),
economics and other multicomponent systems.

The globality of these laws and principles is well-seen if the system where this global law takes place develops as a corollary of a minor impulse and functions adaptively when system elements are in balance with the impact and stops functioning in cases of high values of the impact.

First, we should define the demands, which the parameterized connection should meet during experimental and theoretical modeling of water or other ecological system's adaptive reaction, *S*, on the impact, *r*. To simplify this process, we should analyze an ecosystem, which was affected by one impact, as a whole.

1) Adaptive system should have two thresholds: reaction *S* threshold appearing at the impact with average value, r_0, and system reaction on the impact with an average value, r^*, stopping threshold (Figure 15.1)
2) System reaction should rise at the impact increase with low values ($r \sim r_0$)
3) System reaction should decrease at the impact increase with high values ($r \sim r^*$)
4) Function $S(r)$, characterizing system adaptive reaction on the impact at $r_0 \leq r \leq r^*$, should be the analog function *R*.[29]
5) In order to satisfy these conditions, the diagram showing the system reaction on the impact will look like the function in Figures 15.2-15.4. As the analog $S(r)$ reaction

---

[29] There is an idea (especially amongst psycho-physicians) that the environment and organisms do not react on some insignificant impact increase. First, the system is not influenced by such an increase, but if it increases again, then the reaction also increases abruptly. We should neglect the insignificant increases of the impact and its reaction during mathematical modeling.

goes to zero twice ($S(r_0) = S(r^*) = 0$), the maximum value existence results from the well-known theory of an analog function's mathematical analysis (Rolle's Theorem).

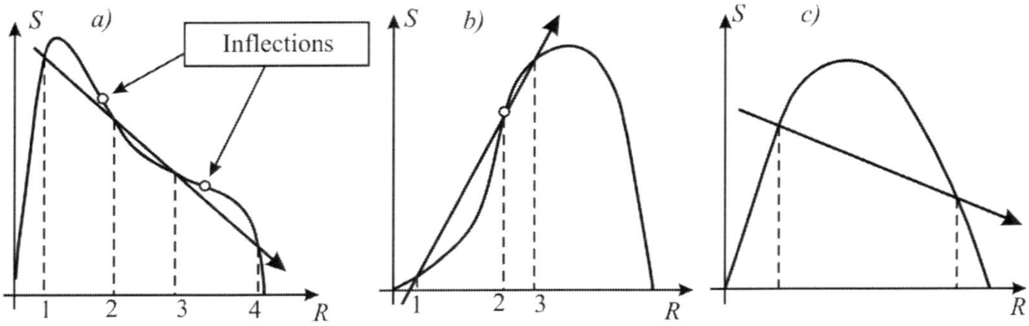

Figure 15.1.

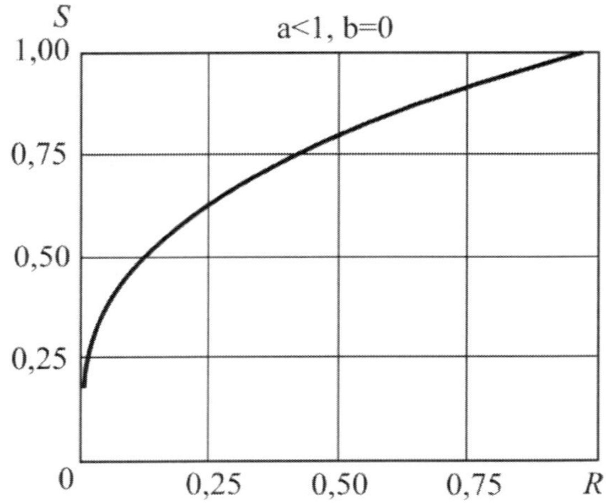

Figure 15.2.

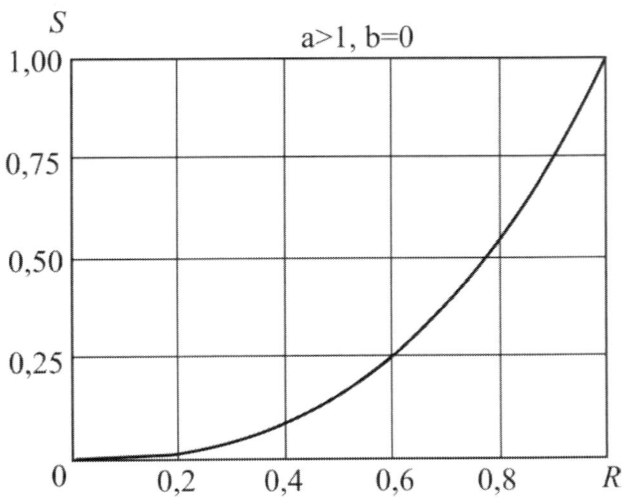

Figure 15.3.

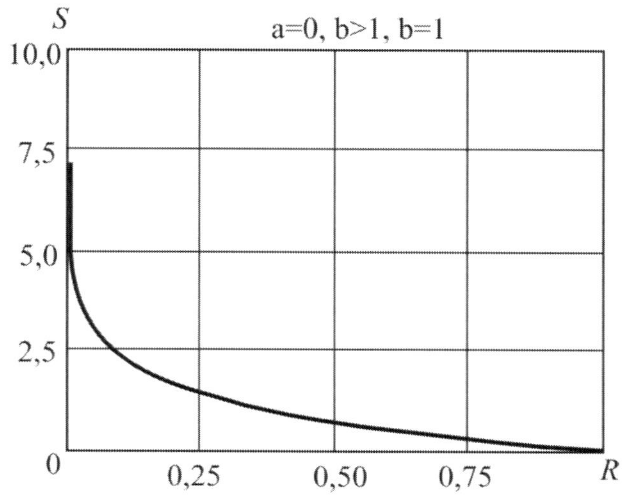

Figure 15.4.

To simplify the calculation's dimensional impact, r is changed into the non-dimensional impact characteristic R:

$$R = \frac{r - r_0}{r^* - r_0} \tag{1}$$

In the case of $r = r_0$, we have $R = 0$. In the case of $r = r^*$, we have $R = 1$. That is why the non-dimensional impact changes in the interval:

$$0 \leq R \leq 1$$

Moreover, the reaction at the ends of the last interval is equal to zero ($S(r_0) = S(r^*) = 0$, see section 5 of Chapter 15):

$$S(0) = S(1) = 0 \tag{2}$$

The first example of parametrization is taken from psycho-physics, and it is a synthesis of Mr. Zabrodin's law [8]:

$$S(R) = CR^a ln^b \left(\frac{1}{R}\right) \tag{3}$$

where C, a and b are the parameters, not dependent on the impact, R, and defining an ecosystem's reaction on the impact. Function (3) meets conditions (2) if $C > 0$, $a > 0$ and $b > 0$. Function S(R) is minimal at $R^* = \exp(-b/a)$.

The diagram of the function S (R) cannot have more than two inflections at different values of a and b parameters, defining this function. The curve concavity direction changes at the inflections. Curves with the inflections corresponding to an aquatic ecosystems adaptive reaction on the impact, however, also have a dynamic feature, except for the abovementioned geometrical one. The fact is that curves with the inflections are hypersensible to the direction.

The curve without inflections is crossed by the straight line, drawn in any direction, in two points (see Figure 15.1*c*). The curves having one inflection can be crossed by the straight line in three points (see Figure 15.1*b*), and curves with two inflections can be crossed by the straight line in even four points (see Figure 15.1*a*).

These features of curves with inflections are used in theory and practice of the dynamic systems because one or several inflections exist at the diagrams, corresponding to the birth and death coefficients; this is one of the features of a dynamic system's loss of stability [9]. Functions (3) described in paper [10], where diagrams have inflections, are also used to define the basic sync frequency of an aquatic ecological systems functioning as impact-reaction.

The function $S(R)$ diagram depends on the parameters a and b values, included in the function (3) definition. Therefore, if $a = 1$ and $b = 0$, the function $S(R)$ diagram is linear. If $a < 1$ and $b = 0$, the function diagram (3) is an exponential function, concave down (see Figure 15.2). If $a > 1$ and $b = 0$, the function diagram (3) is an exponential function, concave up (see Figure 15.3). If $a = 0$ and $b > 1$, the function $S(R)$ is an exponential logarithmic connection, concave up (see Figure 15.4). In the case where $a = 0$ and $b < 1$, the function $S(R)$ is an exponential logarithmic connection with an inflection (Figure 15.5). That is why features of the function diagram $S(R)$ depend on the following inequations to appear in cases of modeling the aquatic ecological system reaction on the impact, R:

$$(a < 1, b < 1); (a < 1, b > 1); (a > 1, b < 1); (a > 1, b > 1) \tag{4}$$

Inequations (4) separate the a and b parameters plane into four areas. We will make diagrams of the $S(R)$ functions for each area (4).

1. ($a < 1$, $b < 1$) area. In this case, the dependence diagram (3) has no inflections if the point with coordinates (a, b) at parameters (a, b) plane is inside of the curvilinear quadrilateral, limited by straight lines $b = 0$, $a = 1$ and $b = 1$ and parabola $b = 4a(1 - a)$ (Figure 15.6). Inside of a curvilinear triangle, limited by straight lines $b = 1$ and $a = 1$ and parabola $b = 4a(1 - a)$, the function diagram $S(R)$ has two inflections, and their ordinates are on the right side of the present function maximum (Figure 15.7).

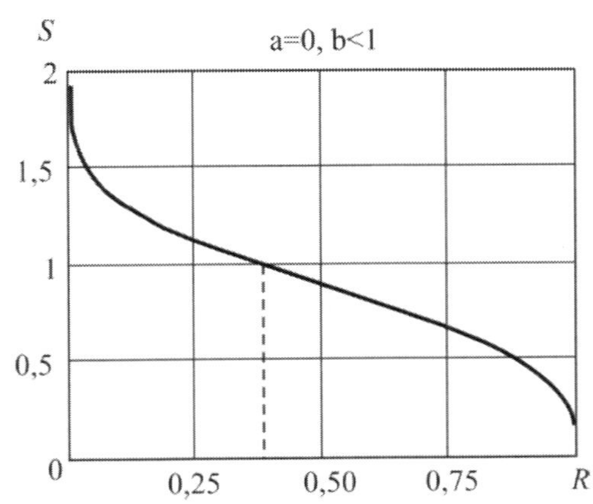

Figure 15.5.

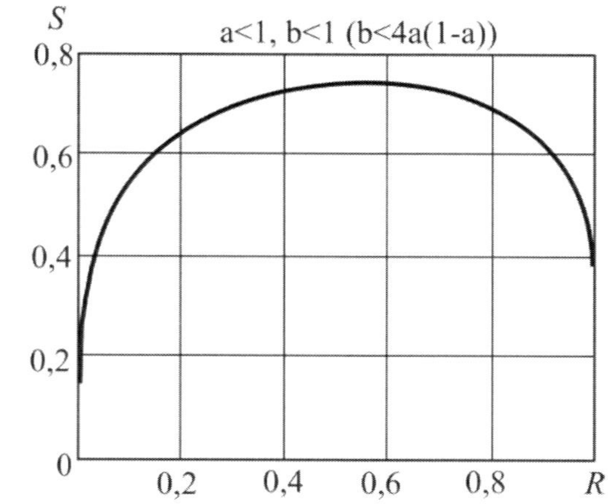

Figure 15.6.

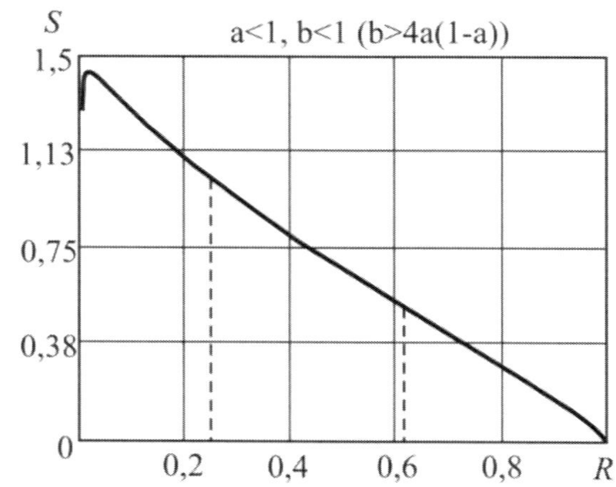

Figure 15.7.

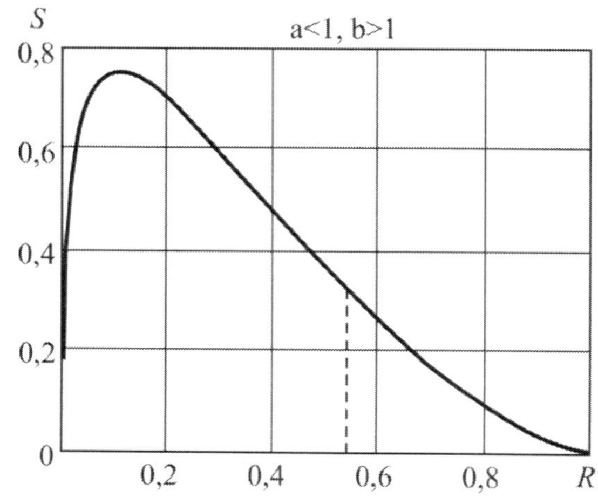

Figure 15.8.

2. Area ($a < 1$, $b > 1$). Here the function diagram (3) has only one inflection and its ordinate is on the right side of the function *S(R)* maximum ordinate (Figure 15.8).

3. Area ($a > 1$, $b < 1$). The function diagram (3) has one inflection, and its ordinate is on the left side of the function *S(R)* maximum ordinate (Figure 15.9).

4. Area ($a > 1$, $b > 1$). The function diagram (3) has two inflections, and their ordinates are on different sides of the function *S(R)* maximum ordinate (Figure 15.10).

After analyzing the information shown in Figures 15.2-15.10, we will make a flat diagram, characterizing the aquatic ecological system's semi-empirical adaptive reaction (3) on the impact (Figure 15.11). We called them the diagrams of "adaptation condition." You can also find more detailed descriptions of function (3) in paper [10].

Straight lines *Ha*, *Pb*, *AB*, *CD* and the parabola interval *OK* in Figure 15.11 are called "undirected arcs" of the diagram. Points *O*, *G*, *E*, *K* and *F* are called the "diagram point-group." Moving around the plane OAB crossing diagram, the undirected arcs and point-group when parameters a and b are changing corresponds to the unsteady-state processes in an aquatic ecological system. HOFA, OGEF, BEGA and other areas are called "diagram sectors."

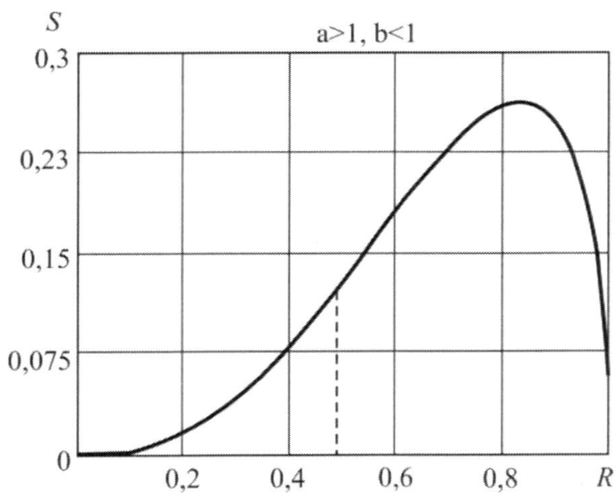

Figure 15.9.

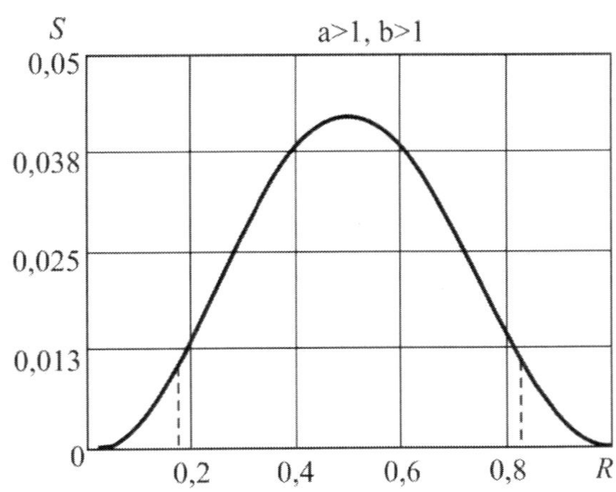

Figure 15.10.

The mathematical expression (3), however, is not general (a general expression of adaptive reaction is its integral approximation through the solution of the equations system, shown in Chapter 14). The approximation form of the adaptive reaction on the impact depends on the results of the experiment and also on experiment itself, only if an accurate definition of the adaptive reaction is not used for this experiment (see Chapter 13). For example, the following types of connections are used in biological research:

$$S(R) = CR^{a}(1\text{-}R)^{b} \tag{5}$$

This connection meets the requirements (2).

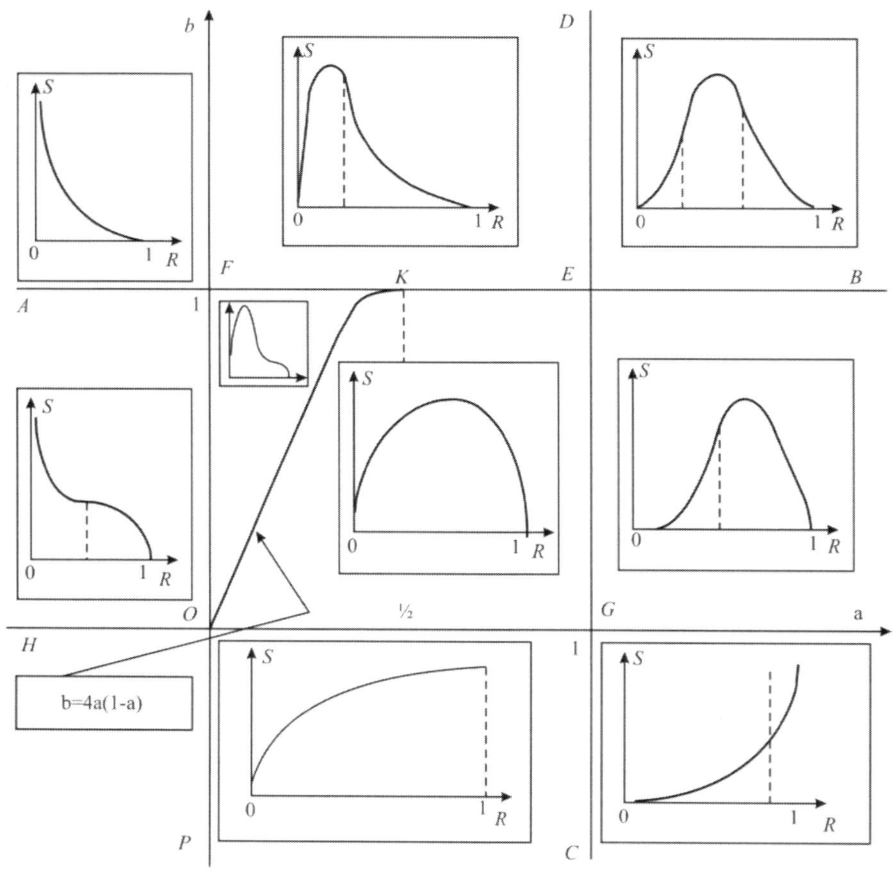

Figure 15.11.

Suppressing the corroboration, we will make the diagram of connection with *a* and *b* parameters featuring (5) (Figure 15.12). This diagram is identical to the diagram in Figure 15.11 in almost all sectors, excluding the sector, where limited by two straight lines, $a = 0$, $b = 0$, $a = 1$ and $b = 1$ inside the area $a < 1$ and $b < 1$. Speaking of these areas, the connection feature (5) varies in the triangles, limited by the straight line $a + b = 1$ (see Figure 15.12). We should also mention that function (5) is in its maximum if:

$$R = \frac{a}{a+b}$$

The diagrams Figure 15.12 features in its left sector are deficient compared to the diagram Figure 15.11 features; both diagrams are good models of equation connection features, shown in Figure 15.1. That is why both functions (3) and (5) can be used as the aquatic ecosystem model for its adaptive reactions upon an impact.

The diagrams, shown in Figures 15.11 and 15.12 are the examples of functions meeting condition (2) and having only one maximum on the interval [0, 1]. For example, the following functions:

$$f(x) = (1-x)^a \sin^b x\,, g(x) = \sin^a x \cos^b(\pi x/2) \qquad (7)$$

and other ones can be expressed by diagrams identical with those shown in Figures 15.11 and 15.12. Such diagrams can differ only in their lower left sector, if $0 \le a \le 1$ and $0 \le b \le 1$. This expands the opportunities of an aquatic ecosystems integral adaptive reaction modeling.

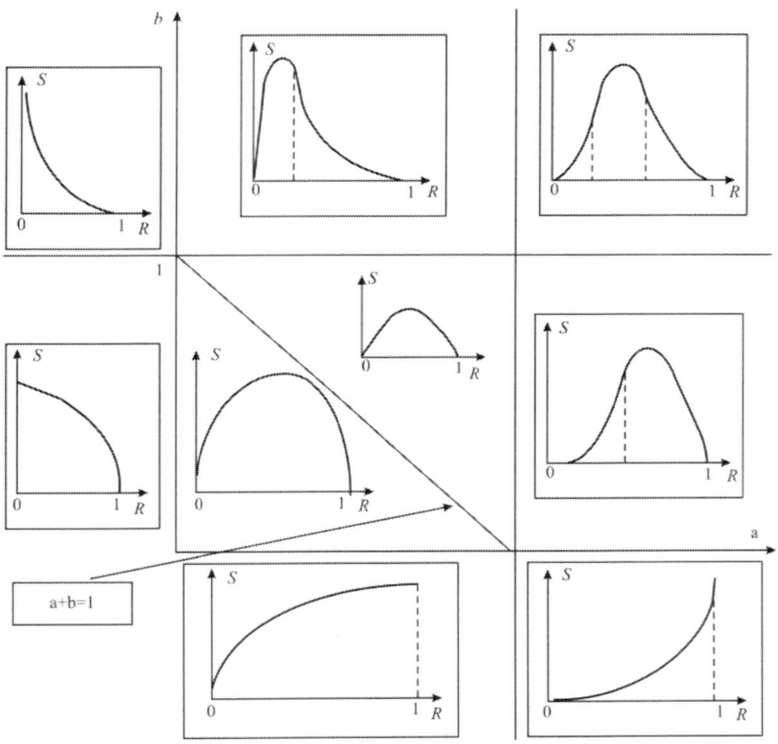

Figure 15.12.

Diagrams of functions $f(x)$ and $g(x)$, if $a = 2$ and $b = 1$, are shown in Figure 15.13. Functions $f(x)$ and $g(x)$ change type if they are analogous to the type of the abovementioned reaction changes.

If two external forces influence an aquatic ecosystem, the reaction dependence on the impact is expressed as follows: $S = S(R_1, R_2)$ (Figure 15.14). We will not show the relatively complex equation of the area, presented on Figure 15.14. We can show the research analogous to this case, define the inflections—changing the area $S = S(R_1, R_2)$ curvature direction—and make a nonplanar diagram of the two-dimensional aquatic ecosystem adaptive reaction on the impact by taking into account the parameters defining this reaction. Speaking about the nonplanar diagram, we should mention that undirected arcs here are the subspaces and surfaces; point-groups here are the lines where these subspaces and surfaces intersect. Now we will describe the adaptive reactions defined by the experimental research of hydrobiological and hydrodynamical processes in the aquatic ecosystems.

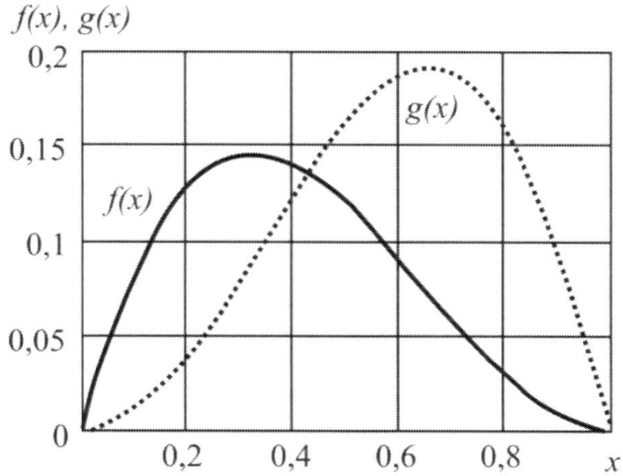

Figure 15.13. Diagrams of $f(x)$ and $g(x)$ functions.

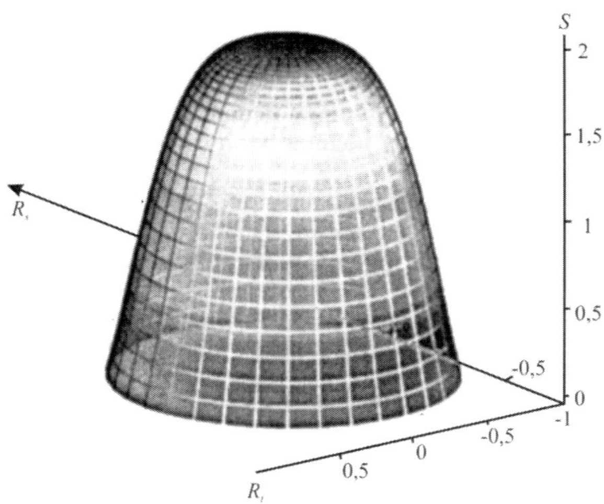

Figure 15.14.

15.1. EXAMPLES OF THE AQUATIC ECOLOGICAL SYSTEM ELEMENT'S ADAPTIVE REACTION UPON IMPACT

The following examples are used to show that the diagrams of the body most adaptive to reactions on the impact are convex functions with one maximum, and the body-impact system functioning in its neighborhood is the most adaptive one.

Example No. 1

S.A. Menzikov and P.A. Gdovskiy, working at the Institute of Biology of Inside Water RAS, analyzed the activity of chloride activated adenosine triphosphoric acid (ATA) taken from the crucian carp osmetic lepidic tissue with a dependence on a magnesium concentration [11]. The results of this experiment are shown in Figure 15.15. The authors of this research did not give any details in a table, because their aim was to find out the optimum level of magnesium impact. Therefore, the maximum was fixed at the magnesium concentration equal to 0.07 moles. ATA activity, shown on Figure 15.15, corresponds to the sector *BED* of the adaptational conditions diagram (see Figure 15.11).

Tabular information absence gives no chance to make a curve, fitting ATA reaction by the analytical connection. The next example shows the simplest process of the reaction fitting and this process of fitting will not be shown in the following examples.

Example No. 2. Experimental Data Fitting

The method of normalizing periodical impact on a one-component oscillatory process taking place in aquatic ecological systems can also be applied for normalizing the impact on a multicomponent oscillatory process. Before giving corresponding integral characteristics of multicomponent oscillatory processes in aquatic ecosystems, however, we should give the examples of scientific observations and experiments in mathematical treatment. Our aim is to make functional dependencies by fitting the experimental data with the planned accuracy.

Before some specific experimental data mathematical treatment, we should review the methods of estimation of tabular or other experimental data fitting errors. There are two basic ways to make such estimations.

The first way is based on comparing fitting results that are maximal in deviation from the specified experimental dependence. If $y(x)$ is the result of the experiment, $y_0(x)$ is the result of the fitting, and the experiment error estimation is defined as an absolute magnitude maximum difference of two (experimental dependence and fitting dependence) ordinates:

$$\Delta_C y = \max_{a \le x \le b} |y(x) - y_0(x)|, \tag{8}$$

in this case $[a, b]$ is a fitting segment. This estimation is usually referred to the B - A length of the fitting segment, where B and A are the borders of function $y_0(x)$ values.

The second method of estimation is based on the integral deviation of the experimental and fitting functions' absolute magnitude:

$$\Delta_L y = \int_a^b |y_0(x) - y(x)|\, dx. \tag{9}$$

In this case, the error is estimated in percents in relation to the area under the experimental function diagram.

Now, we will give an example of experimental data fitting by the elementary functions on the basis of the aquatic ecosystem element's reaction on the external impact experiments results.

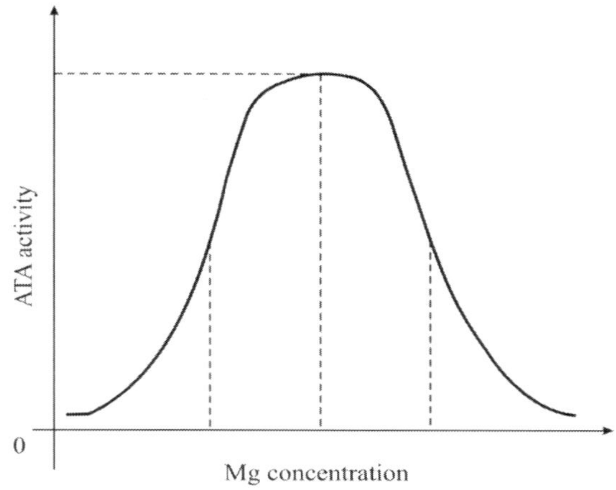

Figure 15.15.

Example No. 3

Let us discuss Table 15.1 where the ribulose-diphosphate-carboxylaze enzyme from lucerne green leaves' dependence on pH is shown. This experiment was run by T.V. Burova and other specialists [12]. The fitting process can be simplified by bringing dimensionless variables. These dimensionless variables should vary on the section [0, 1]. In this case, dimensionless analogs of pH (x variable) and temperature T (function y(x)) are expressed in the formulas:

$$x = \frac{pH-7{,}55}{10{,}50-7{,}55}, y = \frac{T-333{,}0}{342{,}2-333{,}0} \quad (10)$$

Therefore, the dimensionless analogs of pH and temperature, T, are changed at values $0 \le x$ and $y \le 1$. It is necessary to solve the equations (10) for pH and T to revert to dimensional variables.

Table 15.1.

No.	pH	T°C
1	7.50	343.2
2	7.75	343.0
3	8.00	343.0
4	8.25	343.0
5	8.50	342.7
6	8.75	342.2
7	9.00	341.6
8	9.25	341.0
9	9.50	339.8
10	9.75	338.8
11	10.00	336.8
12	10.25	335.6
13	10.50	333.0

Table 15.2.

No.	1	2	3	4	5	6	7	8	9	10	11	12	13
t	0	0.068	0.153	0.237	0.322	0.407	0.492	0.576	0.661	0.746	0.831	0.915	1
y	1	0.980	0.980	0.980	0.980	0.902	0.843	0.784	0.667	0.569	0.373	0.255	0

Now we will convert the values shown in Table 15.1 into dimensionless variables (10). The results of this conversion are shown in Table 15.2. According to these results, we will make an experimental diagram (Figure 15.17) based on the dimensionless variables, conforming to the "dimensional" diagram shown on Figure 15.6.

On this experimental diagram (see Figure 15.16), you can see that the following functions: $y_1(x) = \cos(\pi x/2)$, $y_2(x) = 1 - x^4$, $y_3(x) = 1 - x^3$ and $y_4(x) = 1 - x^2$ are very close to the experimental curve. $y_1(x)$, $y_2(x)$, $y_3(x)$ and $y_4(x)$ function diagrams can be seen in Figure 15.17, where the process of fitting and experimental functions approaching is also very well presented. The experimental diagram is between two ($y_3(x)$ and $y_4(x)$) function diagrams and is closer to the $y_3(x)$ function. That is why function $y_5(x) = 1 - x^{2.8}$ (Figure 15.18) is very close to the experimental diagram. Now, we will count the fitting error estimation for function $y_5(x)$ according to the abovementioned formulae for coordinate-wise and integral error estimations. The area under the experimental curve (see Figure 15.17) equals $S = 0.746$ dimensionless units. The area under the curve $y_5(x) = 1 - x^{2.8}$ (see Figure 15.18), is:

$$\int_0^1 (1 - x^{2.8})\, dx = 0.737, \frac{0.746 - 0.737}{0.746} = 0.0121 \qquad (11)$$

Thus, the relative (area under the experimental function diagram) integral fitting error estimation, calculated for the function $y_5(x)$ according to the abovementioned formulae, is 1.2%.

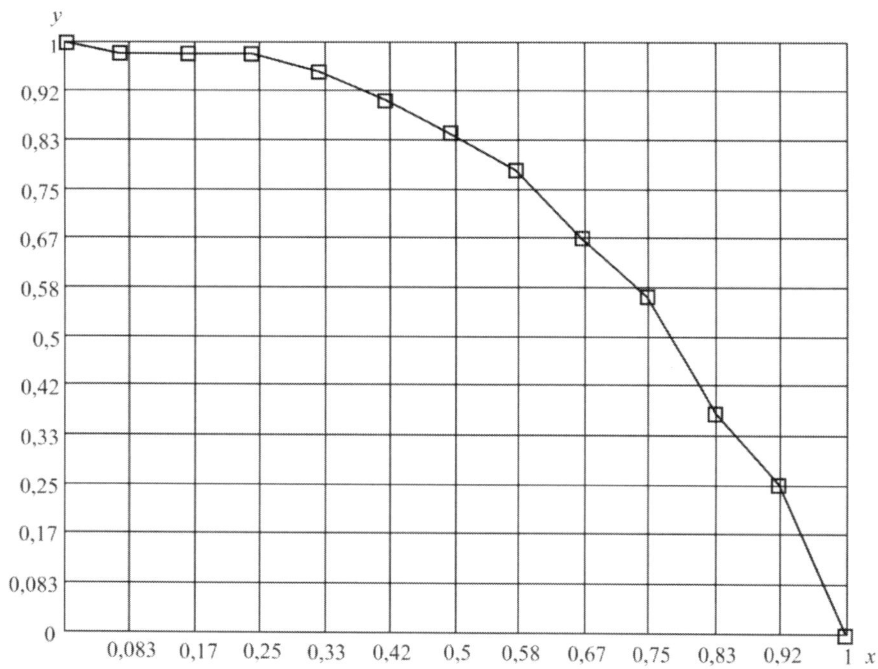

Figure 15.16.

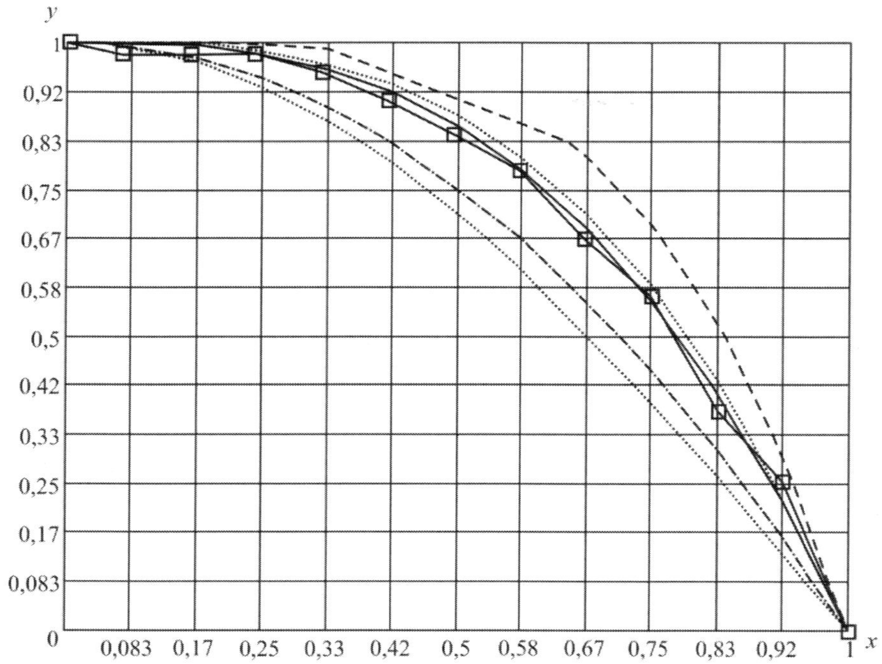

Figure 15.17.

The fitted function ordinates are much more deviated from the function, which is the result of the fitting, if $x \approx 0.45$. This deviation equals the difference 0.893 - 0.875 = 0.018. The ratio error of approximation according to the two functions' ordinates maximal deviation, related to the fitting section, equal 1 and is 1.8%. Moreover, the fitting function value between two points (if x = 0.45), found by the experiment, was defined by the linear interpolation method.

It is very easy to avoid the fitting error while comparing areas (1.2%); it is necessary to count the exponential function $y_0(x) = 1 - x^a$ so that the area under this function diagram concurs with the area $-S$ = 0.746 under the fitted function diagram. The area under the function $y_0(x)$ diagram is:

$$S_0 = \int_0^1 (1 - x_a)\,dx = \frac{a}{a+1}. \qquad (12)$$

That is why we should use the following equation:

$$\frac{a}{a+1} = 0.746 \Rightarrow a = 2{,}937 \qquad (13)$$

to define parameter a.

Thus, the function $y_0(x) = 1 — x^{2.937}$ fits the experimental one so that the areas under both diagrams coincide. Function $y_0(x)$ diagram and the experimental function diagram are shown in Figure 15.18.

We should mention, however, that the full error estimation of the fitting, counted by the areas comparing method, increases the error of fitting counted by the experimental and fitted functions ordinates comparing method. The function $y_0(x)$ average approximation error, relegated to the fitting interval length equal to 1, at two functions ordinates, has a maximal deviation of 2.9% but not less than 1.8% as it is for the $y_5(x)$ function. That is why a very good approximation will have the function with the diagram passing between the $y_0(x)$ and $y_5(x)$ functions. This function is $Y(x) = 1 - x^{2.868}$ (Figure 15.19). Ratio errors of the experimental diagram fitting are 0.7% according to the method of comparing areas and 0.9% according to the method of comparing ordinates. Both errors do not go beyond one percent. That is why experimental diagram fitting (see Figure 15.16) is done by the function:

$$Y(x) = 1 - x^{2.868}, \Delta_C y(x) = 0.9\%, \Delta_L y(x) = 0.7\% \quad (14)$$

(see Figure 15.19) can be taken as suitable, because the errors are within the accuracy of the experiment.[30]

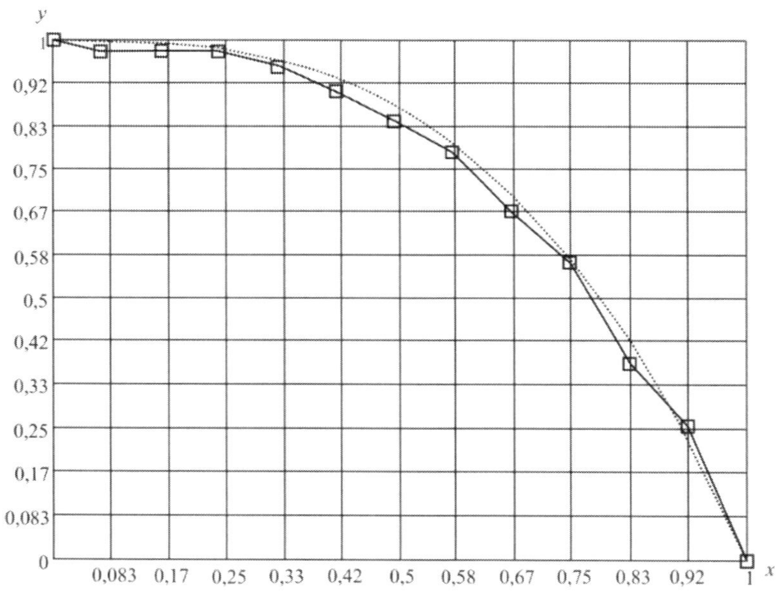

Figure 15.18.

The most important and most difficult question in the course of fitting is the choice of the initial function, because its diagram should qualitatively repeat properties of the experimental function diagram (see Figure 15.15). There are several standard ways of tabular and other data fitting: Fourier's series (intervals of Fourier's series, which is more exact) and orthogonal polynomials (Lezhandr's, Chebyshev's, Ermit's, etc.). Each way of fitting pulls function together with an arbitrary specified accuracy. All known ways of fitting, however, including the one mentioned above, have one drawback: the fitted function is not connected with the experimental one from a physical standpoint. Therefore, now we will just express (without proving) one theorem excluding such a lack of the experimental data fitting by the given set of functions; if the differential equation modeling hydrochemical or other process is known, the experimental diagram of such a process can be fitted by solutions of a nonlinear analog of the initial differential equation with the accuracy of the experiment carried out. In this fitting

[30] Unfortunately, the accuracy of the hydrochemical and hydrobiological experiments cannot be fixed.

method, the fitting function with the same physical significance is assigned to the experimental function.

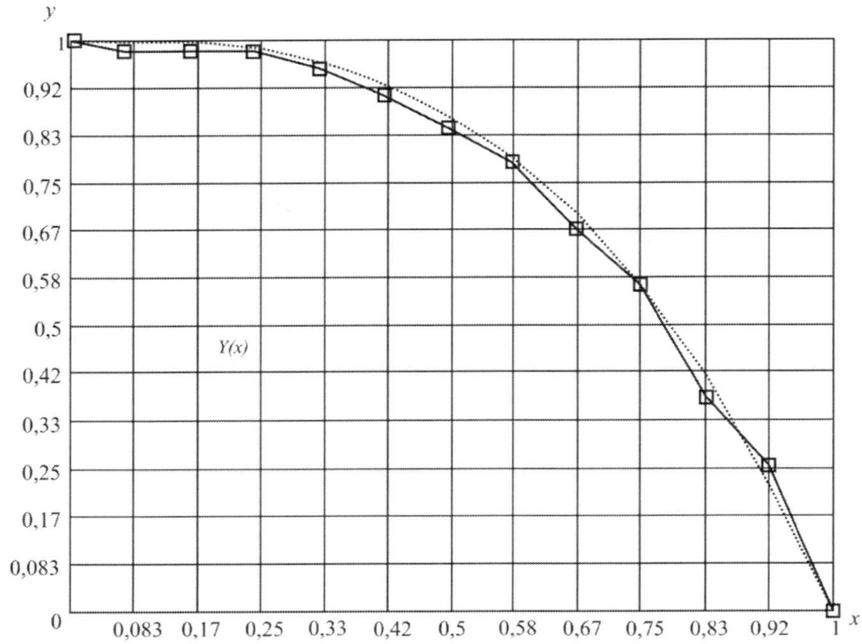

Figure 15.19.

Experimental and fitting curves in Figure 15.19 are the right descending branch of the model adaptation reaction (see the diagram in Figure 15.11, sector *OKEG*). This means that the process described by the experimenters passes the stages beginning with the adaption for impact and finishing the disrupting of tested enzymes. Moreover, the adaption process is only ¼ of the body's environmental capacity (see Figure 15.19).

Let us give more examples of the adaptive reactions, defined in the course of experiments. Thus, if the negative influence of the pollutant or radiation destroying biological bodies was investigated in the course of experiment, the term “adaptation” will be used with the negative meaning.

Example No. 4

R.V. Khimko searched some peculiarities of higher aquatic plants’ reaction on lighting changing [13]. *Vallisneria americana* and *Elodea canadensis* were used as the observational material. The factor influencing the plants was visible light, x. The bioelectric potential, ΔE, oscillatory amplitude was the reaction of aquatic plants on this influence.

The results of the experiment are shown in Figure 15.20.

These plants’ reaction with regards to lighting is shown by curves on the adaptation condition diagram (Figure 15.11) in sector *DEFb*. Adaptation curves of this sector differ from the curves in the *BED* sector by the lower adaptation capability level. This happens because the curves shown in Figure 15.20 have only one flex point. That is why aquatic plants’ reaction to the lighting is not as active as, e.g., the reaction of plants' tumors induction to a dose of radiation.

№	x W/m²	ΔE, B	
		Vallisneria	Elodea
1	3	40	50
2	5	80	56
3	8	90	62
4	12	88	77
5	16,5	80	80
6	18	76	81
7	25	70	86
8	31	64	84
9	46	60	78
10	61	52	70
11	75	48	64
12	91	46	60
13	110	44	58

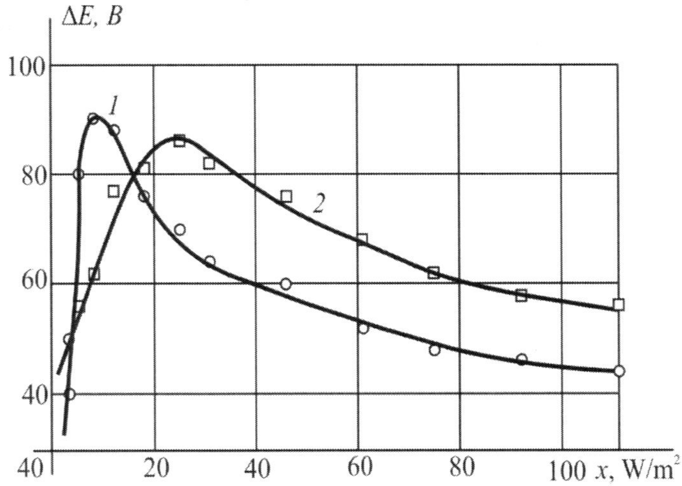

Figure 15.20. Dependence of the bioelectric potential ΔE oscillatory amplitude of Vallisneria americana (*1*) and Elodea canadensis (2) on the radiation dose *x*.

Figure 15.20 analysis shows that *Vallisneria americana*'s adaptation to the illumination intensity is a little bit higher than the adaptation of *Elodea canadensis*. The stability of its interaction with the illumination intensity, however, is lower than that of *Elodea canadensis*. This results from that fact that the peak of the *Vallisneria americana* adaptation curve around the maximum is higher than the peak of the *Elodea canadensis* (the same dimensioning, see Figure 15.20).

Example No. 5.

When M.A. Bolshakov and S.I. Alexeeva worked on the problem explained in Example No. *4*, they studied the Lymnaeidae pacemaker neuron's electric activity depending on their heating rate [14]. Heating temperature was the influencing factor. Changes of the spontaneous neurons' electric activity in a normal saline solution (SS) and ouabain solution (OS) were taken as the reaction of neurons.

The results of these experiments are shown in Figure 15.21.

The adaptation curve, corresponding to the neurons' spontaneous electric activity in the normal saline solution, has only one flex point. It is fitted by the adaptation condition diagram curve (see Figure 15.11), located in the sector *DEFb*. The adaptation curve, corresponding to the neurons' spontaneous electric activity in the ouabain solution, has two flex points and is fitted by the adaptation condition diagram curve located in sector *BED*.

Adaptation curves, shown in Figure 15.21, were plotted in one scale. Adaptation of neurons' spontaneous electric activity in the ouabain solution is 16 units higher than the adaptation of the neurons' spontaneous electric activity in the normal saline solution. The distance between the same line flex points of the adaptation curve, corresponding to the neurons spontaneous electric activity in the ouabain solution, is more than the distance between the same line flex point of adaptation curve corresponding to the neurons' spontaneous electric activity in the normal saline solution. That is why the ouabain solution stimulates adaptation and the neurons' spontaneous electric activity stability.

№	T °C	N	
		SS	OS
1	18,0	6,25	20,00
2	19,0	10,00	21,15
3	20,0	15,00	25,00
4	21,0	20,00	32,00
5	21,5	21,00	35,00
6	22,0	22,50	37,50
7	22,5	23,50	39,00
8	23,0	24,00	40,00
9	23,5	23,50	39,00
10	24,0	21,00	37,00
11	24,5	18,00	36,00
12	25,0	15,00	35,00
13	26,0	5,00	25,00
14	27,0	0,00	10,00
15	28,0	0,00	0,00

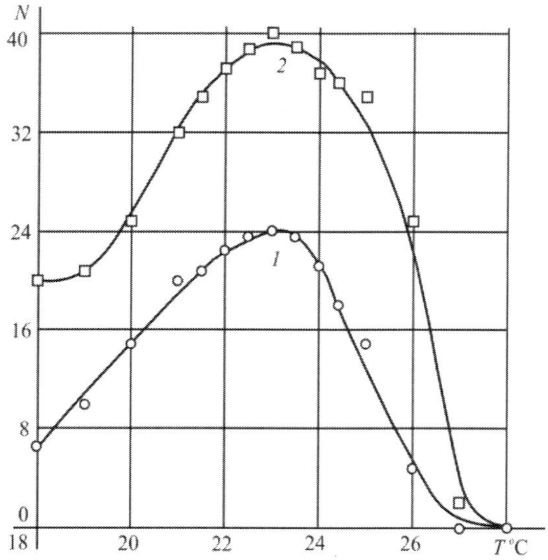

Figure 15.21. Temperature dependence of neurons' spontaneous electric activity in normal saline solution (SS) (*1*) and in an ouabain solution (OS) (*2*).

Example No. 6

This is the example of studying temperature stimulus for bacterial group growth. These bacteria were sampled in the Rybinsk Reservoir during the freezing ice period and put into a special thermostat. This experiment was carried out by V.I. Romanenko, V.A. Gavrilov and L.I. Zakharova [15]. The organic compounds' (protein hydrolisate) absorption by bacterial plankton was taken into account as a group reaction on the temperature action.

The protein hydrolisate absorption by bacterial plankton in the Rybinsk Reservoir experimental dependence on temperature is shown in Figure 15.22. The highest level of protein hydrolisate absorption by bacterial plankton was fixed at a temperature equal to 20°C. This so-called system "protein hydrolisate—bacterial plankton" is much more adaptive to the temperature activity at this very temperature level. The flex point at the adaptation curve diagram (T ≈ 37 °C) is located far from the maximum and means that the protein is stable to the temperature activity over the range of T = 0...40°C. The curve, shown on the diagram, is fitted by one of the curves in the *DEFb* sector of the adaptation conditions diagram (Figure 15.11).

Example No. 7

I.O. Solntseva and G.I. Vinogradova studied salinity resistance of yeast strains, sampled from the freshwater ecosystems [16]. Twenty-three types and 13 kinds of yeasts (total 200 strains) taken from different organs (body surface, branchiaes, intestinal canal) of fish inhabiting the Volga Reservoir, Danube river estuary and Ladoga Lake were used as the experimental material. Wort agar medium in seawater was used during the experiments. The salt content level was an influencing factor. The reaction was expressed by the ratio between the number of strains, capable of normal quantity growth (in percents) and the number of strains growing in normal salinity conditions. It was a double experiment; all the measurements were made before the yeasts adapted to the salt water and after that the fact.

№	T°C	y %
1	1,0	25,00
2	2,5	31,25
3	5,0	43,75
4	7,5	56,25
5	10,0	67,15
6	12,5	78,10
7	15,0	87,50
8	17,5	93,75
9	20,0	100,00
10	22,5	100,00
11	25,0	95,00
12	27,5	87,50
13	30,0	77,00
14	32,5	62,50
15	35,0	43,75
16	37,5	18,75
17	40,0	10,00
18	42,5	7,25
19	45,0	6,25
20	47,5	6,00
21	50,0	5,00

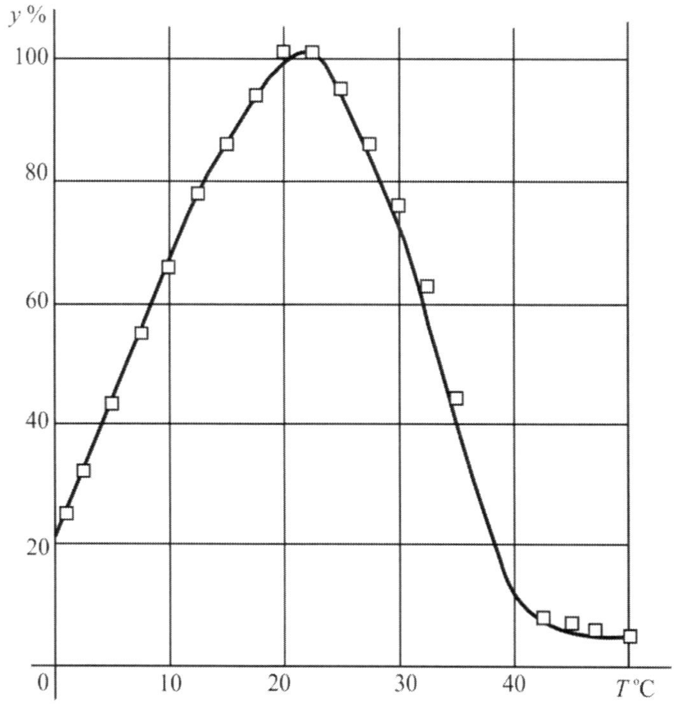

Figure 15.22. Protein hydrolisate absorption by bacterial plankton in the Rybinsk Reservoir, experimental dependence on atmospherical temperature, *T*.

The table and curves with the results of the experiments are shown in Figure 15.23. Both experimental curves are fitted by the connections, including into the adaptation conditions diagram in Figure 15.11. Before the adaptation, the curve is fitted by the right part of the adaptation curve in sector *DEFb*. After yeasts adapted to the seawater, the curve is fitted by the right part of the adaptation curve in sector *GEKO*. Moreover, yeasts adaptation to the seawater is well-seen in the salinity area 50-100%, which is more than one fifth (20%) (see Figure 15.23) of the yeasts in normal condition in seawater.

Example No. 8

V.L. Voronin studied the ultrasonic homogenization impact on the number of colony-forming units in the living pikes body blenna [17]. Different types of fungi were evolved from the blenna affected by the 44 kilohertz ultrasonic waves under a temperature of 28°C and electric potential of 100 Volt. A radiation dose, expressed in duration of ultrasonic impact, was used as an influencing factor. Fungi medium reaction was expressed by the number of colony-forming units (in percents) comparing to the control.

Four types of fungi were used for experiment:

1) Phoma in a culture
2) Phoma
3) Cladosporium
4) Penicillum

To simplify the experiment and because *Penicillum* almost did not react on the impact, we are giving the data only on three fungi types (Figure 15.24).

№	x %	y ‰	
		before adaptation	after adaptation
1	50	100	100
2	60	90	100
3	70	83	100
4	80	75	100
5	90	65	98
6	100	60	97
7	110	50	95
8	120	44	90
9	130	37	86
10	140	31	78
11	150	26	65
12	160	21	56
13	170	17	50
14	180	12	40
15	190	8	30
16	200	5	15

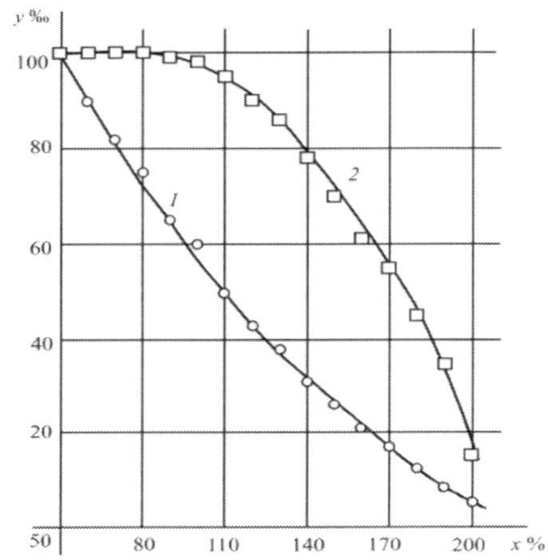

Figure 15.23. Dependence of the yeast strains' abundance, *y*, capable of normal quantity growth on the salinity, *x*, before adaptation (*1*) and after it (*2*).

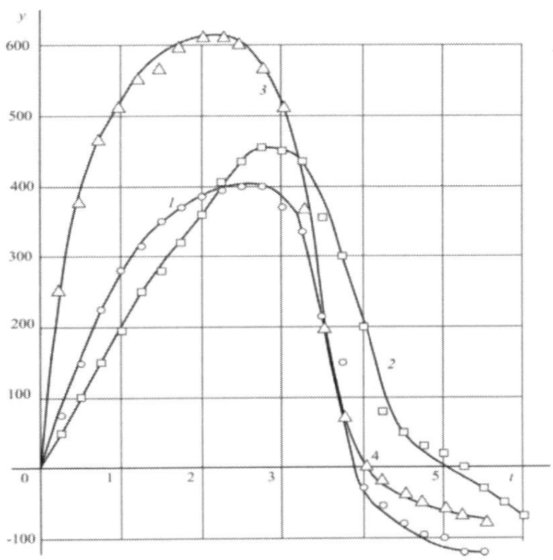

Figure 15.24. Dependence of the *Phoma* in colony (*1*), *Phoma* (2) and *Cladosporium* (3) colony-forming units quantity, *y*, on the ultrasonic impact time, *t*.

In comparing the diagrams in Figure 15.24 with the model diagram (see Figure 15.11), we can see that all fungi experimental reactions can be fitted by the curves of the *DEFb* sector. The most adaptive for ultrasonic impact is *Cladosporium* fungi. The less adaptive one is *Phoma* in culture. The most stable one in these conditions is *Phoma*.

Example No. 9

I.V. Veshnyakova and V.I. Romanenko studied dependence of duckweed (Lemna minor L) CO_2 exhalation while breathing on the temperature in thermostatic valve [18]. Temperature was an influencing factor during this experiment, and the quantity of CO_2 was a reaction.

The duckweed breathing capacity experimental curve is shown in Figure 15.25. This curve contains two flex points, which are situated distantly from the duckweed breathing passages' maximal adaptation area. This curve is situated in a "calm" *BED* area of the adaptation conditions' model diagram in Figure 15.11. The stable dependence of the duckweed breath on the temperature can be well-seen at 25- 42°C, which is about 42% of the temperature range before the breath failure appears. A very hairpin diagram decreasing after the maximal point proves that at 42°C and higher, duckweed breathing passages' adaptation level is too low. Fast aquatic ecological systems' adaptation to an impact reduction can probably be a reason for the ecosystem stability reduction or can even cause its full destruction. Taking care of such processes in the aquatic ecosystems elements is one of the most important aspects of ecological normalization.

№	*y* mkg/l	*T* °C
1	10,0	0,12
2	15,0	0,17
3	20,0	0,22
4	25,0	0,30
5	30,0	0,38
6	35,0	0,47
7	40,0	0,54
8	42,5	0,55
9	43,5	0,20
10	45,0	0,15
11	47,5	0,10
12	50,0	0,04
13	52,5	0,00

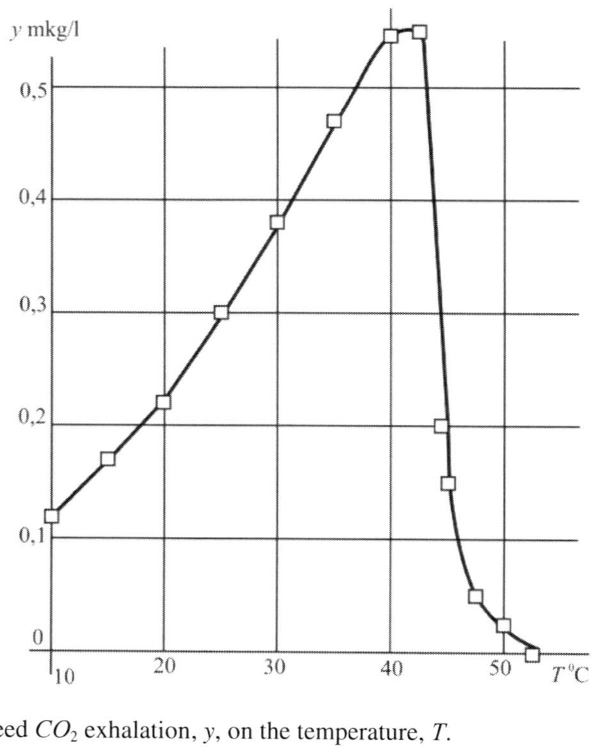

Figure 15.25. Dependence of duckweed CO_2 exhalation, *y*, on the temperature, *T*.

Example No. 10

V.A. Alexeev studied the dependence of some water insects and spider-like species' toxicoresistance on the water temperature [19]. *Chironomus plumosus* slugs, sampled in February in Rybinsk Reservoir inshore waters, were used as material for the experiment. Phenol C was an influencing factor, and the number of surviving slugs was presented as the reaction during this experiment.

Results of the experiment are shown in Figure 15.26 in tabular and diagram forms. The curve of the slugs' reaction on phenol impact is fitted by the adaptation condition left on the diagram in Figure 15.11, located in one of the upper sectors: *BED* or *DEFb*. In case of phenol's weak concentrations, it is not known what is going on with the slugs. That is why the diagram in Figure 15.26 does not give any possibility to view the slugs' stability. We can only say that this stability is very high. The flex point of such an adaptation curve has an abscissa of 1,950 μg /l. That is why stabilization area of the interaction between the slugs and phenol is not less than 70% of the total capability of the slugs living in the phenol.

Example No. 11

A.Y. Karataev studied *Stictochironomus Psammophilus Tscheru* abundance dynamics in heat sink different temperature zones in terms of Lukomskoe Lake littoral in 1971 and 1980 [20].

The dependence, resulting from the Karataev's experiment, is shown in Figure 15.27. *Stictochironomus Psammophilus Tscheru*'s adaptation abilities to the temperature impact in this experiment are similar to those adaptation abilities described in the previous experiment, and we will not show them now.

№	*C* mkg/l	*N* %
1	900	100
2	1050	98
3	1200	96
4	1350	90
5	1500	80
6	1650	50
7	1800	30
8	1950	30
9	2100	12
10	2250	6
11	2400	2
12	2550	0

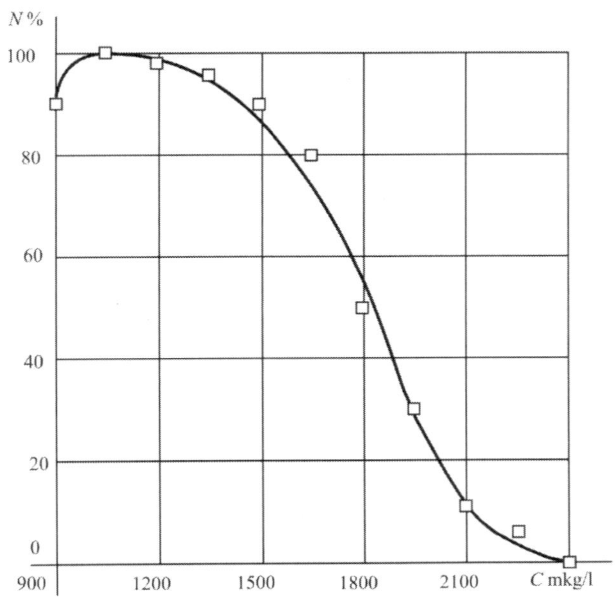

Figure 15.26. Dependence of some water insects and spider-like spices toxicoresistance (*N*) on phenol concentration (*C*) in water.

№	$T\,^0C$	N ex/m^2
1	10,0	3900
2	11,0	3500
3	12,0	3300
4	13,0	2800
5	14,0	2300
6	15,0	1500
7	15,5	1000
8	16,0	700
9	16,5	500
10	17,0	300
11	18,0	200
12	19,0	100
13	20,0	0

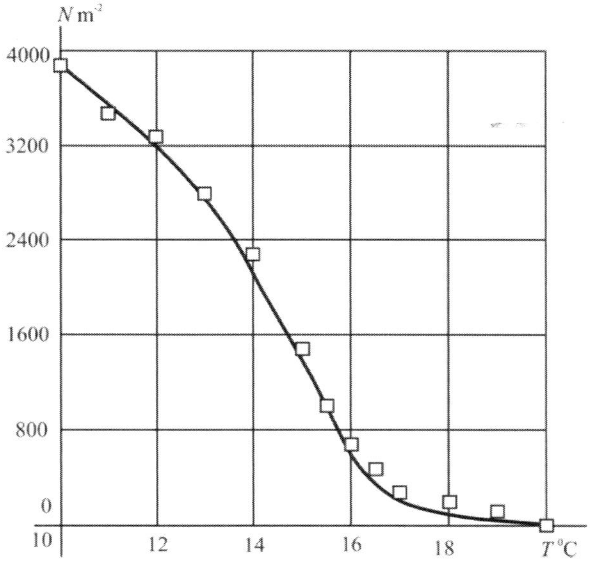

Figure 15.27. Dependence of *Stictochironomus Psammophilus Tscheru* abundance (N) on water temperature (T).

Example No. 12

I.V. Semenov and G.V. Semenova found out the dependence of ammonifying microorganisms' abundance on nitrogen-containing organic substances in the river Kambileevka during their study of the river water self-purification processes [21]. The results of their experiment are shown in Figure 15.28, where the curve is fitted by the adaptation conditions diagram in sector *BED* (see Figure 15.11).

The high peak of the experimental curve maximum indicates the low stability of river-water self-purification stability: ammonifying microorganisms and nitrogen-containing organic substances in river Kambileevka interaction stability is not more than 23% of the general microorganisms' capabilities to resist such kinds of contamination.

Example No. 13

V.I. Romanenko, M.K. Obey-Aberu and L.I. Zakharova described the temperature regime of bacteria abundance growth in silts [22]. The experiment was carried out in the laboratory and at the same time in natural conditions. Water temperature was an influencing factor, and CO_2 thermotrophy assimilation in percents from its maximum is the reaction.

The full-scale natural experiment was carried out in summer in the Rybinsk Reservoir, and the laboratory was carried out into the silt and put into a methane-tank with thermophilic microflora. The dependencies resulting from the experiments are shown in Figure 15.29. Both in the laboratory and full-scale experiments, the stabilizing capabilities of bacteria abundance growth at temperature rise were not too great. Bacteria stabilizing capabilities in full-scale conditions, however, are higher than in laboratory ones.

Moreover, interaction within the stimulation processes in the laboratory were not as intensive as the ones in full-scale conditions, and after the adaptation peak in the laboratory stimulation it quickly recessed so that it could be a reason of bacteria destruction. Both curves, shown in the figure, are situated in sector *BED* of the adaptation conditions diagram (see Figure 15.11).

Example No. 14

K.O. Granstrem and E.V. Vorobeynikov studied viruses and bacteria reaction on the electromagnetic field (EMF) radio-frequency range [23]. The subject of research here is *staphylococcal bacteriophage* influenced by EMF 250 V/m with the excitation time of 3 hours. The radiation frequency was used as the excitation controllable parameter, while bacteriophage inactivation, i.e., its lytic activity, is the reaction.

№	*C* mg/dm³	*B* rel. unit
1	0	0,00
2	2	0,03
3	3	0,06
4	4	0,13
5	6	0,38
6	8	0,68
7	10	0,80
8	11	0,81
9	12	0,80
10	14	0,74
11	16	0,64
12	18	0,56
13	20	0,48
14	22	0,42
15	24	0,36
16	26	0,30
17	28	0,25
18	30	0,20
19	36	0,10
20	44	0,03

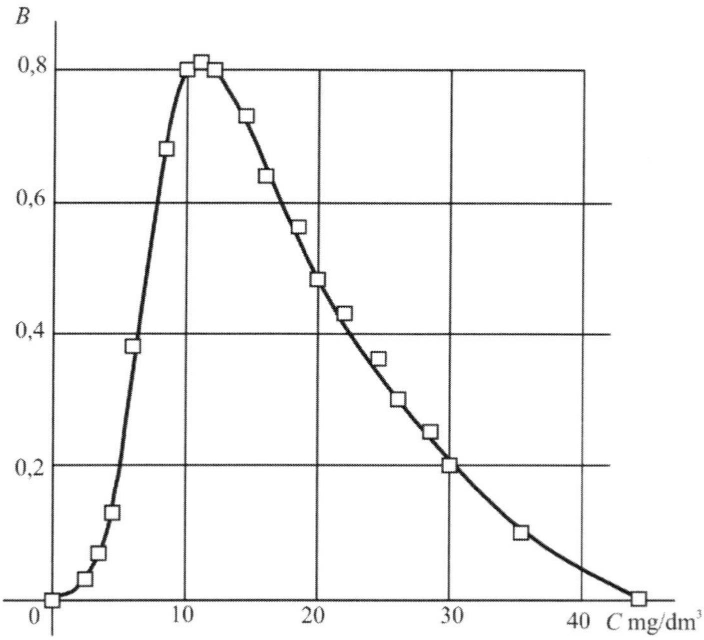

Figure 15.28. Dependence of the ammonifying microorganisms' abundance (*B*) on nitrogen-containing organic substances concentration (*C*).

The results of the experiment are shown in Figure 15.30. The curve, shown in this figure is fitted by the line and situated in the adaptation conditions diagram sector *GEKO* (see Figure 15.11). Viruses and bacteria adaptation to the radiation influence is not too high, and it is impossible to speak about the stabilization of this interaction on the basis of this diagram, because the curve has no flex points.

№	T	E %	
		F	L
1	10,0	25,0	2,0
2	13,3	7,0	2,0
3	16,7	11,0	3,5
4	20,0	17,5	4,0
5	23,3	25,0	5,0
6	26,7	45,0	5,5
7	30,0	98,0	6,5
8	33,3	100,0	7,5
9	36,7	98,0	9,0
10	40,0	40,0	11,0
11	43,3	7,5	14,0
12	46,7	3,5	22,0
13	50,0	2,0	37,5
14	53,3	—	65,0
15	56,7	—	100,0
16	60,0	—	50,0
17	63,3	—	35,0
18	66,7	—	26,0
19	70,0	—	20,0
20	73,3	—	15,0
21	76,7	—	10,0
22	80,0	—	7,0

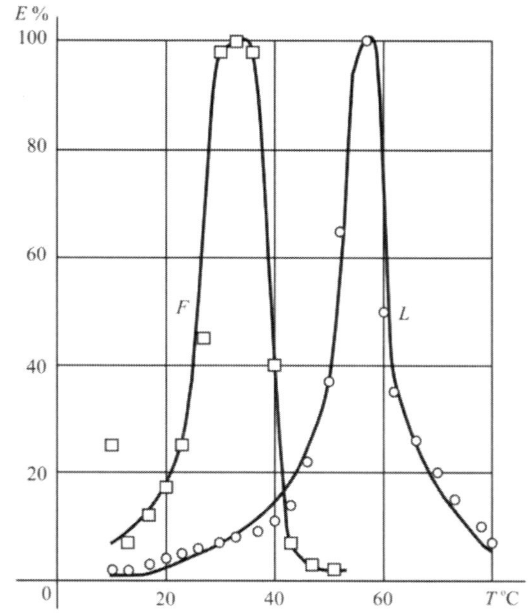

Figure 15.29. Dependence of CO_2 heterotrophic assimilation activity (*E*) on temperature (*T*) in full-scale (F) and laboratory (*L*) conditions.

№	*f* MHz	*A* ‰
1	0,5	5
2	1,0	23
3	1,5	30
4	2,0	40
5	2,5	45
6	3,0	48
7	3,5	49
8	4,0	50
9	4,5	50
10	5,0	49
11	6,0	48
12	8,0	46
13	10,0	43
14	12,0	40
15	20,0	24
16	22,0	21
17	24,0	17
18	26,0	13
19	28,0	8

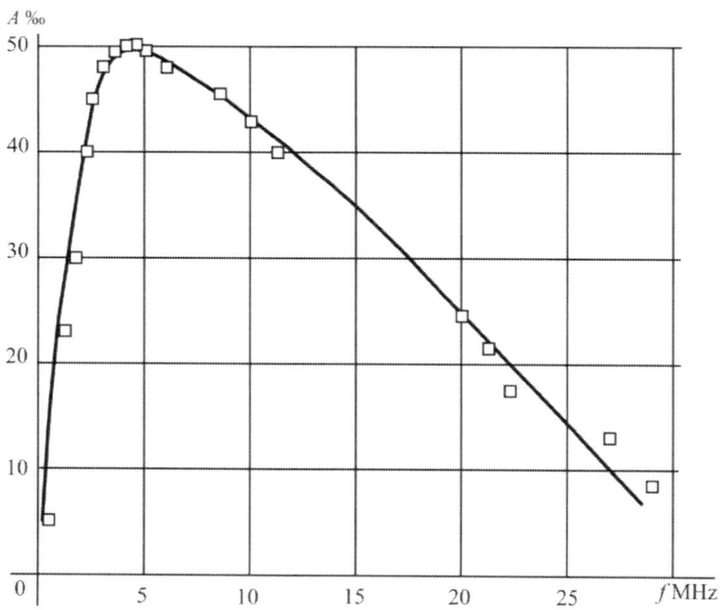

Figure 15.30. Dependence of bacteriophage, *A*, inactivation on the EMF frequency, *f*.

Example No. 15

L.V. Shchedrina, S.N. Kolyubaeva, etc., studied DNA repair synthesis after γ-radiation in stimulated and non-stimulated *phytohemagglutinin lymphocytes* (*PHA-lymphocytes*) [24]. The experimental dependence drawn by the scientists is presented in Figure 15.31. The experimental curve can be seen in the adaptation conditions diagram sector *GEKO* (see Figure 15.11).

Example No. 16

M.G. Beridze, together with other scientists, studied postradiational changes in neural tissue membrane-bound enzymes activity, particularly ionizing radiation impact on the ATA (adenosine triphosphoric acid) and rats encephalon micro-modal fraction n-nitrophenil phosphate hydrolisis [25].

The curve resulting from Beridze's experiment is shown in Figure 15.32. This curve can be fitted by the adaptation dependencies in sectors *aGEB* and *BED* of the diagram (see Figure 15.11). As the experiment was not finished, we cannot fix all the possibilities of the enzyme adaptation to radiation. We can only state that at the beginning (until x = 3.75 Gy) radiation stimulates the enzyme adaptation very slowly, and after that value, radiation stimulates this process very quickly and the stabilization interval of the enzyme interaction with the radiation consequences is wide enough (not lesser than 59% of the enzyme resistance against the radiation consequences ability).

№	*x* Gy	*y*
1	5	28
2	8	37
3	10	63
4	15	90
5	20	118
6	25	120
7	27	119
8	30	118
9	35	107
10	40	93
11	45	80
12	50	70
13	55	60

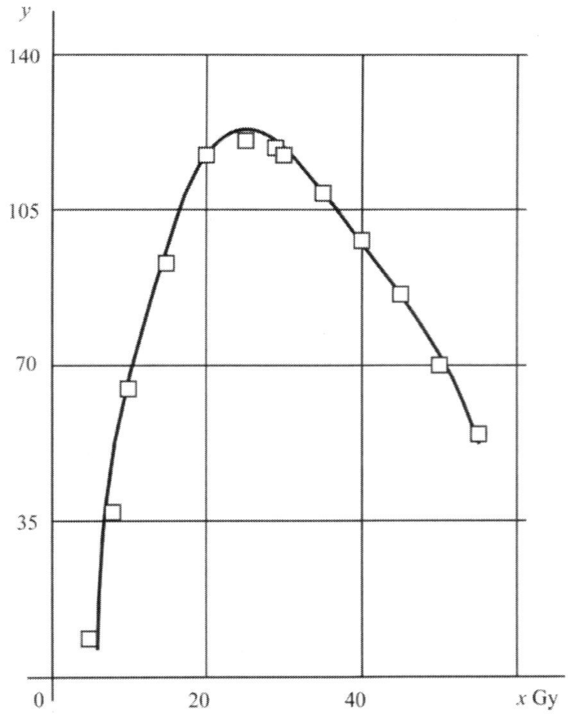

Figure 15.31. Dependence of cells (y) with nuclei number (for 1,000 lymphocytes) on the radiation dose (x).

№	x Gy	y
1	1,75	1,20
2	2,33	1,22
3	2,92	1,25
4	3,50	1,28
5	4,08	1,46
6	4,67	1,64
7	5,25	1,76
8	5,83	1,83
9	6,42	1,87
10	7,00	1,88
11	7,58	1,86
12	8,17	1,82
13	8,75	1,67

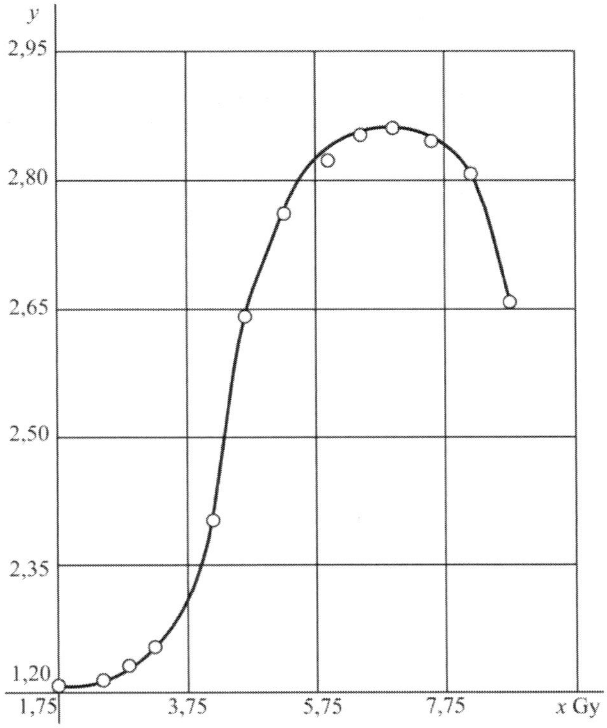

Figure 15.32. Dependence of the enzyme activity in percents of the zymolites (y) hydrolisis on the ionizing radiation dose (x).

Example No. 17

D.T. Zaorgievskaya and A.I. Gaziev studied the dependence of DNA breaks quantity on the UV radiation dose [26]. Cells were irradiated and then incubated for 30 minutes and the single-straight DNA quantity was found out.

The results of the experiment described above are shown in Figure 15.33. The curve $y = y(x)$, resulting from the experiment, is seen in the adaptation conditions diagram sectors *GEKO* and *DEFb* (see Figure 15.11). The results of this experiment as well as the results of the previous experiment do not let us characterize the most important, from the point of view of ecological normalizing, branch of the curve, defining the process of adaptation break.

Example No. 18

L.L. Alievskaya, together with other scientists, studied the phalloidin impact on fish skate cross muscles [27]. Only the descending branch of the fish skate cross muscles adaptation reaction diagram on the phalloidin impact is characterized for this experiment (Figure 15.34). Curve's ascending branch, defining the process of the muscle's adaptation to the impact is out of the scientist's consideration. On the basis of the experiment results, however, we can state that the fish skate cross muscles are highly adaptive to the phalloidin impact, because the adaptation curve maximum area in this case is very wide and is not less than 25% of all the fish skate muscle's abilities to resist the impact. The horizontal distance between the adaptation curve's right inflection point and its relation to the phalloinid changes full is 58%. The left branch of the adaptation curve is not described here. That is why stabilization the interval exceeds 58% of all the fish skate cross muscle's abilities to resist the phalloidin impact. The curve is situated in one of the adaptation conditions diagram sectors (*DEFb* or *BED*) (see Figure 15.11).

№	x J/m²	y %
1	0	0,0
2	10	6,0
3	20	9,5
4	30	13,0
5	40	15,3
6	50	17,2
7	60	19,2
8	70	20,0
9	80	20,4
10	90	20,2
11	100	19,0

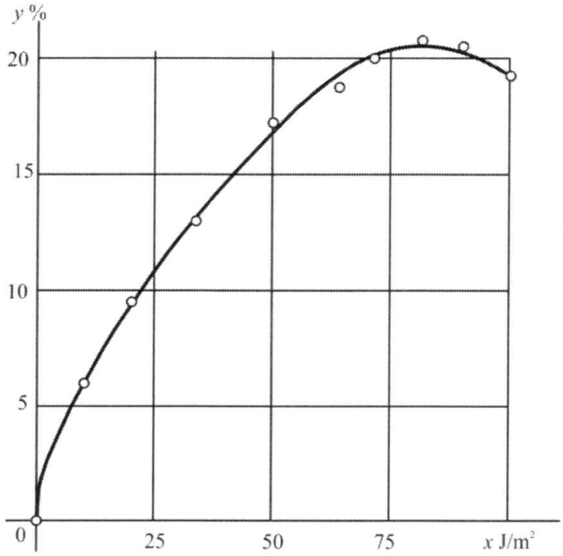

Figure 15.33. Dependence of single-straight DNA quantity (y) in the cell on the radiation dose (x).

№	x relative units	y 10^{-5} H
1	4,25	350
2	4,50	350
3	4,75	340
4	5,00	325
5	5,25	300
6	5,50	217
7	5,75	150
8	6,00	90
9	6,25	40
10	6,50	20
11	6,75	10
12	7,00	5

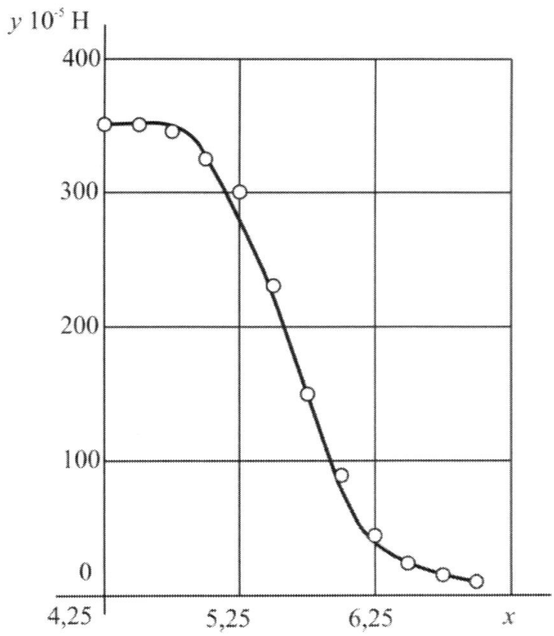

Figure 15.34. Dependence of the fish skate cross muscles strength (y) on phalloidin concentration (x).

Example No. 19

A.M. Emelyanenko and B.L. Bovykin studied the dependence of the bull brain H-octane and H-tetradecanoic solutions lipid bilayers' (LB) electrical capacity on gallium chloride and isonicotinic acid hydrazide complex concentration [28]. The results of the experiment are shown in Figure 15.35. The corresponding adaptation curve is situated in the adaptation conditions diagram sector *aGEB* (see Figure 15.11). Ascending and descending branches of the LB electrical capacity adaptation to the gallium chloride impact are very sharp (because a logarithmic scale is used for the impact axes). That is why the membrane's electrical capacity adaptation to the gallium chloride is high enough (because in case of changing a logarithmic scale into ordinary one, the maximum point area will be wide). Logarithmic characteristics of the stabilization interval (horizontal distance between the curve diagram left point and its right branch, referred to the whole interval of the measurements) are 60%. Logarithmic scale basis in this experiment is not mentioned that is why we can only state that the stabilization of the membrane's adaptation to the chloride impact does not exceed 60%.

№	x relative units	y uF/sm^2
1	3,750	0,525
2	4,000	0,550
3	4,125	0,570
4	4,250	0,610
5	4,375	0,670
6	4,500	0,700
7	4,625	0,695
8	4,750	0,685
9	4,875	0,680
10	5,000	0,670
11	5,125	0,660
12	5,250	0,675
13	5,375	0,630
14	5,500	0,600
15	5,625	0,580
16	5,750	0,550
17	5,875	0,530
18	6,000	0,510

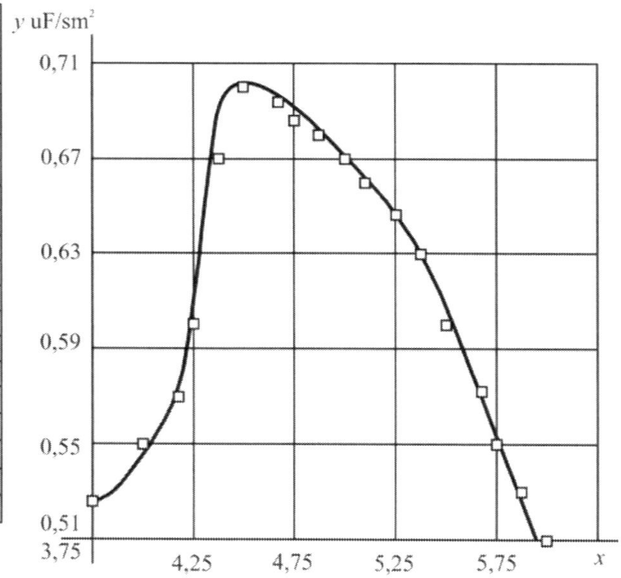

Figure 15.35. Dependence of the LB electric capacity (*y*) on the logarithm of the gallium chloride complex concentration(*x*).

Example No. 20

B.I. Sukhorukov, together with other researchers, studied γ-radiation affecting DNA and RNA exposed in media with different acidity level with and without oxygen [29].

The results of the experiment are shown in Figure 15.36. The experimental dependence is qualitatively expressed by the adaptation curve in the adaptation conditions diagram sector *BED* (see Figure 15.11). DNA destruction is stable at the pH values 2.75-4.75. That is why the stable zone of the DNA destruction in this experiment is 2/3 or 67% of the impact general interval.

Example No. 21

B.I. Davydov, together with other scientists [30], on the basis of the experiments collected data about the biological reaction of mice on the microwave irradiation impact. The total radiation dose (24 J/sm^2) was fixed during these experiments and its power was changed. The results of such experiments are shown in Figure 15.37. Such reactions can be found in the curvilinear sector *OKF*; the curves there have two inflections on the right side of the maximum adaptation (mice anti-adaptation to the impact and death in this case) abscissa. Two (or several) inflections mean that the adaptation reduction interval increases. If several inflections are on the maximum adaptation abscissa left side, the adaption abilities stimulation interval increases.

№	pH	y
1	1,75	0,135
2	2,00	0,150
3	2,25	0,160
4	2,50	0,175
5	2,75	0,200
6	3,00	0,225
7	3,25	0,240
8	3,50	0,250
9	3,75	0,245
10	4,00	0,235
11	4,25	0,215
12	4,50	0,190
13	4,75	0,170
14	5,00	0,150
15	5,25	0,135
16	5,50	0,125
17	5,75	0,115

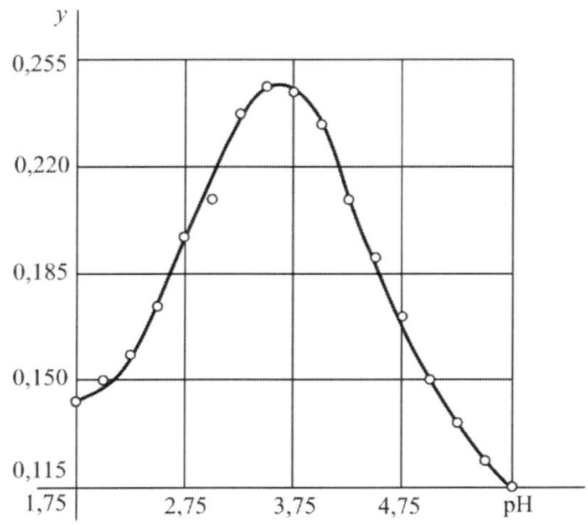

Figure 15.36. Dependence of the DNA and RNA destruction on pH with irradiation dose 600 Gy.

№	pH	y
1	1,75	0,135
2	2,00	0,150
3	2,25	0,160
4	2,50	0,175
5	2,75	0,200
6	3,00	0,225
7	3,25	0,240
8	3,50	0,250
9	3,75	0,245
10	4,00	0,235
11	4,25	0,215
12	4,50	0,190
13	4,75	0,170
14	5,00	0,150
15	5,25	0,135
16	5,50	0,125
17	5,75	0,115

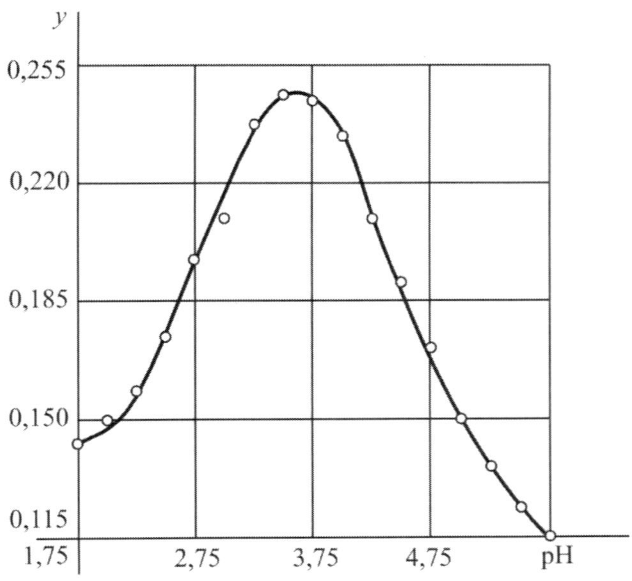

Figure 15.37. Dependence of the dead mice quantity (y) on microwave irradiation.

15.2. Oscillatory Process Adaptation Characteristics. Laboratory Experiment

Now, we will give an example of the oscillatory system's adaptation reaction on the oscillation. This example is based on the laboratory experiment carried out by Hydrochemical Institute specialists to find out the interaction of herbicides with aquatic microorganisms [31, 32]. This experiment's aim was based on that fact that the pesticides are widely used in agriculture and the researchers have been interested in the problem of the environmental pollution by pesticides and, as a result, microflora composition type change (this microflora plays a great role in a water body's self-purification) for a very long time. Some microorganisms produce biocatalysts and adapt to these substances faster using them as feed elements. Researching shows, however, that the kinetics of the pesticide's destruction by microorganisms is typical for bacterial growth during the "unfamiliar" substrate reclamation. Moreover, some microorganisms use organic substances as feed elements, other ones use them as nutritional compounds. Lower organisms are interconnected, first of all, by choronomic organic metabolites, which are likened to the relations of "impact-reaction" and "resource-consumer."

2M-4x herbicide, widely used in agriculture, was chosen for this experiment as an influencing factor (impact). Researchers studied aquatic organisms' biomass reaction on these pesticides.

The experiment was made as follows; natural water and 70-80% 2M-4x herbicide solution (water-soluble powder) were used in this case.

The experiments were made in pathogen-free conditions (herbicide was autoclaved at 1,013 hPa pressure for 20 minutes). The concentration of the applied chemicals calculation was based on the basic solution, containing 100 ng/l of 2M-4x herbicide. The following concentrations were used during experiments: 0.025, 0.050, 0.100, 0.500, 2,500 and 5,000 mg/l. To control the results of the experiments, researchers used natural water. The researchers did not apply any chemicals to this natural water. The experiments were done twice.

Natural water inoculation on agar-meat infusion (AMI) was made as follows:

- 1st inoculation was made while filling bell jars with water
- 2nd one was made after the agent application (two hours later)
- 3rd, 6th, 8th, 11th and 12th ones came 3, 6, 8, 11 and 12 hours later. Bacterial cells grew in thermostat at a temperature equal to 30°C. The bacterial cells' grown colonies were counted in five days.

This data was processed in advance. First of all, corresponding values, resulting from the control experiments, were subtracted from total values. Secondly, bacteria quantity (in this case, it is an aquatic environment adaptation reaction on the herbicide impact) was normalized by the bacteria arithmetic average on each day of the experiment. This normalization was made for all the herbicide concentrations resulting from the experiment. Average normalized values are the most informative ones in the case of adaptation reactions research, because normalization (as the element of the experimental data statistical manipulation) helps to exclude variable errors. The results of the experiments are shown in Table 15.3 and are summarized in Table 15.4.

We should also mention that the data received on the first day of the experiment is less informative. It is natural, because from the point of view of biophysics, the object's reaction on the impact cannot proceed suddenly. Therefore, to describe this experiment and its results, we will use the data beginning with the second day.

Table 15.3. Bacteria quantity in case of 2M-4x different concentrations

Inoculation No.	Control	Concentration, 10-2 mg / dm3						
		0.025	0.050	0.1000	0.5000	1.000	2.500	5.000
1	149	151	147	140	198	148	143	150
2	149	551	547	240	198	326	183	203
3	115	462	687	155	355	427	611	337
4	33	218	164	120	125	250	142	272
5	8	88	463	210	333	62	67	120
6	117	767	1,226	836	380	1,378	347	606
8	7	48	507	102	36	33	88	73
11	63	147	770	110	83	208	94	317
12	19	297	150	73	103	67	107	53

Table 15.4. Bacteria experimental average quantity (average units) after normalization

Day	Concentration, mg / dm3							
	0.000	0.025	0.050	0.100	0.500	1.000	2.500	5.000
2nd	0.00	1.98	1.95	0.45	0.00	1.16	0.17	0.29
3rd	0.00	1.09	1.80	0.13	0.75	0.98	1.56	0.70
4th	0.00	1.22	0.87	0.57	0.61	1.43	0.72	1.56
5th	0.00	0.43	2.47	1.10	1.77	0.29	0.32	0.61
6th	0.00	0.96	1.64	1.07	0.39	1.87	0.34	0.73
8th	0.00	0.32	3.90	0.74	0.23	0.67	0.63	0.51
11th	0.00	0.46	3.84	0.26	0.11	0.79	0.17	1.38
12th	0.00	2.71	1.28	0.98	0.82	0.47	0.86	0.33
Average	0.00	1.15	2.22	0.66	0.78	0.96	0.60	0.76

The microorganism's interaction with the herbicide is an oscillatory process. The character of this interaction is described in the tables shown below. The microorganism's average quantity dependence on the herbicide concentration is shown in Figure 15.38. A nonuniform scale of the concentration axis is used for higher informativeness.

Now we will compare Figure 15.38 with Figures 14.1 and 14.2. The qualitative coincidence of these figures is evident. Bacteria interaction with the herbicides curve is a damping oscillatory process. The same dependence coincides with the adaptation oscillatory processes. Thus, from the point of view of the microorganism's adaptation to the herbicides impact, *the average characteristic of the bacteria interaction with the microorganisms is a typical oscillatory process adaptation curve.*

The analysis made *proves the efficiency of the adaptation normalization method, worked out and described in the present monograph.*

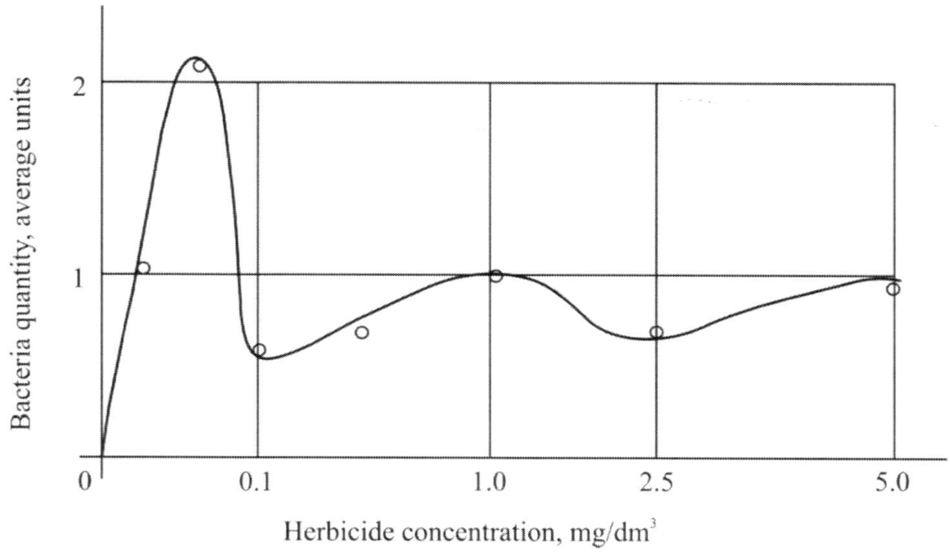

Figure 15.38.

15.3. Full-Scale Study

A. M. Nikanorov, A. V. Zhulidov and other scientists working at the Hydrochemical Institute [33] studied processes connected with the freshwater ecosystems buffer capacity to heavy metals.

The research was carried out at the Usman River, situated on the territory of Voronezh reserve. The abovementioned reserve was chosen for a long-period research because it is situated in the most suitable are for research; it is a forest-steppe zone, characterized by optimum conditions for all ecosystem components' development.

As it was impossible to make a long-term natural experiment (years: 1977-1981, 1983, 1984) that would take into account all hydrobionts' nutritional groups, the scientists decided to concentrate on phytoplankton, which is some kind of "producer," defining a large variety of hydrochemical and hydrobiological criteria.

Cuprum, which is highly active biologically and can be completely dissolved by the organic substance, was used for these experiments.

The aim of the experiments was to study the interconnection of the river Usman phytoplankton biomass' seasonal and annual dynamics and the concentration of cuprum dissolved in the water of the river.

All the data on the phytoplankton biomass' seasonal and annual dynamics and quantity of the dissolved cuprum are shown in Table 15.5.

Now we will not make a detailed analysis of the abovementioned experiments [33]; we will just describe the phytoplankton adaptation characteristics in interaction with cuprum. These characteristics are described by the average values of phytoplankton and cuprum concentrations, shown in Table 15.5.

Table 15.5. Average multiannual changing of the dissolved cuprum and phytoplankton biomass content in the Usman River (years 1977-1981, 1983, 1984)

Month	Dissolved cuprum, mkg/l	Phytoplankton, mg/m^3
January	0.687 ± 0.120	139.8 ± 59.7
February	0.636 ± 0.107	143.7 ± 55.1
March	0.593 ± 0.129	184.8 ± 96.3
April	0.356 ± 0.056	283.5 ± 87.3
May	0.386 ± 0.098	357.4 ± 118
June	0.319 ± 0.079	371.1 ± 91.2
July	0.332 ± 0.084	471.9 ± 77.0
August	0.316 ± 0.070	504.4 ± 79.1
September	0.307 ± 0.091	472.9 ± 82.4
October	0.401 ± 0.123	341.7 ± 103
November	0.545 ± 0.106	252.9 ± 114
December	0.706 ± 0.058	144.0 ± 58.2

To make the mathematical processing of the information, shown in Table 15.5, easy, we will begin by analyzing the cuprum and phytoplankton concentration and will continue with non-dimensional variables. Now, let us define the following units: M is the cuprum concentration, F is the phytoplankton concentration, and cuprum (μ) and phytoplankton (φ) concentration are the non-dimensional variables, defined according to the following formulae:

$$\mu = \frac{M - M_{min}}{M_{max} - M_{min}}, \varphi = \frac{F - F_{min}}{F_{max} - F_{min}} \quad (15)$$

The average values are defined according to Table 15.5:

$M_{min} = 0.307$, $M_{max} = 0.706$, $F_{min} = 139.8$, $F_{max} = 504.4$

Now we have to recalculate the concentration values, shown in Table 15.5, according to the non-dimensional variables, using formulae (15) (see Table 15.6).

Table 15.6.

Month…...	I	II	III	IV	V	VI
μ…………	0.952	0.825	0.717	0.123	0.198	0.030
φ…………	0.000	0.011	0.123	0.394	0.597	0.634
Month…….	VII	VIII	IX	X	XI	XII
μ…………..	0.063	0.023	0.000	0.236	0.596	1.000
φ…………..	0.911	1.000	0.910	0.554	0.310	0.012

The second step is to plot the diagrams of the average multiannual changing of the dissolved cuprum and phytoplankton biomass content in the Usman River time dependencies (years: 1977-1981, 1983, 1984) (Figure 15.39) according to the data from Table 15.6.

The diagrams, shown in Figure 15.39, are the example of cuprum and phytoplankton biomass concentration characteristics' dependence on time period (but not on impact, as all the above mentioned adaptation curves). Nevertheless, the basic data on the phytoplankton adapting to the cuprum impact is shown in Figure 15.39.

Phytoplankton and cuprum adaptation interaction period lasted six months from mid-April until mid-October. This period is estimated by the horizontal distance between the cross points of the dependencies of the cuprum and phytoplankton concentration in the river; this dependence is shown in Figure 15.39.

The period of the phytoplankton adapting to the cuprum impact (April-October) is estimated by the distance between the left and right flex points on the diagram, characterizing the phytoplankton concentration.

The cuprum concentration since May until October is very low and almost equal each month.

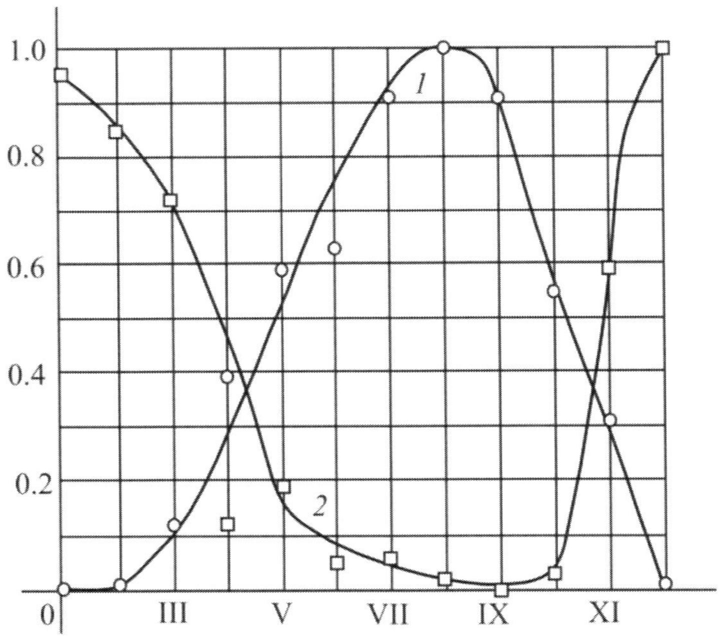

1—phytoplankton, 2—cuprum

Figure 15.39. Cuprum and phytoplankton biomass concentration characteristics' dependence on time period.

We should also mention that the phytoplankton concentration has its maximal value one month earlier than the cuprum concentration.

We suggested a new idea of an adaptation reaction study in paper [33]; the matter was studied not only in terms "impact-reaction," but in terms "impact, reaction and period of time." This is shown in one and the same diagram or in one and the same table. This approach is much more general than studying adaptation reaction only in terms "impact-reaction," because here, we have manifold possibilities to describe this very adaptation reaction.

For example, if we analyze the data shown in Table 15.6, we will understand that the best fitting of a cuprum and phytoplankton concentration in water is a parabolic one. If $Y(t)$ is a phytoplankton concentration and $X(t)$ is a cuprum concentration, the parabolic approximation of both types of concentrations' calculation by the least-squares method will give us the following result:

$$Y(t) = At^2 + Bt + C, X(t) = -at^2 + bt + c \ (a > 0) \tag{16}$$

where A, B, C, a, b, c are known coefficients, which were found out by fitting.

Bringing the parabolas (16) in order, we will have the following equation:

$$Y(t) = Pt^2+Q, X(t) = -pt^2 + q(p>0) \tag{17}$$

where P , Q, p, q are known coefficients.

If we exclude the period of time from the equations (17), however, we will get a plankton concentration adaptation dependence on the cuprum concentration impact:

$$Y = -Mx + N \ (M< 0) \tag{18}$$

where

$$M = \frac{P}{p}, N = \frac{Qp - Pq}{p}$$

The abovementioned equation (18) is a linear dependence of the phytoplankton concentration on the cuprum concentration (Figure 15.40). Thus, it gives no chance to define phytoplankton's adapting characteristics in case of the cuprum impact. The description of this data is based on the phytoplankton concentration parametric dependence on the cuprum concentration (see Figure 15.39). Linear dependence (18) is not described by the above mentioned examples, and its parametric characteristics, shown in Figure 15.39 and fitted by the equations (16), can give us more information connected with the aquatic ecosystem impact.

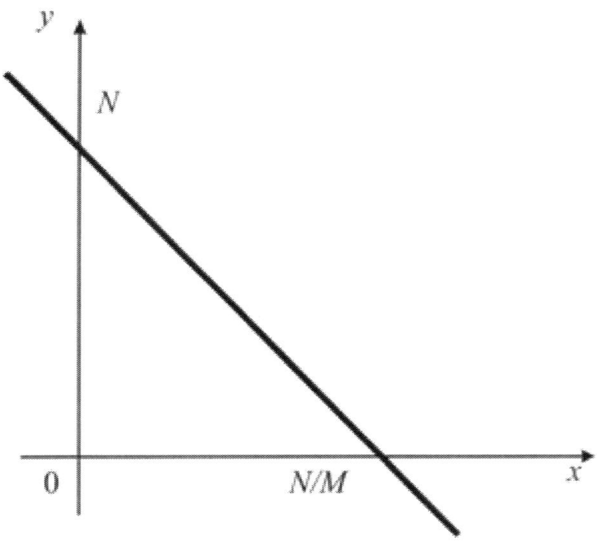

Figure 15.40. Linear dependence of the phytoplankton concentration on the cuprum concentration.

The parametric dependencies are more general than explicit ones, like $y = f(x)$, and these explicit dependencies are included into the parametric dependencies group. The explicit dependencies can be defined by the parametric dependencies by excluding the parameter (as it was made in the equation (18)). In this case, the information defined only by the parametric dependencies can be lost (as it happened in the case of equation (16) transformation into the equation (18)). That is why this work shows a new approach to the aquatic ecosystem adaptation reaction to the external impact.

Paper [33] is also a very good example of the 95% accurate confidential intervals for the average multiannual changing of the dissolved cuprum and phytoplankton biomass content in the Usman River definition (years: 1977-1981, 1983, 1984).

The last example, given in section 15.4, is taken from hydrology. It appears that *the channel's liquid turbulent stream forms a water flow speed profile that looks like a typical dynamic system adaptation reaction on the external impact.*

15.4. Turbulent Stream Channel

K.Ya.Kondratyev, A.M. Nikanorov and their co-authors considered the flat, one-dimensional liquid flow in the direction of an axis x between two fixed parallel plates: $y = 0$ and $y = h$, caused by pressure difference ($-dp/dx$) = const. The boundary problem for the Navier–Stokes equations system with its isotropic turbulent resistance, κ [34], in case of a liquid sticking to the plates after including gravitation into the pressure, is limited to a boundary problem for the simple second-order differential equation ($v_x = v(y)$, $v_y = 0$, $v_z = 0$):

$$\nu \frac{d^2 \upsilon}{dx^2} - \kappa \upsilon = \frac{1}{\rho} \frac{dp}{dx};\ \upsilon(0) = \upsilon(h) = 0 \quad (19)$$

The solution of the problem is expressed by the function:

$$\upsilon = -\frac{1}{\rho\,\kappa} \left\{ 1 - ch\left[\left(y - \frac{h}{2} \right) \sqrt{\frac{\kappa}{\nu}} \right] sch \sqrt{\frac{\kappa h^2}{4\nu}} \right\} \frac{dp}{dx} \quad (20)$$

in the case of $\kappa = 0$, this profile line coincides with the parabolic profile:

$$\upsilon = -\frac{1}{2\mu} \frac{dp}{dx} y(h - y) \quad (21)$$

of the Poiseuille flow between two parallel plates [34]. In case of high values of the turbulent resistance with the κ coefficient, the influence of the second member shown in equation (20) in square brackets implies only next to the two-dimensional channel border. The speed is almost constant in the other flow area (20):

$$\upsilon = -\frac{1}{\rho\kappa} \frac{dp}{dx} \left(1 - sch \sqrt{\frac{\kappa h^2}{4\nu}} \right) \quad (22)$$

In the case of high values of κ,, formula (20) describes a *trapezoid section*, known from research, with a plug flow and an interlayer at the channel walls, which is characteristic for turbulent liquid flow speed between two parallel plates (Figure 15.41). This very section is an adaptation reaction of the water flowing into the channel on the pressure drop impact.

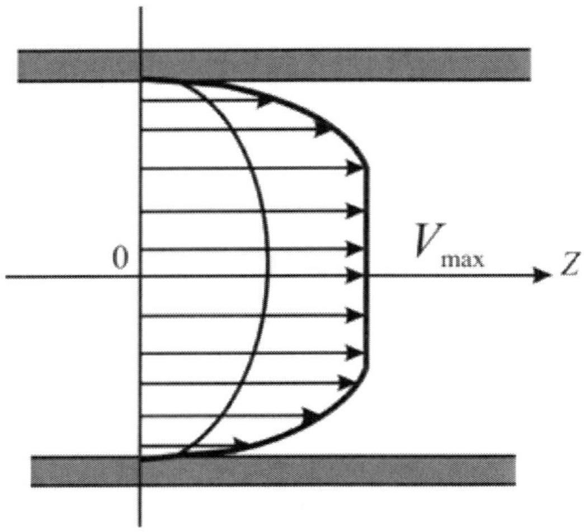

Figure 15.41. Trapezoid section of the turbulent flow in tube. Thin line is a Poiseuille parabola.

A very wide range of impacts on absolutely different objects was described in the present chapter. Examples show that the mathematical characteristic of the biological objects' reaction on impact has a form of the adaption reaction described in the previous chapters. At the beginning of the present chapter we described systems that can be described by different types of the adaptation reaction: tolerance principle, Olly's principle, etc. It is possible to expand the research area and to give a set of similar examples from other branches of knowledge: medicine, psychology, physics, etc. To find out these examples, it is enough for an interested person to read corresponding educational or reference books.

The examples given in the present paper make it possible to find out the following regularity: *the adaptation dynamical (ecological, biological, hydrological, etc.) system reaction on the impact looks like and has characteristics of a curve (or surface) with one or several maxima, and the ecological system is mostly adaptive to the impacts on it in the neighborhood of these maxima.*[31]

In the majority of the given examples, the adaptation curves' maximum points are shifted from the impact jog center. The maximum point shift to the left side means the adaptation process is precipitating; its shift to the right side means the process is slowing down.

Now we will make a conclusion from this chapter, taking into account all the abovementioned examples. It is necessary to define the adaptation process stabilization as the relation of the horizontal distance from the right (left) flex point to the left (right) adaptation curve branch, expressed in percents, to the total impact range.

If the abovementioned diagram has no flex points, the *coefficient of the adaptation process' stabilization is determined by the average width of a horizontal adaptation curve.*

[31] Sometimes such maxima are called "plateau."

Chapter 16

EXAMPLES OF MATHEMATICAL DESCRIPTIONS OF ECOLOGICAL SYSTEMS

A large number of examples of the adaptive response of different ecosystems to exposure were given above. Now, let us consider the examples of mathematical description of dynamical systems. The solutions of such systems as was shown in the previous chapters are the basis for the determination of the adaptive response.

In the modern mathematical ecology and in the dynamics of the biochemical systems and other areas of system analysis of complex multifactorial processes, the multiblock mathematical models have been adopted, each of the blocks of which, in its turn, is a mathematical model of processes occurring in the block. A simplified mathematical description of processes in the block of the system model is constructed as follows. The simplest description of the population within a block is its description through the quantity or mass $M(t)$, where t time, and any physical structure is regarded as the mass. In our case, it is the pollution of the aquatic ecosystem (chemical, biological, anthropogenic, etc.). Alongside with the mass, a number of the resource $N(t)$ is considered. A physical quantity competing with the mass is regarded as the quantity of the resource, for example, as the quality of the water or the environmental well-being of the water body.

The construction of any mathematical model, which is not based on previously known equations, begins with the construction of general laws of conservation (mass, substance, energy, etc.), the laws that model oscillating conditions, if it is known that the system operates under the oscillating conditions with the exposure, as well as the laws of balance, etc. Then, the law of variation of the environmental resource, $N(t)$, of the system can be written as follows (1):

$$\frac{dN}{dt} = Q(M,N) - MV(M,N) \tag{1}$$

where $Q(M, N)$ is the rate of the resource arrival into the system of water quality—pollution type, etc., $V(M, N)$ is the rate of the resource consumption by one "unit" of pollution and the total amount is equal to $M(t)$.

Similarly, the law of the pollution variation $M(t)$ is written as (2):

$$\frac{dM}{dt} = kV(M,N)M - m(M,N)M \tag{2}$$

where $kV(M,N)$ ($0 < k < 1$) is part of the consumable resource that is spent on water purification and $m(M, N)$ is a self-purification factor.

Reasoning during the selection of environmental models, models of dynamic economy, models of global development, etc., are similar to the given ones [10]. The only difference consists in the conceptual sense of the functional parameterization of $Q(M, N)$, $V(M, N)$, $m(M, N)$ and others, taking into account the dimension of the system depending on the number of its blocks. Therefore, in general terms, the dynamics of the considered systems are described by the matrix equation as follows (3):

$$\frac{dM}{dt} = A(M,t) - B(M,t) \tag{3}$$

where A is the rate of the resource arrival, B is the rate of the resource consumption and M is the matrix of the system state.

Sometimes, it is meant that $A—B = G$, and the system (3) is written as (4):

$$\frac{dM}{dt} = G(M,t) \tag{4}$$

Given systems of equations can be used to model various processes in aquatic and other ecological systems, let us show how on the basis of such systems, the simple processes of the entry of hydroecological system into the stationary and synchronous modes of operation can be modeled.

16.1. Entry into the Stationary Mode of Functioning

Ecological systems, including the aquatic ones, are generally in a state of dynamic equilibrium. The conditions of the aquatic ecological systems functioning in the state of dynamic equilibrium can be different: based on stationary (time-independent) processes, oscillating processes, transitional and other processes. In this section, we model the processes of the aquatic ecosystem entry into the stationary mode of functioning. Such processes in the aquatic ecosystems occur after the change in its initial state as a result of pollution or, on the contrary, water purification. In addition, the entry of the aquatic ecosystems into the stationary mode of functioning may occur as a result of transient processes in the ecosystem. To describe such processes, it is necessary to give an answer to the question of whether the organization of water quality changing in system (3) under which the system quality resource - pollution begins to function in the stationary mode for a long period of time (at $t \to \infty$) is possible. The answer to this question is important in analyzing any ecosystems, including the aquatic ones.

Despite the importance of this question, its mathematical solution is rather simple. Let us show that entry of system (3) into the stationary mode of functioning (Figure 16.1) can be described for a simple system of type (3), consisting of a single equation, the right-hand member of which is time independent. Let the water quality, $m(t)$, be described by the equation (5):

$$\frac{dm}{dt} = a(m) - b(m) \tag{5}$$

For the simplicity, we assume that there are no external sources of pollution (or they have ceased their influence) and the rate of water quality improvement is limited by its intrinsic properties. This abstraction is rough, but is still sufficient to model the process under consideration.

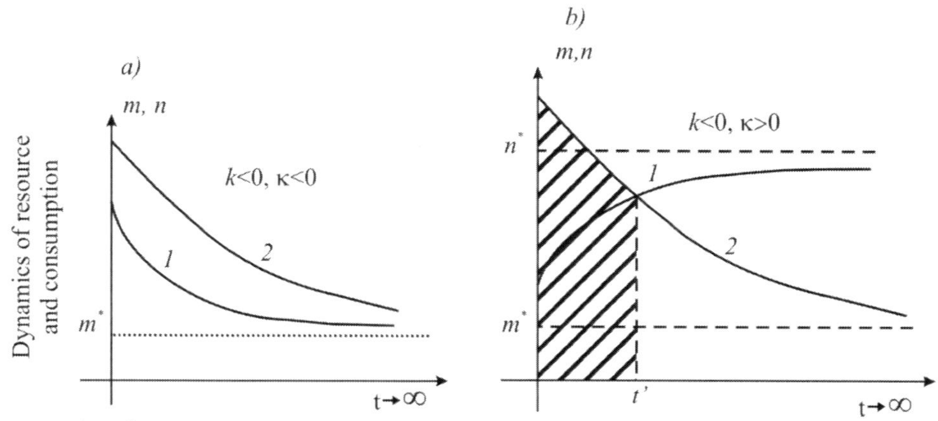

1—consumption, 2—resource

Figure 16.1.

When modeling the entry into stationary dynamics, pollution is described by the equation that is similar to equation (5). We do not give this equation because in the considered process of noninteracting quality and pollution resource, the properties of pollution may be similar to the properties of the resource.

The equivalent mathematical formulation of the question concerning the entry of the system of equations (5) into the stationary mode of functioning is as follows: whether the solutions $m = m(t)$ of the equation (5) may have a horizontal asymptote m^*? More precisely, whether the process:

$$m(t) \to m^* \neq \infty \text{ at } t \to \infty$$

is possible for the solution of equation (5). This process means that in case of large values of time ($t = \infty$), solutions of equation (5) have an extreme. This fact, in its turn, means that the derivative of the left-hand member of equation (5) comes to zero at $t \to \infty$:

$$\lim_{\substack{t \to \infty \\ m \to m^*}} \frac{dm}{dt} = 0$$

Therefore, the sufficient condition for the existence of a horizontal asymptote $m = m^*$ of the solutions $m = m(t)$ of equation (5) is the ratio (6):

$$a(m) - b(m) = 0 \to m = m^* \tag{6}$$

In other words, in case of the entry of the solutions $m = m(t)$ of equation (5) into the stationary mode of functioning, the value $m = m^*$ must be a single-valued root of the right-hand member of equation (5).

If the right-hand member of equation (5) has a single-valued root $m = m^*$, then this equation can be written as:

$$\frac{dm}{dt} = \varphi(m)(m - m^*) \qquad (7)$$

Where:

$$\varphi(m) = \frac{a(m) - b(m)}{m - m^*}$$

Besides, according to L'Hospital rule:

$$\varphi(m^*) = \frac{d}{dm}[a(m) - b(m)]\Big|_{m=m^*} = \mathrm{k} \neq 0$$

so far as $m = m^*$ is a single-valued root of equation (6).

Let us restrict ourselves in the analysis of equation (3) for large values of time ($t \to \infty$). In this case $m \sim m^*$, and instead of equation (7), it is sufficient to consider the equation:

$$\frac{dm}{dt} = k(m - m^*), k = \varphi(m^*) \qquad (8)$$

describing the asymptotic behavior of solutions of equation (5) at large values of time ($t \to \infty$).

The general solution of equation (5) is the one-parameter set of functions:

$$m = m^* + ce^{kt}(t \to \infty) \qquad (9)$$

where c is an invariable characterizing the desired set of solutions. It is determined from the initial conditions under which the system described by equation (5) exists. The symbol ($t \to \infty$) in formula (9) means that this formula describes the behavior of solutions of equation (5) asymptotically for large values of time.

Let us repeat that pollution is described by the similar (5) equation and the similar (9) solution thereof. Let $n(t)$ stand for the pollution of the water. Then at ($t \to \infty$), the behavior of the pollution is described by formula:

$$n = n^* + \sigma e^{\kappa t}(t \to \infty) \qquad (10)$$

where σ and κ are invariables, andσ is an arbitrary invariable while the invariable κ is determined similarly to the invariable k.

Despite the seeming simplicity of expressions (9) and (10), the processes corresponding to them often can be found in the aquatic environment and can be divided into two main classes.

The first class is associated with the processes in which the system resource - consumption (e.g., water quality and its pollution) is not destroyed over time. In such systems, the resource is replenished at the expense of its internal dynamics and consumption does not exceed the capabilities of the resource. The first class is characterized by negative values of parameters ($k < 0$, $\kappa < 0$), determining the simultaneous decrease of resource and consumption in this case. These dynamics are illustrated in Figure 16.1*a*. To simplify the construction of plots, the asymptotic values of the resource (m^*) and the consumption (n^*) are assumed to be equal in this figure. Let us be more precise, in that in Figure 16.1a, $c > 0$ and $\sigma > 0$.

The second class of dynamics under consideration is associated with the processes in which the resources are exhausted over time, and consumption from this moment begins to exceed the capabilities of the resource. Such systems exist for some time, and then, after the resource is exhausted, perish. The second class is characterized by the parameters k and κ, with the different signs: sign($k\cdot$ k) < 0[32]. The dynamics of such processes is shown in Figure 16.1*b*. In this figure, t^* stands for the time of existence of the system resource - consumer, the shaded area corresponds to the field of activity of the system, as well as $c > 0$ and $\sigma < 0$.

We should also consider the case of positive values of parameters k and κ determining the behavior of the systems under consideration of large values of time. In this case, at $k > \kappa > 0$, the growth of the resource exceeds the growth of consumption. Such system exists for an indefinite period of time. In this case, when $\kappa > k > 0$, the growth in consumption exceeds the potential growth of the resource. Such systems become destroyed.

We should also consider the case when equation (6) has once repeated its root. Alongside with equation (6), the following equation should be performed (11):

$$\frac{da(m)}{dm} - \frac{db(m)}{dm} = 0, a(m) = b(m) \tag{11}$$

ensuring that the right-hand member of equation (5) will come to zero by time. The equation $a(m)$ - $b(m)$ in (11) is given for reference.

We consider that equations (11) are compatible and that the common root of these two equations exists.[33] Let μ^* stand for this root. Then equation (5) may be written as (12):

$$\frac{dm}{dt} = (m - \mu^*)^2 \psi(m) \tag{12}$$

Where:

$$\psi(m) = \frac{a(m) - b(m)}{(m - \mu^*)^2}$$

besides:

[32] signa is the sign of number, a. More precisely, signa = 1 at a > 0 and signa = -1 at a < 0.

[33] Usually, the mathematical dynamics of real ecological systems are parameterized. The examples of parameters are the coefficients of interaction (e.g., water quality and its pollution), the viscosity of water, its thermal conductivity, diffusion coefficients, etc. In this case, equation (5) contains at least one parameter, and the system of equations (11) then has two variables, i.e., it is closed.

$$\psi(\mu^*) = \frac{1}{2}\frac{d^2}{dm^2}[a(m) - b(m)]\bigg|_{m=\mu^*} = l \neq 0$$

At large values of time ($t \to \infty$) instead of equation (12), we have the equation (13):

$$\frac{dm}{dt} = \ell(m - \mu^*)^2 \tag{13}$$

the solution of which are the functions:

$$m = \mu^* - \frac{1}{c+lt}(t \to \infty) \tag{14}$$

where c is an arbitrary invariable determined from initial conditions of the system existence which is usually known.

At large values of time ($t \to \infty$), the invariable c in expression (14) may be neglected as compared with the large member, lt. Then, instead of (14) we shall consider the equation:

$$m = \mu^* - \frac{1}{lt}\ (t \to \infty) \tag{15}$$

Similarly, the process of consumption $n(t)$ at large values of time is modeled by the ratio:

$$n = \nu^* - \frac{1}{\lambda t}(t \to \infty) \tag{16}$$

where the invariable λ is determined in the same ways as the invariable l.

At $l < 0$, function (15) describes a monotonic, but a limited, reduction of the resource. If $\lambda > 0$, function (16) describes monotonic, but a limited, increase in the consumption. It should be noted that formulas (15) and (16) model growth of the resource and decrease in the consumption but not exponentially.

The dynamics of system (15) and (16) are shown in Figure 16.2, where the curves *1* characterize decrease of the resource and the curves *2* represent an increase of the consumption. The difference describes the activity of the system.

If $\mu^* - \nu^* > 0$ (see Figure 16.2 *a*), then the activity is sufficient and functioning of the system is stable.

If $\mu^* - \nu^* < 0$ (see Figure 16.2 *b*), then the activity is not sufficient. The system is starving and is on the verge of death. In this case, it is necessary to include the control of the system with additional parameterization of equations (5) with the purpose to change the sign of difference $\mu^* - \nu^*$ from negative to positive. Let us make a note that the exponential nature of the resource and the consumption change parameterization of the system will not save it from starving unless it changes the qualitative behavior of the system by removing the exponents from solutions of the equations that correspond to the system.

Figure 16.2 illustrates the case $\mu^* = \nu^* = \mu$, when the state of the system is optimal. In this case, the system has neither excess nor shortage of resources. This case is also referred to as the managerial, because it requires the additional parameterization of the system.

Even those simple monitorings over the systems, resource and consumer of which they do not interact directly, suggest broad possibilities for the modeling of processes of an ecological system's functioning.

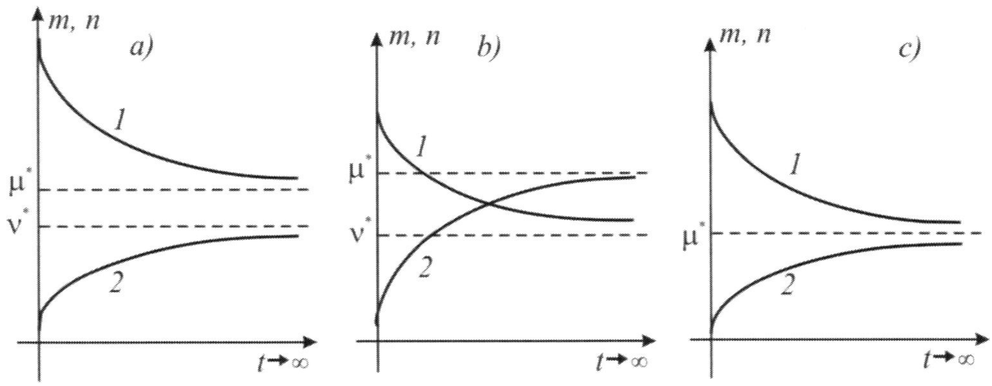

Figure 16.2.

16.2. Further Examples

In view of importance of the processes of the aquatic system's dynamics entry into the stationary mode of functioning (as well as taking into account the examples of destabilization of aquatic ecosystems), let us consider further examples relating to such processes. Let $x(t)$ stand for the combined characteristic of the water properties contributing to the change of its quality.

Function $x(t)$ for brevity will mean the water quality. In the simplest case, the rate of the quality change is proportional to the results of actions aimed at improving the water quality, i.e., proportional to $x(t)$. Then the simplest equation describing the change in the water quality over time has the following form:

$$\frac{dx}{dt} = ax + b \tag{17}$$

Here b is the invariable characterizing the change in the water quality in the case when for the changing of the water quality, no actions are taken. The invariable a determines the degree of improvement ($a > 0$) or deterioration ($a < 0$) of the water quality. All solutions of equation (17) have the form as follows:

$$x = Ce^{at} - \frac{b}{a} \tag{18}$$

The invariable C is determined from the initial state of quality of the water body under consideration.

If at the initial moment of time, the water quality was characterized by the value $x_0 (x(0) = x_0)$, then from equation (18) at $t = 0$, we will find $C = ax_0+b$. Therefore, the change in water quality over time is described by the function:

$$x = \frac{ax_0 + b}{a} e^{at} - \frac{b}{a} \tag{19}$$

The right-hand member of equation (17) goes to zero at $x = -\frac{b}{a}$. Therefore, when ($t \to \infty$) change in the water quality x(t) over time may enter into stationary mode. The entry of water quality into stationary mode will occur only in cases when there is $\lim_{x\to\infty} x(t)$.

The existence of such a limit in this case is determined by the sign of parameter a, being part of equation (17).

At $a > 0$ and at $ax_0 + b > 0$ the function $x(t)$ increases with no limit with growth in time t, and the limit of water quality improving does not exist. Such dynamics of reservoirs is ideal and is rarely observed in the natural environment.

At $a > 0$ and at $ax_0 + b < 0$, the function $x(t)$ decreases with no limit over time, and the limit of the water quality deteriorating does not exist. Such reservoirs perish.

At $a < 0$, the exponent in expression (19) decreases with no limit over time. Therefore, the water quality enters into a stationary mode over time and takes a value of (20):

$$x(\infty) = -\frac{b}{a} = x_\infty \tag{20}$$

The limiting value $x = x_\infty$ is that unknown value that determines the state of water in the reservoirs with unstable dynamics of its quality, because the water quality does not decrease lower than level x_∞ over time.

If value x_∞ is positive, then the water quality enters to level x_∞ over time and remains at this level. Let min x stand for the minimally permissible quality of water that is necessary for the functioning of the reservoir. Then at $x_\infty >$ min x, there is a buffer capacity B of the water quality in the reservoir:

$$B = x_\infty - \min x$$

At $0 < x_\infty <$ min x, the objective of water protection measures is to maintain the rather high level x_∞. This can be achieved by means of increasing the parameter b, being a part of equation (17) and characterizing the rate of change of the water quality due to its internal capabilities, i.e., due to the process of water self-purification.

If the value x_∞ is negative, then the quality of water decreases to the level x_∞ over time. In this case, the reservoir may perish unless the water capability for its quality restoration is changed due to the control b.

The dynamics of the water quality change (see formulas (17) and (19)) at $a < 0$ are shown in Figure 16.3.

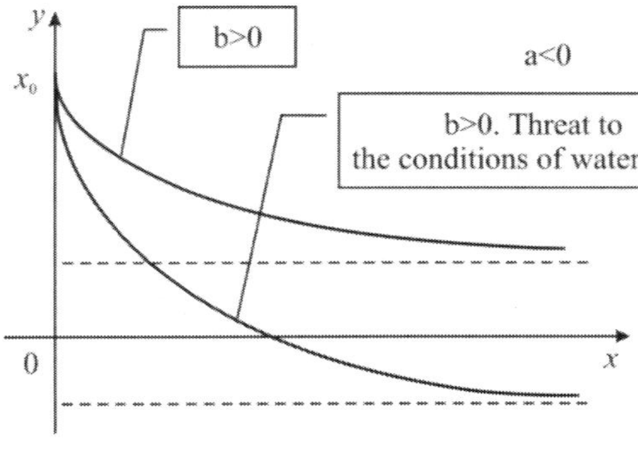

Figure 16.3.

Chapter 17

SYNCHRONIZATION OF PROCESSES IN AQUATIC ECOSYSTEMS

Let us consider the examples of aquatic adaptive systems' dynamics entry into the synchronous modes of functioning of water quality and its pollution (as well as the examples of oscillatory destabilization of aquatic ecosystems). Again, let us denote the unlimited combined parameters of the water composition characterizing its quality through $x(t)$. The value $y(t)$ will stand for the unlimited combined characteristic of the water pollution. The function $x(t)$, as in the previous chapter, will stand for the water quality. For the function $y(t)$, we will use the term "pollution." In this chapter, we will assume the simplest conditions to describe the dynamical system $\{x(t), y(t)\}$ for the quality – water pollution. There are two such conditions.

1. The rate of water quality improvement (the rate of self-purification of the water) decreases with the increase of pollution, therefore:

$$\frac{dx}{dt} = -ay + F(t) \tag{1}$$

where $F(t)$ is the function contributing to the improvement of water quality and $a > 0$ is the factor characterizing an increase of pollution.

2. The rate of increase of pollution increases with the deterioration of the water. Therefore:

$$\frac{dy}{dt} = bx + G(t) \tag{2}$$

where $G(t)$ is the function characterizing the pollution accumulation and $b > 0$ is the factor of water quality improvement.

In order to investigate the changes in the water quality $x(t)$ depending on external factors $F(t)$ and $G(t)$, we will exclude the pollution $y(t)$ from equations (1) and (2). To do this, we will differentiate equation (1) over time t and put the rate dy/dt from equation (2) into the received ratio. Then, in the simplest case, the water quality depending on external characteristics of pollution $G(t)$ and the performance of water protection measures $F(t)$ is modeled by the equation:

$$\frac{d^2x}{dt^2} + \omega^2 x = f(t) \quad (3)$$

where $f(t) = \frac{dF}{dt} + G(t)$ is the summary of the external influence on the water taking into account the pollution and saving of the water quality and $\omega = \sqrt{ab}$ is the positive parameter of the dynamical system quality - pollution of the water. The physical sense of the parameter ω will be cleared in the process of analysis of equation (3).

At the initial moment of time, the water quality and the rate of its change over time should be known:

$$x(0) = x_0, \left.\frac{dx}{dt}\right|_{t=0} = y_0 \quad (4)$$

From equation (1), it follows that the parameter y_0 in the second condition (4) characterizes the number of pollutants in the water at the initial moment of time. The ratios (4) are initial conditions for equation (3) or the conditions of the Cauchy problem (3) and (4).

First, we will describe the properties of the water quality modeled by equation (3) in the case when $f = 0$, i.e., the aquatic system is not influenced externally at all. In this case, the general solution of homogeneous [34] equation (3) is expressed by the dependence:

$$x = Pcos(\omega t) + Qsin(\omega t) = Acos(\omega t + \alpha) \quad (5)$$

of the water quality x on time t. The invariables P, Q, A and α are arbitrary and should be determined from the initial conditions (4). Putting the function $x(t)$ from formula (5) into the boundary conditions (4), we will determine the invariables P and Q and introduce them into the expressions (5). Then, we will get the dependence of the water quality on time in the case when pollution is absent and water protection measures are not taken:

$$x = x_0 \cos(\omega t) + \frac{y_0}{\omega}\sin(\omega t) = \frac{\sqrt{\omega^2 x_0^2 + y_0^2}}{\omega} cos\left(\omega t - arctg\frac{y_0}{\omega x_0}\right) \quad (6)$$

Therefore, in the case of the absence of input of pollutants and water purification, the water quality varies with the frequency determined by internal properties of the water body (as the frequency $\omega = \sqrt{ab}$ is determined by the characteristics a and b of the water quality change and its deterioration is due to internal reservoir processes). The amplitude and phase of oscillations of the water quality are determined by initial characteristics of the water body. In mode (6) at rough approximation. the terrestrial reservoirs function in the absence of man-induced impact; the quality of water of such reservoirs varies with respect to the average level.

The oscillations (6) are shown in Figure 17.1 at $x_0 = 3$ and $y_0 = -\omega$. The phase of such oscillations is equal to arctg (l/3) ≈ 0.322 ≈ 18.4°.

Its amplitude oscillates around the equilibrium position. The amplitude is equal to $\sqrt{10}$ = 3.162 of conventional quality units.[35]

---

34 The equation is named homogeneous when there are no known functions in it or in this case, the external force is absent.

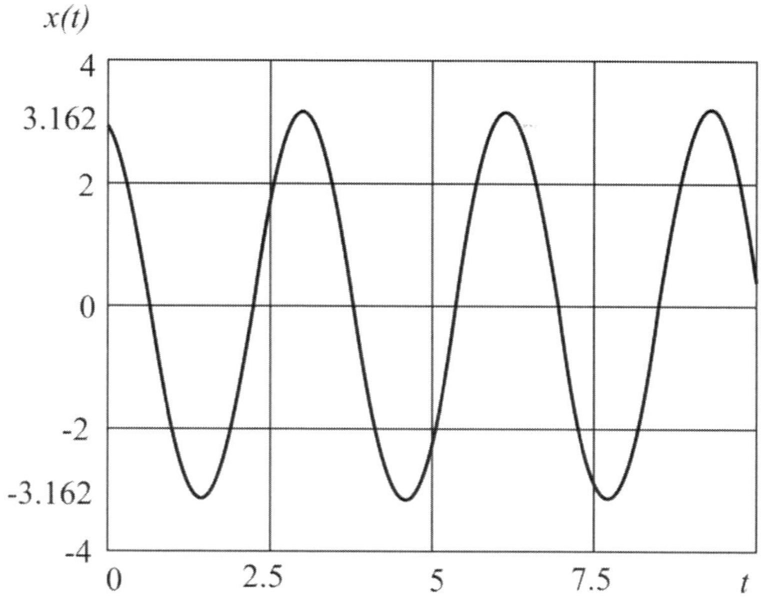

Figure 17.1.

The conclusion about this tranquil nature of the change in the water quality in the absence of external forces having an influence on the aquifer system can be made without giving a full solution to the homogeneous equation (3). To do this at $f(t) = 0$ let us multiply both members of equation (3) by the derivative dx/dt. Then equation (3) will take a form (7):

$$\frac{dx}{dt}\cdot\frac{d^2x}{dx^2} + \omega^2 x \cdot \frac{dx}{dt} = \frac{1}{2}\frac{d}{dt}\left(\frac{dx}{dt}\right)^2 + \frac{\omega^2}{2}\frac{dx^2}{dt} = 0 \tag{7}$$

This equation is easily integrated. Its integral is as follows:

$$x^2 + \omega^2 x^2 = C^2 \tag{8}$$

where the point about the symbol denotes the derivative of the function corresponding to the symbol over time t. The invariable of integrating in formula (8) is expressed through C^2.

Ratio (8) is the first integral of equation (3) or the energy integral of this equation. This name is given due to the fact that the first member in the left-hand part of ratio (8) is determined by the rate of the process and, therefore, is associated with its kinetic energy. The second member in the left-hand part of the same ratio is determined by the position of the system and is associated with the potential energy of the system. The energy integral is a family of curves depending on one parameter C^2. The family of curves of the first integral of the differential equation is located in the plane of coordinates $(x,\dot{x})$. Such a family is named *the phase portrait* of the differential equation, and the plane coordinates $(x,\dot{x})$ is named *the phase plane* of the same equation. Each curve of the phase portrait is named *the phase trajectory*.

[35] Units (physical, chemical and biological parameters) that determine the quality of water are the essential indicators of balance of the substance ⇔ the energy ensuring relatively stable operation of the water body for a long period of time. To determine them is the objective of the full-scale environmental modeling.

From formula (8), it follows that the phase portrait of equation (8) is a family of confocal ellipses with semi-axes *C* and *C*/ω located in the phase plane (Figure 17.2). The ellipses are the phase trajectories of the development of water quality in this situation. The development (movement) in the system water quality – pollution takes place tangentially to the phase trajectories. The arrows in Figure 17.2 show the direction of development. They are arranged taking into account the directions of rates {*-ay*, *bx*} determined in equations (1) and (2). The figure shows that the phase trajectory of the system under consideration does not go away from the equilibrium position. The functioning of such systems can be stable unless the external forces are applied to the system water quality – pollution violate (see below) the balance that was created in their absence.

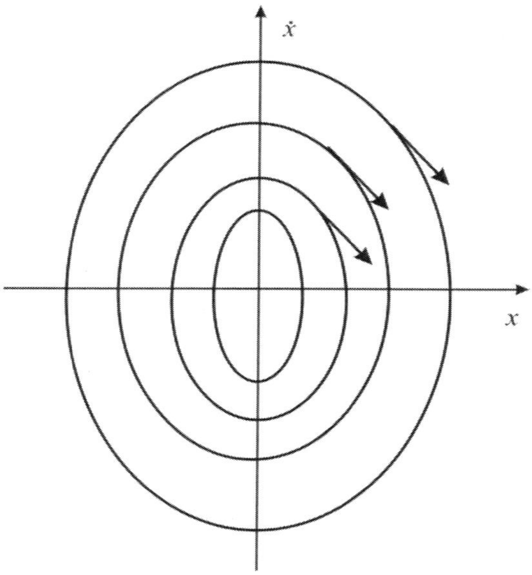

Figure 17.2. Phase portrait of equation (3).

Without giving full information on the solution to the Cauchy problem, the phase plane gives an idea about the general nature of the development of processes in the system under consideration. In the case of a fixed value of the invariable C^2, development of the system does not go away from the phase trajectory. Therefore, the processes corresponding to the phase portrait shown in Figure 17.2 have a periodic nature and any value both of the function itself (water quality), and its derivative (rate of the quality change), does not deviate from the equilibrium position ($x = 0$, $\dot{x} = 0$) at a distance exceeding major semi-axes of the ellipse corresponding to the phase trajectory, and over time, they return to the initial position. The invariable C^2 in equation (7) of the family of phase trajectories can be determined from the initial conditions (4) of the Cauchy problem:

$$C^2 = \dot{x}_0^2 + \omega^2 y_0^2 \quad (9)$$

In terms of analyzing the processes development character in this system, however, the determination of the invariable C in formula (9) is not necessary.

Equation (3) is met in the most varied branches of theory and practice of science. Chemistry, biology, electromagnetic dynamics, astronomy, theory and practice of oscillatory processes and many other branches of science usually use equation (3) in their studies. It is not accidental; oscillating processes are known both in nature and in technology.

Let us make a note that equation (3) in the mechanics and physics is known as the "mathematical pendulum."

Let us give an example of the phase plane that is different from the phase plane of equation (3) (Figure 17.3), which describes the steady flow of the oscillating process. Let us consider the system of two differential equations modeling the process in the absence of competition. Let $x(t)$ be the quality of the water and $y(t)$ the factor purifying the water (self-purification or water treatment), the effect of which increases with deterioration of the water quality. Then instead of equation (1) for modeling two interrelated factors, in the absence of its competition, we have the simplest system of equations:

$$\frac{dx}{dt} = ay + F(t), \frac{dy}{dt} = bx + G(t) \tag{10}$$

Removing the factor $y(t)$ from equations (10) with the help of differentiation of the first equation, we will get the equation ($\omega = \sqrt{ab}$)

$$\frac{d^2x}{dt^2} - \omega^2 x = f(t);\ x(0) = x_0, \dot{x}(0) = y_0 \tag{11}$$

describing the quality of water in the simplest, two-component, non-competitive system. The energy integral to equation (10) has the following form:

$$\dot{x}^2 - \omega^2 x^2 = C^2 \tag{12}$$

Therefore, the phase portrait of equation (10) is the system of confocal hyperbolas located in the plane of coordinates $(x, \dot{x})$ (see Figure 17.3).

The point with coordinates $(x, \dot{x})$ fixing the state of the system moves along the hyperbole in the direction of the arrows in the figure. The direction of the arrows corresponds to the direction of the vector $\{ax, by\}$, which determines the rate in system (10). In contrast to the dynamics described by equation (3) and shown in Figure 17.2, the point on the phase trajectory in Figure 17.3 can deviate from the initial state of the system located on the hyperbole at any amount. The movement of the system away from the equilibrium position may mean the destruction of the system, unless such a movement is not directed toward the improvement of the water quality.

Let us continue with the consideration of the water properties described by equation (3).

To analyze the dependence of water quality on man-induced or other impacts on the aquifer system, let us determine the common dependence of $x = x(f(t))$ of the state of water quality from any external forces $f(t)$. For linear differential equations, there is a relatively simple way of such relations construction—the Lagrange method of variation of arbitrary variables. Omitting the conclusion, let us give a solution of equation (3) that already takes into account the initial conditions (4):

$$x = x_0 \cos(\omega t) + \frac{y_0}{\omega} \sin(\omega t) + \frac{1}{\omega} \int_0^t f(\tau) \sin(\omega(t - \tau))\, d\tau \quad (13)$$

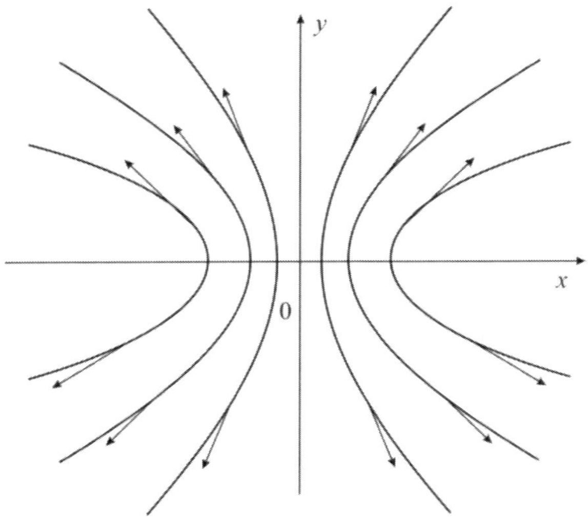

Figure 17.3.

For simplicity, at the first step of water quality study determined by equation (12), we assume that at the initial moment the aquifer system is in balance, i.e., the initial conditions (4) for the aquifer systems have the form as follows:

$$x(0) = 0, \dot{x}(0)=0 \quad (14)$$

If the external force $f(t)$ has an impact on the water ecosystem being at the initial moment in balance, then formula (13) is simplified:

$$x = \frac{1}{\omega} \int_0^t f(\tau) \sin(\omega(t - \tau))\, d\tau \quad (15)$$

If in the future it is necessary to take the initial state of the system, which is different from the balanced system, into account, you will need to impose oscillations determined by formula (6) (or the sum of the first two members in equation (13)) on the received solution, which determined by the external influences for any initial conditions.[36]

Let us restrict ourselves with the description of water quality change under the influence of the most typical factors. Let us consider the following cases.

1. Let the forces having an impact on the aquifer system have an exponential nature: $f(t) = e^{at}$. Let us introduce this expression of force into formula (15). Then, we will find out that the water quality regulated in the reservoir by the factor has an exponential growth ($a > 0$) or decrease ($a < 0$):

---

[36] The imposition of the initial state on the current one caused by the influence of external forces is valid only for linear systems which includes equation (3). Nonlinear dynamics require the simultaneous consideration of the initial and current states of the system (see below).

$$x(t) = \frac{\omega e^{at} - \omega \cos(\omega t) - a\sin(\omega t)}{a^2 + \omega^2} \tag{16}$$

From formula (16), it follows that the oscillating nature of the change in the water quality over time may turn out to be different only at positive values of the parameter a. In this case over time, formula (16) will be dominated by the exponential member and the quality of water will improve. If the positive parameter $a \ll 1$, the water quality in average increases, but the oscillating nature of the quality change remains for some time. Such dynamics of increase in the water quality are shown in Figure 17.4 at $a = 0.02$ and $\omega = 5$.

From formula (16), it follows that the oscillating growth of the function $x(t)$ may be changed for its oscillating attenuation, if the parameter sign is changed for the positive and if the function $x(t)$ models, for example, the process of changes of the incoming pollutants. Therefore, the given example demonstrates capabilities of control of dynamical systems. At present, such a control is widely used only in electrical engineering.

If the parameter a is negative, then the change of the water quality (16) returns over time to oscillations around the equilibrium position. Such dynamics of change in the water quality are shown in Figure 17.5 at a = -0.02 and ω = 5. At $t \sim 100$, the water quality varies practically around the equilibrium position. The amplitude of oscillations of the water quality in Figure 17.4 and 17.5 is close to five relative (dimensionless) units of quality.

Now, let us show how, with application of model (3), it is possible to model the process of water quality entry into the stationary mode that is different from the initial equilibrium position characterized by ratio (14). Let an external force changing the quality of the water in the reservoir have a form $f(t) = 1 - e^{-at}$, where $0 < a < 1$. Over time ($t \to \infty$) the value of force $f(t)$ will take a value of one. Without writing an analytical expression of the water quality (15) depending on this force $f(t)$, we present a plot (Figure 17.6) depending on quality of water on time at a = -0.005 and $\omega^2 = 0.5$. From equation (3), it follows that at $t \to \infty$, the quality of water $x(t)$ takes the value $x(\infty) = 20$. This follows from the plot (see Figure 17.6) of the water quality change. In this case, however, the transition to another stationary mode is accompanied with the change of the water quality of the oscillating nature.

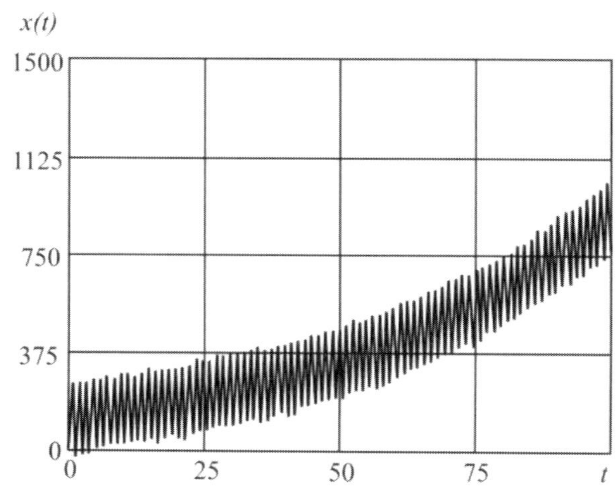

Figure 17.4.

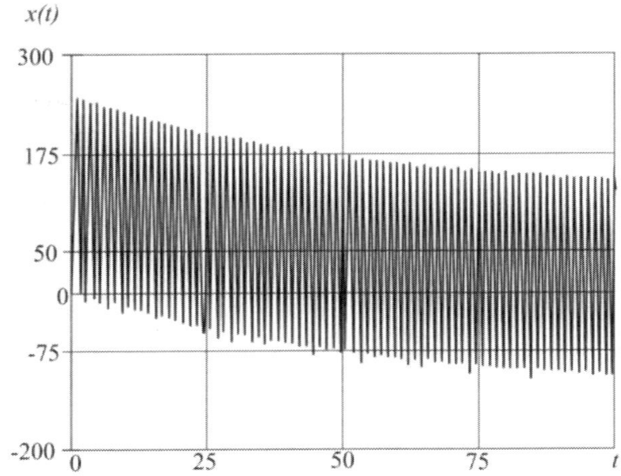

Figure 17.5.

2. The second example of the water quality study described by formula (15), refers to the case of oscillating nature of the change of external influence on water. Let the force having an influence on water be recurrent over time *t* and vary with the frequency $\Omega: f(t) = \cos(\Omega t)$. Introducing this expression of force into formula (15), we will get a formula determining the change in the water quality depending on the time and frequency on the applied force action:

$$x(t) = \frac{a(\cos(\omega t) - \cos(\Omega t))}{\omega^2 - \Omega^2} \tag{17}$$

As an example, we provide the dependence diagram (17) at $\omega = 1$ and $\Omega = 9$ (Figure 17.7). This figure clearly shows how the higher overtones of the external force are imposed on the tranquil 2π- periodicity of changes in the water quality.

At $\omega = \Omega$ there occurs an uncertainly in formula (17). By expanding it according to L'Hospital's rule (differentiating the numerator and the denominator as to the parameter Ω), we obtain the dependence of the water quality on time in the case when frequency of the external influence on water is equal to the frequency of characteristic oscillation of the water quality:

$$x(t) = -\frac{at}{2\omega}\sin(\omega t) \tag{18}$$

Expression (18) describes the case of resonant interaction determining the parameters of water quality with the external influence on water. The resonant case is qualitatively different from the tranquil pattern determined by dependence (17) and shown in Figure 17.6. In the resonant case, the amplitude of the water quality increases over time (Figure 17.8, where $\omega = \Omega = 2$).

The resonance phenomena can be both useful in terms of increasing the function, modulating oscillations in the water quality, and dangerous, as sharp resonance may even destroy the dynamical system external influence - water quality.

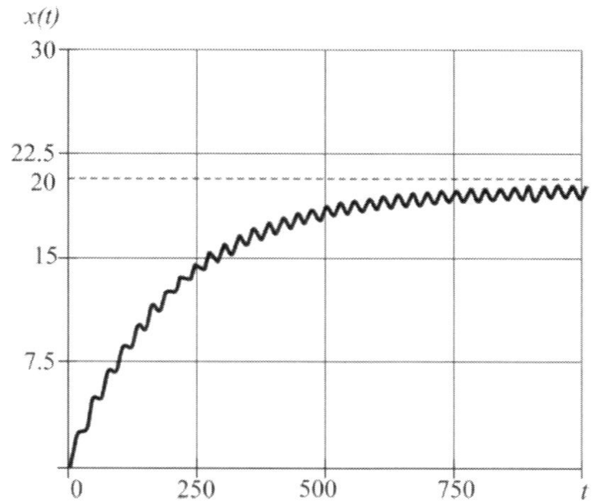

Figure17.6.

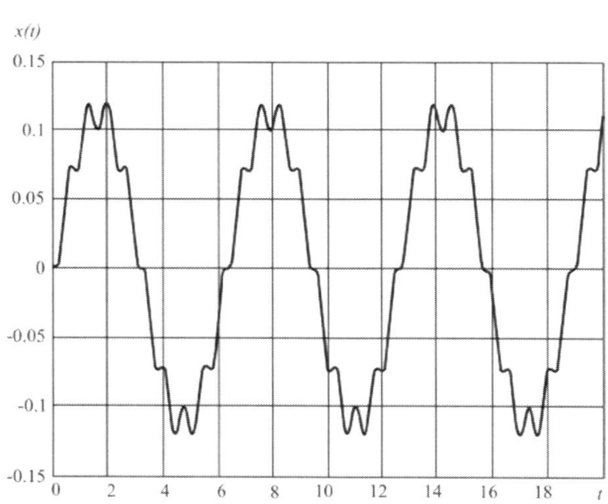

Figure 17.7.

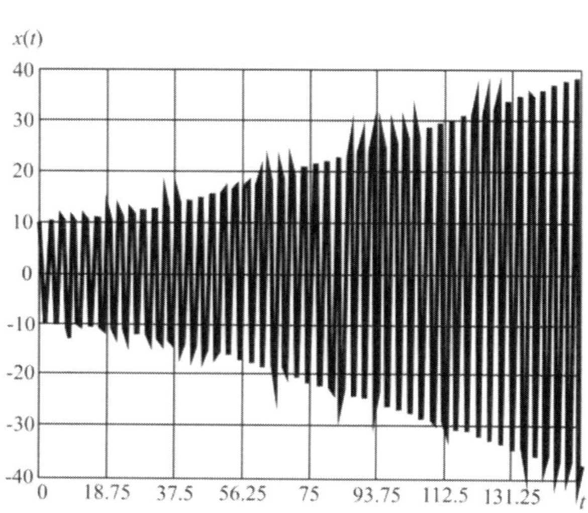

Figure17.8.

These examples demonstrate the ways of water quality regulation in the reservoirs in cases when the general nature of changes in the water quality is oscillating. To model the quality of water, the simplest linear equation (3) describing oscillating processes, was used. The examples show that equation (3) has broad possibilities for describing oscillating processes in the reservoirs.

The processes discussed in this chapter described by two-component oscillating systems definitely occur in the natural water bodies. The problem of the future is to fill such oscillating systems with the real ecological content—chemical characteristics and physical parameters that are included in the given mathematical expressions.

The only threshold situation in a two-component oscillating system is resonance. The resonance phenomena are dangerous if, for example, the resonance is related to the increase in the number of pollutants and the factors of influence on the ecological system.

The prevention of the resonance in multicomponent oscillating systems is a relatively simple task. It is enough to change the frequency of external oscillations regulating the state of an aquatic ecosystem away from its resonance value. For example, it is enough to change the frequency of the pollutants incoming or the frequency of water purification in the ecological system.

The next step in the development of mathematical methods for the description of aquatic ecological systems will be the attempt to appreciate the interaction and interdependence of three parameters of freshwater objects to the study of which the experiments in the mesocosms were devoted, namely the concentration of dissolved organic substances, the number of biological components and the concentration of heavy metals. In perspective, bottom sediments will be added to this list when its real role in the formation of water quality is determined.

Theoretically, it is possible to identify the critical points of interaction between biotic and abiotic components, analyzing the stability of the process and appreciating the methods for prevention of catastrophic phenomena in aquatic ecological systems. Preliminary analysis indicates that a three-component system, in the case of the changing of the mentioned three parameters, goes through a complex system of unstable stationary conditions resulting in a strange attractor that eventually may cause chaotic processes and the destruction of the ecosystem (in the sense that dynamical equilibrium between key parameters determining the conditions of its functioning will be violated).

REFERENCES

[1] Yegorov V.A., Kallistov Yu.N., Mitrofanov V.V., Piontkovsky A.A. Mathematical models of global development. – L.: *Gidrometeoizdat,* 1980. –192 p.

[2] Murry J. Nonlinear differential equations in biology. Lectures on the models. - M.: *Mir,* 1983. —397 p.

[3] Prigozy I., Strebgers I. Order from chaos.—M.: *Progress,* 1986. —432 p.

[4] Ebeling V. Formation of structures during irreversible processes. — M.: *Mir,* 1979. — 279 p.

[5] Kaznacheev V.P. Modern aspects of adaptation. — Novosibirsk: *Nauka,* 1980.

[6] Blekhman I.I. Synchronization in nature and tecnique. — M.: *Nauka,* 1981.

[7] Odum Yu. Fundamentals of ecology. — M.: *Mir,* 1975. —305 p.
[8] Bardin K.V. A basic psychophysical law and its variants. — Kazan, 1981. – P. 14—34.
[9] Moiseev N.N. Mathematical problems of system analysis. — M: *Nauka,* 1981. —396 p.
[10] Nikanorov A.M., Pantykhin Ya.V., Zavolzhensky M.V., Kokhanaya S.I. Mathematical description of adaptive capabilities of ecological systems // In the collection: Environmental regulation and modeling a man impact on aquatic ecosystems. Edition 1. — L.: *Gidrometeoizdat,* 1988. – P. 32—57.
[11] Menzikov S.A., Gdovsky P.A. Chlor-activated ATAF from olfactic lining of crucian carp // *Biology of inner waters.* — 1989. — No. 83. – P. 62—65.
[12] Burova T.V., Sainsky A.A., Danilenko A.N. etc. Conformational stability of ribulose biphosphate carboxylase of the green leaves according to differential scanning microcalorimetry // *Biophysics.* —1989. — Volume XXXIV, edition 4. – P. 751—752.
[13] Khimko P.V. Some features of bioelectric reaction of higher aquatic plants in case of changing the lightning conditions // *Hydrobiological journal.* — 1985. — Volume 21, No. 4. – P. 99—103.
[14] Bolshakov M.A., Alexeeva S.I. Change in the electrical activity of racemaker neutrons of pond snails depending on the rate of their heating // *Biophysics.* — 1986. — Volume XXXI, edition 3. – P. 521—523.
[15] . Romanenko V.I., Gavrilov V.A., Zakharova L.I. The temperature optimum for the development of bacterial communities in the Rybinsk Reservoir during the subglacial period // *Biology of internal waters.* — 1987. — No. 79. – P. 17—19.
[16] Solntseva I.O., Vinogradova G.I. Salt tolerance of natural yeast strain from freshwater bodies // *Biology of internal waters.* —1988. — No. 79. – P. 22—24.
[17] Voronin V.L. Basic groups of saprophytic fungi on the surface of the pike body // *Biology of internal waters.* — 1989. — No. 83. – P. 16—19.
[18] Veshnyakova I.V., Romanenko V.I.. Emission of CO_2 by the duckweed (*Lemna minor a*) in the process of breathing depending on the temperature // *Biology of internal waters.* — 1987. — No. 75. – P. 10—13.
[19] Alekseev V.A. The dependence of toxicoresistence of some aquatic insects and arachnids on temperature // *Biology of internal waters.* — 1974. — No. 21. – P. 48—50.
[20] Karataev A. Yu. The dynamics of numerosity *Stictochironomus Psammophilus Tscheru* in different temperature zones of the heat sink of thermal power plant // *Biology of internal waters.* —1986. — No. 71. – P. 25—28.
[21] Semyonov I.V., Semyonova G.V. On the use of some mathematical models when studying the processes of water self-purification // *Hydrochemical materials.* — 1972, volume 28. – P. 205—211.
[22] Romanenko V.I., Obei-Aberu M.K., Zakharova L.I. Determination of thermal optimum for the development of bacteria in the silt // *Biology of internal waters.* —1988. — No. 80. – P. 76—77.
[23] Granstrem K.O., Vorobeynikov Ye.V. Reaction of viruses and bacteria to the exposure of the electromagnetic field of radio-frequency range // *Works of Leningrad Sanitary-Hygiene Medical Institute.* — 1985. – P. 57—63.
[24] Schedrina L.V., Kolyubaeva S.N., Komar V.Ye. Correlation of repair synthesis of DNA and the occurrence of cells with micronuclei during γ-irradiation of stimulated and

unstimulated phytohemagglutinin-lymphocyte of the peripheral blood of a man // *Radiobiology.* — 1985. — Volume XXV, edition 6. – P. 733—734.

[25] Beridze M.G., Nadoreishvili K.Sh., Chiloshvili D.V. Postradiation changes of the activity of membrane-related enzymes of the nerve tissue // *Radiobiology.* —1989. Volume XXIX, edition 5. – P. 176—178.

[26] Zaorgievskaya D.T., Gaziev A.I. The influence of protein inhibitors on the level of degradation of replanned DNA synthesis in UV- and γ- irradiated cells // *Radiobiology.* — 1989. — Volume XXIX, edition 6. – P. 23—29.

[27] Alievskaya L.L., Bukatina A.Ye., Graph N.A. etc. Change of the characteristic of the fource-pCa of the cross-striated muscle of the shark-ray under the influence of phalloidin // *Biophysics.* — 1987. — Volume XXXII, edition 1. – P. 67—70.

[28] Yemelyanenko A.M., Bovykin B.L. On the possibility of existence of the hydrocarbon solvent layer between the layers of belayed lipid membrane // *Biophysics.* — 1988. — Volume XXXIV, edition 3. – P. 117—120.

[29] Sukhorukov B.I. etc. Influence of the oxygen on radiation damages of nucleic acids irradiated in the medium with a different pH // *Radiobiology.* — 1988. — Volume XXVIII, No. 1. – P. 78—81.

[30] Davydov B.I., Antipov V.V., Tikhonchuk V.S. Biological effect, regulation and protection from electromagnetic radiation. — M.: *Energoatomizdat,* 1984. —161 p.

[31] Nikhanorov A.M., Pantyukhin Ya.V., Zavolzhensky M.V., Kokhanaya S.I., Botsenyuk K.L. Adaptive characteristics of aquatic ecological systems // In the collection: Environmental regulation and modeling of a man impact on aquatic ecosystems Edition 2. — St. Petersburg: *Hydrometeoizdat,* 1999. – P. 6—54.

[32] Pantyukhin Ya.V., Zavolzhensky M.V., Simonovich A.I. Determination of ecological capacity of biological objects with respect to herbicides. // *Materials XXIX of hydrochemical meeting.* Rostov-on-Don. 1987. 13 p.

[33] Nikanorov A.M., Zhulidov A.V., Dubova N.A., Gekov V.Ph., Kamov I.Yu. Buffering capacity of freshwater ecosystems to heavy metals and hydro-biological factors determining it // In the collection: Environmental regulation and modeling of a man impact on aquatic ecosystems. Edition 1. — L.: *Hydrometeoizdat,* 1988. – P. 58—69.

[34] Kondratyev K.Ya., Nikanorov A.M., Pantyukhin Ya.V., Zavolzhensky M.V. Some turbulent flows in water bodies. — *Report of the Academy of Sciences of Russia.* — 1992. — Volume 324, No. 3. – P. 676—680.

ABBREVIATIONS

AC	aircrafts
ACS	automatic control station
AES	artificial Earth satellite
AF	accumulation factor
AFSRI	Azov Fisheries Scientific and Research Institute
AHEM	Administration for Hydrometeorology and Environmental Monitoring
AMI	agar-meat infusion
ANKOS-VG	automated subsystem of observations and control over environmental pollution surface waters
AOS	aeronautic observation stations
	automatic operation system
ASCSW	automated system for control of surface waters
ASIL	approximately safe impact level
ASS	anionic synthetic surfactants
ATA	adenosine triphosphoric acid
BBI	Belgian Biotic Index
BOD	biochemical oxygen demand
BR	brightness ratios
BS	bottom sediments
CC	complex formation capacity
CCP	critical parameters of water pollution
CEM	Center for Environmental Monitoring
COD	chemical oxygen demand
D	ground resolution
DBWMA	Don Basin Water Management Association
DNA	deoxyribonucleic acid
DOM	dissolved organic matter
DP	dispatching post
	disastrous phenomenon
EIFAC	European Inland Fishery Advisory Commission
EHP	extremely high pollution level
EMF	electromagnetic field

EPA	Environmental Protection Agency
FA	fulvic acid
FEC	fuel and energy complex
GC	generalized coordinates
GEMS/Water	Global Environmental Monitoring System
H	flight altitude
HA	humic acid
HCI	Hydrochemical Institute
HCL	hydrochemical laboratory
HM	heavy metals
HP	high pollution level
HPP	heat and power plants
ICC	index of chemical contamination
IPC	information processing center
IR	infrared
ISW	inland surface water
LB	lipid bilayers
M	mineralization
MD	measuring devices
MHBL	mobile hydrobiological laboratory
MHCL	mobile hydrochemical laboratory
MNR	Ministry of Natural Resources of the RF
MPC	maximum permissible concentration
MPEL	maximum permissible ecological loads
MR	microwave radiation
NAWQA	National Water Quality Assessment Program
NCAHEM	North Caucasus Administration of Hydrometeorology and Environmental Monitoring
NWC	National Water Council
OAC	optically active components
ODMS	onboard data measurement system
OECD	Organization for Economic Cooperation and Development
OS	ouabain solution
OSS	observation space station
PAH	polyclic aromatic hydrocarbons polynuclear aromatic hydrocarbons
PC	personal computer
PCP	physicochemical parameters
PH	pump and hydraulics
RC DAP	Regional Center for Data Acceptance and Processing
RD	Regulatory Document
RF	Russian Federation
RLSBO	side-looking radar
RNA	ribonucleic acid
Roshydromet	Russian Federal Service for Hydrometeorology and Environmental Monitoring

RP	radiation parameters
RREPC	Rostov Regional Environment Protection Committee
S	satellite
SAF	sediment accumulation factor
SFEAE	subsystem of freshwater ecosystems anthropogenic eutrophication
SHBL	stationary hydrobiological laboratory
SHCL	stationary hydrochemical laboratory
SMS	state monitoring system
SMSTSW	subsystem for monitoring of state of transboundary surface waters
SRLI	safe reference level of impact
SS	saline solution
SSEM	State Sanitary and Epidemiological Surveillance
SSEPC	State Sanitary and Epidemiological Supervision Center of the Rostov Region
SSMEC	State Service for Monitoring of Environmental Conditions
SSOMPN	State Service for Observation and Monitoring of Pollution of the Objects of the Nature
SSWM	System of the Surface Water Monitoring
SWPLCI	specific combinatoric index of the water pollution level
T	temperature
TOC	total organic carbon
UNEP	United Nations Environmental Program
UNESCO	United Nations Educational, Scientific, and Cultural Organization
US/USA	United States (of America)
USGS	United States Geological Survey
USSR	Union of Soviet Socialist Republics
WHO	World Health Organization
WMO	World Meteorological Organization
WOB	water oxygen balance
WPLCI	combinatoric index of the water pollution level
WQI	water quality index

INDEX

A

B

C

D

E

F

J

K

L

M

N

O

P

Q

R

S

W

X

Y

Z